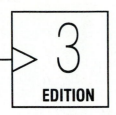

PRACTICAL DIGITAL DESIGN USING ICs

Joseph D. Greenfield
Rochester Institute of Technology

REGENTS/PRENTICE HALL, Englewood Cliffs, N.J. 07632

Greenfield, Joseph D. 1930–
    Practical digital design using ICs / Joseph Greenfield.—3rd ed.
        p.   cm.
    Includes bibliographical references and index.
    ISBN 0-13-689894-7
    1. Digital integrated circuits.   2. Digital electronics.   3. Logic
circuits.   I. Title.
TK7874.65.G74   1994
621.3815—dc20

93-4991
CIP

Editorial/production supervision: **Marcia Krefetz**
Interior design: **Kathy Kasturas**
Cover design: **Mike Fender**
Manufacturing buyer: **Ed O'Dougherty**
Prepress Buyer: **Ilene Sanford**
Editorial Assistant: **Melissa Stephens**

 © 1994 by REGENTS/PRENTICE HALL
Prentice-Hall, Inc.
A Division of Simon & Schuster
Englewood Cliffs, New Jersey 07632

Printed in the United States of America
10 9 8 7 6 5 4 3 2 1

ISBN 0-13-689894-7

PRENTICE-HALL INTERNATIONAL (UK) LIMITED, LONDON
PRENTICE-HALL OF AUSTRALIA PTY. LIMITED, SYDNEY
PRENTICE-HALL CANADA INC., TORONTO
PRENTICE-HALL HISPANOAMERICANA, S.A., MEXICO
PRENTICE-HALL OF INDIA PRIVATE LIMITED, NEW DELHI
PRENTICE-HALL OF JAPAN, INC., TOKYO
SIMON & SCHUSTER ASIA PTS. LTD., SINGAPORE
EDITORA PRENTICE-HALL DO BRASIL, LTDS., RIO DE JANEIRO

Contents

Chapter 15   Arithmetic Circuits   487

Chapter 16   The Basic Computer   517

References   661

Appendices

Index   715

Preface

The purpose of this book is to introduce the world of digital electronics to the student of the mid-1990s, to enable him or her to analyze and design these circuits, and to allow the student to use the laboratory to experiment and create.

Since the publication of the second edition of *Practical Digital Design Using ICs*, there have been many changes in digital electronics, however, much is still the same. What has remained constant is the use of the basic gates, ANDs, ORs, NANDs, etc. However, their operation must still be explained to the beginning student. As in the previous editions, I strive to start simply and proceed to the more complex topics. I also strive to illustrate every concept with an example, and to identify every gate I use in my circuits so they can be checked out in the laboratory.

One of the major changes in the past few years has been the introduction of new families of ICs. There are now many varieties and it is impossible to tell which one the student will encounter in his or her college or in the workplace. Fortunately, the same **7400** number has the same basic function in all families; I have tried to indicate this by giving most of my ICs a **74x** number, except when the IC is only available in one family.

A second major change is the introduction of PLDs, particularly PALs. I devote Chapter 12 to these devices, and endeavor to cover them as thoroughly as possible, given the constraint of space.

Digital electronics has also become more tightly entwined with microprocessors and personal computers. I have emphasized this intermarriage in this book, and I have used some of the circuitry in the IBM-PC as examples of good digital design. In addition, I have devoted two chapters to interfacing ICs and other devices with microprocessors.

Finally, I have introduced more modern test instruments, such as logic analyzers, digital oscilloscopes and workstations. The discussion of these instruments can be found in the last chapter of this book because the student must attain some sophistication before he can use these instruments properly.

Chapter one is an introduction to the arithmetic used in digital electronics and computers. The binary system, hexadecimal system, and 2's complement arithmetic are presented. Arithmetic using other bases has been ignored because it is irrelevant—even octal is obsolete.

Chapter 2 introduces the basic operations and theorems of Boolean algebra. One of the problems encountered when teaching these topics is the illustration of how to use the laboratory effectively. I have added several sections oriented toward laboratory procedures. Hence, the student will be able to test and verify the circuits and the theorems discussed.

The first references to data books were also made in this chapter. Each student should have at least one manufacturer's data book to use in conjunction with the text book. When I taught the course, the 1988 Logic Data Book (Texas Instruments) was used as the major reference.

Chapter three presents SOP and POS forms; an orderly way of expressing Boolean equations. It also leads to systematic reduction by Karnaugh maps. Several practical examples using these methods of simplification are given. These circuits can be tested in the laboratory.

The various logic families and their characteristics are presented in Chapter four. Due to their proliferation, it is difficult to condense all of the information about them into one chapter, however, I have tried to give the students a reasonable overview of the field.

The basic logic gates are introduced in Chapter five. I have placed much more emphasis on CMOS, 3-state gates, and line drivers than in the previous edition because they are used so often.

Chapters six, seven, and eight present the basic sequential gates of digital electronics: FFs, one-shots, and counters. All the chapters have been modernized. In Chapter six, the master-salve FF has been deleted as obsolete, however, more modern applications, such as an NRZ-to-Biphase converter were added. Chapter seven covers more modern one-shots and crystal oscillators. Chapter eight covers currently used counters and frequency dividers.

State tables have become increasingly important as PALs have come into common use. Therefore, Chapter nine is basically a new chapter on state tables, although some of the material from the previous edition using state tables with counters has been included. The emphasis is on using state tables to design digital circuits.

Chapters ten and eleven introduce shift registers and multiplexers. They are updated versions of the chapters in the second edition. In Chapter ten there is an emphasis on the design of shift register timing sequences. Johnson counters and Pseudo-Random Sequence Generators have also been added to the chapter. Chapter eleven places more emphasis on the design of decoders, because they are used in keyboard decoding and in memory address decoding. Display drivers have also been added to Chapter eleven.

Chapter twelve is a new chapter on programmable logic, primarily PALs. PALASM, which I believe is the most commonly-used and readily available language has been used, but if the student learns to use any PAL language, he should be able to readily adjust to other languages. The step-by-step design of a circuit for a PAL is covered, culminating in the blowing of the PAL. Both combinatorial and registered PALs are covered as well as simulation. Examples are presented. Finally, gate arrays are introduced.

Chapter thirteen is an updated version of the chapter on memories in the second edition. This chapter has been moved up because memories have become more important. Memory bus decoding has been emphasized. Examples of the use of memories in the IBM-PC and the APPLE computer are presented.

Chapter fourteen discusses EXCLUSIVE-OR circuits and their use in comparators, parity checkers and generators, and Gray code converters. Parity checking is emphasized and the parity circuit used in the IBM-PC is presented. Chapter fifteen is an updated version of the chapter on arithmetic units. ALUs and APUs are discussed.

Chapter sixteen explains the operation of a computer and shows how a computer can be designed and built using the circuits discussed in this book. The latter part of the chapter discusses modern microprocessors and their function. Only an overview can be presented because a deeper understanding would involve a knowledge of how to program the microprocessor, which space limitations preclude explaining. This course is often taught in conjunction with a course on a specific microprocessor where these topics can be presented.

Interface ICs and their connections to µPs and peripheral devices are covered in Chapter seventeen. Intel port addressing and Motorola memory-mapped addressing are explained. Examples of serial and parallel interfacing are presented. An overview of personal computers is also presented. The student is now capable of understanding the components that comprise a personal computer and how they function.

Chapter eighteen discusses A/D and D/A converters. The circuitry of these converters is explored and the important parameters are discussed. Applications such as their use in a digital voltmeter and a digital storage oscilloscope are explained.

Chapter nineteen covers the methods of constructing digital circuits, including wiring and printed circuit boards. It then goes into debugging these circuits, where I try to include methods of debugging that I have used in both industry and academia. Finally, modern test instruments such as logic analyzers, digital storage oscilloscopes, and workstations are introduced. Sections of this chapter may be used earlier in the course, depending on the instructor's preference. I have presented this chapter as the last chapter because I wanted to develop the basic circuits used in logic analyzers and DSOs before discussing them, so the students could understand how they worked.

The author would like to acknowledge the cooperation of the Tektronix Corporation and the Hewlett-Packard Corporation for supplying me with photographs and materials I used in the book. I would also like to commend their sales engineers in their Philadelphia offices for helping me. In addition I would like to acknowledge Bruce J. Liban of R.F. Products, Inc., Camden NJ, for his help in procuring permissions.

Joseph D. Greenfield
Feasterville, PA.

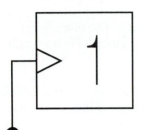

Binary Numbers

1-1 Introduction to Digital Electronics

This book introduces the reader to digital electronics and enables him or her to understand, design, and construct digital circuits. Digital circuits are the "brains" of the technological world; they are used to *control* many things, from highly complex industrial processes, to robots, to automobiles (the electronic control module, EMC, in your automobile is primarily digital), to household appliances such as microwave ovens and VCRs. Computers and microprocessors are built from digital circuits, and in Chapter 16 we will design a rudimentary computer based on the circuits covered in the intervening chapters.

Digital circuits consist of a group of logic *gates* that perform useful tasks. They are described throughout this book, beginning in Chapter 2. These gates and functions have been incorporated into several series of *integrated circuits* (ICs). The **7400** series of ICs is the predominant series and it will be covered in this book. These ICs have experienced a phenomenal growth in the past several years, and millions of them have been manufactured. They can be purchased from manufacturers, electronics distributors, electronics stores such as Radio Shack, many computer stores, and by mail or phone from discount houses.[1] Their low price and wide availability have caused them to be designed into almost all digital circuits produced in the last 10 years.

Microprocessors (μPs), which are small computers packaged in an IC and

[1]These discount houses often advertise in magazines such as *Popular Electronics* or *Byte*. Jameco is such a discount house (See References).

are the computing element in personal computers (PCs), have also enjoyed phenomenal growth recently. Some engineers believed that μPs would replace ICs, but this has not happened. Rather, μPs work in *conjunction* with digital ICs to provide a high level of intelligence and sophistication to digital hardware. Figure 1-1 is a photograph of the system board of the IBM personal computer (PC). This is the printed circuit board that controls the PC. The small, black modules mounted on the board are the ICs.

Figure 1-2 is a *module chart* of the same systems board. The module chart shows which ICs are used on the board and where they are placed. We have subdivided the board into four areas:

1. The *basic logic area*. This consists of the basic logic gates (often called "glue chips") used in the PC. Almost all of these are **7400** series ICs. They are covered in Chapters 2 to 15.
2. *The memory area*. This consists of both RAMs and ROMs. They are covered in Chapter 13.
3. *The microprocessor itself*. This is the Intel **8088** μP, and there is a socket for a *coprocessor*, the Intel **8087**. An **8087** coprocessor aids the μP by performing sophisticated mathematical operations when required. Many PCs function without coprocessors, but if mathematical calculations are to be performed frequently, a coprocessor should be added to increase the speed of these operations.
4. *Peripheral ICs*. These ICs are primarily involved with *Input/Output* (I/O). Some of these ICs are discussed in Chapters 17 and 18.

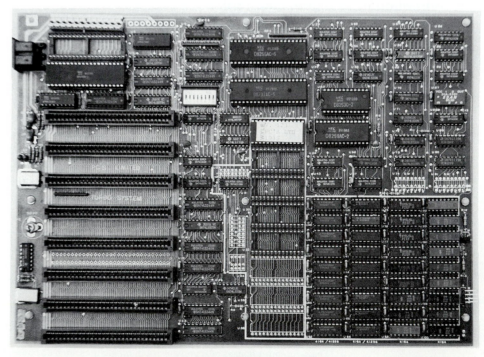

Figure 1-1 The IBM PC system board.

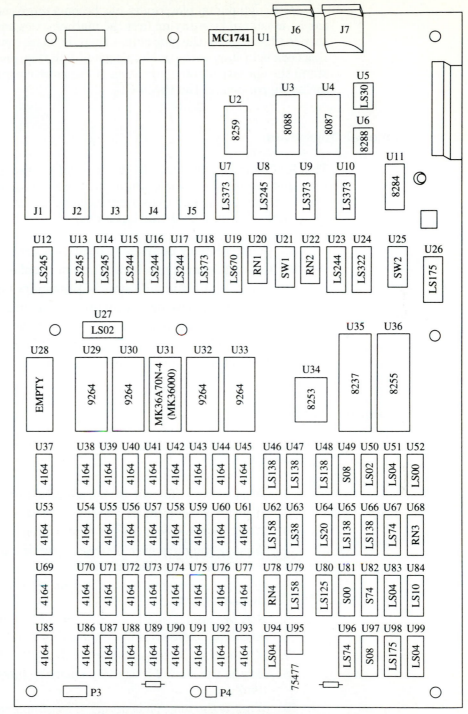

Figure 1-2 Module chart of the IBM PC. (Courtesy of International Business Machines Corporation.)

Although digital electronics has become more complex in recent years, it can be mastered by studying first the mathematics and logic of digital devices, then the basic gates (the glue chips), then the more complex devices. The goal of this book is to start with the simple gates, and then enable the reader to understand the operation and application of the more complex devices and systems, including portions of the IBM PC.

1-2 Instructional Objectives

This first chapter introduces the basic number systems used with digital systems and computers. It shows the student how to use the binary and hexadecimal number systems and to perform 2s complement arithmetic. After reading the chapter the student should be able to:

1. Convert binary numbers to decimal numbers.
2. Convert decimal numbers to binary numbers.
3. Find the sum and difference of two binary numbers.
4. Convert negative binary numbers to their 2s complement form.
5. Add and subtract numbers in 2s complement form.
6. Use hexadecimal numbers instead of binary numbers.
7. Add, subtract, and negate hexadecimal numbers.

1-3 Self-Evaluation Questions

As the student reads the chapter, he or she should be able to answer the following questions:

1. What is the difference between digital and analog circuits?
2. How are the outputs of a circuit defined to make the output digital?
3. What are the advantages of digital circuits?
4. What is the difference between a *bit* and a *decimal digit*? In what respects are they similar?
5. In a *flow chart*, what is the function of the rectangular and diamond-shaped boxes?
6. How are binary addition and subtraction different from decimal addition and subtraction? How are they similar?
7. How are bits, bytes, and words related?
8. How is the sign of a 2s complement number determined by inspection?
9. What is the major advantage of 2s complement notation?
10. How can the positive equivalent of 2s complement negative numbers be found?
11. What is the advantage of hexadecimal arithmetic?
12. How can binary numbers be converted to hexadecimal and vice versa?

To start the study of digital electronics, one must be able to distinguish between *analog* and *digital* quantities.

1-4-1   Analog Voltages

The output of most electronic circuits is an *analog quantity*, typically a voltage. An analog quantity is a quantity that may assume *any* numerical value within the range of possible outputs. An electronic circuit is capable of producing many outputs in response to different inputs. For example, 5.12 volts (V) might be one output, 3.76 V another; any voltage within the precision of the voltmeter is a legitimate output. The speedometer of an automobile is another example of an analog quantity; the needle indicates the speed of the vehicle, which can be any velocity from 0 mph up to the speed limit.

Figure 1-3 shows an analog voltmeter and a digital voltmeter. The meter of Fig. 1-3a is analog because the indicator needle points to the voltage being measured. The voltmeter of Fig. 1-3b is commonly called a digital voltmeter because it presents its outputs as a series of digits. It is actually an analog meter, however, because it can measure *any* voltage that appears at its input terminals. The output appears to be digital because the output is displayed as a series of decimal digits. These type of digital voltmeters are quite simple to design and build. They basically consist of an analog-to-digital converter (discussed in Chapter 18) and a digital readout (discussed in Chapter 11). By the end of this book, the reader will understand how these meters work.

1-4-2   Digital Voltages

Digital circuits differ from analog circuits by providing only two voltage values as an output. An output can be represented as either a logic one (1) or a logic zero (0), *and nothing else*. Of course, digital engineers are still dealing with electronic circuits whose actual outputs are voltages. What they have done is to *define* a certain range of voltages as a logic 1 and another range of voltages as a logic 0. Typically, the 1 and 0 ranges are separated by a *forbidden* range of voltages. TTL integrated circuits define a 1 output as any voltage between 2 V and 5 V and a 0 as any voltage between ground (zero) and 0.8 V. But what if the actual output of the circuit is in a forbidden or undefined range, 1 V, for instance? "Then," says the digital engineer succinctly, "the circuit is malfunctioning. Fix it." Methods of diagnosing and repairing malfunctions in digital circuits are discussed in Chapter 19.

There are several advantages gained by restricting the output of an electronic circuit to one of two possible values. First, it is rarely necessary to make fine distinctions. Whether an output is 3.67 V or 3.68 V no longer matters; in both cases it is a logic 1. Since well-designed logic circuits produce voltages near the middle of the range defined for 1 or 0, there is no difficulty in distinguishing between them. In addition, a digital circuit is very tolerant of any drift in the output caused by component aging or changes. A change in a component would almost have to be catastrophic to cause the output voltage to drift from a 1 to a 0 or an undefined value. Another advantage is that it is far easier for electronic circuits to remember a 1 or a 0 than to remember an analog quantity like 3.67 V.

(a)

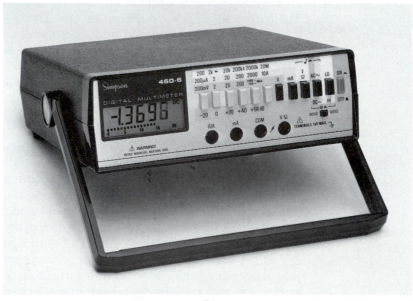

(b)

Figure 1-3 Photograph of multimeters: (a) The Simpson 260-8 Analog Voltmeter; (b) The Simpson 460-6 Digital Voltmeter. (*Courtesy of Simpson Electric Company.*)

Since all but the simplest digital circuits require the ability to remember the value of a voltage *after the conditions that caused that voltage have disappeared,* this is a very important consideration.

Advances in digital technology have been spectacular. The first computers were built in the early 1950s using large vacuum tubes that had poor reliability and consumed enough power to heat the building. Tubes were replaced by transistors as soon as the latter became available and reliable. Engineers quickly found ways to build small, efficient, and inexpensive digital circuits using a single transistor, a couple of diodes, and a few resistors. Now, however, these *discrete* components (the resistors, transistors, and diodes) have all disappeared inside the IC. Recently, ICs have become slightly larger and far more complex. The microprocessor itself is an example of a very large scale (VLSI) integrated circuit containing several thousand gates.

The computer of Fig. 1-1 might be used to illustrate the difference between the analog and the digital engineer. The analog engineer would view the signals passing between the ICs as a continuous voltage wave form; the digital engineer would view them as a *pulse train* of 1s and 0s, with the exact voltages unspecified. These two viewpoints are illustrated in Fig. 1-4. When computers or digital circuits are being considered, the digital engineer's visualization is more pertinent.

1-5   Uses of Binary Bits

The output of a digital circuit is one of two values represented by a single binary digit (a 1 or a 0) commonly called a *bit*. A single bit is enough to answer any question that has *only two* possible answers. For example, a typical job application might ask "What is your sex?" A 1 could arbitrarily be assigned to a male and a 0 to a female, so a single bit is enough to describe the answer to this question. A single bit is all the space a programmer needs to reserve in his computer for this answer.

However, another question on the job application might be "What is the color of your hair?" If the possible answers are black, brown, blonde, and red, a single bit cannot possibly describe them all. Now *several* bits are needed to describe all possible answers. We could assign one bit to each answer (i.e., brown = 0001, black = 0010, blonde = 0100, red = 1000), but if there are many possible answers to the given question, many bits are required. The coding scheme presented above is not optimum; it requires more bits than are really necessary to answer the question.

It is most economical to use *as few bits as possible* to express the answer to a question, or a number, or a choice. The crucial question is: *"What is the minimum number of bits required to distinguish between n different things?"*

Whether these *n* things are objects, possible answers, or *n* numbers is immaterial. To answer this question we realize that each bit has two possible values. Therefore, k bits would have 2^k possible values. This fact is the basis for Theorem 1.

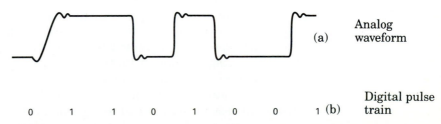

Figure 1-4   Visualization of a wave form: (a) analog viewpoint; and (b) digital viewpoint.

Theorem 1.

The minimum number of bits required to express n different things is k, where k is the smallest number such that $2^k \geq n$.

A few examples should make this clear.

EXAMPLE 1-1

What is the minimum number of bits required to answer the hair color question, and how could they be coded to give distinct answers?

SOLUTION   There are four possible answers to this question, therefore, $2^k = 4$. Since 2 is the smallest number such that $2^2 \geq 4$, $k = 2$ and 2 bits are needed. One way of coding the answers is $00 = $ brown, $01 = $ black, $10 = $ blonde, and $11 = $ red.

EXAMPLE 1-2

How many bits are needed to express a single decimal digit?

SOLUTION   There are 10 possible values for a single decimal digit (0 through 9); therefore, $2^k \geq 10$. Since $k = 4$ is the smallest integer such that $2^k \geq 10$, 4 bits are required.

EXAMPLE 1-3

A computer must store the names of a group of people. If we assume that no name is longer than 20 letters, how many bits must the computer reserve for each name?

SOLUTION   To express a name, we need only the 26 letters of the alphabet, plus a space and perhaps a period, or a total of 28 characters for each letter or character of the name. Here $2^k \geq 28$ so that $k = 5$. Therefore, 5 bits are required for *each* character and, since space must be reserved for 20 such characters, 100 bits are needed for each name.

1-6   Binary to Decimal Conversion

In the early chapters of this book only two number systems, *binary* (base 2) and *decimal* (base 10) are considered. The decimal system contains the numbers 0–9 and is the one we commonly use. The binary system must be mastered because it is used extensively in digital engineering. The *hexidecimal* (base 16) number system is an extension of the binary system. It and the 2s complement system are used by computer engineers. Since we will be describing components of computers starting in Chapter 13, these systems are also covered in this introductory chapter.[2] To eliminate any possible confusion, a subscript is used to indicate which number system is employed. Thus, 101_{10} is the decimal number whose value is one hundred and one, while 101_2 is a *binary* number whose deci-

[2]This material will not be used until Chapter 13. Therefore, the reader may prefer to skip the rest of this chapter on first reading. He or she must understand it, however, before reading Chapter 13.

mal value is five. Of course, *any* number containing a digit from 2 to 9 is a decimal number because binary numbers can only contain the digits 0 and 1.

The value of a decimal number depends on the magnitude of the decimal digits expressing it and on their *position*. A decimal number is equal to the sum $D_0 \times 10^0 + D_1 \times 10^1 + D_2 \times 10^2 + \cdots$, where D_0 is the least significant digit (the rightmost digit in the number), D_1 the next significant, and so on.

EXAMPLE 1-4

Express the decimal number 7903 as a sum to the base 10.

SOLUTION   Here D_0, the least significant digit, is 3, $D_1 = 0$, $D_2 = 9$, and $D_3 = 7$. Therefore, 7903 equals:

$$
\begin{array}{llll}
  & 3 \times 10^0 = 3 \times 1 & = & 3 \\
+ & 0 \times 10^1 = 0 \times 10 & = & 0 \\
+ & 9 \times 10^2 = 9 \times 100 & = & 900 \\
+ & 7 \times 10^3 = 7 \times 1000 & = & 7000 \\
\hline
  & & & 7903
\end{array}
$$

Similarly, a group of *binary* bits can represent a number in the *binary* system. The binary base is 2; therefore the digits can only be 0 or 1. However, a binary number is also equal to a sum, namely $B_0 \times 2^0 + B_1 \times 2^1 \cdots$, where B_0 is the least significant bit, B_1 the next significant bit, and so on. The powers of 2 are given in the "Binary Boat" or table of Appendix A. In this table, n is the exponent and the corresponding positive and negative powers of 2 are listed to the left and right of n, respectively.

A binary number is a group of ones (1s) and zeros (0s). To find the equivalent decimal number, we simply add the powers of 2 that correspond to the 1s in the number and omit the powers of 2 that correspond to the 0s of the number.

EXAMPLE 1-5

Convert 100011011_2 to a decimal number.

SOLUTION   The first bit to the left of the binary point corresponds to $n = 0$, and n increases by one (increments) for each position further to the left. The number 100011011 has 1s in positions 0, 1, 3, 4, and 8, so the conversion is made by obtaining those powers of 2 corresponding to these n values (using Appendix A, if necessary) and adding them:

n	2^n
0	1
1	2
3	8
4	16
8	256
	283

Therefore, $100011011_2 = 283_{10}$.

EXAMPLE 1-6

A *word* is a basic unit of computer information and consists of several bits. In the Intel **8086** μP computer each word consists of 16 bits, that is $k = 16$. How many numbers can be represented by a single **8086** word and what are they?

SOLUTION   Since 16 bits are available, any one of 65,536 (2^{16}) numbers can be expressed. These numbers range from a minimum of sixteen 0s to a maximum of sixteen 1s, which is the binary equivalent of 65,535. Therefore, the 65,536 different numbers that can be expressed by a single word are the decimal numbers 0 through 65,535.

1-6-1   Conversion of Binary Fractions to Decimals

Decimal fractions can be expressed as a sum of digits times 10 to *negative* powers. For example, 0.3504 equals:

$$
\begin{array}{rcl}
3 \times 10^{-1} &=& 0.3 \\
+ 5 \times 10^{-2} &=& 0.05 \\
+ 0 \times 10^{-3} &=& 0 \\
+ 4 \times 10^{-4} &=& \underline{0.0004} \\
&& 0.3504
\end{array}
$$

Similarly, binary fractions can be expressed as sums of *negative* powers of two. The table of Appendix A can again be used if $n = 1$ is taken as the first position to the *right* of the decimal[3] point and n increases as the position moves to the right. Here the *negative* powers of 2 are added up.

EXAMPLE 1-7

Convert the binary fraction 0.11010001 to a decimal fraction.

SOLUTION   In this example, the 1s appear in the 1, 2, 4, and 8 positions (reading toward the right). From Appendix A we find:

n	2^{-n}
1	0.5
2	0.25
4	0.0625
8	0.00390625
	0.81640625

Therefore, $0.11010001_2 = 0.81640625_{10}$.

1-7   Decimal to Binary Conversion

It is often necessary to convert decimal numbers to binary. Humans, for example, supply and receive decimal numbers from computers that work in binary;

[3]Strictly speaking, a "decimal point" in a binary number should be called a binary point.

consequently, computers are continually making *binary to decimal* and *decimal to binary* conversions.

To convert a decimal number to its equivalent binary number, the following algorithm or procedure may be used, where **K** is the position of the bit:[4]

1. Obtain **N** (the decimal number to be converted).
2. Determine if **N** is odd or even.
3. a. If **N** is odd, write 1 and subtract 1 from **N**. Go to step 4.
   b. If **N** is even, write 0.
4. Obtain a new value of **N** by dividing the **N** of step 3 by 2.
5. a. If **N** > 1, go back to step 2 and repeat the procedure.
   b. If **N** = 1, write 1. The number written is the binary equivalent of the original decimal number. The number written first is the least significant bit (lsb), and the number written last is the most significant bit (msb).

This procedure can also be implemented by following the *flow chart* of Fig. 1-5. Computer programmers often use flow charts to describe their programs

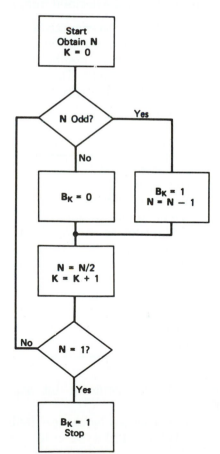

Figure 1-5 Flow chart for decimal-to-binary conversion of whole numbers.

[4]This procedure is essentially the same as repetitive division by 2.

graphically. For the rudimentary flow charts drawn in this book, the *square* box is a *command*, which must be obeyed *unconditionally*. The *diamond-shaped* box is a *decision* box. If the answer to the question within the decision box is YES, the YES path must be followed; otherwise the NO path is followed. The flow chart of Fig. 1-5 starts with the given number **N**; since **K** equals 0, initially we are writing B_0, the least significant digit. Note that equations in a flow chart are programmer's equations, not algebraic equations. The "equation" $N = N - 1$ makes no sense mathematically. What it means here is that **N** is *replaced* by $N - 1$.

On the initial pass through the flow chart, B_0, the *least* significant bit, is written as 0 or 1, depending on whether **N** is even or odd. Next **N** is divided by 2 and **K** is incremented so that on the following pass B_1, the second least significant digit will be written. We continue looping through the flow chart and repeating the procedure until $N = 1$. Then the most significant bit is written as a 1, and the process stops. The bits written are the binary equivalent of the decimal number.

EXAMPLE 1-8

Find the binary equivalent of the decimal number 217.

SOLUTION The solution proceeds according to the algorithm or flow chart. When an odd number is encountered, a 1 is written as the binary digit and subtracted from the remaining number; when the remaining number is even, 0 is written as the binary digit. The number is then divided by 2. The process continues until the number is reduced to 1.

Remaining Number		Binary Digit or Bit
217	Odd—subtract 1	1
216	Divide by 2	
108	Even—divide by 2	0
54	Even—divide by 2	0
27	Odd—subtract 1	1
26	Divide by 2	
13	Odd—subtract 1	1
12	Divide by 2	
6	Even—divide by 2	0
3	Odd—subtract 1	1
2	Divide by 2	
1	Finish	1

Note that the least significant bit was written first. Therefore, $217_{10} = 11011001_2$.

This result can be checked by converting back from the binary to the decimal number. $11011001 = 128 + 64 + 16 + 8 + 1 = 217_{10}$.

1-7-1 Converting Decimal Fractions to Binary Fractions

Decimal fractions must also be converted to binary fractions in certain applications. Decimal fractions may not have an *exact* binary equivalent, and the decimal to binary conversion often produces a *repetitive sequence* of binary bits.[5] The decimal to binary conversion procedure starts with the *most significant* binary bit, the bit immediately to the right of the decimal point, and then proceeds to the right, one bit at a time. Each bit has only *half* the value of the preceding bit, and the engineer stops the conversion after obtaining a sufficiently precise binary representation of the decimal fraction.

The following procedure can be used to convert a decimal fraction to a binary fraction:

1. Obtain **N**.
2. Double **N**.
3. a. If the new value of **N** is greater than 1, write 1 as the next most significant bit, subtract 1 from **N**, and go back to step 2.
   b. If the new value of **N** is less than 1, write 0 as the next most significant bit and go back to step 2.

An equivalent procedure is given by the flow chart of Fig. 1-6. Notice that we start with the most significant fractional bit, the bit immediately to the right of the decimal point, and proceed one bit to the right each time we loop through the flow chart.

EXAMPLE 1-9

Convert 0.78125_{10} to binary.

SOLUTION   Follow the procedure or the flow chart:

Number		Binary Bit
0.78125		
	Double **N**	
1.5625		
	Subtract 1	1
0.5625		
	Double **N**	
1.125		
	Subtract 1	1
0.125		
	Double **N**	
0.25		
	Double **N**	0
0.5		
	Double **N**	0
1.0		
	Subtract 1	1
0.0		

---

[5]This also happens with decimal fractions such as $\frac{1}{3}$ or $\frac{1}{9}$, etc.

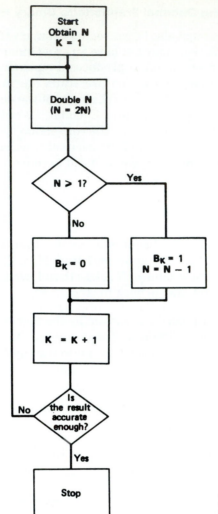

Figure 1-6 Flow chart for decimal-to-binary conversion of fractions.

Each binary bit is written immediately after **N** is doubled. If **N** is greater than or equal to 1, the bit is a 1 and 1 is subtracted from **N**, but if **N** is less than 1, the bit written is a 0 and no subtraction is performed.

Here we find that $0.78125_{10} = 0.11001_2$. This is one of the happy examples where the decimal fraction has an exact binary equivalent, as indicated by the fact that **N** is exactly 0 after the last subtraction.

EXAMPLE 1-10

Convert 0.85 to binary.

SOLUTION Again the flow chart or algorithm procedure is followed.

Number	Binary Bit
0.85	
Double **N**	

1.7		1
	Subtract 1	
0.7		
	Double **N**	
1.4		1
	Subtract 1	
0.4		
	Double **N**	
0.8		0
	Double **N**	
1.6		1
	Subtract 1	
0.6		
	Double **N**	
1.2		1
	Subtract 1	
0.2		
	Double **N**	
0.4		0

Repetitive Sequence

The results indicate that to 5 binary places 0.85 = 0.11011. The repetitive sequence is 0110 because **N** = 0.4 at the beginning and at the end of this sequence. Therefore, we can immediately write the equivalent of 0.85 to 9 or 13 bits by just appending repetitive sequences; that is,

$$0.85 = 0.11011001100110011(0) \text{ etc.}$$

Repetitive sequences

This result was checked with a hand calculator using the Binary Boat (Appendix A).

0.11011	= 0.84375
0.110110011	= 0.84960937
0.1101100110011	= 0.8499755859

Notice that each result is less than 0.85 because the bits of lesser significance were not included; however, as more bits are used the result approaches 0.85.

1-7-2 Mixed Numbers

Mixed numbers, which consist of both integers and decimals, can be converted from binary to decimal or decimal to binary by working on the integer and fraction portions separately.

EXAMPLE 1-11

Convert 10011.101_2 to a decimal number.

SOLUTION   The integer part, 10011, converts to 19, and the decimal portion converts to 0.625; therefore, $10011.101_2 = 19.625$.

EXAMPLE 1-12

Convert the decimal number 33.3 to binary.

SOLUTION   33 converts to 100001 and 0.3 becomes 0.0100110011, where the 0011 pattern is repetitive. Therefore, $33.3_{10} = 100001.0100110011_2$.

1-8   Addition and Subtraction of Binary Numbers

The binary number system is a valid mathematical system, and operations such as addition, subtraction, multiplication, and division can be performed on binary numbers in a manner similar to those operations on decimal numbers. In this section, the most commonly performed arithmetic operations, addition and subtraction, will be discussed. The reader should consult more specialized texts (see the References at the end of the book) for multiplication, division, squares, square roots, and other more complex arithmetic operations.

1-8-1   Addition of Binary Numbers

The addition of binary numbers is similar to the addition of decimal numbers except that $1 + 1 = 0$ with a carry out to the *next significant* place. A carry into a *more significant* position acts like an additional 1.

EXAMPLE 1-13

Add the binary numbers $A = 11101100$ and $B = 1100110$.

SOLUTION

Column	9	8	7	6	5	4	3	2	1	(Decimal Addition)
Carry		⌒	⌒	⌒		⌒	⌒			
A		1	1	1	0	1	1	0	0	(236)
B			1	1	0	0	1	1	0	(102)
Sum	1	0	1	0	1	0	0	1	0	(338)

The carry row shows where binary carries were generated. The addition proceeded as follows:

1. Column 1 (least significant digit) $0 + 0 = 0$.
2. Column 2 $0 + 1 = 1$.
3. Column 3 $1 + 1 = 0$ plus a carry output.
4. Column 4 $0 + 1$ plus a carry input from Column 3 sums to a 0 and produces a carry out to Column 5.
5. Column 5 $0 + 0 = 0$, but the carry input from Column 4 makes the sum 1.
6. Column 6 $1 + 1 = 0$ and a carry output to Column 7.

7.	Column 7	$1 + 1$ plus a carry input results in a sum of 1 and a carry output.
8.	Column 8	B does not have an eighth bit; therefore, a leading 0 can be assumed. Here $0 + 1$ plus a carry input yields a 0 sum plus a carry output.
9.	Column 9	Since neither A nor B has a ninth digit, leading 0s are written for both. In Column 9 we have $0 + 0$ plus a carry in from Column 8, which gives a sum of 1. Since there is no carry out of Column 9, the addition is complete.

The sum of Example 1-13 can be checked by converting the numbers to their decimal equivalents. These numbers are shown in parentheses beside the sum.

1-8-2 Subtraction of Binary Numbers

The rules for the subtraction of binary numbers are:

1. $1 - 1 = 0$
2. $0 - 0 = 0$
3. $1 - 0 = 1$
4. $0 - 1 = 1$ with a borrow out

In order to borrow, change the next 1 in the minuend to a 0 and change all intervening 0s to 1s.

EXAMPLE 1-14

Subtraction 101101001 from 100011010011.

SOLUTION

Column	12 11 10 9 8 7 6 5 4 3 2 1	(Decimal Subtraction)
Borrow	⌒ ⌒ ⌒ ⌒⌒ ⌒	
Minuend	1 0 0 0 1 1 0 1 0 0 1 1	(2259)
Subtrahend	1 0 1 1 0 1 0 0 1	−(361)
Difference	1 1 1 0 1 1 0 1 0 1 0	(1898)

The borrow row indicates where borrows have occurred. The subtraction proceeds as follows:

1. Column 1 $1 - 1 = 0$.
2. Column 2 $1 - 0 = 1$.

3. Column 3 $0 - 0 = 0$.
4. Column 4 $0 - 1 = 1$. The 1 in Column 5 is changed to a 0 because of the borrow out generated in Column 4.
5. Column 5   This is now $0 - 0 = 0$.
6. Column 6 $0 - 1 = 1$. The 1 in Column 7 is changed to a 0.
7. Column 7   Because of the borrow from Column 6 this now becomes $0 - 1$ or 1 with a borrow out that changes the 1 in Column 8.
8. Column 8   This becomes $0 - 0 = 0$.
9. Column 9 $0 - 1 = 1$. Since Columns 10 and 11 are 0, the borrow must be from Column 12. Columns 10 and 11 contain intervening 0s so they change to 1s and Column 12 changes to a 0.
10. Column 10   This is now $1 - 0 = 1$.
11. Column 11   This is now $1 - 0 = 1$.
12. Column 12   This is now $0 - 0 = 0$.

The results were checked by converting the binary numbers to their decimal equivalents, which are shown in parentheses beside the numbers.

1-9   2's Complement Arithmetic

When building hardware to accommodate binary numbers, two problems arise:

1. The number of bits in a hardware register is limited.
2. Negative integers must also be represented.

These problems do not arise in conventional pencil-and-paper arithmetic. If additional bits are needed, the number can always be extended to the left and negative numbers can always be represented by a minus sign.

In a digital system, such as a computer or a controller, numbers are stored in a digital circuit called a *hardware register* (to be covered in Section 6-13-1).

Since a hardware register consists of a finite number of bits, the range of numbers that can be represented is constrained. Each bit of the register can assume one of two values (0 or 1) and an n-bit register can therefore contain one of 2^n numbers. If positive binary numbers are used, the 2^n numbers that can be represented are 0 through $2^n - 1$ (a string of n 1s represents number $2^n - 1$).

EXAMPLE 1-15

What range of decimal numbers can be represented by a 12-bit register?

SOLUTION   Since 2^{12} equals 4096, the range of numbers that can be represented by a 12-bit number is from 0 to $2^{12} - 1$, or 4095. A string of twelve 1s can represent the number 4095.

TABLE 1-1

Register Lengths in Bits of Various Microprocessors

Microprocessor	Register Length in Bits
Intel 8080, 8085; Motorola 6800, 6809, 68H011; Rockwell 6502; Zileg Z80	8
Intel 8086, 8088	16
Motorola 68000, 68010, 68020; Intel 80386, 80486	32

The range of numbers that can be represented by a single computer word is also restricted by the *word length* of the computer word. This is one reason why *larger* computers and more powerful microprocessors have *longer* word lengths. Table 1-1 lists some common microprocessors and the size of the registers that contain their numbers.

1-9-1   2's Complementing Numbers

The 2's complement number system is used in all modern computers to express numbers and must be understood by digital engineers. It is similar to the binary number system, but both positive and negative numbers can be represented. The MSB (Most Significant Bit, the leftmost bit of the number) of a 2's complement number *denotes the sign* (0 means the number is positive; 1 means the number is negative), but the *MSB is also part of the number. In 2's complement notation, positive numbers are represented as simple binary numbers with the restriction that the MSB is 0.* Negative numbers are somewhat different. To obtain the representation of a negative number, use the following algorithm:

1. Represent the number as a positive binary number.
2. Complement it. (Write 0s where there are 1s and 1s where there are 0s in the positive number.)
3. Add 1.
4. Ignore any carries out of the MSB.

EXAMPLE 1-16

Given 8-bit words, find the 2's complement representation of:

(a) 25
(b) −25
(c) −1

SOLUTION   (a) The number +25 can be written as 11001. Since 8 bits are available, there is room for three leading 0s, making the MSB 0.

$$+25 = 00011001$$

(b) To find −25, complement +25 and add 1:

$$+25 = 00011001$$
$$\overline{(+25)} = 11100110 \quad \text{(complement)}$$
$$+1$$
$$\overline{}$$
$$-25 = 11100111$$

Note that the MSB is 1.

(c) To write -1, take the 2's complement of $+1$.

$$+1 = 00000001$$
$$\overline{(+1)} = 11111110 \quad \text{(complement)}$$
$$+1$$
$$\overline{}$$
$$-1 = 11111111$$

From this example, we see that a solid string of 1s represents the number -1 in 2's complement form.

To determine the magnitude of any *unknown negative number*, simply take its 2's complement as described above. The result is a *positive number whose magnitude equals that of the original number*.

EXAMPLE 1-17

What decimal number does the 8-bit, 2's complement number 11110100 represent?

SOLUTION   The number is negative because its MSB is 1. Complementing the given number, we obtain

$$00001011$$
$$\text{Adding 1} \qquad \underline{+1}$$
$$00001100$$

This is the equivalent of $+12$. Therefore, $11110100 = -12$. If $+12$ were complemented, *using the same rules*, it will give the number -12.

1-9-2   The Range of 2's Complement Numbers

The maximum positive number that can be represented in 2's complement form is a single 0 followed by all 1s, or $2^{n-1} - 1$ for an n-bit number. The most negative number that can be represented has an MSB of 1 followed by all 0s, which equals -2^{n-1}. Therefore, an n-bit number can represent any one of $2^{n-1} - 1$ positive numbers, plus 2^{n-1} negative numbers, plus 0, which is 2^n total numbers. Every number has a unique representation.

Other features of 2's complement arithmetic are:

1.   Even numbers (positive or negative) have an LSB of 0.
2.   Numbers divisible by 4 have the 2 LSBs equal to 0 (see Example 1-18).
3.   In general, numbers divisible by 2^n have n LSBs of 0.

EXAMPLE 1-18

What range of numbers can be represented by an 8-bit word using 2's complement representation?

SOLUTION   The most positive number that can be represented by 8 bits is a 0 followed by seven 1s or $01111111 = (127)_{10}$ or $(7F)_{16}$.
The most negative number in 8 bits is $10000000 = -128$.

Therefore, any number between $+127$ and -128 can be represented by an 8-bit number in 2's complement form. There are 256 numbers in this range, as expected, since $2^8 = 256$. Note also that the 7 LSBs of -128 are 0, as required, since -128 is divisible by 2^7.

In the BASIC programming language, *integer* variables are restricted to numbers between -32768 and $+32767$. This is because BASIC uses the 2's complement system and reserves 16 bits for each integer variable.

1-9-3   Adding 2's Complement Numbers

Consider the simple equation $C = A + B$. Although it seems clear enough, we cannot immediately determine whether an addition or subtraction operation is required. If A and B are both positive, addition is required. But if one of the operands is negative and the other is positive, a subtraction operation must be performed.

The major advantage of 2's complement arithmetic is: *If an addition operation is to be performed, the numbers are added regardless of their signs. The answer is in 2's complement form with the correct sign.* Any carries out of the MSB should be ignored.

EXAMPLE 1-19

Express the numbers 19 and -11 as 8-bit, 2's complement numbers, and add them.

SOLUTION   The number $+19$ is simply 00010011. To find -11, take the 2's complement of 11.

$$
\begin{aligned}
11 &= 00001011 \\
(\overline{11}) &= 11110100 \\
-11 &= 11110101
\end{aligned}
$$

Now $+19 + (-11)$ equals:

$$
\begin{array}{r}
00010011 \\
+11110101 \\
\hline
00001000
\end{array}
$$

Note that there is a carry out of the MSB, which is ignored. The 8-bit answer is simply the number $+8$.

EXAMPLE 1-20

Add -11 and -19.

SOLUTION First -19 must be expressed as a 2's complement number:

$$\begin{aligned} 19 &= 00010011 \\ (\overline{19}) &= 11101100 \\ -19 &= 11101101 \end{aligned}$$

Now the numbers can be added:

$$\begin{array}{rr} (-19) & 11101101 \\ +\ (-11) & \underline{11110101} \\ \text{Answer}\ (-30) & 11100010 \end{array}$$

Again, a carry out of the MSB has been ignored.

1-9-4   Subtraction of Binary Numbers

Subtraction of binary numbers in 2's complement form is also very simple and straightforward. *The 2's complement of the subtrahend is taken and added to the minuend.* This is essentially subtraction by changing the sign and adding. As in addition, the signs of the operands and carries out of the MSB are ignored.

EXAMPLE 1-21

Subtract 30 from 53. Use 8-bit numbers.

SOLUTION   Note 30 is the subtrahend and 53 the minuend.

$$\begin{aligned} 53 &= 00110101 \quad \text{(Minuend)} \\ 30 &= 00011110 \quad \text{(Subtrahend)} \end{aligned}$$

Taking the 2's complement of 30 and adding, we obtain

$$\begin{array}{rl} (\overline{30}) &= 11100001 \\ -30 &= 11100010 \\ +53 &= \underline{00110101} \\ & \ \ \ 00010111 \ = 23 \end{array}$$

EXAMPLE 1-22

Subtract -30 from -19.

SOLUTION   Here $-19 = 11101101$ (see Example 1-20).

$$-30 = 11100010 \ \text{(Subtrahend)}$$

Note: -30 is the subtrahend. 2's complementing -30 gives $+30$ or 00011110.

$$\begin{array}{rl} -19 & 11101101 \\ +30 & \underline{00011110} \\ & 00001011 \ = \ +11 \end{array}$$

The carry out of the MSB is ignored and the answer, $+11$, is correct.

A problem associated with binary arithmetic should now be apparent: there are too many 1s and 0s, and a shorthand notation for expressing them is needed. In most literature and documentation today, the convention of using *hexadecimal notation* has been adopted. The manufacturers of all the modern microprocessors use hexadecimal arithmetic in their literature to condense the 1s and 0s required to express addresses or data.

The hexadecimal system is a *base 16* arithmetic system. Since such a system requires 16 different digits, the letters A through F are added to the 10 decimal digits (0–9). The advantage of having 16 hexadecimal digits is that each digit can represent a unique combination of 4 bits, and that any combination of 4 bits can be represented by a single hex digit.[6] Table 1-2 gives both the decimal and binary values associated with each hexadecimal digit.

In many computers, the terms *nibble, byte,* and *word* are used with specific meanings. These are listed in Table 1-2.

TABLE 1-2

Bit Quantities

4 bits = One *nibble* or *hexidecimal digit.*
8 bits = Two nibbles or one *byte.*
16 bits = Two bytes or one *word.*
32 bits = Two words or one *longword* (Motorola's term) or one *Doubleword* (Intel's term).

From Table 1-2 we can see that one byte can be expressed by two hex digits, one word will require 4 hex digits, and a long or double word will require 8 hex digits to express it.

A byte is the word size for many microprocessors, and memories (see Chapter 13) are often described by the number of bytes they contain.

1-10-1   Conversions between Hexadecimal and Binary Numbers

To convert a binary number to hexadecimal, start at the least significant bit (LSB) and divide the binary number into groups of four bits each. Then replace each 4-bit group with its equivalent hex digit obtained from Table 1-3.

EXAMPLE 1-23

Convert the binary number 110000010111111101 to a hex number.

SOLUTION   We start with the LSB and divide the number into 4-bit nibbles. Each nibble is then replaced with its corresponding hex digit as shown:

0011	0000	0101	1111	1101
3	0	5	F	D

When the most significant group has less than 4 bits, as in this example, leading 0s are added to complete the 4-bit nibble.

[6]The word *hex* is often used as an abbreviation for hexadecimal.

TABLE 1-3

Table of Hexadecimal Digits

Hexadecimal Digit	Decimal Value	Binary Value
0	0	0000
1	1	0001
2	2	0010
3	3	0011
4	4	0100
5	5	0101
6	6	0110
7	7	0111
8	8	1000
9	9	1001
A	10	1010
B	11	1011
C	12	1100
D	13	1101
E	14	1110
F	15	1111

To convert a hex number to binary, simply replace each hex digit by its 4-bit binary equivalent.

EXAMPLE 1-24

Convert the hex number 1CB09 to binary.

SOLUTION   We simply expand the hex number:

1	C	B	0	9
0001	1100	1011	0000	1001

Thus, the equivalent binary number is:

$$11100101100001001$$

It is not necessary to write the leading 0s.

1-10-2   Conversion of Hex Numbers to Decimal Numbers

The hex system is a base 16 system; therefore, any hex number can be expressed as:

$$H_0 \times 1 + H_1 \times 16 + H_2 \times 16^2 + H_3 \times 16^3 \cdots$$

where H_0 is the least significant hex digit, H_1 the next, and so on. This is similar to the binary system of numbers discussed in section 1-6.

EXAMPLE 1-25

Convert 2FC to decimal.

SOLUTION The least significant hex digit, H_0, is C or 12. The next digit (H_1) is F or 15. This must be multiplied by 16 giving 240. The next digit, H_2, is 2, which must be multiplied by 16^2, or 256. Hence, 2FC = 512 + 240 + 12 = 764.

An alternate solution is to convert 2FC to the binary number 1011111100 and then perform a binary to decimal conversion.

Decimal numbers can be converted to hex numbers by repeatedly dividing them by 16. After each division, the remainder becomes one of the hex digits in the final answer.

EXAMPLE 1-26

Convert 9999 to a hex number.

SOLUTION Start by dividing by 16 as shown in the table below. After each division, the quotient becomes the number starting the next line and the remainder is the hex digit with the least significant digit on the top line.

Number	Quotient	Remainder	Hex Digit
9999	624	15	F
624	39	0	0
39	2	7	7
2	0	2	2

This example shows that $(9999)_{10} = (270F)_{16}$. The result can be checked by converting 270F to decimal, as shown in Example 1-25. By doing so we obtain:

$$(2 \times 4096) + (7 \times 256) + 0 + 15 = 9999$$
$$8192 + 1792 + 0 + 15 = 9999$$

1-10-3 Hexadecimal Addition

When working with μPs, it is often necessary to add or subtract hex numbers. They can be added by referring to hexadecimal addition tables, but we suggest the following procedure:

1. Add the two hex digits (mentally substituting their decimal equivalent).
2. If the sum is 15 or less, it can be directly expressed in hex.
3. If the sum is greater than or equal to 16, subtract 16 and carry 1 to the next position.

The following examples should make this procedure clear.

EXAMPLE 1-27

Add D + E.

SOLUTION D is the equivalent of decimal 13 and E is the equivalent of decimal 14. Together they sum to 27 = 16 + 11. The 11 is represented by B and there is a carry. Therefore, D + E = 1B.

EXAMPLE 1-28

Add B2E6 and F77.

SOLUTION The solution is shown below.

Column	4	3	2	1	Decimal
Carry		⌒	⌒		
Augend	B	2	E	6	(45798)
Addend		F	7	7	(3959)
Sum	C	2	5	D	(49757)

- Column 1   6 + 7 = 13 = D. The result is less than 16 so there is no carry.
- Column 2   E + 7 = 14 + 7 = 21 = 5 + a carry because the result is greater than 16.
- Column 3   F + 2 + 1 (the carry from Column 2) = 15 + 2 + 1 = 18 = 2 + a carry.
- Column 4   B + 1 (the carry from Column 3) = C.

Like addition, *hex subtraction* is analogous to decimal subtraction. If the subtrahend digit is larger than the minuend digit, one is borrowed from the next most significant digit (MSD). If the next MSD is 0, a 1 is borrowed from the next digit and the intermediate digit is changed to an F.

EXAMPLE 1-29

Subtract 32F from C02.

SOLUTION The subtraction proceeds as follows:

Column	3	2	1
Minuend	C	0	2
Subtrahend	3	2	F
Difference	8	D	3

- Column 1   Subtracting F from 2 requires a borrow. Because a borrow is worth 16, it raises the minuend to 18. Column 1 is therefore 18 − F = 18 − 15 = 3.

- Column 2   Because Column 2 contains a 0, it cannot provide the borrow out for Column 1. Consequently, the borrow out must come from Column 3, while the minuend of Column 2 is changed to an F. Column 2 is therefore $F - 2 = 15 - 2 = 13 = D$.
- Column 3   Column 3 can provide the borrow out needed for Column 1. This reduces the C to a B and $B - 3 = 8$.

As in decimal addition, the results can be checked by adding the subtrahend and difference to get the minuend.

1-10-4   Negating Hex Numbers

The negative equivalent of a positive hex number can always be found by converting the hex number to binary and taking the 2s complement of the result (section 1-9-1). A shorter method exists, however.

1. Add to the least significant hex digit the hex digit that makes it sum to 16.
2. Add to all other digits the digits that make it sum to 15.
3. If the least significant digit (LSD) is 0, write 0 as the LSD of the answer and start at the next digit.
4. The number written is the negative equivalent of the given hex number.

This procedure works because the sum of the original number and the new number is always 0.[7]

EXAMPLE 1-30

Find the negative equivalent of the hex number 20C3.

SOLUTION   The least significant digit is 3. To make 16, D must be added to 3. The other digits are 2, 0, and C. To make 15 in each case, we add D, F, and 3, respectively. The negative equivalent of 20C3 is therefore DF3D. This example can be checked by adding the negative equivalent to the positive number. Since X plus $-X$ always equals 0, the result should be 0.

$$
\begin{array}{r}
2\ 0\ C\ 3 \\
+\ D\ E\ 3\ D \\
\hline
0\ 0\ 0\ 0
\end{array}
$$

The carry out of the most significant digit is ignored.

1-10-5   Arithmetic Using Hexadecimal Numbers

One must often convert decimal numbers to hex and then add or subtract them. If the numbers are relatively small, there are shortcuts that make this procedure very simple, as Example 1-31 illustrates.

[7]This procedure is equivalent to taking the 16's complement of the number.

EXAMPLE 1-31

Convert the decimal numbers -43 and -48 to hex and add them.

SOLUTION Perhaps the simplest way to convert a negative decimal number to hex is to find the positive hex equivalent to the decimal number and then negate it.

$$+43_{10} = 2 \times 16 + 11 = 2B_{16}$$

Negating 2B, as in section 1-10-4, we obtain D5.

$$+48_{10} = 3 \times 16 + 0 = 30_{16}$$

Negating 30 (see line 3 of section 1-10-4), we obtain D0.

The sum of D0 and D5 is A5. This is the correct answer. The carry out of the MSB is ignored.

CHECK Because A5 contains two nibbles, and because the most significant nibble is between 8 and F, A5 is an 8-bit negative number. To find its positive equivalent, A5 can be negated to find its positive equivalent, which is 5B. But if $5B_{16} = 91_{10}$, then $A5_{16}$ must be equivalent to -91_{10}.

SUMMARY

This chapter introduced the student to the arithmetic used by digital and computer engineers. The binary system is extensively used by digital engineers and the hexadecimal system is universally used by computer engineers. The 2's complement number system, which is also universally used for computer arithmetic, was also discussed. The operations of addition, subtraction, and negation in these number systems were discussed and examples were presented.

GLOSSARY

Analog quantity: A continuously variable quantity; one that may assume any value usually within a limited range.

Binary system: A number system with 2 as the base.

Bit: A single digit quantity: a 1 or a 0.

Byte: A group of 8 bits.

Digital quantity: A variable that has one of two possible values.

Hexadecimal Digit: A single digit from 0–F. It can be represented by four binary bits.

Integrated circuit (IC): A small electronic package usually containing several circuits.

Least Significant Bit (LSB): The rightmost bit in a number.

Microprocessor (µP): A small computer in an IC package.

Most Significant Bit (MSB): The leftmost bit in a number.

Nibble: See Hex digit.

Word: A group of bits that constitutes the basic unit of information within a computer.

Section 1-5.

1-1. How many bits are required to distinguish between 100 different things?

1-2. A major league baseball team plays 162 games a year. Before the season starts, how many bits must be reserved to express the number of games the team will win and how many bits to express the number of games the team will lose?

1-3. A line printer is capable of printing 132 characters on a single line, and each character is 1 of 64 symbols (26 alphabetics plus 10 numbers plus punctuation). How many bits are needed to print an entire line?

Section 1-6.

1-4. Express the following decimal numbers as a sum.
- **(a)** 4507
- **(b)** 137,659
- **(c)** 889.7061

1-5. Convert the following binary numbers to decimal.
- **(a)** 10111
- **(b)** 110101
- **(c)** 110001011
- **(d)** 11011.10111
- **(e)** .00010011
- **(f)** 110001111.011101

Section 1-7.

1-6. Convert the following decimal numbers to binary.
- **(a)** 66
- **(b)** 252
- **(c)** 5795
- **(d)** 106,503
- **(e)** 0.3
- **(f)** 0.635
- **(g)** 0.82
- **(h)** 67.65
- **(i)** 3,477.842

Section 1-8.

1-7. For each of the following pairs of numbers, find $A + B$ and $A - B$, completing the third and fourth columns below.

	A	B	$A + B$	$A - B$
(a)	11011	1001		
(b)	111001	101010		
(c)	111000111	100101		
(d)	101111011	1000101		

1-8. Find $A + B$ and $A - B$ by converting each number to binary and doing the additions and subtractions in binary. Check the results by converting back to decimal.

	A	B	$A + B$	$A - B$
(a)	67	39		
(b)	145	78		
(c)	31,564	26,797		

1-9. How high can you count using only your fingers?

Section 1-9.

1-10. Find the 8-bit 2's complement of the following numbers:
(a) 99
(b) -7
(c) -102

1-11. Determine by inspection which of the following 2's complement numbers are divisible by 4:
(a) 11011010
(b) 10011100
(c) 01111000
(d) 00001010
(e) 01000001
(f) 01010100

1-12. Express each of the following numbers in 9-bit, 2's complement form, and add them:

(a)   85
  $+37$

(c) -85
  $+37$

(b)   85
  $+(-37)$

(d) -85
  $+(-37)$

1-13. Do the following subtractions after expressing the operands in 11-bit, 2's complement notation:

(a)   36
  $-(23)$

(c) -450
  $-(-460)$

(b)   835
  $-(214)$

(d)   310
  $-(-579)$

1-14. A number in 2's complement form can be inverted by subtracting it from -1. Invert 25 using this procedure and 8-bit numbers.

Section 1-10.

1-15. Convert the following binary numbers to hexadecimal:
(a) 11111011
(b) 1011001
(c) 10000011111100
(d) 10010101100011101

1-16. Convert the following hex numbers to binary:
   (a) 129
   (b) 84C5
   (c) 5CF035
   (d) ABCDE2F
1-17. Convert the numbers in Problem 1-16 to decimal numbers.
1-18. Convert the following numbers to hex:
   (a) 139
   (b) 517
   (c) 2,000
   (d) 105,684
1-19. Perform the following hex additions:
   (a)   99   **(b)** CB
        +89   DD

   (c) 15F02   **(d)** 2CFB4D
       3C3E   5DC98B

1-20. Perform the following hex subtractions:
   (a) 59   **(b)** 1CC
        F   DE

   (c) 1002   **(d)** 5F306
       5F8   135CF
1-21. Find the negative equivalent of the following hex numbers:
   (a) 23
   (b) CB
   (c) 500
   (d) 1F302
   (e) F5630

Answers to selected problems are given in Appendix F.

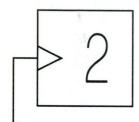

Boolean Algebra

2-1 Instructional Objectives

This chapter presents the rules, theorems, and operations of Boolean algebra so that they may be applied to digital design. The reader will learn to:

1. Construct truth tables.
2. Simplify Boolean expressions.
3. Analyze electronic circuits using Boolean algebra.
4. Design and construct electronic circuits that physically implement Boolean expressions.
5. Take the complement of a given Boolean expression.
6. Analyze and design relay circuits using Boolean algebra.

2-2 Self-Evaluation Questions

Watch for the answers to these questions as you read the chapter. They should help you to understand the material. When you have completed the chapter, return to this section and be sure you can answer the questions below:

1. Given N variables, how many entries must a truth table contain?
2. How are the outputs of a truth table determined?

3. How is addition similar to OR operation and multiplication to AND operation?
4. How are the theorems of Boolean algebra used?
5. What arithmetic operations are not allowed in Boolean algebra?
6. What is the relationship between a Boolean expression and its complement?
7. What is the significance of a bubble in a logic circuit?
8. How do you obtain the alternate representation for a gate?
9. Why should logical expressions be simplified before attempting to build them?
10. What is the advantage of selecting gates so that bubbles are connected to bubbles?
11. Are normally open (n.o.) and normally closed (n.c.) relay contacts complements of each other? Explain.

2-3 Introduction to Boolean Algebra

Boolean algebra is a branch of mathematics that is directly applicable to digital design because the equations of Boolean algebra can be physically implemented by using electronic gates.

The basic Boolean equation is simply:

$$G = f(w, x, y, z, \cdots) \tag{2-1}$$

Equation (2-1) states that the output, G, is a *function* of the input variables w, x, y, z, and so on. Stated another way, the value of G is *determined by the values of the input variables*. The unique feature of Boolean algebra is that all the variables in a Boolean equation, *outputs as well as inputs*, may only assume the values 0 or 1—that is, true or false. Numbers such as 2, -1, or 0.5 are *excluded* from Boolean equations. There is, however, a direct correspondence between digital circuits, also restricted to 0s and 1s on both their inputs and outputs, and Boolean equations.

2-3-1 Word Problems

When a supervisor first presents a design problem to an engineer, he or she does so by giving the engineer a *verbal description* of what he *feels* he wants designed. All too often, the descriptions are vague in many respects and fail to define the output desired in response to many significant input combinations. Boolean algebra can help clarify problems stated in words as well as electronic problems. To apply Boolean algebra to a *word problem*, a two-step procedure must be followed:

1. The number of variables implied by the definition of the problem must be determined.
2. The significance of each of the two values of each variable must be defined.

An example should make this clear.

EXAMPLE 2-1

Given the following statement, identify the variables and assign a value to each:

The president of a company and 3 assistants are voting on whether to accept a contract. If the president votes for it, then 2 yes votes (including the president's) are enough; but if the president votes no, all 3 assistants must vote yes in order for the contract to be accepted.

SOLUTION   After some thought we realize that the result or output of this statement is the decision that determines the acceptance or rejection of the contract. Let us call the output C, and assign C a value of 1 if the contract is accepted, and a value of 0 if it is rejected.

There are 4 people whose votes determine the result. Let us call the president P, and the 3 assistants A_1, A_2, and A_3. We can assign a 1 to each individual if he votes to accept the contract and a 0 if he votes to reject it. This statement may be described by the equation:

$$C = f(P, A_1, A_2, A_3) \tag{2-2}$$

This is a Boolean equation; all input variables and the output, C, have a value of either 0 or 1. Only the function itself remains to be defined.

2-3-2   Truth Tables

Truth tables are often constructed to relate the output of a circuit to its inputs. *A truth table lists every possible combination of inputs and the output corresponding to each combination of inputs.* Therefore, if a function has N inputs, there are 2^N possible combinations of these inputs and there will be 2^N entries in the truth table. If the 2^N entries are listed in a haphazard or arbitrary manner, one might inadvertently list some entries twice and omit others. To avoid this, the following entry-listing procedure is recommended:

1.   List the input variables in a binary sequence. If there are four variables, for example, the list should run from 0000 to 1111.
2.   List the output corresponding to each input combination. The output must be determined by examining the statement of the problem and the value of each input variable. A clear problem statement allows the designer to determine the proper output for each combination of input variables.

EXAMPLE 2-2

Construct the truth table for Example 2-1.

SOLUTION   There are four input variables in this example, so we must have $2^4 = 16$ lines in the truth table. Arbitrarily, A_3 is chosen as the least significant variable, A_2 next, then A_1, and the president, P, as the most significant variable. The truth table is shown in Fig. 2-1. First the columns for the 4 input variables are listed in accordance with the above procedure, giving 16 lines in the table, each with a different set of inputs. The output, C, for each line is determined by examining the inputs and the statement of the problem. For example, the first line of the truth table reads 0000, which means that everyone votes against the contract. Of course it cannot be accepted and $C = 0$.

The sixth line of the truth table is 0101, which indicates that the president and the second assistant are against the contract but the first and third assistants are for it. An examination of the word statement indicates the contract is not accepted and again $C = 0$.

The tenth line, 1001, indicates the president and the third assistant are for the contract, which is enough to have the contract accepted, and $C = 1$.

From the truth table we can see whether the contract is accepted for all possible combinations of votes. Thus, the truth table implicitly specifies the function, f, in Eq. (2-2).

2-4 Operations with Boolean Variables

Truth tables are a very precise and comprehensive way of describing a function. Unfortunately, they may quickly become large and unwieldy. For example, 8 input variables require a 256 line truth table, which we might have to write on a scroll. In Boolean algebra, certain operations and manipulations are permissible and they allow us to simplify expressions without the use of truth tables.

P	A_1	A_2	A_3	C
0	0	0	0	0
0	0	0	1	0
0	0	1	0	0
0	0	1	1	0
0	1	0	0	0
0	1	0	1	0
0	1	1	0	0
0	1	1	1	1
1	0	0	0	0
1	0	0	1	1
1	0	1	0	1
1	0	1	1	1
1	1	0	0	1
1	1	0	1	1
1	1	1	0	1
1	1	1	1	1

Figure 2-1 Truth table for Example 2-2.

2-4-1 Complementation

To *complement* a variable is to *reverse* its value. A complemented variable is represented by placing a bar over the variable. Thus, if $x = 1$, $\bar{x} = 0$; conversely, if $x = 0$, $\bar{x} = 1$.[1] An electronic gate whose function is to change logic 1s to 0s (and vice versa) is called an *inverter* (see section 2-5-2).

In technical literature, complementation is often referred to as the NOT of a variable. Thus, the term $\bar{A}$ is often called A-bar or NOT A.

2-4-2 Addition—The OR Operation

Boolean *addition* is equivalent to a logical OR. The *plus* symbol $(+)$ is the symbol used to indicate addition or ORing. As in ordinary arithmetic, $0 + 0 = 0$ and $0 + 1 = 1$. The difficult question is "How much is $1 + 1$ in a system where the number 2 is not allowed?" By definition, $1 + 1 = 1$. Thus, the expression $s = x_1 + x_2$ means that $s = 0$ only if *both x_1 and x_2 are 0*, but $s = 1$ if *either x_1 or x_2* **or** both equal 1. In general, the equation

$$s = x_1 + x_2 + \cdots + x_n$$

means $s = 1$ *if any x is 1* and $s = 0$ *only if all the x's are 0.*

EXAMPLE 2-3

You want to go to the movies but refuse to go alone. So you send a note to each of your four friends, Alice, Betty, Cindy, and Doris, asking them to join you. You will go to the movies if *one or more* of them say yes. Express this situation as a Boolean equation.

SOLUTION Let us choose the variables *A, B, C,* and *D* to represent the girls, and assign a value of 1 to each variable if that girl says yes. *M* is chosen as the variable "movies," and $M = 1$ means you will go. Equation (2-3) represents the situation.

$$M = A + B + C + D \qquad (2\text{-}3)$$

Arithmetically, if any one or more of the girls say yes, *M* is a 1 and you go to the movies. Logically we can say you will go to the movies if Alice OR Betty OR Cindy OR Doris OR any combination of the girls say yes. This demonstrates the equivalence of Boolean algebra and the logical OR.

Another example of the OR function can occur when using a personal computer (PC) with DOS. If the operator types an erroneous command, the computer often responds with the message BAD COMMAND OR FILE NAME. This means the command cannot be executed because the computer cannot understand the command OR it cannot find the file. Possibly both are in error.

---

[1]Some authors use x' instead of $\bar{x}$ to denote the complement of x.

2-4-3 Multiplication—The AND Operation

Boolean multiplication is equivalent to a logical AND operation. The rules for multiplication are the same as in ordinary arithmetic. If $s = x \cdot y$,[2] $s = 1$ only if *both x and y are* 1, and $s = 0$ *if either or both x and y are* 0. In general the equation

$$s = x_1 x_2 \cdots x_n$$

means that $s = 1$ *only if all the x values are 1* or that $s = 0$ if **any x** *has a value of 0.*

EXAMPLE 2-4

The girls (named in Example 2-3) are thinking about going to Rochester. They each have $3.00 and their car needs $11.00 worth of gas to get there. Let R be a variable that equals 1 if they do go to Rochester and express this situation as a Boolean equation.

SOLUTION They can go to Rochester only if they all agree to go and pay their share of the gas. Thus the Boolean equation is:

$$R = ABCD \tag{2-4}$$

Logically, Eq. (2-4) says they will go to Rochester only if Alice AND Betty AND Cindy AND Doris all agree to go and pay for the gas.

2-4-4 Other Permissible Boolean Operations

The Boolean operations of *addition* and *multiplication* (ANDing and ORing) are both *commutative* and *associative*. Therefore:

1. $xy = yx$
2. $x + y = y + x$
3. $(x + y) + z = x + (y + z) = x + y + z$
4. $(xy) z = x (yz) = xyz$

This allows us to manipulate Boolean expressions *without* concern for the order in which variables appear.

Factoring is another permissible operation in Boolean algebra. For example, $xy + xz = x (y + z)$.

2-4-5 Prohibited Operations

The operations of *subtraction* and *division* are *not* permitted in Boolean algebra. Consequently, if *identical terms* appear on either side of an equation, they *cannot* be cancelled (by subtraction) as in ordinary algebra (see Example 2-19). It also means both sides of an equation cannot be divided by a common variable. Fractions such as x/y do *not* occur in Boolean algebra.

---

[2]As in ordinary algebra, the dot signifying a product is usually omitted. Thus, $s = x \cdot y$ is the same as $s = xy$; the *latter* form is *preferred* and used throughout this book.

The COMPLEMENT, OR, and AND functions that were discussed in section 2-4 can all be implemented using *logic gates*. *A logic gate is an electronic circuit that has one or more inputs and only one output.* All the inputs and the output are voltages at either the 0 or 1 levels. The output therefore is a logical function of the inputs. Logic gates are packaged inside Integrated Circuits (ICs) with several gates within each package. The simplest gates, those discussed here, all come in 14-pin Dual-In-line Packages (DIPs). The packages are called dual-in-line because they have two rows of pins for their connections. Figure 2-2a shows the actual IC and Fig. 2-2b shows the pin numbering for the 14-pin package.

Figure 2-2 (a) A 14-pin DIP package; and (b) Pinout and dimensions (Reprinted by permission of Texas Instruments.)

ICs get their power from a single supply of voltage and power, labeled V_{CC}. For almost all ICs, V_{CC} is +5 Volts. Power, and its return ground, are generally applied to the *corner pins* of an IC. For most 14-pin packages, V_{CC} is placed on pin 14 and ground is connected to pin 7.

2-5-1 IC Numbering and Lettering

Digital ICs are labeled starting with the number **74**. This is usually followed by some letters, then another two- or three-digit number, and another letter. The **74** at the start indicates that it is a member of the **7400** *series* of ICs. This is the series that is almost universally used and is considered in this book.

The **7400** series consists of several *families* of ICs. The next set of letters indicates which family the IC belongs to. Families need not concern us now; they are discussed in Chapter 4.

The next number indicates the *function* (AND, OR, NAND, etc.) of the chip or IC. *ICs from all families with the same number perform the same function.* The last letter indicates the *type of package* the IC comes in. Most commonly, this is the inexpensive plastic package denoted by the letter N.

In most cases, the family of the IC and the package type are unimportant because *they do not affect the logic function*. Therefore, ICs will usually be identified in this book as **74x** where **x** is the general term for the family. The numbers for an inverter, for example, are **74** at the start and **04** at the end. Table 2-1 gives the numbering for the inverters in some of the families. These can be referred to, collectively, as **74x04**.

2-5-2 Inverters

Three of the simplest and most commonly used logic gates are the inverter, the OR gate and the AND gate. They will be discussed in the following sections. The function of an *inverter* is either to change a 1 input to a 0 output, or a 0 input to a

TABLE 2-1

Nomenclature for inverters in various families

IC Number	Letters	Family
7404N	None	Standard
74LS04N	LS	Low-power Schottky
74HCT04N	HCT	High-speed, TTL compatible CMOS

1 output. Hence it *inverts* or *complements* the signal. An inverter has only one input and one output. The symbol for an inverter, a small triangle with a bubble, is shown in Fig. 2-3. Both Figs. 2-3a and 2-3b are valid representations of an inverter. The standard TTL inverter is the **7404** IC.

The small circle shown at the *output* of Fig. 2-3a (or the input of Fig. 2-3b) is called a *bubble*, and is used to indicate the parts of the circuit where the asserted or true level of a signal is LOW. If a signal is actually present (asserted), *we expect those points in the signal path that are not connected to bubbles to be logic 1s, and those points that are connected to bubbles to be logic 0s.* Conversely, if the same signal is not present or asserted (logic 0) the points connected to bubbles should be 1, and the points wihout bubbles should be 0. Thus, in Fig. 2-3a, if the input signal to the inverter is asserted (1), the output will be 0. This is indicated by the bubble on the output. If the signal is not asserted, the input, A, will be LOW and the output ($\bar{A}$) will be HIGH. In Fig. 2-3b, the asserted level of the signal on the input is LOW and the asserted level of the output is HIGH, as indicated by the absence of a bubble. If the input A is asserted or present, the input to the gate of Fig. 2-3b should be a 0.

(a) Inverter-bubble on output

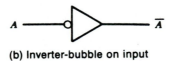

(b) Inverter-bubble on input

Figure 2-3 Two equivalent representations for an inverter.

Engineers obtain their information about specific ICs from *data books*, such as *The TTL Data Book* published by Texas Instruments, a leading manufacturer of ICs. Figure 2-4 shows a page of the Data Book that pertains to the **74x04** inverters. Looking at the left column we find:

1. The package options: a description of the various packages in which this IC is available.
2. A description of the **74x04**.
3. A function table: It shows how the IC functions, and gives the outputs for the various inputs. For the **74x04** it simply shows that the output (Y) is HIGH if the input (A) is LOW, and vice-versa.
4. The IEEE (Institute of Electrical and Electronic Engineers) Standard logic symbol from Standard 91-1984: This standard has not proved to be popular and we do not use it in this book.

- **Package Options Include Plastic "Small Outline" Packages, Ceramic Chip Carriers and Flat Packages, and Plastic and Ceramic DIPs**

- **Dependable Texas Instruments Quality and Reliability**

description

These devices contain six independent inverters.

The SN5404, SN54LS04, and SN54S04 are characterized for operation over the full military temperature range of −55°C to 125°C. The SN7404, SN74LS04, and SN74S04 are characterized for operation from 0°C to 70°C.

FUNCTION TABLE (each inverter)

INPUTS A	OUTPUT Y
H	L
L	H

logic symbol†

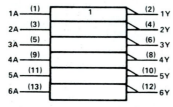

†This symbol is in accordance with ANSI/IEEE Std. 91-1984 and IEC Publication 617-12.
Pin numbers shown are for D, J, and N packages.

logic diagram (positive logic)

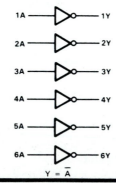

$$Y = \overline{A}$$

SN5404 . . . J PACKAGE
SN54LS04, SN54S04 . . . J OR W PACKAGE
SN7404 . . . N PACKAGE
SN74LS04, SN74S04 . . . D OR N PACKAGE
(TOP VIEW)

```
1A  [1    14] VCC
1Y  [2    13] 6A
2A  [3    12] 6Y
2Y  [4    11] 5A
3A  [5    10] 5Y
3Y  [6     9] 4A
GND [7     8] 4Y
```

SN5404 . . . W PACKAGE
(TOP VIEW)

```
1A  [1    14] 1Y
2Y  [2    13] 6A
2A  [3    12] 6Y
VCC [4    11] GND
3A  [5    10] 5Y
3Y  [6     9] 5A
4A  [7     8] 4Y
```

SN54LS04, SN54S04 . . . FK PACKAGE
(TOP VIEW)

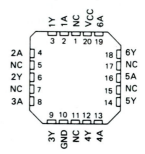

NC - No internal connection

TEXAS INSTRUMENTS
POST OFFICE BOX 655012 • DALLAS, TEXAS 75265

Figure 2-4 Data book description of the **74x04** (Reprinted by permission of Texas Instruments.)

5. The positive logic diagram: This is similar to Fig. 2-3 and is the type most commonly used, and it is used throughout this book. It shows that the **74x04** IC contains six inverters. It also associates each pin on the IC with its function.

6. The logic equation for the gate, $Y = \overline{A}$ is shown below the logic diagram.

The right side of the page shows the pinouts for several types of packages. In general, we will use the N package because it is the least expensive and most commonly available package. The pinout shows, for example, that the input to gate 4 of the **74x08** is input 4A, which is on pin 9, and the corresponding output is 4Y, which is on pin 8.

EXAMPLE 2-5

The input shown in Fig. 2-5a is applied to pin 3 of a **74x04**. What is its output and where can it be found?

SOLUTION   The output of an inverter is the complement of its input. Figure 2-3 shows that the output of the gate whose input is on pin 3 appears on pin 4. The output is shown in Fig. 2-5b.

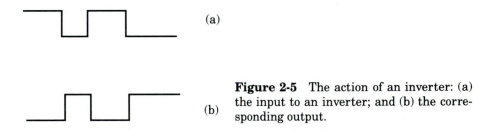

(a)

(b)

Figure 2-5   The action of an inverter: (a) the input to an inverter; and (b) the corresponding output.

2-5-3   OR Gates

An OR gate electrically performs the logical OR function; if any input is HIGH (logic 1), the output is also HIGH. A two input OR gate and its truth table are shown in Fig. 2-6. The truth table shows that if either input is 1, the output will be 1.

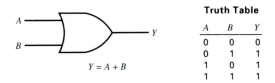

Truth Table

A	B	Y
0	0	0
0	1	1
1	0	1
1	1	1

$Y = A + B$

Figure 2-6   A 2-input OR gate.

EXAMPLE 2-6

The two inputs to an OR gate are shown in Fig. 2-7b. Sketch the output.

SOLUTION   The output is also shown in Fig. 2-7b. Whenever either (or both) inputs are HIGH, the output is HIGH.

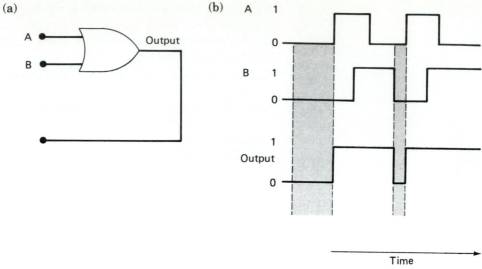

Figure 2-7 Solution to Ex. 2-6: (a) OR gate; and (b) Outputs. (Ronald J. Tocci, DIGITAL SYSTEMS: Principles and Applications, 5e, © 1991. Reprinted by permission of Prentice-Hall, Englewood Cliffs, New Jersey.)

Conceptually, three or more input OR gates are possible. Three- and four-input OR gates are shown in Fig. 2-8. Example 2-3 of section 2-4-2 could be implemented electrically using the 4-input OR gate. If the A, B, C, and D represent Alice, Betty, Cindy, and Doris, then a HIGH on the Y output indicates that one or more of the girls has agreed to go with you, so you are going to the movies.

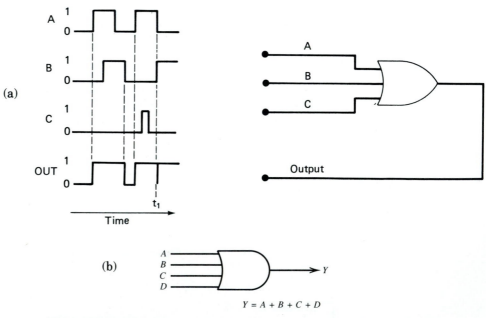

Figure 2-8 Multiple input OR gates: (a) 3-input OR gate; and (b) 4-input OR gate.

(Ronald J. Tocci, DIGITAL SYSTEMS: Principles and Applications, 5e, © 1991. Reprinted by permission of Prentice-Hall, Englewood Cliffs, New Jersey.)

Unfortunately, the gates shown in Fig. 2-8 do not exist. In the **7400** series, only 2-input OR gates are manufactured. These are the **74x32** ICs, whose pin-out is shown in Fig. 2-9. A 3-input OR gate can be constructed using two **74x32** gates, as shown in Fig. 2-10. Its output will be HIGH if any one of the three inputs is HIGH. Four or more input AND gates can be constructed by extending the number of 2-input OR gates shown in Fig. 2-10.

FUNCTION TABLE (each gate)

INPUTS		OUTPUT
A	B	Y
H	X	H
X	H	H
L	L	L

(a)

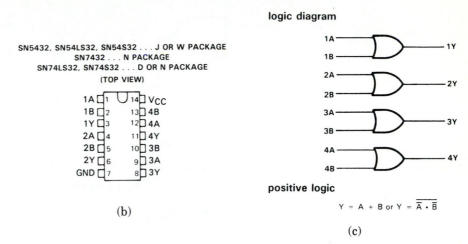

SN5432, SN54LS32, SN54S32 . . . J OR W PACKAGE
SN7432 . . . N PACKAGE
SN74LS32, SN74S32 . . . D OR N PACKAGE
(TOP VIEW)

1A	1	14	V_{CC}
1B	2	13	4B
1Y	3	12	4A
2A	4	11	4Y
2B	5	10	3B
2Y	6	9	3A
GND	7	8	3Y

(b)

logic diagram

positive logic

$$Y = A + B \text{ or } Y = \overline{\overline{A} \cdot \overline{B}}$$

(c)

Figure 2-9 The **74x32**: (a) function table; (b) Pinout; and (c) logic diagram. (Reprinted by permission of Texas Instruments.)

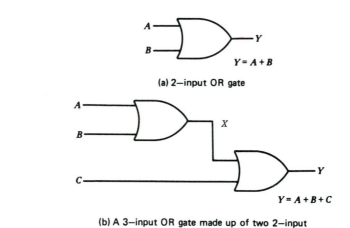

$$Y = A + B$$

(a) 2—input OR gate

$$Y = A + B + C$$

(b) A 3—input OR gate made up of two 2—input OR gates

Figure 2-10 A 3-input OR gate made up of two 2-input OR gates.

EXAMPLE 2-7

In Fig. 2-10b, assume that input A is HIGH. Show that the output, Y, will also be HIGH.

SOLUTION If the A input to OR gate 1 is HIGH, its output, X, must also be HIGH. But X is an input to OR gate 2, so its output, Y, must also be HIGH. This conforms to the OR function.

2-5-4 AND Gates

The AND gate is similar to the OR gate and conceptually may have any number of inputs. Its output is the logical AND of the inputs (i.e., all inputs must be 1 in order for the output to be a 1). The standard symbol for an AND gate is shown in Fig. 2-11, where both 2- and 4-input AND gates are shown along with their logical equations.

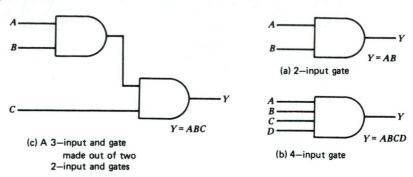

Figure 2-11   AND gates.

The basic TTL AND gate in the standard TTL series is the **74x08** quad 2-input AND gate. Triple 3- and dual 4-input AND gates are also available. If only **74x08**s are used, however, 3- and 4-input AND gates can be constructed from **74x08** gates as shown in Fig. 2-11c.

The manufacturer's literature must be consulted to properly use any TTL gate. Figure 2-12 is taken from the TTL Logic Data Book, 1988, published by Texas Instruments, Inc. It shows the pinout, function table, and the logic diagram of the **74x08** AND gate.

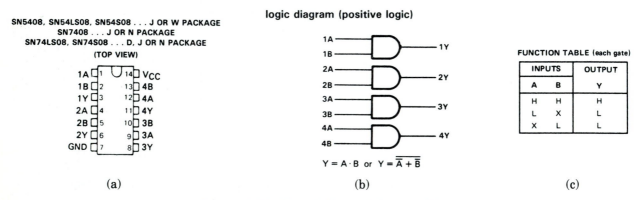

Figure 2-12   The **74x08** quad 2-input AND gate: (a) pinout; (b) logic diagram; and (c) function table. (Reprinted by permission of Texas Instruments.)

2-6   Circuit Construction and Testing

This section covers the simplest methods of constructing and testing a circuit in the laboratory. They allow the user to go into the laboratory and test the ICs discussed in the previous section and to verify the circuits and theorems presented in the following sections.

2-6-1 Construction

Perhaps the simplest way to construct an IC is to use a *superstrip* as shown in Fig 2-13. A superstrip has a series of holes in it at 0.1 inch centers, and ICs can be plugged directly into them. Wires can also be plugged into connecting holes to make any necessary connections. This is the easiest way to build up an experimental digital circuit. More permanent digital construction is discussed in Chapter 19.

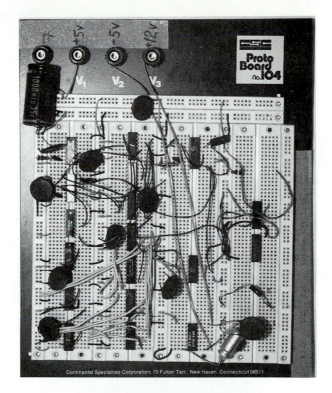

Figure 2-13 A superstrip

2-6-2 Testing

At its simplest level, testing is a matter of connecting logic 1s and 0s to the input of a digital circuit and monitoring the outputs to determine if they conform with the function table for the IC or the expected results for the circuit as a whole. Logic inputs can be taken from a switch as shown in Fig. 2-14. If the switch is open, the output is *pulled-up* (see section 5-5-1) to +5 V and is a logic 1. If the switch is closed, the output is connected directly to ground and is a logic 0. If several inputs are needed, several switches must be used.

There are several ways to determine if the voltage at a point in the circuit is a logic 1 or 0:

- Indicator Lights: An indicator light or Light-Emitting Diode (LED; see section 11-10-1) can be connected to the point. The front panel of many digital devices uses these lights to indicate their status.
- Oscilloscopes: The probe of an oscilloscope can be connected to the point of interest. If the oscilloscope is set on dc (direct current), the

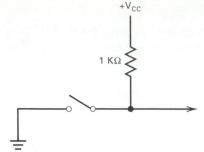

Figure 2-14 Connecting a switch for a Logic 1 or 0.

trace will go up when the point is at a logic 1. Oscilloscopes will also show if there is any variation in the levels at the point to which it is connected.

• Logic Probes: A logic probe is shown in Fig. 2-15. It is often the student's favorite way of testing. A logic probe has three lights on it: one indicates a 0, another indicates a 1, and the third light flashes whenever the voltage changes states.

Figure 2-15 A Logic Probe. Courtesy of Hewlett-Packard, Inc.

2-6-3 Light and Switch Panels

The use of a light and switch panel is very helpful in testing digital circuits, especially those built on superstrips. Figure 2-16 is a photograph of a light and switch panel used at the Rochester Institute of Technology. It consists of three parts:

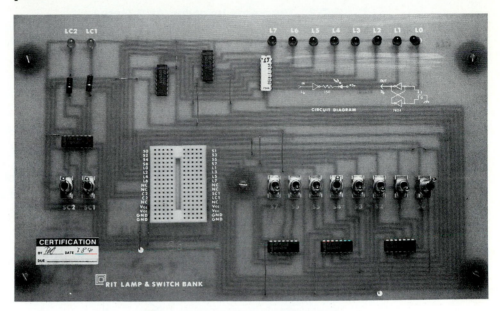

Figure 2-16 A Light and Switch Panel

- Ten LEDs to indicate the logic levels at 10 places in a circuit.
- Ten switches to connect 1s or 0s into the circuit. The switches are debounced (see section 7-8).
- A superstrip that is connected to the lights and switches. This allows the user to run wires from the switches to the circuit and run wires from the circuit outputs to the LEDs.

These panels have been used continuously to test both digital and microprocessor circuits.

2-7   Theorems

It is advantageous to reduce logic expressions to their simplest forms. The reasons for this are discussed in section 2-8. Several *basic theorems* in Boolean algebra are used to *simplify* expressions and equations. These are listed here along with pertinent explanations and derivations as necessary.

2-7-1   Theorems Involving OR Gates

1. $x + 0 = x$

ORing a variable with 0 does *not* change the value of that variable.

2. $x + 1 = 1$

This theorem should be interpreted as the OR of anything (a variable or a combination of several variables forming an expression) with a logic 7 equals 1. This is extremely useful; it often allows the engineer to delete some variables from an expression (see Example 2-9).

2-7-2 Theorems Involving AND Gates

3. $x \cdot 0 = 0$
4. $x \cdot 1 = x$

As in ordinary algebra, ANDing with 0 gives a 0 result and the ANDing of a variable with 1 does not change the value of the variable.

Theorems 1 and 2 can be demonstrated in the laboratory by the circuit of Fig. 2-17. Input A of the **74x32** OR gate is connected to a square wave generator, and the square wave can be observed on an oscilloscope. Input B is connected to a switch. If the switch is in the UP position, it is connected to V_{CC} through a resistor. This makes the input HIGH, or a logic 1. If the switch is in the DOWN position it is connected to ground, or a logic 0.

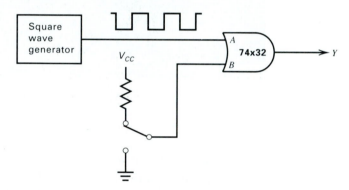

Figure 2-17 A laboratory circuit to demonstrate Theorems 1 and 2.

EXAMPLE 2-8

If an oscilloscope is connected to the output of Fig. 2-17, what will it show

    a) if the switch is down?
    b) if the switch is up?

SOLUTION

    a) If the switch is down, input B is a 0. The output will then be the same as input A. Thus the student should observe a square wave. This is an illustration of theorem 1 ($x + 0 = x$).

    b) If the switch is up, the B input will always be 1, and the output will always be 1. The student should observe a straight line on the oscilloscope at a level somewhere between 3 and 5 volts.

Theorems 3 and 4 can be demonstrated similarly in the laboratory by replacing the OR gate with an AND gate.

2-7-3　Other Theorems Involving a Single Variable

5. $x \cdot x = x$
6. $x + x = x$
7. $x \cdot \bar{x} = 0$
8. $x + \bar{x} = 1$

These theorems are best demonstrated by simply substituting 0s and 1s for x. Theorems 7 and 8 are apparent when we remember that if $x = 1, \bar{x} = 0$ and vice versa.

Theorem 7 can be demonstrated in the laboratory by using the circuit of Fig. 2-18. The inputs to the AND gate will always be complementary and the output should always be a logic 0 (ground potential).

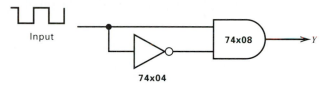

Figure 2-18　A demonstration of Theorem 7.

Theorem 8 can also be demonstrated in the laboratory, by replacing the AND gate with an OR gate. Now the output should always be HIGH.

These theorems can be used to *simplify* more complex Boolean expressions, as the following examples show.

EXAMPLE 2-9

Simplify the logic expression $x + \bar{x} + y$.

SOLUTION　We note that $x + \bar{x} = 1$ by Theorem 8. Therefore,

$$x + \bar{x} + y = 1 + y = 1 \qquad \text{(Theorem 2)}$$

EXAMPLE 2-10

Simplify the logic expression $x(y + z)\bar{x} + w$.

SOLUTION　This expression can be rewritten as

$$x \cdot \bar{x} \cdot (y + z) + w$$

but $x \cdot \bar{x} = 0$, which eliminates the first term entirely. Therefore,

$$x(y + z)\bar{x} + w = w$$

2-7-4   Theorems Involving More Than One Variable

The theorems of the previous section involve only one Boolean variable. Theorems also exist to simplify expressions involving *several* Boolean variables.

$$9. \quad x + xy = x$$

Proof
The expression $x + xy$ can be factored

$$x + xy = x(1 + y) = x$$

Since $1 + y = 1$ by Theorem **2**.

After a theorem has been proven, it is wise to go back and examine the original equality. This verifies the theorem and gives one a feeling for it and its uses. When the expression $x + xy$ is examined, we note that it equals 1 if $x = 1$ and its equals 0 if $x = 0$, regardless of the value of y.

$$10. \quad x(x + y) = x$$

Proof
By multiplying, the left side becomes

$$x(x + y) = x \cdot x + xy = x + xy$$

This is identical to the left side of Theorem **9**.

$$11. \quad x + \bar{x}y = x + y$$

Proof
Here a bit of imagination helps. Consider the expression $(x + \bar{x})(x + y)$.

$$(x + \bar{x})(x + y) = x \cdot x + xy + \bar{x}x + \bar{x}y$$
$$= x + xy + \bar{x}y$$
$$= x + \bar{x}y$$

But $(x + \bar{x}) = 1$ so it is also apparent that

$$(x + \bar{x})(x + y) = x + y$$

This proves Theorem **11**.

Examining both sides of the original equality we see that they are surely 1 if $x = 1$, but if $x = 0$ both sides of the expression are equal to y.

$$12. \quad xy + \bar{y}z + xz = xy + \bar{y}z$$

Proof

This theorem can be proved by properly expanding the right side.

$$\begin{aligned}
xy + \bar{y}z &= xy(1 + z) + \bar{y}z(1 + x) \\
&= xy + xyz + \bar{y}z + x\bar{y}z \\
&= xy + \bar{y}z + xz(y + \bar{y})
\end{aligned}$$

The situations where Theorem **12** can be applied are not always obvious. One must generally look for a *variable multiplied by a second variable*, and the *complement of the first variable multiplied by a third variable*. Then, *if the product of the second and third variables is present, it can be eliminated*.

EXAMPLE 2-12

Simplify $f(w, x, y, z) = \bar{x}y + wy\bar{z} + xw\bar{z}$.

SOLUTION In this problem one must be careful. The x, y, and z of the specified function do not correspond directly to the x, y, and z variables in Theorem **12**. Some engineers find it clearer to change the variables to a, b, c, and d.

$$f(w, x, y, z) = f(a, b, c, d) = \bar{b}c + ac\bar{d} + ba\bar{d}$$

Examining this expression, we find a variable (b) times a second variable $(a\bar{d})$ and its complement $(\bar{b})$ times a third variable (c). Looking at Theorem 12, the following correspondence can be established:

$$y = b, \quad x = a\bar{d} \quad and \quad z = c$$

Therefore,

$$\bar{b}c + ac\bar{d} + ba\bar{d} = \bar{b}c + ba\bar{d}$$

or returning to the original function:

$$f(w, x, y, z) = \bar{x}y + wy\bar{z} + xw\bar{z} = \bar{x}y + xw\bar{z}$$

The term $wy\bar{z}$ can be deleted without changing the value of the expression.

$$13. \quad (x + y)(x + \bar{y}) = x$$

Proof

$$\begin{aligned}
(x + y)(x + \bar{y}) &= x + xy + x\bar{y} + \bar{y}y \\
&= x + x(y + \bar{y}) + 0 \\
&= x + x \\
&= x
\end{aligned}$$

$$14. \quad (y + z)(\bar{y} + x) = xy + \bar{y}z$$

Proof
Direct expansion gives

$$(y + z)(\bar{y} + x) = xy + \bar{y}z + xz$$

By Theorem **12** this reduces to $xy + \bar{y}z$.

2-8   Logical Manipulation

In the practical world, each logic expression is translated into electronic gates or relays and their interconnections. Therefore, the *simplest* and *most compact* Boolean statement results in the *fewest* gates and wires. Gates themselves are relatively inexpensive. But when the additional costs, in both time and money, of mounting extra circuits, debugging them, and repairing them when they fail are considered, appreciable savings accrue to people who make an effort to eliminate unneeded circuits.

2-8-1   Reduction of Logical Expression

Often Boolean expressions an engineer sees in industry (and these expressions may describe actual circuits as built) are *not* written in their simplest form. By using the theorems of section 2-7-5, many of these expressions can be reduced, which leads to the elimination of useless circuitry. The following examples demonstrate some of the reduction techniques used in simplifying and reducing logical expressions.

EXAMPLE 2-13

Simplify the expression

$$f(x, y, z) = \bar{x}(\bar{y} + \bar{z}) + yz + x\bar{z}$$

SOLUTION   It is usually best to begin by breaking parentheses open.

$$f(x, y, z) = \bar{x}(\bar{y} + \bar{z}) + yz + x\bar{z} = \bar{x}\bar{y} + \bar{x}\bar{z} + yz + x\bar{z}$$

Examining this expression, we find that

$$\bar{x}\bar{z} + x\bar{z} = \bar{z}(x + \bar{x}) = \bar{z} \qquad \text{(Theorem 8)}$$

Therefore,

$$\begin{aligned}
f(x, y, z) &= \bar{x}\bar{y} + \bar{x}\bar{z} + yz + x\bar{z} \\
&= \bar{x}\bar{y} + yz + \bar{z} \\
&= \bar{x}\bar{y} + y + \bar{z} \qquad &\text{(Theorem **11**)} \\
&= \bar{x} + y + \bar{z} \qquad &\text{(Theorem **11**)}
\end{aligned}$$

The last two simplifications ($yz + \bar{z} = y + \bar{z}$) and ($\bar{x}\bar{y} + y = y + \bar{x}$) used Theorem **11**. No further simplifications can be made.

EXAMPLE 2-14

Simplify the expression

$$f(A, B, C) = (A + B)(A + BC) + \bar{A}\bar{B} + \bar{A}\bar{C}$$

SOLUTION

$$\begin{aligned}
f(A,B,C) &= (A + B)(A + BC) + \bar{A}\bar{B} + \bar{A}\bar{C} \\
&= A + AB + ABC + BC + \bar{A}\bar{B} + \bar{A}\bar{C} \\
&= A(1 + B + BC) + BC + \bar{A}\bar{B} + \bar{A}\bar{C}
\end{aligned}$$

The term in parentheses equals 1 and can be dropped:

$$f(A,B,C) = A + \bar{A}\bar{B} + \bar{A}\bar{C} + BC$$

Since A is a term already in the expression, another identical term (another A) can be added without changing the value of the expression (by Theorem **6**). Therefore,

$$\begin{aligned}
f(A,B,C) &= (A + \bar{A}\bar{B}) + (A + \bar{A}\bar{C}) + BC \\
&= A + \bar{B} + A + \bar{C} + BC \\
&= A + \bar{B} + \bar{C} + B
\end{aligned}$$

But $B + \bar{B} = 1$. Therefore, the value of the function $f(A,B,C) = 1$.

Logically this means there is no combination of variables A, B, and C one can choose that makes $f(A,B,C) = 0$. Physically it means no gates are actually needed to implement f, other than a conductor. If f is needed it can be wired directly to a voltage in the logic 1 range.

EXAMPLE 2-15

Simplify

$$f(x,y,z) = x + \bar{y}z + (x + \bar{y}z) \text{ times } Q$$

where Q = any expression.

SOLUTION The expression $x + \bar{y}z$ can be factored from the given expression

$$\begin{aligned}
f(x,y,z) &= x + \bar{y}z + (x + \bar{y}z) Q \\
&= (x + \bar{y}z)(1 + Q) \\
&= x + \bar{y}z \qquad\qquad\qquad \text{(Theorem 2)}
\end{aligned}$$

since the last term in parentheses equals 1.

EXAMLE 2-16

Simplify

$$f(W,X,Y,Z) = \bar{W}XY\bar{Z} + \bar{W}XYZ + W\bar{X}\bar{Y}Z + W\bar{X}YZ + WX\bar{Y}Z + WXY\bar{Z}$$

SOLUTION

$$\begin{aligned}
f(W,X,Y,Z) &= \bar{W}XY\bar{Z} + \bar{W}XYZ + W\bar{X}\bar{Y}Z \\
&\quad + W\bar{X}YZ + WX\bar{Y}Z + WXY\bar{Z} \\
&= \bar{W}XY(\bar{Z} + Z) + W\bar{X}Z(\bar{Y} + Y) \\
&\quad + WX\bar{Y}Z + WXY\bar{Z}
\end{aligned}$$

Before the indicated simplifications are performed, it is best to rewrite the third and first terms near the last two terms. This allows us to simplify the last two terms.

$$\begin{aligned}
f(W,X,Y,Z) &= \bar{W}XY(\bar{Z} + Z) + W\bar{X}Z(\bar{Y} + Y) + W\bar{X}\bar{Y}Z \\
&\quad + WX\bar{Y}Z + WXY\bar{Z} + \bar{W}XYZ \\
&= \bar{W}XY(\bar{Z} + Z) + W\bar{X}Z(\bar{Y} + Y) \\
&\quad + W\bar{Y}Z(\bar{X} + X) + (W + \bar{W})XY\bar{Z} \\
&= \bar{W}XY + W\bar{X}Z + W\bar{Y}Z + XY\bar{Z}
\end{aligned}$$

EXAMPLE 2-17

Show that the following expression is an equality by adding a term to each side:

$$ab + \bar{a}\bar{b} + bc = ab + \bar{a}\bar{b} + \bar{a}c$$

SOLUTION   By using Theorem **12** we can expand the last two terms on the left side to

$$\bar{a}\bar{b} + bc = \bar{a}\bar{b} + bc + \bar{a}c$$

By using Theorem **12** on the first and third terms of the right side, we find that

$$ab + \bar{a}c = ab + \bar{a}c + bc$$

When we make these substitutions

$$ab + \bar{a}\bar{b} + bc + \bar{a}c = ab + \bar{a}\bar{b} + \bar{a}c + bc$$

A term-by-term comparison now shows that the two sides are equal.

2-8-2   Logical Equivalence[3]

Two Boolean expressions are equivalent *only* if they are equal for *all possible values of the variables* in *both expressions*. To *prove two expressions unequal, it is only necessary to find a single set of values for the variables that makes the expressions unequal*. But if two expressions are equal, this must be proved by logical manipulation or, as a last resort, by truth tables.

EXAMPLE 2-18

Does $X + WZ = WX + \bar{W}X\bar{Y} + W\bar{X}Z + \bar{W}Y$?

SOLUTION   We start by manipulating the right-hand side in an attempt to reduce it.

$$\begin{aligned}
WX &+ \bar{W}X\bar{Y} + W\bar{X}Z + \bar{W}Y \\
&= WX + \bar{W}(Y + X\bar{Y}) + W\bar{X}Z \\
&= WX + \bar{W}(Y + X) + W\bar{X}Z &&\text{(Theorem 11)} \\
&= (W + \bar{W})X + \bar{W}Y + W\bar{X}Z &&\text{(Theorem \ 8)} \\
&= X + W\bar{X}Z + \bar{W}Y &&\text{(Theorem 11)} \\
&= X + WZ + \bar{W}Y
\end{aligned}$$

[3]This section contains advanced material and may be omitted on first reading.

Since no further reductions seem possible, the question is now:

$$\text{Does } X + WZ \stackrel{?}{=} X + WZ + \bar{W}Y$$

The two sides do not look alike; but to prove it, a set of values must be found to make the sides unequal. The right side of the equation contains all the terms in the left side, so if either term (X or WZ) equals 1 boths sides must be 1. However, if we choose values such that the left side is 0, it may be possible to make the right side equal to 1. If we choose $X = 0$, $W = 0$, and $Y = 1$, we get $\bar{W}Y = 1$. Now the left side equals 0 and the right side equals 1. This proves the terms are unequal.

EXAMPLE 2-19

Does $bc + abd + a\bar{c} \stackrel{?}{=} bc + a\bar{c}$

SOLUTION   First we try to make the term abd equal 1 while making the right side equal 0. Unfortunately, we do not succeed (if a and b both equal 1, the right side equals 1), so perhaps the terms are equal. Manipulating the left side we obtain

$$
\begin{aligned}
bc &+ abd + a\bar{c} \\
&= bc + abd + a\bar{c} + ab &&\text{(Theorem } \mathbf{12}) \\
&= bc + a\bar{c} + ab &&\text{(Theorem } \mathbf{9}) \\
&= bc + a\bar{c} &&\text{(Theorem } \mathbf{12})
\end{aligned}
$$

Now we have demonstrated that the two sides *are* equal. This explains why it was *not* possible to find a set of values to make them different.

EXAMPLE 2-20

A misguided mathematician would like to subtract the term $a\bar{c}$ from both sides of the *equality*:

$$bc + abd + a\bar{c} = bc + a\bar{c}$$

Would they still be equal if he did so?

SOLUTION   With the term $a\bar{c}$ removed, the expressions would be

$$bc + abd = bc$$

Now we find that if $a = 1, b = 1, c = 0,$ and $d = 1$, the left side equals 1 and the right side equals 0; therefore, the terms are unequal. This example demonstrates why subtraction is *not* allowed in Boolean algebra.

2-8-3   Implementing Logic Expressions in the Laboratory

The expressions and results of all the previous examples can be verified by simulating the expressions in the laboratory using AND gates, OR gates, and inverters. Example 2-21 shows that it is not difficult to implement logical expressions in the laboratory.

EXAMPLE 2-21

Use AND and OR gates to implement the function $Y = AB + BCD$.

SOLUTION   The first term requires A and B to be ANDed, and the second term requires that B, C, and D be ANDed. These two partial results must then be ORed together. The circuit using only 2-input gates is shown in Fig. 2-19.

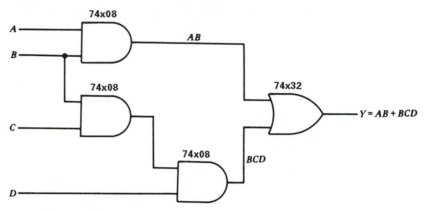

Figure 2-19   Circuit to produce the function $Y = AB + BCD$.

2-9   Complementation of Functions

Two functions, F and $\bar{F}$, are complementary if they depend on the same set of input variables and if, *for every combination of values of the input variables*, the values of F and $\bar{F}$ are *inverse*. One way to determine the complement of a function, F, is to draw up a *truth table*. $\bar{F}$ is then a 1 whenever F is a 0, and 0 whenever F is a 1.

EXAMPLE 2-22

Draw up the truth table for $F = X + \bar{Y}Z$. From it find F and $\bar{F}$ and simplify $\bar{F}$ if possible.

SOLUTION   The truth table of Fig. 2-20 was constructed according to the procedure of section 2-3-2. The value of F is determined from the equation and $\bar{F}$ is then written simply as the inverse of F. Since upon examining the truth table we find $\bar{F}$ to be 1 on lines 0, 2, and 3, we can write the following equation:

$$\begin{aligned}
\bar{F} &= \bar{X}\bar{Y}\bar{Z} + \bar{X}Y\bar{Z} + \bar{X}YZ \\
&= \bar{X}(\bar{Y}\bar{Z} + Y\bar{Z} + YZ) \\
&= \bar{X}(\bar{Y}\bar{Z} + Y) && \text{(Theorem 8)} \\
&= \bar{X}(\bar{Z} + Y) && \text{(Theorem 11)}
\end{aligned}$$

X	Y	Z	F	$\bar{F}$
0	0	0	0	1
0	0	1	1	0
0	1	0	0	1
0	1	1	0	1
1	0	0	1	0
1	0	1	1	0
1	1	0	1	0
1	1	1	1	0

Figure 2-20 Truth table for $F = X + \bar{Y}Z$.

2-9-1 DeMorgan's Theorem

Fortunately the complement of a Boolean expression can be found without using truth tables by **DeMorgan's theorem**, which states the following.
  The complement of a function f can be found by:

1. *Replacing each variable by its complement.*
2. *Interchanging all the* **AND** *and* **OR** *signs.*

To apply DeMorgan's theorem, the given expression is first investigated to find all the AND and OR signs. These are changed to OR and AND signs, respectively, and then the variables are complemented.

┌ **EXAMPLE 2-23**

If $F = X + \bar{Y}Z$, as in Example 2-22, find $\bar{F}$.

SOLUTION The $+$ sign between X and $\bar{Y}$ must be changed to a multiplication sign, and the implied multiplication sign between Y and Z must be changed to a $+$ sign. The variables must also be complemented. This procedure is illustrated below.

$$F = X + \bar{Y} \cdot Z$$
$$\downarrow \downarrow \downarrow \downarrow \downarrow$$
$$\bar{F} = \bar{X} \cdot (Y + \bar{Z})$$

The result, $\bar{F} = \bar{X}(Y + \bar{Z})$, agrees with the results obtained from the truth table.

Two important corrolaries of DeMorgan's theorem are:

$$\overline{(X_1 + X_2 + \cdots + X_N)} = \bar{X}_1\bar{X}_2 \cdots \bar{X}_n \qquad (2\text{-}5)$$

$$\overline{(X_1 X_2 X_3 \cdots X_n)} = \bar{X}_1 + \bar{X}_2 + \bar{X}_3 + \cdots + \bar{X}_n \qquad (2\text{-}6)$$

EXAMPLE 2-24

For Examples 2-3 and 2-4, find $\bar{M}$ and $\bar{R}$ and explain their meaning physically.

SOLUTION In Example 2-3 we found that

$$M = A + B + C + D$$

Using Eq. (2-5) we obtain

$$\bar{M} = \bar{A}\bar{B}\bar{C}\bar{D}$$

Physically this means you are *not* going to the movies ($\bar{M} = 1$) if all the girls say no ($\bar{A} = \bar{B} = \bar{C} = \bar{D} = 1$).

In Example 2-4 we found that

$$R = ABCD$$

Here using Eq. (2-6) we obtain

$$\bar{R} = \bar{A} + \bar{B} + \bar{C} + \bar{D}$$

This means that the girls are *not* going to Rochester ($\bar{R} = 1$) if any one of them doesn't want to go. The rest of the girls, collectively, cannot afford the cost of the gasoline.

Occasionally, it is necessary to simplify expressions that already contain *complemented subexpressions*. The best procedure:

1. Take the complement of that portion of the expression that is complemented.
2. Simplify the remainder of the expression.

EXAMPLE 2-25

Simplify the expression:

$$F = (X + \bar{Z})(\overline{Z + WY}) + (VZ + W\bar{X})(\overline{Y + Z})$$

SOLUTION A complement sign outside a parentheses refers to everything within the parentheses. By DeMorgan's theorem:

$$(\overline{Z + WY}) = \bar{Z}(\bar{W} + \bar{Y})$$

While

$$(\overline{Y + Z}) = \bar{Y}\bar{Z}$$

Then

$$
\begin{aligned}
F &= (X + \bar{Z}) \cdot \bar{Z} \cdot (\bar{W} + \bar{Y}) + (VZ + W\bar{X})\bar{Y}\bar{Z} \\
&= (X\bar{Z} + \bar{Z})(\bar{W} + \bar{Y}) + W\bar{X}\bar{Y}\bar{Z} \\
&= \bar{Z}(\bar{W} + \bar{Y}) + W\bar{X}\bar{Y}\bar{Z} \\
&= \bar{Z}\bar{W} + \bar{Z}\bar{Y} + W\bar{X}\bar{Y}\bar{Z} \\
&= \bar{Z}\bar{W} + \bar{Z}\bar{Y}(1 + W\bar{X}) \\
&= \bar{Z}\bar{W} + \bar{Z}\bar{Y}
\end{aligned}
$$

When expressions to be complemented become complex, the place to substitute signs is not always obvious. Any possible confusion can be eliminated by breaking the function up into subfunctions and simplifying them.

EXAMPLE 2-26

(a) Find the complement of the expression of Example 2-25 without simplifying.

(b) Simplify the complement.

SOLUTION The function

$$F = (X + \bar{Z})(\overline{Z + WY}) + (VZ + W\bar{X})\overline{(Y + Z)}$$

looks formidable. Let us write

$$F = F_1 F_2 + F_3 F_4$$
$$F = F_5 + F_6$$

where

$$F_1 = X + \bar{Z}$$
$$F_2 = \overline{(Z + WY)}$$
$$F_3 = (VZ + W\bar{X})$$
$$F_4 = \overline{(Y + Z)}$$
$$F_5 = F_1 F_2$$
$$F_6 = F_3 F_4$$

Now proceeding a step at a time we obtain:

$$F = F_5 + F_6$$
$$\bar{F} = \bar{F}_5 \cdot \bar{F}_6$$
$$\bar{F}_5 = \bar{F}_1 + \bar{F}_2$$
$$\bar{F}_6 = \bar{F}_3 + \bar{F}_4$$

then

$$\bar{F} = (\bar{F}_1 + \bar{F}_2)(\bar{F}_3 + \bar{F}_4)$$

and

$$\bar{F}_1 = \bar{X}Z$$
$$\bar{F}_2 = [\overline{(\overline{Z + WY})}] = Z + WY$$
$$\bar{F}_3 = (\bar{V} + \bar{Z})(\bar{W} + X)$$
$$\bar{F}_4 = Y + Z$$

Simplifying, we have

$$\bar{F} = (\bar{X}Z + Z + WY)[(\bar{V} + \bar{Z})(\bar{W} + X) + Y + Z]$$
$$= (Z + WY)(\bar{F}_3 + Y + Z)$$
$$= Z + ZY + Z\bar{F}_3 + WY\bar{F}_3 + WYZ + WY$$
$$= Z(1 + Y + \bar{F}_3) + WY(1 + Z + \bar{F}_3)$$
$$= Z + WY$$

We note that the simplified expression for $\bar{F}$ is indeed the complement of the simplified expression for F found in Example 2-25. Thus, we can be confident that both of these difficult examples were done correctly.

Besides inverters, many gates perform inversion as a normal part of their function. The simplest inverting gates are NAND and NOR gates.

A NAND gate takes the AND of its inputs, and, in the process, inverts the output. Because common-emitter transistor circuits normally invert, NAND gates are easier to build and more commonly used than noninverting gates, such as AND gates. The standard symbol for a 3-input NAND gate is shown in Fig. 2-21. Note that it is simply an AND gate symbol with a bubble on the output

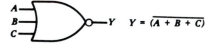

$$Y = \overline{(A\ B\ C)}$$

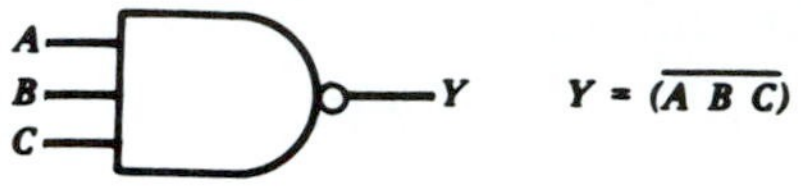

Figure 2-21   A 3-input NAND gate.

to indicate inversion. The process is AND first, then invert; therefore, the output equation for the NAND gate of Fig. 2-21 is

$$Y = \overline{(ABC)}$$
$$= \bar{A} + \bar{B} + \bar{C} \qquad \text{(by DeMorgan's theorem, section 2-8)}$$

A NOR gate similarly takes the OR of its inputs and then inverts it. The symbol for a NOR gate is shown in Fig. 2-22. Note that it is an OR gate with a bubble on the output. The logic equation for the NOR gate is

$$Y = \overline{(A + B + C)}$$
$$Y = \bar{A}\bar{B}\bar{C}$$

$$Y = \overline{(A + B + C)}$$

Figure 2-22   A 3-input NOR gate.

2-10-1   NAND Gates

The standard NAND gate in the **7400** series is the **74x00** IC. It is a quad 2-input NAND gate.

In addition to the **74x00**, other NAND gates are available as shown in Fig. 2-23:

1. The **74x04** hex inverter (an inverter can be viewed as a 1-input NAND gate).
2. The **74x10** triple 3-input NAND gate.
3. The **74x20** dual 4-input NAND gate.
4. The **74x30** 8-input NAND gate.

All of the preceding NAND gates are in 14-pin packages (V_{CC} is on pin 14, ground is on pin 7), and are readily available from most suppliers.

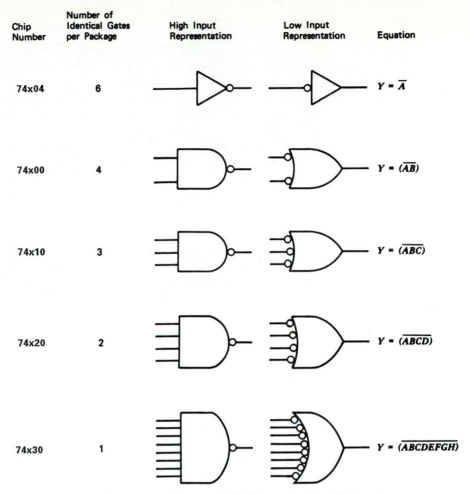

Chip Number	Number of Identical Gates per Package	High Input Representation	Low Input Representation	Equation
74x04	6			$Y = \overline{A}$
74x00	4			$Y = \overline{(AB)}$
74x10	3			$Y = \overline{(ABC)}$
74x20	2			$Y = \overline{(ABCD)}$
74x30	1			$Y = \overline{(ABCDEFGH)}$

Figure 2-23   Common NAND gates of the **7400** series.

EXAMPLE 2-27

Design a 30-input NAND gate using only the NAND gates of Fig. 2-23.

SOLUTION   There are many possible designs that satisfy the problem requirement. One solution, shown in Fig. 2-24, is to connect the inputs to **74x30**s in groups of 8. The last **74x30** only requires 6 inputs so that the unused inputs on the **74x30** are tied to a working input (see section 5-6). The outputs of the **74x30** are then inverted and go to a **74x20**, 4-input NAND gate. The final output will be low only if *all* 30 inputs are HIGH.

2-10-2   TTL NOR Gates

The **74x02** is the basic quad 2-input NOR gate. Other NOR gates available in the standard TTL line are the **74x27** triple 3-input NOR gate and the **74x25** dual 4-input NOR gate. The **74x25** is a strobed gate (see section 5-8-1).

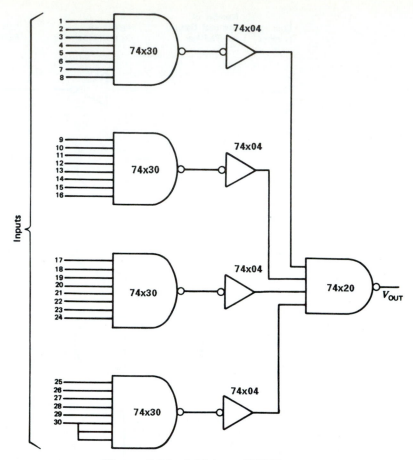

Figure 2-24 A 30-input NAND gate.

2-10-3 Equivalence of Gates and Inversion of Gates

Any gate of the AND-OR-NAND-NOR group may be represented in one of two ways. The two ways to symbolize an inverter were shown in Fig. 2-3. The alternate symbol for a gate is obtained if we:

1. Change the function of the gate (from AND to OR or OR to AND).
2. Change all the bubbles on both the inputs and outputs. (That is, delete bubbles when present and add them when absent.)

Both a symbol and its alternate represent the *same physical gate*. Alternate symbols are often used to clarify signal flow in a circuit (section 2-10-1).

EXAMPLE 2-28

Find the alternate symbol for the 3-input NAND gate of Fig. 2-25a.

SOLUTION The alternate symbol is shown in Fig. 2-25b. It was obtained by:

1. Changing the function of the gate from AND to OR.
2. Adding bubbles to the input.
3. Removing the bubble from output.

Figure 2-25a indicates the output is LOW only if *all* (logic AND) *the inputs are HIGH*. Figure 2-25b indicates the output is HIGH if *any input* (logic OR) *is LOW*. With a little thought, we realize that these are two different ways of saying the same thing!

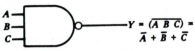

(a) Original 3-input NAND gate

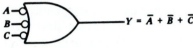

(b) Alternate representation

Figure 2-25 Equivalent representations of a 3-input NAND gate.

EXAMPLE 2-29

Find the alternate representation of the 2-input OR gate of Fig. 2-26a.

SOLUTION The alternate representation, shown in Fig. 2-26b, is obtained by replacing the OR symbol with an AND symbol and adding bubbles on both the inputs and the output.

Figure 2-26a indicates the output is HIGH if *A or B* is high, and Fig. 2-26b indicates the output is LOW only if *A and B* are low. Again, these two statements are equivalent. DeMorgan's theorem shows that the output equations for each gate of Fig. 2-26 are equivalent, that is, $A + B = \overline{\overline{A}\overline{B}}$.

(a) 2-input OR gate

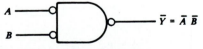

(b) Equivalent representation

Figure 2-26 Equivalent representations of a 2-input OR gate.

NAND and NOR gates are basically inverting gates with bubbles on *either* the inputs *or* the output, *but not both* (Fig. 2-26). AND and OR gates are basically noninverting and either have bubbles on both inputs and outputs, or have no bubbles at all, as Fig. 2-26 shows.

All the logic expressions discussed previously can be implemented and built in the laboratory using AND gates, OR gates, and inverters. AND gates are simply placed wherever two variables are *multiplied* together, and OR gates are placed wherever two variables are *added*. The resulting circuit is the hardware implementation of the given Boolean expression, as shown by Examples 2-30 through 2-32 below.

EXAMPLE 2-30

Implement the expression $Y = (A + \bar{B})C + \bar{A}B\bar{C}$ if:

(a)   Both complemented and uncomplemented input variables are available.

(b)   Only uncomplemented input variables are available.

SOLUTION   (a) The term $A + \bar{B}$ suggests an OR gate. The output of this OR gate must be ANDed with C to get the first term in the expression. To obtain the term $\bar{A}B\bar{C}$ a 3-input AND gate is required. Finally, since the output is a sum, the two terms must be ORed together to produce Y. The results are shown in Fig. 2-27a.

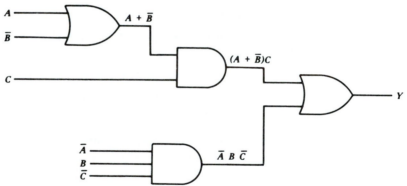

(a) Both complemented and uncomplemented variables available

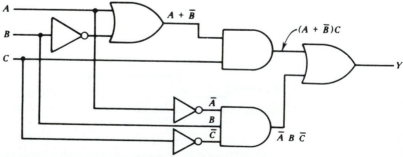

(b) Only uncomplemented inputs are available

Figure 2-27   Two implementations of the expression $Y = (A + \bar{B})C + \bar{A}B\bar{C}$.

(b) If only *uncomplemented* inputs are available, complemented variables can be obtained through the use of inverters. Figure 2-27b is essentially the same as Fig. 2-27a, except that inverters have been added to convert uncomplemented inputs A, B, and C to $\bar{A}$, $\bar{B}$, and $\bar{C}$ where needed.

EXAMPLE 2-31

Using only the gates discussed previously, implement the expression:

$$Y = AB(\overline{\overline{C}D}) + \bar{B}CD + (\bar{A} + \bar{C})(B + D)$$

(a)   Do not simplify the above expression before implementing.
(b)   Simplify first, then implement the expression.

SOLUTION   For part (a) the expression is implemented just as stated. The term $(\overline{\overline{C}D})$ implies an AND gate and an inverter, or more simply a NAND gate. The output of this NAND gate is ANDed with A and B to produce the first term in the expression. The term $\bar{B}CD$ is a 3-input AND gate, and the term $(\bar{A} + \bar{C})(B + D)$ is the AND of two OR gates. Finally, the three terms have to be ORed together because the output is the sum of the terms. The solution is shown in Fig. 2-28a. For simplicity it was assumed that both complemented and uncomplemented variables were available.

The solution of Example 2-31a demonstrates that any combinatorial expression can be implemented using only ANDs, ORs, and so forth.

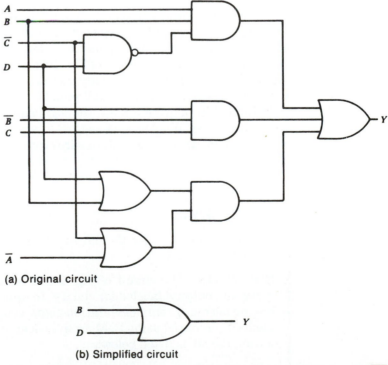

(a) Original circuit

(b) Simplified circuit

Figure 2-28   Two circuits to implement the expression $Y = AB(\overline{\overline{C}D})$ $+ \bar{B}CD + (\bar{A} + \bar{C})(B + D)$.

A better circuit is obtained by *simplifying the expression before* implementing it. The first step in any simplification is to use DeMorgan's theorem to eliminate the complemented form.

$$Y = AB(\overline{\overline{C}\overline{D}}) + \bar{B}CD + (\bar{A} + \bar{C})(B + D)$$
$$= AB(C + \bar{D}) + \bar{B}CD + (\bar{A} + \bar{C})(B + D)$$

This expression can be simplified further either by drawing Karnaugh maps (see section 3-5) or by algebraic manipulation. Trying algebraic manipulation, we obtain:

$$Y = ABC + AB\bar{D} + \bar{B}CD + \bar{A}B + \bar{A}D + B\bar{C} + \bar{C}D$$
$$= B(AC + \bar{A}) + AB\bar{D} + D(\bar{B}C + \bar{C}) + \bar{A}D + B\bar{C}$$
$$= B(C + \bar{A}) + AB\bar{D} + D(\bar{C} + \bar{B}) + \bar{A}D + B\bar{C}$$
$$= BC + B\bar{C} + B\bar{A} + AB\bar{D} + D\bar{C} + D\bar{B} + \bar{A}D$$
$$= B + D\bar{B} + D\bar{C} + \bar{A}D$$
$$= B + D$$

Thus, the expression which originally required 7 gates to build (in Fig. 2-28a) reduces to a single OR gate as shown in Fig. 2-28b. This example clearly demonstrates why it is wise to first simplify expressions, instead of immediately plunging in and building the circuit. The circuit of Fig. 2-28a would have been even more complicated if only uncomplemented variables were available. In that case four additional inverters would have been required while the circuit of Fig. 2-28b would still remain unchanged.

We should also be able to reverse the procedure and analyze a circuit that already exists to determine its logic equation. The equation is needed in order to predict its behavior for all combinations of input variables, because occasionally such poor designs as the circuit of Fig. 2-28a actually are built and incorporated into working equipment. Naturally they should be reduced and replaced with a simpler equivalent circuit, if at all possible.

To analyze a given circuit, the logic equations for each gate are written first. Then gates are chained together in accordance with the interconnecting wires until an expression for the entire circuit is achieved, as shown in Example 2-32.

EXAMPLE 2-32

Find the expression for the output of Fig. 2-29. Design a simpler circuit if possible.

SOLUTION   The circuit of Fig. 2-29a is redrawn in Fig. 2-29b with the gates and outputs labeled for clarity. In order to prevent confusion by introducing too many complementation signs, complemented expressions are immediately reduced using DeMorgan's theorem. The output is obtained by following the steps listed below:

1. The output of AND gates 1 and 2 are *AB* and *BC*. These form inputs to gates 7 and 4, respectively.

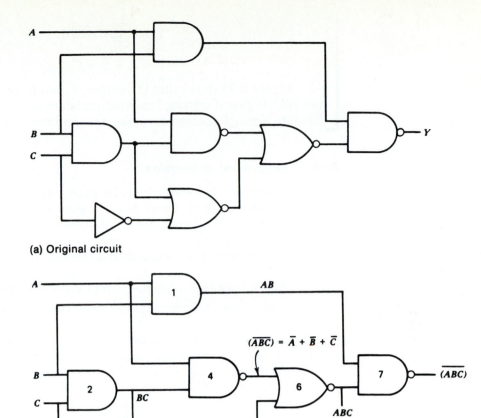

(a) Original circuit

(b) Circuit used for analysis

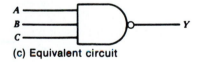

(c) Equivalent circuit

Figure 2-29   Circuit for Example 2-32.

2. Gate 4 is a NAND gate with inputs A and BC. Since NAND gate inverts, the output of gate 4 is the complement of the AND of A and BC or $(\overline{ABC}) = \bar{A} + \bar{B} + \bar{C}$.

3. Gate 5 is a NOR gate with inputs BC and $\bar{C}$. The output is therefore $(\overline{BC + \bar{C}}) = C(\bar{B} + \bar{C}) = C\bar{B}$

4. Gate 6 NORs together the output of gates 4 and 5. Its output is therefore $[\overline{C\bar{B} + \bar{A} + \bar{B} + \bar{C}}] = (\overline{\bar{A} + \bar{B} + \bar{C}}) = ABC$.

5. The output of gate 7 is the NAND of the outputs of gates 1 and 6. Thus the final output is:

$$Y = [\overline{(\overline{ABC})(\overline{AB})}]$$
$$= (\overline{\overline{ABC}})$$
$$= \bar{A} + \bar{B} + \bar{C}$$

Figure 2-29 shows that the output, Y, can be produced by a single 3-input NAND gate if uncomplemented variables are available, or by a 3-input OR gate if complemented variables are available.

2-11-1 Placement of Bubbles

With two alternate representations for each gate, the question arises: Which representation should be used? Many engineers solve the problem by using positive logic, which means placing bubbles on outputs only, and never on inputs. However, we feel it is easier to trace the signal flow, and the operation of a circuit becomes clearer, if we can choose the gate representations so that:

1. Output bubbles are connected to input bubbles.
2. Outputs without bubbles are connected to inputs without bubbles.

Gate representations that satisfy these requirements cannot be selected in all cases, but they can in most cases. They help the engineer "see" the signal flow and facilitate trouble-shooting for the technician.

Consider the problem of finding an expression for the circuit of Fig. 2-30a where bubbles are placed only on the outputs. The output of gates 1 and 2 are $(\bar{A} + \bar{B})$ and $(\bar{C} + \bar{D})$, respectively. Here Y is the output of a NAND gate:

$$Y = \overline{(\bar{A} + \bar{B})(\bar{C} + \bar{D})} = AB + CD$$

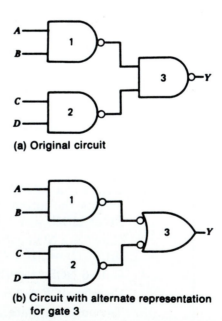

(a) Original circuit

(b) Circuit with alternate representation
for gate 3

Figure 2-30 Two identical circuits with different representations.

Even for this simple circuit, visualization of the signal flow is somewhat difficult, and the answer was achieved by algebraic manipulation.

Now consider the circuit of Fig. 2-30b, which is the same as Fig. 2-30a, but NAND gate 3 has been replaced by its equivalent representation with negative inputs. In Fig. 2-30b, we can see that if A and B or C and D are *both* HIGH, the output of gates 1 or 2 (or both) is LOW. This LOW output is fed to the input of gate 3, causing its output to be HIGH. Therefore,

$$Y = AB + CD$$

The signal flow is traced by noting simply that two HIGH inputs cause a LOW output between the gates, and this LOW level causes the output of gate 3 to be HIGH.

EXAMPLE 2-33

Find an expression for the circuit of Fig. 2-31a. Change the bubbles to make the circuit clearer.

SOLUTION The representation of the 2-input NAND gate can be changed so that all inputs to the second stage NAND gate are positive, as shown in Fig. 2-31b. Now, however, the circuit output is negative. It is the product of the 3 inputs, or:

$$\bar{Y} = (\bar{A} + \bar{B})(CD)(E + F)$$

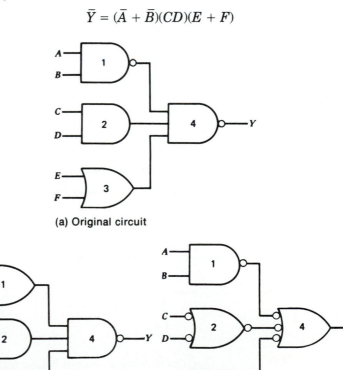

(a) Original circuit

(b) An equivalent representation

(c) A second equivalent representation

Figure 2-31   Circuit for Example 2-33.

If a positive output is needed, an alternate solution is to change the 3-input NAND gate to its alternate representation, as shown in Fig. 2-31c. Since in this case it is convenient to have all the inputs to this NAND gate negative, the first level AND and OR gates are changed to their alternate representations. From Fig. 2-31c it is easy to see that the output is HIGH for *any* combination of inputs that makes the output of one or more of the first level gates LOW. Therefore,

$$Y = AB + \bar{C} + \bar{D} + \bar{E}\bar{F}$$

Note that the outputs of Fig. 2-31b and Fig. 2-31c can be shown to be identical by DeMorgan's theorem.

2-12   Relays

Algebraic expressions may be implemented using relays. Relays are still widely used in innumerable devices (the pinball machine is one common example of a device using relays), and the functions they perform are described by Boolean equations.

Physically a relay consists of a coil and a set of switch contacts or closures. A specified current in the coil causes the switch contacts to open or close and make or break another electric circuit in which the contacts are connected.

There are two types of relay contacts; *normally open* and *normally closed*. For a relay, the normal condition means its deenergized state, that is, *no* current flow in the relay coil. When current does flow the relay is said to be *energized*; normally open contacts close and normally closed contacts open. Figure 2-32 shows the circuit of a relay and the symbols for normally open and normally closed contacts. We see that a current in the coil magnetically attracts the armature of the normally open contact, causing it to close. The same current and mechanical motion pull down the normally closed contact, causing it to open.

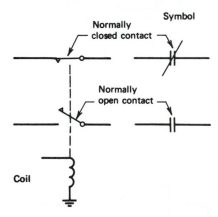

Figure 2-32   Diagram of a relay.

Circuits involving *several* relays quickly become complicated. It is advantageous to use Boolean logic to simplify and reduce them. When Boolean algebra is applied to a circuit, a 1 for the output indicates there is a current path through the circuit. But if the output = 0, the circuit is *open*. For a single relay let us define a closed contact as a 1 and an *energized* relay as a 1. Therefore, normally open contacts are written as unprimed variables. For example, a set of normally open contacts on relay A are written as A. If relay A is not energized, there is *no* current path through the contacts and $A = 0$. Conversely, a set of normally closed contacts is written as $\bar{A}$ because when $A = 0$, there *is* a path through the contacts, but when the relay is energized, $A = 1$ and the path opens ($\bar{A} = 0$).

Typical relay problems involve determining whether a path exists between two points in a relay network, and whether that network can be simplified.

EXAMPLE 2-34

Write the expression for the circuit of Fig. 2-33.

SOLUTION   Because there is more than one possible path between two points in a circuit, as between points A and B in Fig. 2-33, an OR symbol is appropriate since there may be continuity through one path *or* another. Whenever there is *only one possible path*, as between point B and the output, an AND symbol is correct. In the circuit of Fig. 2-33, relay Z must also be closed to provide continuity between the input and the output.

Examining the circuit with these facts in mind, we write the final expression as:

$$\text{Output} = (W + XY)Z$$

In Fig. 2-33, either W or XY provides a path from A to B, but contact Z must provide the path from B to the output.

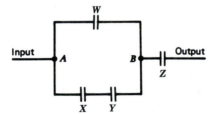

Figure 2-33   Relay circuit for Example 2-34.

EXAMPLE 2-35

Design a relay circuit to implement the function

$$\text{Output} = A(B + C) + \bar{B}(\bar{C} + DE)$$

SOLUTION   The solution is shown in Fig. 2-34. The equation $A(B + C) + \bar{B}(\bar{C} + DE)$ consists of two terms ORed together. This indicates that there are two major paths between the input and output. The term $A(B + C)$ is implemented by the top path and the term $\bar{B}(\bar{C} + DE)$ is implemented by the lower path.

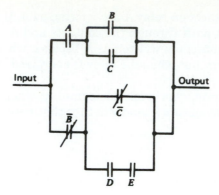

Figure 2-34 Circuit for Example 2-35.

EXAMPLE 2-36

Write the expression for the relay circuit of Fig. 2-35a. Simplify it and find a simpler relay circuit that implements the *same* function.

SOLUTION

1. The expression is written by following every possible path from input to output.
2. Then, the expression is simplified. Thus:

1. Output $= AB + A\bar{C}\bar{B} + A\bar{C}D + CD + C\bar{B} + C\bar{C}B$
2. $= AB + \bar{B}(A\bar{C} + C) + A\bar{C}D + CD$
   $= AB + \bar{B}A + \bar{B}C + A\bar{C}D + CD$ (Theorem **11**)
   $= A + \bar{B}C + A\bar{C}D + CD$ (Theorem **8**)
   $= A + \bar{B}C + CD$ (Theorem **9**)
   $= A + C(\bar{B} + D)$

The logical equations indicate that if relay A is closed, there is always a path between the input and the output. This is correct, but it is *not* obvious from the illustration. A simpler relay circuit that only uses four contacts and conforms to the simplified final equation is shown in Fig. 2-35b.

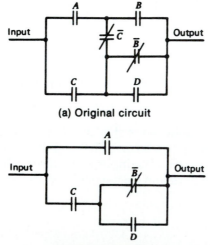

(a) Original circuit

(b) Reduced circuit   **Figure 2-35** Relay circuit for Example 2-36.

SUMMARY

In this chapter the techniques of manipulating, complementing, and reducing Boolean expressions were explained and demonstrated. DeMorgan's theorem for complementing functions was also discussed. These techniques are often used by design engineers to eliminate unnecessary circuitry.

The AND, OR, NAND, and NOR gates available in the **7400** series were introduced, and methods of producing the Boolean functions using those gates were explained. Relay logic, still used in some applications, was also explained.

GLOSSARY

Boolean algebra: An algebra of two valued variables.

Bubble: A symbol (a small circle) indicating where the asserted level of the signal is LOW.

Complement: Invert the value of a variable or expression.

Inverter: An electronic gate whose output is the complement of its input.

NAND gate: A gate combining the functions of ANDing followed by inversion.

NOR gate: A gate combining the functions of ORing followed by inversion.

Normally closed contact: A relay contact that is closed when there is no current in the relay coil and opens when current is applied (relay energized).

Normally open contact: A relay contact that is open when there is no current in the relay coil (relay deenergized).

PROBLEMS

Section 2-3.

2-1. A machine operates with four essential variables controlling its operation. For the machine to be operating properly at least two of these control variables must be present at the same time. However, when the machine is not operating correctly we wish to have some signal to alert us to the problem. Draw up a truth table and find an expression for the alarm signal.

Section 2-4.

2-2. Refer to Ex. 2-4. If it only cost $9.00 for gas to get to Rochester, draw up a truth table to show when the girls can go.

Section 2-5.

2-3. Sketch a 4-input OR gate using a single **74x32**. Show the pin numbers.

2-4. Sketch a 6-input AND gate using two **74x08**s. Show the pin numbers.

Section 2-7-4.

2-5. Sketch a circuit that can be built in the laboratory to verify
    (a) Theorem 9
    (b) Theorem 11
    (c) Theorem 12

Section 2-8-1.

2-6. An engine (cooled by water and lubricated by oil under pressure) has a warning signal light that turns ON when one or both the following conditions are present: (1) engine temperature is *high*; (2) engine temperature is *low* but both the water level and the oil pressure are inadequate.

    Let x, y, and z denote, respectively, engine temperature, water level, and oil pressure. Assume that these are measured by sensors that put out either a 0 or a 1 signal. That is, $x = 0$ means temperature is low, $x = 1$ temperature high, and so on.

    (a) Set up a truth table with x, y, and z as the inputs and $T(x, y, z)$ as the output of the warning light control circuit. $T = 1$ indicates that the light goes on.

    (b) Obtain a minimal form of T using algebraic manipulation.

2-7. The conditions under which an insurance company will issue a policy are (1) a married female 25 years old or older, or (2) a female under 25 years, or (3) a married male under 25 with no accident record, or (4) a married male with an accident record, or (5) a married male 25 years or older with no accident record. Obtain a simplified logic expression stating to whom a policy can be issued.

2-8. Simplify each of the following expressions:
    (a) $XY(X + Y\bar{Z})$
    (b) $T(x,y,z) = (x + \bar{x} + \bar{y})x\bar{z} + x\bar{z}(y + \bar{y})$
    (c) $T(x,y,z) = xy + x\bar{y} + \bar{x}z$
    (d) $T(w,x,y,z) = (w + x)(x + y)(w + \bar{x} + y + \bar{z}) + \bar{x} + \bar{y} + \bar{w}$
    (e) $T(a,b,c,d) = \bar{a}\bar{d}(\bar{b} + \bar{c}) + (\bar{b} + c)(b + \bar{c})$

Section 2-8-2.

2-9. Show that each of the following identities is true:
    (a) $\bar{a}b + ac = ac + bc + \bar{a}\bar{c}b + b\bar{a}\bar{c}d$
    (b) $a\bar{b} + \bar{a}\bar{c} + \bar{a}b + \bar{c}b = \bar{c} + a\bar{b} + \bar{a}b$
    (c) $(X + \bar{Z})(X + Y + \bar{Z})(Y + \bar{Z}) = \bar{Z} + XY$
    (d) $AB + (\overline{AB + \bar{A}\bar{B}}) = A + B$
    (e) $XY\bar{Z} + (\overline{XZ + \bar{X}YZ + \bar{Y}Z}) = \bar{Z}$

2-10. Determine whether each of the following equations is true. If any are false, find a set of values that makes the two sides unequal.
    (a) $(X + \bar{Y} + XY)(X + \bar{Y})\bar{X}Y = 0$
    (b) $xyz + w\bar{y}\bar{z} + wxz = xyz + w\bar{y}z + wx\bar{y}$
    (c) $ab + \bar{a}\bar{b} + a\bar{b}c = ac + \bar{a}\bar{b} + \bar{a}bc$
    (d) $X + YZ = XY + \bar{X}\bar{Y}W + \bar{X}\bar{Y}Z + \bar{Y}W$

2-11. Sketch a laboratory circuit to verify the identity of Problem 2-9d.

Section 2-9.

2-12. Complement the following expressions.
    (a) $T(a,b,c,d) = \bar{a}d(\bar{b} + c) + \bar{a}\bar{d}(b + c) + (\bar{b} + \bar{c})$
    (b) $T(x,y,z) = (x + y)(\bar{x} + z)(y + z)$
    (c) $T(a,b,c,d,e) = a\bar{b}c + (\bar{a} + b + d)(ab\bar{d} + \bar{e})$
    (d) $(a + b\bar{c})(c + \bar{d}(e + f))$

Section 2-10-1.

2-13. Given $X\bar{Y} + \bar{X}Y = Z$, show that $X\bar{Z} + \bar{X}Z = Y$.

2-14. Find the alternate representation for the gates of Fig. P2-14.

2-15. Using only NOR gates, design:
    (a) A 2-input NAND gate.
    (b) A 2-input AND gate.
    (c) A 2-input OR gate.

2-16. If only inverters and 3-input NAND gates are available, design:
    (a) A 3-input AND gate.
    (b) A 3-input OR gate.
    (c) A 3-input NOR gate.

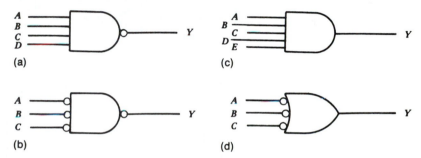

Figure P2-14

Section 2-11.

2-17. Implement the following expressions. Do not simplify or take complements.
    (a) $(a + b\bar{c})(c + \bar{d}(e + f))$
    (b) $AB + C(\bar{B} + CA) + \bar{A}B\bar{D}$
    (c) $\overline{WXY(Z + \bar{W}\bar{X})} + WX(\bar{Y} + \bar{Z})$
    (d) $[(X + \bar{\bar{Y}}Z) + W(XY + \bar{Z})]$
    (e) $[(A + B\bar{C})(D + \bar{A}\bar{C})] + CD$
    (f) $[X(\bar{Y} + Z) + (\overline{W}\overline{X}\overline{Z})]$

2-18. **(a)** Find the logic expression for the circuits of Fig. P2-18 without simplifying.
    (b) See if you can draw a simpler circuit.

2-19. When approaching an intersection, an automobile driver may make a right turn if the traffic light is green, or if it is red and no car is approaching on the intersecting road. Put this statement into Boolean algebra, simplify it, and restate it more simply. (*Hint*: Red = $\overline{\text{Green}}$.)

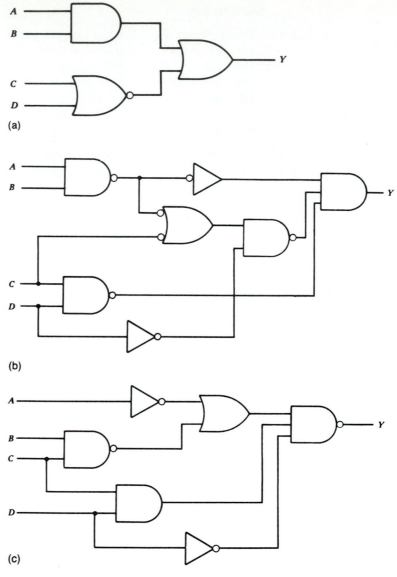

Figure P2-18

Section 2-12.

2-20. For the circuits of Fig. P2-20, find the logic equations without simplifying; then simplify the circuits.

2-21. Use relays to implement the following logic expressions. Do not simplify.
   (a) $BD + \bar{B}(A\bar{C} + \bar{A}C)$
   (b) $XY\bar{Z} + XZ(YW + \bar{X}V)$
   (c) $AB(C\bar{B}D + ACD + BDE)$

To be sure you understand this chapter, return to section 2-2 and review the questions. If you cannot answer certain questions, review the appropriate sections of the text to find the answers.

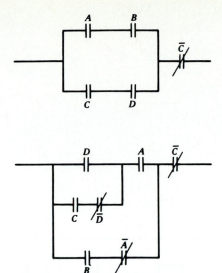

Figure P2-20 Circuits for Problem 2-8.

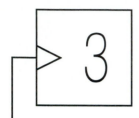

Systematic Reduction
of Boolean Expressions

3-1   Instructional Objectives

This chapter introduces the Karnaugh map method of simplifying logic expressions. After reading this chapter the student should be able to:

1. Express a given Boolean function in standard SOP and POS form.
2. Find the numerical equivalent of SOP and POS terms.
3. Construct Karnaugh maps for given functions.
4. Use Karnaugh maps to obtain the minimal form for given expressions.
5. Implement SOP and POS expressions using logic gates.

3-2   Self-Evaluation Questions

Watch for the answers to the following questions as you read the chapter. They should help you understand the material presented. When you have finished the chapter, return to this section and be sure you can answer all of the questions.

1. Define an SOP term. What is the difference between an SOP and a standard SOP term?
2. Define a POS term. What is the difference between a POS and a standard POS term?

3. If a term has k literals missing, how many terms does its expansion to a standard POS or SOP form contain?
4. If k literals are missing from an SOP or POS expression, what specific terms does the expression represent?
5. Define subcubes. How are they used in minimizing functions?
6. Why is the largest essential subcube selected in minimizing functions?
7. How many literals are missing when a 2, 4, or 8 cell subcube is written?
8. How do you check a simplified expression to be sure all the specified terms are covered?
9. Why is it wise to start simplifying Karnaugh maps by selecting essential subcubes?
10. How do you plot a POS Karnaugh map?
11. When do "don't cares" typically arise? Why are they advantageous to the designer?

3-3   Standard Forms — Introduction

The expressions considered in Chapter 2 were simplified by applying the rules and theorems we had developed. It requires a generous helping of inspiration and experience, however, to decide *which* theorems may be applied usefully to a particular expression. An ounce of inspiration is often worth a pound of logic, but unfortunately there is no formula for inspiration. Consequently, a more systematic way of reducing Boolean expressions must be examined so that the pound of logic is there when the ounce of inspiration is missing! Furthermore, a systematic approach prevents the designer from deluding himself into thinking he has the minimal form of an expression when he has only achieved a partial simplification.

For example, if we try to simplify the expression

$$f(W,X,Y,Z) = \bar{W}\bar{X} + \bar{X}\bar{Y}\bar{Z} + \bar{W}\bar{Z} + YZ$$

we can try a few theorems, but we will achieve no reduction. Does this mean the expression cannot be simplified at all or merely that we have not learned to simplify it? The application of systematic logic answers this type of question.

In order to develop a systematic method of reducing Boolean expressions, the following definitions are required:

- **Literal.** The occurrence of a variable in either its complemented or uncomplemented state in an expression. The term $A\bar{B}C$ *consists of three literals: A, $\bar{B}$, and C.*
- **Product term.** A product term consists solely of the product of literals. $A\bar{B}C$ is a product term, while $A(\bar{B} + C)$ is not a product term because of the + sign.
- **Sum term.** A sum term consists solely of the sum of literals. $A + \bar{B} + C$ is a sum term; similarly $A(B + C)$ is not a sum term because of the implied multiplication.

- **Domain.** The domain of a function is the set of variables for which that function exists. This may be defined *explicitly* or *implicitly*. Explicitly, the expression $f(x,y,z)$ means the function exists for the variables x, y, and z. Implicitly, domains are implied by word statements of a problem. Such statements must be examined to determine what the pertinent variables are; these form the domain of the function. In Example 2-1, the domain of the contract problem proved to be the president and three assistants, or the variables P, A_1, A_2, A_3.

- **Standard product term.** A product term that contains *one* literal from *every* variable in the domain of the function. For example, if the domain is A, B, C, and D, the term $A\bar{B}\bar{C}D$ is a standard product term while the term ACD is not because the literal B is missing.

- **Standard sum term.** A *sum* term that contains *one* literal from every variable in the domain. $\bar{A} + \bar{B} + \bar{C} + \bar{D}$ is a standard sum term, but $A + B + D$ is not.

- **Sum of products form.** An expression is in a sum of products (SOP) form if it is a sum of product terms. For example, $A\bar{B} + ABC$ is in SOP form. An expression is in *standard SOP form* if *each term* in the sum is a *standard product term*.

- **Product of sums form.** An expression is in the product of sums (POS) form if it consists only of the product of a group of sums. The expression $(A + B)(A + B + \bar{C})$ is in POS form. If the expression is the product of a group of *standard* sum terms it is in *standard POS form*.

The sum of products (SOP) form is very useful. The input to most Karnaugh maps (see section 3-5) is in SOP form. Even more important, Programmable Array Logic (PALs—see Chapter 12) is ideally suited to expressions in SOP form. SOP and POS forms can be evaluated as a Boolean expression as Example 3-1 shows.

EXAMPLE 3-1

What is the value of the SOP expression $A\bar{B} + ABC$ if $A = 1$, $B = 0$, and $C = 1$?

SOLUTION   The term ABC is 0, but the term $A\bar{B}$ is 1. Therefore, the value of the expression is 1.

3-3-1   Conversion of Product Terms to Standard SOP Form

A product term that is not a standard product term because it has several literals missing can be expanded into the sum of several standard product terms. Consider a product term from a domain of n variables. If the term has k literals missing and n-k literals specified, the term can be expanded into the standard SOP form where each standard product term has the n-k literals that are specified and the terms consist of all combinations of the k literals that are missing. Therefore, a product term with k variables missing expands into 2^k standard terms to include all possible combinations of the k variables. To demonstrate how the conversion to standard SOP form works, consider:

$$f(V,W,X,Y) = V\bar{W}\bar{X} \qquad (3\text{-}1)$$

There are four variables in the domain of Eq. 3-1 ($n = 4$). One of the literals, Y, is missing ($k = 1$) and the other three are specified as V, W, and X. Equation 3-1 can therefore be expanded to two terms. Both terms will contain V, $\bar{W}$, and $\bar{X}$. One term will contain Y and the other term will contain $\bar{Y}$.

To expand Eq. 3-1, it can be multiplied by $Y + \bar{Y}$ (which equals 1), however, to give:

$$V\bar{W}\bar{X}(Y + \bar{Y}) = V\bar{W}\bar{X}Y + V\bar{W}\bar{X}\bar{Y} \qquad (3\text{-}2)$$

The function is now in standard SOP form because each term contains one literal for each variable in the domain.

EXAMPLE 3-2

If $f(V,W,X,Y)$ in Eq. (3-1) is 1, what are the values of V, W, and X?

SOLUTION   The function can only be 1 if V, $\bar{W}$, and $\bar{X}$ all equal 1. Therefore, $V = 1$, $W = 0$, and $X = 0$. Note that Eq. (3-2) must now equal 1 regardless of the value of Y.

If another variable, Z for instance, were included in the domain, Eq. (3-2) could be multiplied by $Z + \bar{Z}$. The result would be 4 terms. The terms would include all possible combinations of the missing variables (Y and Z) and each term would include $V\bar{W}\bar{X}$.

EXAMPLE 3-3

Given $f(a,b,c,d,e) = \bar{a}bd$, write f in standard SOP form.

SOLUTION   The missing variables are c and e. Because 2 variables are missing the solution will contain 4 terms. Each term will contain 1 combination of c and e ($\bar{c}\bar{e}$, $c\bar{e}$, $\bar{c}e$, or ce) and the specified variables $\bar{a}bd$. Therefore,

$$f(a,b,c,d,e) = \bar{a}bd$$
$$= \bar{a}b\bar{c}d\bar{e} + \bar{a}bcd\bar{e} + \bar{a}b\bar{c}de + \bar{a}bcde$$

EXAMPLE 3-4

Given $f(a,b,c,d) = a\bar{b} + ac\bar{d}$, write it in standard SOP form.

SOLUTION   The given function consists of 2 product terms. Each can be expanded into standard SOP form. The first term has 2 literals missing and will result in 4 product terms. The second term has only 1 literal missing and will result in 2 SOP terms. Thus, the answer is assumed to consist of 6 SOP

---

terms. However, some of the 6 terms may occur more than once. In that case, the *redundant* terms may be eliminated (Theorem **6**).[1]

$$f(a,b,c,d) = a\bar{b} + ac\bar{d}$$
$$= \underbrace{a\bar{b}\bar{c}\bar{d} + a\bar{b}\bar{c}d + a\bar{b}c\bar{d} + a\bar{b}cd}_{\text{Expansion of the term } a\bar{b}} + \underbrace{a\bar{b}c\bar{d} + abc\bar{d}}_{\substack{\text{Expansion of} \\ \text{the term } ac\bar{d}}}$$

$$= a\bar{b}\bar{c}\bar{d} + a\bar{b}\bar{c}d + a\bar{b}c\bar{d} + a\bar{b}cd + abc\bar{d}$$

The term $a\bar{b}c\bar{d}$ occurred in both expansions, but it need only be written once, as the final answer shows.

3-3-2   Expansion of Sum Terms to Standard POS Form

Sum terms can be expanded into standard POS form in a manner similar to the expansion of product terms into standard SOP form. Expanding a sum term with k missing literals results in the product of 2^k POS terms and the terms contain all the 2^k possible combinations of the unspecified terms.

To demonstrate the expansion, consider:

$$f(X,Y,Z) = X + \bar{Y}$$
$$= X + \bar{Y} + \bar{Z}Z \qquad \qquad \text{(Theorem } \mathbf{5})[2]$$
$$= (X + \bar{Y} + Z)(X + \bar{Y} + \bar{Z}) \qquad \text{(Theorem } \mathbf{13})[2]$$

Since the term $\bar{Z}Z = 0$, it may be added to an expression whenever it is helpful and it does not change the value of the expression.

EXAMPLE 3-5

Express $f(A,B,C,D) = (A + \bar{C} + \bar{D})(A + \bar{B})$ in standard POS form.

SOLUTION   The first term in the product, $A + \bar{C} + \bar{D}$, has the variable B missing and expands to $(A + B + \bar{C} + \bar{D})(A + \bar{B} + \bar{C} + \bar{D})$. The second term has the variables C and D missing. Because this POS form has two literals missing, it will expand to four terms ($k = 2$, $2^k = 4$). Each term will contain the literals A and $\bar{B}$, and the four terms will contain all combinations of the literals C and D. The term expands to

$$A + \bar{B} = (A + \bar{B} + \bar{C} + \bar{D})(A + \bar{B} + \bar{C} + D)$$
$$\cdot (A + \bar{B} + C + \bar{D})(A + \bar{B} + C + D)$$

Therefore,

$$f(A,B,C,D) = (A + \bar{C} + \bar{D})(A + \bar{B})$$
$$= (A + B + \bar{C} + \bar{D})(A + \bar{B} + \bar{C} + \bar{D})$$
$$\cdot (A + \bar{B} + \bar{C} + \bar{D})(A + \bar{B} + \bar{C} + D)$$
$$\cdot (A + \bar{B} + C + \bar{D})(A + \bar{B} + C + D)$$

[1] See section 2-7-3.
[2] See section 2-7.

$$= (A + B + \bar{C} + \bar{D})(A + \bar{B} + \bar{C} + \bar{D})$$
$$\cdot (A + \bar{B} + \bar{C} + D)(A + \bar{B} + C + \bar{D})$$
$$\cdot (A + \bar{B} + C + D)$$

As before, the term $A + \bar{B} + \bar{C} + \bar{D}$ appeared twice and is written only once.

3-3-3 Numerical Representation of SOP Forms

Functions in standard SOP form are often represented by the sum (Σ) of a group of numbers, where each number corresponds to a particular SOP term. This allows us to write each term more concisely, and is used by digital engineers for this reason. The numeric form is found by assuming that the *first* variable in the domain is the *most significant* bit of binary number and proceeding until the *last* variable, which is the *least significant* bit of the number. In standard SOP form, uncomplemented literals are assigned the value of 1 and complemented literals are assigned a value of 0. We arrive at a numerical result simply by reading each term as a binary number and writing its decimal equivalent.

EXAMPLE 3-6

Write the equation $f(X,Y,Z) = \bar{X}YZ + X\bar{Y}Z$ in numeric form.

SOLUTION   In the first term, $\bar{X}YZ$, X is complemented but Y and Z are uncomplemented. In numeric form this reads as 011 or a binary 3. In the second term only Y is complemented, so this term is read as 101 or binary 5. Therefore the equation can be written as

$$f(X,Y,Z) = \Sigma(3,5)$$

Notice that this is more concise. The advantage is even greater when more variables are involved as Ex. 3-7 shows.

EXAMPLE 3-7

Write the numerical form of the standard SOP function:

$$f(a,b,c,d) = \bar{a}\bar{b}\bar{c}\bar{d} + \bar{a}\bar{b}cd + \bar{a}b\bar{c}d + ab\bar{c}d + abc\bar{d}$$

SOLUTION   We can attach numbers to each term as described above.

$$f(a,b,c,d) = \bar{a}\bar{b}\bar{c}\bar{d} + \bar{a}\bar{b}cd + \bar{a}b\bar{c}d + ab\bar{c}d + abc\bar{d}$$

0000	0011	0101	1101	1110
0	3	5	13	14

Therefore,

$$f(a,b,c,d) = \Sigma(0,3,5,13,14)$$

EXAMPLE 3-8

If $f(W,X,Y,Z) = \Sigma(1,5,13,15)$, write f in standard SOP form.

SOLUTION   In this example, we must write the literals to correspond to the given numbers.

$$f(W,X,Y,Z) = \Sigma(1,5,13,15)$$
$$= 0001 + 0101 + 1101 + 1111$$
$$= \overline{W}\overline{X}\overline{Y}Z + \overline{W}X\overline{Y}Z + WX\overline{Y}Z + WXYZ$$

With a little practice one can go directly from the numerical representation to the SOP form and back without the intervening step.

If a product term is not in standard form, because one or more literals are missing, it becomes the sum of several terms or several numbers. It is possible to go from the literal form directly to the numbers by placing an X wherever a literal is missing and then writing numbers corresponding to the given literals and all possible combinations of X values.

EXAMPLE 3-9

If $f(a,b,c,d) = a\bar{c}\bar{d}$, find the numeric SOP representation of the function.

SOLUTION   The function $a\bar{c}\bar{d}$ corresponds to 1X00, where X is used in place of the missing literal b. Substituting 0 and 1 for X, we get the numbers 8 and 12, respectively. Therefore,

$$f(a,b,c,d) = a\bar{c}\bar{d} = \Sigma(8,12)$$

Indeed, 8 and 12 are the only two 4-bit numbers where the most significant bit is 1 and the least significant bits are both 0.

EXAMPLE 3-10

If $f(a,b,c,d,e) = \bar{b}e$, express f in numerical form.

SOLUTION   Since 3 variables are missing, we expect the result to contain $2^3 = 8$ terms. We can write

$$f(a,b,c,d,e) = \bar{b}e = \text{X0XX1}$$

where the Xs correspond to the missing variables a, c, and d, the 0 corresponds to $\bar{b}$, and the 1 in the least significant bit position corresponds to e. The results are shown in Fig. 3-1. First the X_1, X_2, and X_3 values are listed simply as the binary numbers from 0 to 7. This assures us 8 different combinations for the Xs. Then the X_1, X_2, and X_3 values are substituted into the term X0XX1 and the numerical results are simply the decimal equivalents of the resulting binary numbers. Therefore,

$$f(a,b,c,d,e) = \bar{b}e = \Sigma(\mathbf{1,3,5,7,17,19,21,23})$$

X_1 X_2 X_3	X 0 X X 1	Numerical Result
0 0 0	0 0 0 0 1	1
0 0 1	0 0 0 1 1	3
0 1 0	0 0 1 0 1	5
0 1 1	0 0 1 1 1	7
1 0 0	1 0 0 0 1	17
1 0 1	1 0 0 1 1	19
1 1 0	1 0 1 0 1	21
1 1 1	1 0 1 1 1	23

Figure 3-1 Expansion of the function $\bar{b}e$.

3-3-4 Numerical Representation of POS Forms

At this point, an important distinction between SOP and POS forms must be made:

> *If a function is expressed in SOP form, it equals 1 if any term in the sum equals 1 and 0 otherwise. But if a function is expressed in POS form, it equals 0 if any term in the product is 0 and 1 otherwise.*

Functions in the standard POS form are listed as a product (Π) of a group of numbers. *These numbers correspond to input values that cause the value of the function to be 0. A sum term can be 0 only if all the literals making up that sum are 0.* Therefore, a term such as $W + \bar{X} + Z$ equals 0 only if $W = 0, X = 1$, and $Z = 0$. For sum terms, a numerical representation is obtained by assigning a *value of 0 to each uncomplemented variable and a value of 1 to each complemented variable.* This is the reverse of the procedure for obtaining a numerical representation of a product term.

EXAMPLE 3-11

What is the numerical representation of the sum term $\bar{W} + X + Y + \bar{Z}$?

SOLUTION   By assigning a 0 to the uncomplemented variables and a 1 to the complemented variables, we get

$$\bar{W} + X + Y + \bar{Z}$$
$$1 \quad\; 0 \quad 0 \quad\; 1$$

Since this is the binary equivalent of 9, the numerical representation of $\bar{W} + X + Y + \bar{Z}$ is 9.

Note that if $f(W, X, Y, Z) = \bar{W} + X + Y + \bar{Z}$, f is 0 only if $W = 1, X = 0$, $Y = 0$, and $Z = 1$, which corresponds to a numeric value of 9. For any other combination of input variables, at least one of the literals has a value of 1 and f equals 1.

EXAMPLE 3-12

(a) Explain the meaning of

$$f(a,b,c,d) = \Pi(0,4,7,11,14,15)$$

(b) Express it as literals in standard POS form.

SOLUTION (a) Because of the product (Π) sign the function equals 0 when the numeric value of the input is 0, 4, 7, 11, or 15. By using the rules developed in this section we can translate f into literals:

(b) $f(a,b,c,d) = \Pi(0,4,7,11,14,15)$
$= (a + b + c + d)(a + \bar{b} + c + d)(a + \bar{b} + \bar{c} + \bar{d})$
$\cdot (\bar{a} + b + \bar{c} + \bar{d})(\bar{a} + \bar{b} + \bar{c} + d)(\bar{a} + \bar{b} + \bar{c} + \bar{d})$

where each sum term corresponds to one of the numbers in the original expression.

The numeric representation of sum terms that have one or more literals missing can be obtained by using the given literals and every combination of the unspecified literals. This is similar to finding the numeric value for SOP terms.

EXAMPLE 3-13

If $f(a,b,c,d) = a + \bar{c}$, find its numeric representation in POS form.

SOLUTION The term $a + \bar{c}$ can be written as 0X1X, where the 0 in the most significant position represents a, and the first X represents the missing b literal, the 1 represents $\bar{c}$, and the second X represents the missing d literal. All four combinations of Xs will be included in the numerical representation. Therefore,

$$f(a,b,c,d) = \Pi(0010, 0011, 0110, 0111)$$
$$= \Pi(2,3,6,7)$$

EXAMPLE 3-14

If

$$f(W,X,Y,Z) = (\bar{W} + Y)(W + \bar{Y} + Z)(X + Y + \bar{Z})$$

find the numeric representation of f.

SOLUTION The first term, $\bar{W} + Y$, expands into 1X0X or 8, 9, 12, and 13. The term $W + \bar{Y} + Z$ expands to 0X10 or 2 and 6, and the term $X + Y + \bar{Z}$ expands to X001 or 1 and 9. Therefore,

$$f(W,X,Y,Z) = \Pi(1,2,6,8,9,12,13)$$

The number 9 occurred twice but should only be written once.

3-3-5 AND-OR and NAND Implementation of SOP Forms

Any expression in SOP form can be implemented by using AND gates followed by OR gates. This AND-OR implementation is accomplished by ANDing together the literals that comprise each term and ORing the output of each AND gate.

SOP expressions can also be implemented by using two levels of NAND gates. The first level NANDs the literals of each term producing a low output, and a second level NAND gate, serving as an OR for negative inputs, ORs together the outputs of the first level NAND gates. In practice, the two level NAND gate implementation is used more often because it requires only one type of gate.

EXAMPLE 3-15

The expression for segment a of a 7-segment display (see section 3-8-3) is:

$$\text{Segment a} = A + BD + \bar{B}\bar{D} + CD$$

Design a circuit to implement this expression:
(a) Using AND-OR logic.
(b) Using NAND logic.

SOLUTION (a) The AND-OR implementation of an SOP expression can always be designed by first ANDing together each term and then ORing the results. In this expression the terms B and D must be put into an AND gate. So must the terms $\bar{B}$ and $\bar{D}$ and the terms C and D. The term A can be left alone. The outputs of these four gates must be put into an OR Gate. The final circuit is shown in Fig. 3-2a. Because 4-input OR gates are not manufactured, the function is shown implemented by three 2-input OR gates as explained in Chapter 2. If only uncomplemented variables are available, $\bar{B}$ and $\bar{D}$ can be obtained from inverters. An alternate way to obtain $\bar{B}\bar{D}$ is to use a NOR gate (if available) in place of the AND gate since $\bar{B}\bar{D} = \overline{(B + D)}$. This circuit is shown in Fig. 3-2b.

(b) The solution using only NAND gates is shown in Fig. 3-2c. Each term in the original expression is NANDded together in the set of first level NAND gates. The outputs of each of these NAND gates is then fed to the second level NAND gate to provide the correct output. For this solution the A variable must be inverted because a LOW input to the second NAND gate is needed to produce a HIGH output. Any circuit in SOP form can be implemented by NAND gates by following this procedure. The NAND gate circuit is easily built in the laboratory. The ICs required to implement the circuit are identified in Fig. 3-2c.

3-3-6 NOR Implementation of POS Expressions[3]

Expressions in POS form can be implemented by either a two level OR-AND circuit or a two level NOR circuit. For the two level OR-AND circuit, the first level gates OR together the literals comprising each term. The second level gate ANDs together the outputs of the OR gates. In the NOR implementation, the first level gates NOR together the literals for each term and the second level gate

[3]On first reading, the user may want to skip to section 3-5.

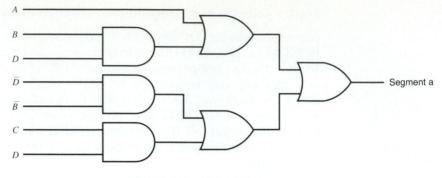

(a) AND-OR implementation

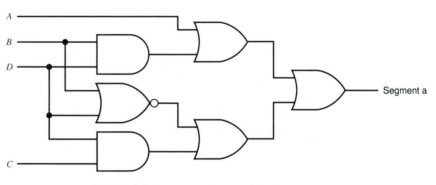

(b) AND-OR implementation (modified)

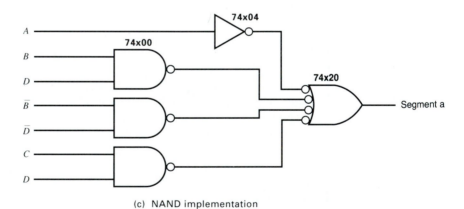

(c) NAND implementation

Figure 3-2 Three ways of building segment A of a 7-segment decoder.

functions as an AND gate for LOW inputs to AND together the outputs of the first level gates.

EXAMPLE 3-16

The POS expression for segment a of a 7-segment decoder is:

$$\text{Segment a} = (\bar{B} + D)(A + B + C + \bar{D})$$

Implement this expression:
(a) Using OR-AND logic.
(b) Using NOR logic.

SOLUTION   For the OR-AND implementation, the first term becomes a 2-input OR gate and the second term becomes a 4-input OR gate. The outputs of the two OR gates are then ANDed to produce the final output as shown in Fig. 3-3a.

For the NOR implementation the first level gates consist of a 2-input NOR and a 4-input NOR, and the second level gate is another 2-input NOR gate. The three NOR gates that implement the function are shown in Fig. 3-3b.

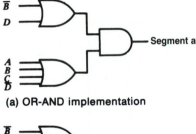

(a) OR-AND implementation

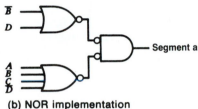

(b) NOR implementation

Figure 3-3   Two implementations of the POS form for segment a of a 7-segment decoder.

3-4   Equivalence of SOP and POS Standard Forms

The SOP and POS numeric representations for a function are complementary. *The numbers that appear in the SOP representation are not in the POS representation and all numbers that are not in the SOP representation appear in the POS representation.*

EXAMPLE 3-17

If $f(a,b,c,d) = \Sigma(1,2,5,6,7,10,11,12,13)$, find its numeric and literal POS representation.

SOLUTION   The domain of the function contains 4 variables, which can be described by the numbers 0 through 15. All the numbers not in the SOP form appear in the POS representation. Therefore,

$$
\begin{aligned}
f(a,b,c,d) &= \Pi(\mathbf{0,3,4,8,9,14,15}) \\
&= (a + b + c + d)(a + b + \bar{c} + \bar{d})(a + \bar{b} + c + d) \\
&\quad \cdot (\bar{a} + b + c + d)(\bar{a} + b + c + \bar{d}) \\
&\quad \cdot (\bar{a} + \bar{b} + \bar{c} + d)(\bar{a} + \bar{b} + \bar{c} + \bar{d})
\end{aligned}
$$

3-4-1   Truth Table Verification

To verify the equivalence of the SOP and POS numeric forms, consider the function defined by the truth table of Fig. 3-4. In SOP form:

$$f(X,Y,Z) = \Sigma(0,2,4,5,7)$$

The SOP form lists the numbers for which the value of the function is 1.

To express the same function in POS form, the numbers where the function is 0 are used:

$$f(X,Y,Z) = \Pi(1,3,6)$$

X	Y	Z	F(X, Y, Z)
0	0	0	1
0	0	1	0
0	1	0	1
0	1	1	0
1	0	0	1
1	0	1	1
1	1	0	0
1	1	1	1

Figure 3-4 Truth table for the function considered in section 3-4-1.

EXAMPLE 3-18

Show that the SOP and POS forms for $f(X,Y,Z)$ are equivalent using algebraic manipulation.

SOLUTION   The SOP form for the function can be obtained directly from the truth table by writing an SOP term corresponding to each 1 in the table. The SOP form can then be simplified.

$$
\begin{aligned}
f(X,Y,Z) &= \Sigma(0,2,4,5,7) \\
&= \bar{X}\bar{Y}\bar{Z} + \bar{X}Y\bar{Z} + X\bar{Y}\bar{Z} + X\bar{Y}Z + XYZ \\
&= \bar{X}\bar{Z}(\bar{Y} + Y) + X\bar{Y}(\bar{Z} + Z) + XZ(Y + \bar{Y}) \\
&= \bar{X}\bar{Z} + X\bar{Y} + XZ
\end{aligned}
$$

Starting with the POS form and simplifying, we find that

$$
\begin{aligned}
f(X,Y,Z) &= \Pi(1,3,6) \\
&= (X + Y + \bar{Z})(X + \bar{Y} + \bar{Z})(\bar{X} + \bar{Y} + Z) \\
&= (X + \bar{Z})(\bar{X} + \bar{Y} + Z) && \text{(Theorem \textbf{13})[4]} \\
&= X\bar{Y} + XZ + \bar{X}\bar{Z} + \bar{Y}\bar{Z} \\
&= X\bar{Y} + XZ + \bar{X}\bar{Z} && \text{(Theorem \textbf{12})[4]}
\end{aligned}
$$

Thus, the POS and SOP forms reduce to identical terms.

---

[4]See section 2-7.

3-4-2 Verification Using DeMorgan's Theorem

For a function with a domain of N variables there are 2^N standard SOP terms. *For any given combination of input variables, however, only one of the 2^N terms can equal 1.* If an SOP function contains all 2^N standard product terms, it will equal 1 because one of the terms must equal 1 and the rest must equal 0 for *any and every* combination of the input variables. Therefore, if a function f is defined in numeric SOP form, all the standard SOP terms that were not included in f must together comprise $\bar{f}$ because if the term that equals 1 is in f, it is not in $\bar{f}$, and vice versa. In Example 3-18,

$$f(X,Y,Z) = \Sigma(0,2,4,5,7)$$

Therefore, the SOP form of

$$\bar{f}(X,Y,Z) = \Sigma(1,3,6)$$
$$= \bar{X}\bar{Y}Z + \bar{X}YZ + XY\bar{Z}$$

If, for example, $\bar{X}\bar{Y}Z = 1$, then $X = 0$, $Y = 0$, and $Z = 1$. This combination of variables makes $\bar{f} = 1$ and $f = 0$ since for these variables every term in f will be 0. To verify the above statements, let us take the complement of $\bar{f}$.

$$\overline{\bar{f}(X,Y,Z)} = f(X,Y,Z)$$
$$\overline{\bar{X}\bar{Y}Z + \bar{X}YZ + XY\bar{Z}} = (X + Y + \bar{Z})(X + \bar{Y} + \bar{Z})(\bar{X} + \bar{Y} + Z)$$

But this is exactly $f(X,Y,Z)$ in POS form, which verifies our statement.

Similarly, *only one of all possible POS terms can equal 0; the rest must equal 1. Therefore, $\bar{f}$ can also be expressed in numeric POS form as the product of all the numbers not included in f.* Again referring to Example 3-18:

$$f(X,Y,Z) = \Pi(1,3,6)$$

Therefore,

$$\bar{f}(X,Y,Z) = \Pi(0,2,4,5,7)$$
$$= (X + Y + Z)(X + \bar{Y} + Z)(\bar{X} + Y + Z)$$
$$\cdot (\bar{X} + Y + \bar{Z})(\bar{X} + \bar{Y} + \bar{Z})$$

If the complement of $\bar{f}$ is taken, the result will be $f(X,Y,Z)$ in SOP form.

3-5 Karnaugh Maps

Karnaugh mapping is a graphic technique for reducing functions to their simplest terms. Once a function is expressed in standard SOP or POS form, it can be plotted on a Karnaugh map, which leads to a systematic minimization of the function. Karnaugh mapping is a viable reduction technique for functions of 3, 4, and 5 variables. For functions of more than 5 variables, more advanced techniques such as the Quine-McClusky algorithm are used.[5] However, these are

---

[5]For a discussion of the Quine-McClusky techniques, see F.J. Hill and G.R. Peterson, *Introduction to Switching Theory and Logical Design*, 3rd ed. (New York: Wiley, 1981), Chapter 7.

more elaborate and cumbersome techniques and only Karnaugh mapping will be discussed in this book.

3-5-1 Construction of the 3-Variable Karnaugh Maps

A Karnaugh map for a function of 3 variables, $f(X,Y,Z)$, is shown in Fig. 3-5a. There are 8 squares within the map, one for each possible number in the decimal representation of the equation. The squares are arranged in a 4-by-2 matrix as shown. The two possible values for the most significant variable (X in this case) head the two columns. The four possible values for the two remaining variables are listed alongside the map. Note that the 11 coordinate is listed above the 10 coordinate. This is deliberately done so that *no coordinate differs from its adjacent coordinate in more than one bit position.*

Figure 3-5b shows the decimal value for each box. This value corresponds to the X, Y, and Z coordinates for that box. As an example, consider the box in the upper right-hand corner. Its X coordinate is 1 and its Y and Z coordinates are 00. As a binary number it can be written as 100 or 4.

To construct a Karnaugh map for a function specified in SOP form:

a) Convert the SOP form to numeric form as shown in Sec. 3-3-3.

b) Put a 1 in each box of the Karnaugh map that corresponds to a 1 in the expression.

The procedure is shown in Ex. 3-19.

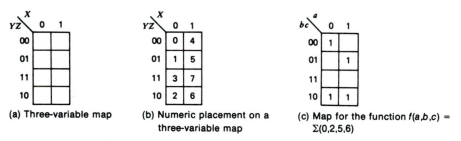

(a) Three-variable map   (b) Numeric placement on a three-variable map   (c) Map for the function $f(a,b,c) = \Sigma(0,2,5,6)$

Figure 3-5   Three-variable Karnaugh maps.

EXAMPLE 3-19

Construct the Karnaugh map for the function

$$f(a,b,c) = \bar{a}\bar{b}\bar{c} + a\bar{b}c + b\bar{c}$$

SOLUTION   The solution is shown in Fig. 3-5c. The first term, $\bar{a}\bar{b}\bar{c}$ corresponds to $a = 0, b = 0, c = 0$, or the number 0. Similarly, the term $a\bar{b}c$ corresponds to 5, and the term $b\bar{c}$ corresponds to the numbers 2 and 6. Therefore

$$f(a,b,c) = \Sigma(0,2,5,6)$$

1s are then placed in the Karnaugh map at the 0,2,5, and 6 locations and the map is complete, as shown.

3-5-2 Simplification of 3-Variable SOP Expressions

A Boolean expression can be simplified using Karnaugh maps by the following procedure:

1. Write the expression in standard SOP form.
2. Plot the Karnaugh map.
3. Form subcubes that cover all the 1s in the expression.
4. Write the resulting expression.

If the subcubes are properly chosen, the result is the expression in minimum SOP form.

Subcubes are groups of 1s that are adjacent to each other on a Karnaugh map. A 2-cell subcube consists of two adjacent 1s, which means the cells containing the 1s border each other in either a vertical or horizontal line. Cells next to each other diagonally are not adjacent. For any 2-cell subcube we find that one variable changes values, but the rest of them have a common value. *The SOP expression for the subcube consists of those variables with a common value. The variable that changes values is not needed in the expression.*

EXAMPLE 3-20

Given $f(a,b,c) = \Sigma(0,2,6)$, draw the Karnaugh map and simplify the function.

SOLUTION The Karnaugh map for the given function is shown in Fig. 3-6a, where three 1s are written in the squares whose coordinates are 0, 2, and 6. The 1s in the 2 and 6 squares are adjacent and form a 2-cell subcube. The coordinates corresponding to 2 and 6 are 010 and 110, so that a (the most significant variable) changes, while $b = 1$ and $c = 0$ for both squares. Therefore, the subcube is written as $b\bar{c}$.

The 1 in the 0 square seems all alone. Anytime a 1 is indeed all alone and cannot be combined with any other 1 to form a subcube, it is a *1-cell subcube* and is represented by a literal for each variable in the domain. The 1 in the 0 square is written as $\bar{a}\bar{b}\bar{c}$ and

$$f(a,b,c) = \bar{a}\bar{b}\bar{c} + b\bar{c}$$

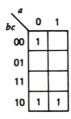

Figure 3-6a Original Karnaugh map for Example 3-20.

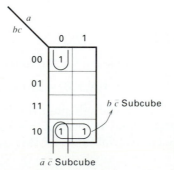

Figure 3-6b Map with subcubes displayed.

After examining the solution of Example 3-20, we realize it can be simplified.

$$f(a,b,c) = \bar{c}(\bar{a}\bar{b} + b)$$
$$= \bar{c}(\bar{a} + b) \qquad \text{(Theorem 11)[6]}$$
$$= \bar{a}\bar{c} + b\bar{c}$$

It appears the Karnaugh map didn't yield the simplest expression after all, but this is incorrect. Heretofore we have ignored the fact that *Karnaugh maps are spherical*; that is, *they wrap around themselves*. For an 8-cell map this means that *the top row is adjacent to the bottom row*. In this example a second subcube can be formed by the 0 and 2 cells. The 0 cell is numerically 000 while the 2 cell is 010. Therefore only the b variable changes; it is excluded and the subcube is represented by $\bar{a}\bar{c}$. The subcubes are shown in Fig. 3-6b.

EXAMPLE 3-21

Find the minimal expression for the function

$$f(a,b,c) = \Sigma(0,2,5,6)$$

SOLUTION   This function was plotted in Fig. 3-5c. As in Example 3-20, the 0, 2, and 6 cells can be combined into 2-cell subcubes. Since the 1 in the 5 cell is truly alone and cannot be combined, it is expressed as $a\bar{b}c$. Therefore,

$$f(a,b,c) = \Sigma(0,2,5,6)$$
$$= \bar{a}\bar{c} + b\bar{c} + a\bar{b}c$$

Note that this expression is simpler than the original expression given in Example 3-19, from which the Karnaugh map was drawn.

3-5-3   Four-Cell Subcubes

The size of a subcube must be a *power* of 2 (i.e., only 2, 4, 8, 16, etc. subcubes exist). For example, a 3-cell subcube is not allowed. If a row of three adjacent 1s appears on a map, they must be combined into two 2-cell subcubes.

Four-cell subcubes are also allowed in a Karnaugh map. A 4-cell subcube appears on a map as either a row of four 1s or a two-by-two square of 1s. For a 4-cell subcube, 2 variables will change and they are omitted. A 4-cell subcube is expressed by 2 fewer literals than the domain of the expression. The maps of Figs. 3-5 or 3-6 are for a domain of three variables. If a 4-cell subcube were to appear in such a map, it would therefore be expressed by only one literal.

EXAMPLE 3-22

Find the minimal expression for the function

$$f(X,Y,Z) = \Sigma(0,1,2,3,4,6)$$

SOLUTION   The Karnaugh map for the function is shown in Fig. 3-7. There is a column of four 1s in the $X = 0$ column. This forms a 4-cell subcube that is described by $\bar{X}$. (Note that for this 4-cell subcube the variables Y and Z

[6] See section 2-7.

changed and became the omitted literals.) The two other 1s at 100 and 110 are adjacent to each other because the top row is adjacent to the bottom row, and can be covered by a 2-cell subcube. However, they can also be covered by a 4-cell subcube consisting of the 000, 100, 010, and 110 cells. These form a two-by-two square because the top and bottom rows are adjacent. When a choice exists, it is wisest to use the *largest* subcube possible, because this will result in an expression with the fewest literals. The 4-cell subcube here is written as $\bar{Z}$ since each cell in the subcube has a Z value of 0. Therefore,

$$f(X,Y,Z) = \Sigma(0,1,2,3,4,6)$$
$$= \bar{X} + \bar{Z}$$

As a check, $\bar{X} = \Sigma(0,1,2,3)$ and $\bar{Z} = \Sigma(0,2,4,6)$. The total sum covers all the terms in f, and no terms that are not in f.

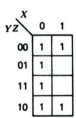

Figure 3-7 Karnaugh map for Example 3-22.

3-5-4   Construction of a 4-Variable Karnaugh Map

Functions of four variables require a map with 16 entries. A 4-variable Karnaugh map is shown in Fig. 3-8a. The four possible combinations for the two most significant variables (W and X) are listed along the top of the map and the four combinations for the least significant variables (Y and Z) are listed along the side. Again the order of listing is 00, 01, 11, and 10, for both the top and side variables. The numeric terms for each of the 16 squares in the map is shown in Fig. 3-8b.

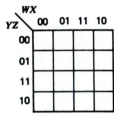

(a) The general 4-variable map

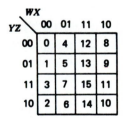

(b) The map with numeric terms entered

Figure 3-8   The 4-variable Karnaugh map.

EXAMPLE 3-23

Construct the Karnaugh map for the function

$$f(W,X,Y,Z) = \Sigma(0,3,5,9,12,15)$$

SOLUTION   The solution is shown in Fig. 3-9 where 1s are written in the Karnaugh map for each term in the specification. For example, the 1 for the number 12 (binary 1100) is written in the square that is the intersection of 11 for the most significant variables and 00 for the least significant variables.

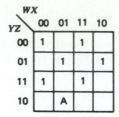

Figure 3-9 Karnaugh map for Examples 3-23 and 3-24.

EXAMPLE 3-24

What combination of variables does the square with the letter A in it in Fig. 3-9 represent?

SOLUTION   The letter A is in the square at the intersection of the 01 and 10 coordinates. Therefore, this square is designated by the number 0110 or decimal 6.

3-5-5   Simplification of 4-Variable Maps for SOP Expressions

Four-variable maps are also simplified by following the procedure developed for 3-variable maps in section 3-5-2. Four-variable maps are spherical also. This means the top row is adjacent to the bottom row and the leftmost column is adjacent to the rightmost column. Eight-cell subcubes may exist in a 4-variable map; if so, they must form a 2-by-4 matrix, containing any 2 complete adjacent rows or columns.

EXAMPLE 3-25

Find the minimal expression for

$$f(W,X,Y,Z) = \Sigma(0,1,2,3,4,8,9,10,15)$$

SOLUTION   First the Karnaugh map for the function is plotted as shown in Fig. 3-10a. Several subcubes are required to cover all the 1s in the map and the expression will be correct when every 1 in the map is covered by at least one subcube. Figure 3-10b shows the subcubes used to cover this function.

Starting on the left, the column $W = 0, X = 0$, is completely filled with 1s. This forms a 4-cell subcube, subcube 1 in Fig. 3-10b, and can be expressed by two literals, as explained in Sec. 3-5-3. The expression for this subcube is $\overline{W}\overline{X}$.

Next consider the 1 in the 0100 cell. This cell can only be combined with the 1 in the 0000 cell to form a 2-cell subcube, which is subcube 2. A 2-cell subcube eliminates one literal and the expression for this particular subcube is $\overline{W}\overline{Y}\overline{Z}$. The X variable changes; consequently it is not included in the expression.

Now let us examine the two 1s in the upper right hand corner; in the 1000 and 1001 cells. They form a 2-cell subcube and can be expressed as $W\overline{X}\overline{Y}$. It is wisest, however, to cover a clee by the largest subcube possible. This eliminates the most variables, and if the circuit is actually to be built, it will lead to the implementation requiring the smallest number of gates, because every literal in an expression requires a gate input. This subcube can be combined into a 4-cell subcube with the cells in 0000 and 0001. This is subcube 3 and its expression is $\overline{X}\overline{Y}$, which is certainly simpler and easier to build than the circuit for the expression $\overline{W}\overline{X}\overline{Y}$.

The 1 in the 1010 is part of an interesting subcube. Each corner cell is adjacent to two other corner cells (but not to the corner cell diagonally opposite it). When all 4 corner cells are filled, however, they form a 4-cell subcube. Their coordinates are 0000, 1000, 1010, 0010. Note that the second and fourth variables are always 0; thus the subcube is described as $\overline{X}\overline{Z}$.

Finally, since the 1 in the 1111 cell cannot be combined with any other 1 on the map, it must be described by using all 4 literals as $WXYZ$. Therefore,

$$f(W,X,Y,Z) = \Sigma(0,1,2,3,4,8,9,10,15)$$
$$= \overline{W}\overline{X} + \overline{X}\overline{Y} + \overline{X}\overline{Z} + \overline{W}\overline{Y}\overline{Z} + WXYZ$$

This covers all the 1s on the map and is the minimal SOP expression for the function.

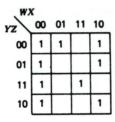

Figure 3-10a Karnaugh map for Example 3-25. a) Original map b) (see below)

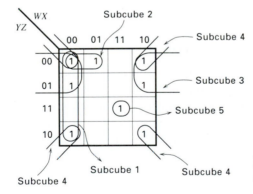

Figure 3-10b Karnaugh map with subcubes displayed.

EXAMPLE 3-26

Find the minimal expression for

$$f(W,X,Y,Z) = \Sigma(0,1,2,3,4,6,8,9,10,11,15)$$

SOLUTION The Karnaugh map is shown in Fig. 3-11. The four 1s in the left-most column are adjacent to the four 1s in the right-most column. Together they form an 8-cell subcube, and 3 literals must disappear. In this case, $X = 0$ for each cell in the subcube while W, Y, and Z all change. Therefore, this subcube is designated as $\bar{X}$. The 2 cells in the upper left row are adjacent to the 2 cells in the lower left row, and form a 4-cell subcube, $\bar{W}\bar{Z}$. The remaining 1 in cell 1111 can be combined with the 1 in cell 1011 to form a 2-cell subcube, WYZ. This covers all the 1s on the map. Therefore,

$$f(W,X,Y,Z) = \Sigma(0,1,2,3,4,6,8,9,10,11,15)$$
$$= \boldsymbol{\bar{X} + \bar{W}\bar{Z} + WYZ}$$

The easiest way to check expressions of this kind is to make a check table as shown below:

Term	Expression	Numbers Covered
$\bar{X}$	X 0 X X	0,1,2,3,8,9,10,11
$\bar{W}\bar{Z}$	0 X X 0	0,2,4,6
WYZ	1 X 1 1	11,15

All the terms of the original specifications have been covered and no terms that are *not* in the specification have been included. This checks the solution.

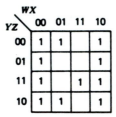

Figure 3-11 Karnaugh map for Example 3-26.

3-5-6 Essential Subcubes[7]

The simplest circuit realization results from choosing the best set of subcubes to cover all the 1s on a given Karnaugh map. To select the best set of subcubes, we first select all the *essential subcubes*, rather than the largest one. *An essential subcube is one that must be chosen to cover a certain 1 on the map because no other subcubes can cover that particular 1.* After the essential subcubes have been chosen, the remaining 1s on the map should be covered as simply as possible.

In Fig. 3-10, for example, we should not start out by deciding how to cover the 1 in cell 0000. This cell can be part of three different 4-cell subcubes. Since a *choice* exists, this cell is *not* essential. The 2-cell subcube, $\bar{W}\bar{Y}\bar{Z}$ which covers the 0000 and 0100 cell, is essential because there is no other way to cover the 0100 cell. The 0001 and 0010 cells are each part of two 4-cell subcubes and are therefore not essential, but the 1 in the 0011 cell can only be covered by a 4-cell subcube, $\bar{W}\bar{X}$; therefore this subcube is essential. (Essential subcubes are not necessarily 2-cell subcubes.) Similarly, the subcube $\bar{X}\bar{Z}$ is essential because it is the only way to cover the 1010 cell and the subcube $\bar{X}\bar{Y}$ is essential because it is the only way to cover the 1001 cell. Since all the terms needed to cover this map, as found in Example 3-25, are essential, the solution is minimal.

---

[7]This section may be omitted on first reading.

EXAMPLE 3-27

Find the minimal expression for f if

$$F(W,X,Y,Z) = \Sigma(1,4,5,6,8,12,13,15)$$

SOLUTION The Karnaugh map for the function is shown in Fig. 3-12a. It is tempting to grab the big 4-cell subcube in the center, $X\bar{Y}$. However, this would leave four 1s uncovered, and they must be covered by four 2-cell subcubes. The resulting expression is

$$F(W,X,Y,Z) = X\bar{Y} + W\bar{Y}\bar{Z} + \bar{W}\bar{Y}Z + \bar{W}X\bar{Z} + WXZ$$

Unfortunately this procedure does not result in a minimum expression for the function. This is because none of the cells in the 4-cell subcube are essential, whereas all the 2-cell subcubes are essential. If we start correctly by taking the four essential 2-cell subcubes, as shown in Fig. 3-12b, we find that each essential subcube covers a 1 in the 4-cell subcube, and together they cover all the 1s. Consequently, the term $X\bar{Y}$ for the 4-cell subcube covers no new 1s and should be omitted. Therefore,

$$F(W,X,Y,Z) = \mathbf{W\bar{Y}\bar{Z} + \bar{W}\bar{Y}Z + WXZ + \bar{W}X\bar{Z}}$$

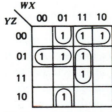

(a) Karnaugh map

(b) Karnaugh map with the essential subcubes selected

Figure 3-12 Karnaugh maps for Example 3-27.

Some functions do not contain any essential cells. Consider $F(W,X,Y,Z) = \Sigma(0,4,5,8,10,11,13,15)$ whose Karnaugh map is shown in Fig. 3-13a. Since each cell in this map can be part of two 2-cell subcubes, no cells are essential. When this occurs, the following general rules are helpful.

1. For each new subcube, try to cover as many previously uncovered 1s as possible.
2. Try to select subcubes so that no 1s are isolated.
3. If each 1 in the map has been covered only once, a minimal expression can be obtained from this cover. It is not always possible to achieve this result.

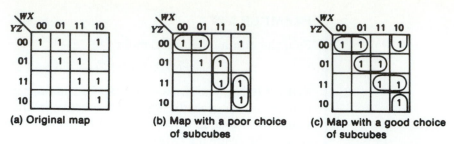

(a) Original map

(b) Map with a poor choice of subcubes

(c) Map with a good choice of subcubes

Figure 3-13 A Karnaugh map containing no essential subcubes.

If we started to cover the function by selecting subcubes as shown in Fig. 3-13b, rule 2 is violated because the 1s in cells 0101 and 1000 are isolated. Two additional subcubes would be needed to cover them and the resulting expression would contain five 2-cell subcubes.

A minimal cover is shown in Fig. 3-13c, where the eight 1s in the map are covered by four 2-cell subcubes and each 1 is covered only once. Therefore,

$$F(W,X,Y,Z) = \Sigma(0,4,5,8,10,11,13,15)$$
$$= \overline{W}\overline{Y}\overline{Z} + X\overline{Y}Z + WYZ + W\overline{X}\overline{Z}$$

is a minimal cover for the expression. Note that this minimal cover is not unique. The expression

$$\overline{X}\overline{Y}\overline{Z} + \overline{W}X\overline{Y} + WXZ + W\overline{X}Y$$

also covers the function with four 2-cell subcubes (See Problem 3-13.)

EXAMPLE 3-28

Find a minimal expression for the Karnaugh map of Fig. 3-14a.

SOLUTION Inspection of the Karnaugh map reveals that each 1 is part of two 4-cell subcubes and there are no essential subcubes. If we select the two columns as 4-cell subcubes, as shown in Fig. 3-14b, we isolate the remaining 1s; this leads to a solution where four 4-cell subcubes are needed to cover the map. Selecting the subcubes as shown in Fig. 3-14c does not isolate any 1s and results in covering the twelve 1s with three 4-cell subcubes. This results in a minimal cover of

$$\overline{A}\overline{B} + BC + A\overline{C}$$

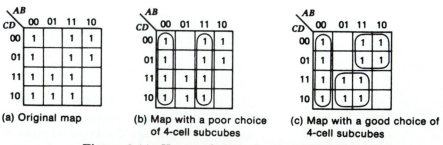

(a) Original map

(b) Map with a poor choice of 4-cell subcubes

(c) Map with a good choice of 4-cell subcubes

Figure 3-14 Karnaugh maps for Example 3-28.

3-5-7 Simplification of 4-Variable Maps for POS Expressions

Karnaugh maps for functions in POS form can be plotted by placing a 0 in each cell where the value of the function is 0. Adjacent cells can then be combined into subcubes in a manner similar to that for SOP expressions. It must be remembered that the coordinates for POS maps are inverted because complemented variables represent 1s and uncomplemented variables represent 0s in POS expressions.

EXAMPLE 3-29

Plot the Karnaugh map for the function

$$f(W,X,Y,Z) = (W + \bar{Z})(\bar{W} + \bar{X} + \bar{Y} + \bar{Z})$$

and find a simpler expression from the map.

SOLUTION   The term $W + \bar{Z}$ can be plotted by placing a 0 on the map at all cells where $W = 0$ and $Z = 1$, or, using the methods of section 3-3-4,

$$W + \bar{Z} = 0XX1 = \Pi(1,3,5,7)$$

so that 0s are placed at the cells in locations 1, 3, 5, and 7. Similarly the term $\bar{W} + \bar{X} + \bar{Y} + \bar{Z}$ causes us to place a 0 in the 1111 cell. The resulting Karnaugh map is shown in Fig. 3-15. From it we see that the 0 in the 1111 cell can be combined as a 2-cell subcube with the 0 in the 0111 cell. For this 2-cell subcube, the W variable changes so it is not in the final expression, while X, Y, and Z are 1. Since 1s are represented by complemented variables for POS functions, this subcube is $\bar{X} + \bar{Y} + \bar{Z}$. Therefore, the function is covered by a 4-cell subcube and a 2-cell subcube and

$$f(W,X,Y,Z) = (W + \bar{Z})(\bar{X} + \bar{Y} + \bar{Z})$$

This expression is simpler than the original expression because it contains one less literal.

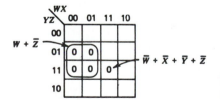

Figure 3-15   Karnaugh map for Example 3-29.

EXAMPLE 3-30

(a) Plot the Karnaugh map and obtain the POS expression for

$$f(a,b,c,d) = \Pi(1,5,6,7,10,12,13,15)$$

(b) Verify your solution using a check table.

SOLUTION   Since the function is specified by a product, 0s are written into each cell of the map that corresponds to a number in the specification, as shown in Fig. 3-16a. Next the essential subcubes are selected as shown in Fig. 3-16b. The 0 in cell 1010 cannot be combined with any other cell. Therefore it must be represented by a term containing all 4 literals, $\bar{a} + b + \bar{c} + d$. The three 2-cell subcubes are all essential, and each covers a term in the 4-cell subcube. The 1111 cell, however, can only be covered by the 4-cell subcube; therefore, the 4-cell subcube is also essential. Writing the proper expression for each subcube in the cover, we obtain

$$f(a,b,c,d) = \Pi(1,5,6,7,10,12,13,15)$$
$$= (\bar{a} + b + \bar{c} + d)(\bar{b} + \bar{d})(a + c + \bar{d})$$
$$(a + \bar{b} + \bar{c})(\bar{a} + \bar{b} + c)$$

A check table can be made for POS as well as SOP expressions.

Term	Expression	Numbers Covered
$\bar{a} + b + \bar{c} + d$	1 0 1 0	10
$\bar{b} + \bar{d}$	X 1 X 1	5,7,13,15
$a + c + \bar{d}$	0 X 0 1	1,5
$a + \bar{b} + \bar{c}$	0 1 1 X	6,7
$\bar{a} + \bar{b} + c$	1 1 0 X	12,13

Together they cover all the numbers (0s) in the original specification and no other numbers. This checks the solution.

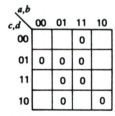

(a) Original map

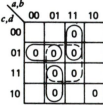

(b) Map with subcubes selected

Figure 3-16 Karnaugh maps for Example 3-30.

EXAMPLE 3-31

Given the function

$$f(W,X,Y,Z) = \Sigma(4,6,7,12,14)$$

(a) Find the minimal SOP expression.
(b) Find the minimal POS expression.
(c) Show by algebraic manipulation that the two expressions are equal.

3 / Systematic Reduction of Boolean Expressions

SOLUTION The SOP Karnaugh map is drawn as shown in Fig. 3-17a. It simplifies to a 4-cell subcube and a 2-cell subcube. Therefore,

$$f(W,X,Y,Z) = X\bar{Z} + \bar{W}XY$$

The POS map is drawn simply by placing 0s in the map wherever there is not a 1, as shown in Fig. 3-17b. It simplifies into an 8-cell subcube and two 4-cell subcubes. In POS form,

$$f(W,X,Y,Z) = X(Y + \bar{Z})(\bar{W} + \bar{Z})$$

To show the two functions are identical, we operate on the POS form:

$$
\begin{aligned}
X(Y + \bar{Z})(\bar{W} + \bar{Z}) &= X(\bar{Z} + \bar{Z}Y + \bar{Z}\bar{W} + \bar{W}Y) \\
&= X(\bar{Z} + \bar{W}Y) \qquad\qquad \text{(Theorem \textbf{9})[8]} \\
&= X\bar{Z} + X\bar{W}Y
\end{aligned}
$$

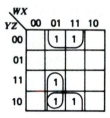

(a) SOP map

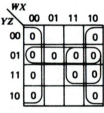

(b) POS map

Figure 3-17 Karnaugh maps for $f(w,x,y,z) = \Sigma(4,6,7,12,14)$.

3-5-8 Optimum Selection of Gates

Heretofore we have developed logic that minimizes the number of literals in a Boolean expression, and we have also developed a procedure for implementing any expression in POS or SOP form. For some logic expressions the SOP form is simpler and easier to implement. For other expressions, the POS form is preferable. There is no a priori way of knowing which form yields the simpler circuit. Consequently, the wise designer develops both the SOP and POS expressions before building the circuit. Implementation then depends on:

1. Which form requires fewer gates.
2. The availability of the required gates. (It makes no sense to use a gate that is not available from the manufacturer, or a gate that is not in stock and may require a long time for delivery when, as is typical in industry, the circuit is needed immediately.)
3. The number of wires required.

[8]See section 2-7.

Often a mixture of ANDs, ORs, NANDs, and so on produces the optimum circuit. As in other areas of engineering, experience, inspiration, and, above all, a little thought help the engineer make the best decision.

EXAMPLE 3-32

Given:

$$f(W,X,Y,Z) = \Sigma(5,6,7,9,10,11,13,14,15)$$

(a) How would the SOP and POS functions be implemented?
(b) Which form is preferable?

SOLUTION (a) The Karnaugh map, shown in Fig. 3-18, is a 3-by-3 rectangle that is covered by four 4-cell subcubes. The SOP representation is

$$f(W,X,Y,Z) = WY + WZ + XY + XZ$$

For NAND implementation, this requires four 2-input gates and a 4-input gate.

The POS representation is obtained by using the top row and left column as two 4-cell subcubes. This gives

$$f(W,X,Y,Z) = (W + X)(Y + Z)$$

(b) The POS form can be implemented with three 2-input NOR gates. Here the POS form has the clear advantage of requiring fewer and less complex gates.

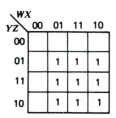

Figure 3-18 Karnaugh map for the function of Example 3-32.

3-5-9 Multiple Outputs[9]

Many circuits require several outputs from the same set of inputs. One common example is 7-segment display drivers where the outputs for each of the 7 segments are derived from the same four inputs (see section 3-8-3).

Karnaugh maps can and should be drawn for each output. Unfortunately, *minimizing the map for each output individually does not necessarily minimize the circuit as a whole.* If the circuit is to be built, the goal is to minimize the entire circuit, not just the circuit for each individual function. As a simple example, consider two functions that must be generated on the same set of inputs:

$$f_1 = \Sigma(4,5,12,13,15)$$
$$f_2 = \Sigma(6,14,15)$$

[9]This section can be omitted at first reading.

Their Karnaugh maps are shown in Fig. 3-19. Simplifying each individually yields:

$$f_1 = X\bar{Y} + WXZ$$
$$f_2 = XY\bar{Z} + WXY$$

The functions can also be expressed as:

$$f_1 = X\bar{Y} + WXYZ$$
$$f_2 = XY\bar{Z} + WXYZ$$

This expression requires one more literal in each output, but the same term, $WXYZ$, can be used to contribute to *both* outputs. The NAND or AND-OR implementation would require one 4-input gate instead of two 3-input gates and is generally preferred (see Problem 3-26).

The theory of minimizing multiple output circuits is complex and not totally developed. We offer the following suggestions.

1. Start by covering all the subcubes of one function that have *no common points* with the other function. For the above example these were the subcube $X\bar{Y}$ for f_1 and $XY\bar{Z}$ for f_2.

2. Where common points exist, examine the maps carefully to try to determine where gates used for one function can also be used for the other function.

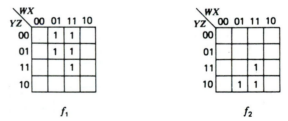

Figure 3-19 Karnaugh maps for f_1 and f_2.

EXAMPLE 3-33

Given:

$$f_1(WXYZ) = \Sigma(0,1,8,9,10,13)$$
$$f_2(WXYZ) = \Sigma(0,1,5,7,9,13,15)$$

Implement f_1 and f_2 using the minimum number of gates.

SOLUTION   The Karnaugh maps are shown in Fig. 3-20. If we were minimizing each function individually, we would certainly use the subcubes $\bar{X}\bar{Y}$ for f_1 and $\bar{Y}Z$ and XZ for f_2. Individual simplification would lead to three terms for each function.

To minimize the functions together, we start by observing that the subcube $WX\bar{Z}$ (the right side corners) is essential for f_1 and has no commonality with f_2. We also see that the subcube XZ is essential for f_2. The subcube $\bar{W}\bar{X}Y$ is

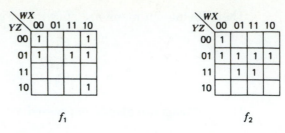

Figure 3-20 Karnaugh maps for Example 3-33.

essential for f_2 but can also be used to cover two terms in f_1. At this point we have:

$$f_1 = W\bar{X}\bar{Z} + \bar{W}\bar{X}\bar{Y}$$
$$f_2 = XZ + \bar{W}\bar{X}\bar{Y}$$

We still must cover the terms 13 and 9 in f_1 and 9 in f_2. For f_1 this requires $W\bar{Y}Z$, but this term can also be used in f_2 instead of the subcube $\bar{Y}Z$, and eliminates the need for that subcube. Therefore, the final expressions become:

$$f_1 = W\bar{X}\bar{Z} + \bar{W}\bar{X}\bar{Y} + W\bar{Y}Z$$
$$f_2 = XZ + \bar{W}\bar{X}\bar{Y} + W\bar{Y}Z$$

As with the minimal expression, there are three terms for each expression, but now two of them are common so the expressions can be implemented with four first level gates instead of six. The implementation is shown in Fig. 3-21.

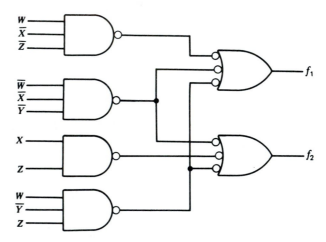

Figure 3-21 Circuit for Example 3-33.

3-6   Don't Cares

For some functions the outputs corresponding to certain combinations of input variables do not matter. This usually occurs because some combinations of the input variables should not or do not exist. A common example is the Binary-Coded-Decimal (BCD) representation of numbers. This code is shown in Fig. 3-22 and uses four bits to represent a single decimal digit, from 0 through 9. The hex digits 0–9 represent the corresponding decimal digit, but the hex digits A–F are not used in this system. Therefore, bit combinations with a value of 10 through 15 should not occur. On a truth table or Karnaugh map, outputs corresponding to these inputs are listed as *don't cares*.

*Don't cares are written as small d*s on Karnaugh maps, and increase the designer's versatility. As on any map, the 1s must be covered and the 0s must be left uncovered. The don't cares, however, may or may not be covered. *Don't cares should be covered if their inclusion simplifies the final expression by creating larger subcubes*, which are expressed by terms with fewer literals. Otherwise there is no need to cover them. Don't cares can be used on an "as needed" basis on both SOP and POS maps.

Binary — Coded Decimal Representation				Decimal Digit Equivalent
0	0	0	0	0
0	0	0	1	1
0	0	1	0	2
0	0	1	1	3
0	1	0	0	4
0	1	0	1	5
0	1	1	0	6
0	1	1	1	7
1	0	0	0	8
1	0	0	1	9

Figure 3-22 The BCD code conversion table.

EXAMPLE 3-34

Given:

$$F(W,X,Y,Z) = \Sigma(0,1,2,5,8,14) + d(4,10,13)$$

Find:

(a) The minimum SOP expression.

(b) The minimum POS expression.

(c) Whether the SOP and POS expressions can be shown to be equal by algebraic manipulation.

SOLUTION (a) The SOP Karnaugh map is shown in Fig. 3-23a. The map may be covered by two 4-cell subcubes (consisting of the upper left-hand corner and the four corners) and a 2-cell subcube. All three subcubes are essential and the minimal SOP expression is

$$f(W,X,Y,Z) = \bar{W}\bar{Y} + \bar{X}\bar{Z} + WY\bar{Z}$$

Note that the don't care in the 1101 cell added nothing to the SOP expression and was not included. Note also that the minimal SOP expression would have been longer if the don't cares in 0100 and 1010 were not available.

(b) The POS Karnaugh map for the function is shown in Fig. 3-23b. It was drawn simply by writing 0s in all the vacant cells on the SOP map and leaving the don't cares as they were. On the POS map the two 4-cell subcubes $\bar{Y} + \bar{Z}$ and $\bar{W} + \bar{Z}$ are essential. The two remaining 0s must be covered by

2-cell subcubes that are not essential. One particular cover is shown in Fig. 3-23c. This leads to the expression:

$$F(W,X,Y,Z) = (\bar{W} + \bar{Z})(\bar{Y} + \bar{Z})(\bar{X} + Y + Z)(W + \bar{X} + \bar{Y})$$

In the POS map the d in cell 1010 did not result in any simplification and consequently was not included.

(c) The SOP and POS expressions are *not* equivalent because they both covered the same subcube (0100); for $W = Y = Z = 0$ and $X = 1$ the SOP expression is 1 while the POS expression equals 0. Nevertheless, they are both valid solutions to the problem because we *don't care* what the result is in response to an input of 0100.

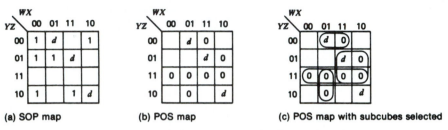

(a) SOP map   (b) POS map   (c) POS map with subcubes selected

Figure 3-23   Karnaugh map for Example 3-34.

3-7   Five-Variable Maps

Karnaugh maps can be used to simplify functions of 5 and 6 variables, although the problem of selecting the proper subcubes becomes more difficult as the number of variables increase. Figure 3-24 shows a 5-variable map with its numeric representation for $f(V,W,X,Y,Z)$. Actually two 4-variable Karnaugh maps are drawn side-by-side; one for $V = 0$ and one for $V = 1$. SOP terms from 0 through 15 are plotted on the $V = 0$ map, and SOP terms from 16 through 31 are plotted on the $V = 1$ map. However, the $V = 0$ map must be considered to be above the $V = 1$ map so that a cell on the $V = 0$ map is adjacent to the corresponding cell on the $V = 1$ map (i.e., the 3 and 19 cells are adjacent, as are the 13 and 29 cells, etc.) This leads to the formation of three-dimensional subcubes and tests a person's powers of visualization.

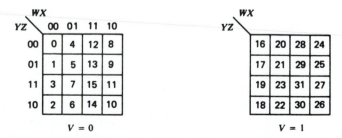

Figure 3-24   Numeric representation of a 5-variable Karnaugh map.

EXAMPLE 3-35

Simplify

$$f(V,W,X,Y,Z) = \Sigma(2,3,4,7,10,11,13,16,18,19,23,24,26,27,29,31)$$

using 5-variable Karnaugh maps.

SOLUTION The Karnaugh map for this function is plotted in Fig. 3-25. The 1-cell 00010 is essential and is covered by the 8-cell subcube consisting of the 2, 3, 10, 11, 18, 19, 26, and 27 squares. An 8-cell subcube deletes 3 literals and leaves 2. The expression for this particular subcube is $\overline{X}Y$ since the variables V, W, and Z change. The 1 in the 00100 cell has no adjacent 1s and must be described by an SOP term containing all 5 literals, $\overline{V}\overline{W}X\overline{Y}\overline{Z}$. The 1 in the 00111 cell forms part of an essential 4-cell subcube consisting of the 3, 7, 19, and 23 cells. For this subcube V and X change values; the subcube is therefore $\overline{W}YZ$. The 1 in the 1101 cell forms an essential 2-cell subcube with the corresponding cell on the $V = 1$ map, $WX\overline{Y}Z$. This completes the cover for all the cells on the $V = 0$ map.

Many of the 1s on the $V = 1$ map were covered when the $V = 0$ map was covered, but the two in the upper corners were not. They can best be covered by the 4-cell subcube $V\overline{X}\overline{Z}$. The only remaining 1 not yet covered is in the 11111 cell. This can be covered either by the 2-cell subcube $VWXZ$ or the 4-cell subcube VYZ. (This subcube is not drawn on the Karnaugh map for clarity.) We prefer using the 4-cell subcube since it contains fewer literals and would be easier to implement.

Gathering our subcubes together we find that

$$F(V,W,X,Y,Z) = \Sigma(2,3,4,7,10,11,13,16,18,19,23,24,26,27,29,31)$$
$$= \overline{X}Y + \overline{V}\overline{W}X\overline{Y}\overline{Z} + \overline{W}YZ + WX\overline{Y}Z + V\overline{X}\overline{Z} + VYZ$$

The answer can be checked by using the check table below:

Term	Expression	Numbers Covered
$\overline{X}Y$	X X 0 1 X	2,3,10,11,18,19,26,27
$\overline{V}\overline{W}X\overline{Y}\overline{Z}$	0 0 1 0 0	4
$\overline{W}YZ$	X 0 X 1 1	3,7,19,23
$WX\overline{Y}Z$	X 1 1 0 1	13,29
$V\overline{X}\overline{Z}$	1 X 0 X 0	16,18,24,26
VYZ	1 X X 1 1	19,23,27,31

Together these terms cover all the numbers in the specification, and no other terms. Therefore, we can be sure that our answer is correct.

Six-variable Karnaugh maps can also be constructed using four 4-variable maps. Of course, it becomes more difficult to select the subcubes correctly as the number of variables increases.

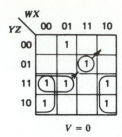
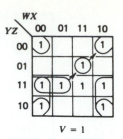

Figure 3-25 Karnaugh map for Example 3-35.

3-7-1 Other Methods of Implementing Logic Expressions

Designers often use other methods besides small scale ICs for implementing logic expressions, especially complex logic expressions. They generally use larger scale ICs, which have more logic in the IC and simplify the designer's role. They will be described briefly. More detailed descriptions can be found in some of the references.

1. **ROMs or PROMs.** Read Only Memories (discussed in Chapter 15) can be used to implement logic expressions. The input variables simply become memory addresses, and the output corresponding to each set of inputs is burned into the ROMs. If the logic requires several outputs based on the same set of input variables (a topic not discussed here), they can easily be accommodated by a ROM.

2. **Multiplexers.** Multiplexers in IC packages can be used to implement logic functions. A single 16-pin IC can handle any function of 4 variables and a 24-pin IC can handle any function of 5 variables. Multiplexer logic is discussed in section 11-8.

3. **Programmable Array Logic (PALs).** PALs are ICs that contain a group of AND gates connected to OR gates. They are designed for SOP inputs. The connections between the gates are made via fuses, which can be selectively blown so the user can implement the required functions.

   PALs are probably the best way to implement complex logic functions, because each PAL can replace four or five ICs. PALs are discussed in Chapter 12.

3-8 Practical Examples

The techniques of logical simplification discussed in this chapter can be applied to many practical problems. Once the simplest logical expression is obtained it can be implemented using relays (section 2-12) or using ICs, (section 2-11). Four practical simplification examples are presented in this section.

A typical series of steps for solving a practical problem is:

1. Draw the truth table.
2. Determine the numeric SOP function.
3. Draw the Karnaugh map.

4. Minimize the Karnaugh map for both the SOP and POS forms of the function.
5. Select the form of implementation most suitable for hardware implementation, as discussed in section 3-5-8.
6. Build the circuit.

3-8-1 Control Problems

Logic circuits are often used to control machinery or processes. The logic inputs come from sensors (e.g., voltage, temperature, and pressure) that monitor the process, and the output of the logic circuit determines the next step in the procedure.

EXAMPLE 3-36

A step in a space vehicle checkout depends on four sensors. Every circuit is functioning if sensor 1 and at least two of the other 3 sensors are also 1s.

(a) Find the minimal SOP expression for the output if the circuit is working properly.

(b) Find the minimal POS expression for the output if the circuit is working properly.

(c) If the circuit is not working properly, an alarm should sound. Find the minimal SOP expression needed to activate the alarm.

SOLUTION   The truth table for the example is shown in Fig. 3-26a, where S_1 is sensor 1, S_2 is sensor 2, and so on. From it we find that

$$f(S_1S_2S_3S_4) = \Sigma(11,13,14,15)$$

The Karnaugh map is drawn in Fig. 3-26b and simplifies into three 2-cell subcubes.

$$f(S_1S_2S_3S_4) = S_1S_2S_3 + S_1S_3S_4 + S_1S_2S_4 = S_1(S_2S_3 + S_3S_4 + S_2S_4)$$

The POS map of Fig. 3-26c simplifies into an 8-cell subcube and three 4-cell subcubes.

$$f(S_1S_2S_3S_4) = S_1(S_3 + S_4)(S_2 + S_3)(S_2 + S_4)$$

The alarm must sound whenever there is a failure. The Karnaugh map for the alarm can be plotted simply by putting 1s in whenever there are 0s on the success map. This simplifies to

$$\tilde{f}(S_1S_2S_3S_4) = \bar{S}_1 + \bar{S}_2\bar{S}_3 + \bar{S}_2\bar{S}_4 + \bar{S}_3\bar{S}_4$$

Note that f could also be obtained by complementing the POS form of f.

3-8-2 Adders

An adder is the most common arithmetic circuit built. Usually an adder is built of many identical stages. A typical stage of an adder that sums two binary numbers, A and B, has inputs A_n, B_n and C_{n-1}, where A_n and B_n are the Nth bits of A and B (the augend and addend) and C_{n-1} is the carryout of the $N-1$ stage (see section 1-7-1). The outputs required are the sum and carry. Such a stage is

Sensors				Output
S_1	S_2	S_3	S_4	
0	0	0	0	0
0	0	0	1	0
0	0	1	0	0
0	0	1	1	0
0	1	0	0	0
0	1	0	1	0
0	1	1	0	0
0	1	1	1	0
1	0	0	0	0
1	0	0	1	0
1	0	1	0	0
1	0	1	1	1
1	1	0	0	0
1	1	0	1	1
1	1	1	0	1
1	1	1	1	1

(a) Truth table

SOP Karnaugh map ($S_3 S_4$ rows, $S_1 S_2$ columns):

$S_3 S_4 \backslash S_1 S_2$	00	01	11	10
00				
01			1	
11			1	1
10			1	

(b) SOP Karnaugh map

POS Karnaugh map ($S_3 S_4$ rows, $S_1 S_2$ columns):

$S_3 S_4 \backslash S_1 S_2$	00	01	11	10
00	0	0	0	0
01	0	0		0
11	0	0		
10	0	0		0

(c) POS Karnaugh map

Figure 3-26 Truth tables and maps for the space vehicle checkout system of Example 3-36.

called a *full-adder*, to distinguish it from a *half-adder*, which has A and B inputs but no carry input. A half-adder can be used to add the two least significant digits, A_0 and B_0, where there is no carry in.

EXAMPLE 3-37

Design a full-adder.

SOLUTION First the truth table is plotted as shown in Fig. 3-27a. Since two outputs, sum and carry, must be generated, two 3-variable Karnaugh maps are drawn as shown in Fig. 3-27b and 3-27c. The Karnaugh map for the sum resembles a checkerboard. There are no 1s that can be combined with others; therefore, no simplifications are possible[10] and the expression for the sum is

$$\text{Sum} = \bar{A}B\bar{C}_{\text{IN}} + A\bar{B}\bar{C}_{\text{IN}} + \bar{A}\bar{B}C_{\text{IN}} + ABC_{\text{IN}}$$

The Karnaugh map for the carry out can be simplified into three 2-cell subcubes. Therefore,

$$\text{Carry Out} = AB + AC_{\text{IN}} + BC_{\text{IN}}$$

[10]No further simplifications are possible using AND-OR logic. When EXCLUSIVE-OR gates are used, they greatly simplify checkerboard Karnaugh maps. The adder is discussed further in Chapter 15.

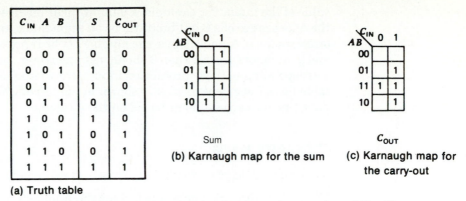

C_{IN}	A	B	S	C_{OUT}
0	0	0	0	0
0	0	1	1	0
0	1	0	1	0
0	1	1	0	1
1	0	0	1	0
1	0	1	0	1
1	1	0	0	1
1	1	1	1	1

(a) Truth table

(b) Karnaugh map for the sum

(c) Karnaugh map for the carry-out

Figure 3-27 Truth tables and Karnaugh maps for a full-adder.

3-8-3 Seven-Segment Displays

Seven-segment displays are used to convert a 4-bit BCD number into a visible readout. A 7-segment display consists of 7 light-emitting diodes (LEDs) arranged as shown in Fig. 3-28a. Modern hand calculators use 7-segment displays for their readouts.

Logically we require that those segments turn on or light up that most closely approximate the shape of the decimal digit equivalent to the binary

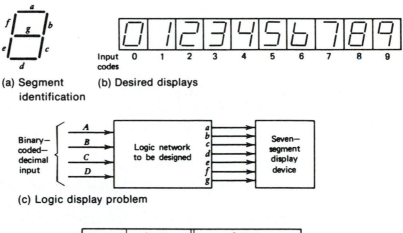

(a) Segment identification

(b) Desired displays

(c) Logic display problem

Decimal Displayed	Inputs				Outputs						
	A	B	C	D	a	b	c	d	e	f	g
0	0	0	0	0	1	1	1	1	1	1	0
1	0	0	0	1	0	1	1	0	0	0	0
2	0	0	1	0	1	1	0	1	1	0	1
3	0	0	1	1	1	1	1	1	0	0	1
4	0	1	0	0	0	1	1	0	0	1	1
5	0	1	0	1	1	0	1	1	0	1	1
6	0	1	1	0	0	0	1	1	1	1	1
7	0	1	1	1	1	1	1	0	0	0	0
8	1	0	0	0	1	1	1	1	1	1	1
9	1	0	0	1	1	1	1	0	0	1	1

(d) Truth table

Figure 3-28 Seven-segment displays. (Reprinted by permission of Texas Instruments.)

value of the input. For example, a 0 input turns on all segments except g to give the appearance of a 0, a 1 input turns on segments b and c, and a 7 input turns on segments a, b, and c. Since the numbers to be represented must be between 0 and 9, numbers greater than 9 should not appear on the inputs, and the outputs corresponding to these prohibited inputs are written as don't cares. The truth table for a 7-segment display is shown in Fig. 3-28d. Note that there are 7 outputs, one for each segment based on the 4 inputs.

EXAMPLE 3-38

Design the logic to display segment a on a 7-segment display.

SOLUTION   Segment a is ON for the numbers 0, 2, 3, 5, 7, 8, 9. The SOP Karnaugh map is shown in Fig. 3-29a. This leads to the simplification:

$$\text{Segment a} = A + BD + \bar{B}\bar{D} + CD$$

Note also that the POS Karnaugh map of Fig. 3-29b contains three 0s and two of them can be combined with don't cares to form a 4-cell subcube. This may be simpler to analyze and gives

$$\text{Segment a} = (\bar{B} + D)(A + B + C + \bar{D})$$

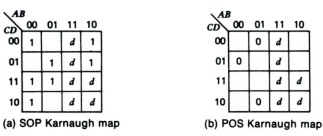

(a) SOP Karnaugh map   **(b) POS Karnaugh map**

Figure 3-29   Karnaugh maps for segment a of a 7-segment display.

3-8-4   Code Conversion

Logic circuits can be used to convert from one code or representation to another. In general, an N-bit input can be converted into an M-bit output. (M is often, but not necessarily, equal to N.) To accomplish this, an N-variable truth table is set up, and M Karnaugh maps (one for each output variable) are drawn. We can then write the equations necessary to construct the code converter.

EXAMPLE 3-39

In the *Excess-3 code* each decimal digit is represented by its binary equivalent plus 3. The truth table is shown in Fig. 3-30. Design a code converter to convert from *BCD* to Excess-3 representation.

SOLUTION   This code converter has 4 inputs (the *BCD* representation of a decimal digit) and 4 outputs. The outputs for inputs greater than 9 are not specified and are assumed to be don't cares. Four Karnaugh maps, one for each output bit are drawn up as shown in Fig. 3-30b. The simplifying subcubes are also shown and the SOP form of the results are listed beneath the maps. A circuit for each output now can be built.

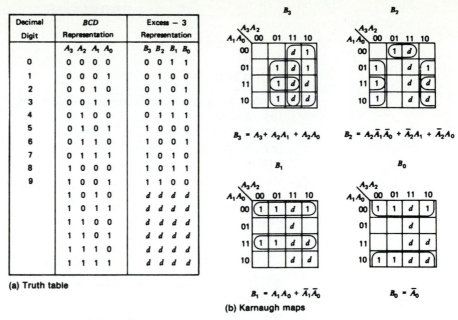

| Decimal | BCD | | | | Excess − 3 | | | |
Digit	Representation				Representation			
	A_3	A_2	A_1	A_0	B_3	B_2	B_1	B_0
0	0	0	0	0	0	0	1	1
1	0	0	0	1	0	1	0	0
2	0	0	1	0	0	1	0	1
3	0	0	1	1	0	1	1	0
4	0	1	0	0	0	1	1	1
5	0	1	0	1	1	0	0	0
6	0	1	1	0	1	0	0	1
7	0	1	1	1	1	0	1	0
8	1	0	0	0	1	0	1	1
9	1	0	0	1	1	1	0	0
	1	0	1	0	d	d	d	d
	1	0	1	1	d	d	d	d
	1	1	0	0	d	d	d	d
	1	1	0	1	d	d	d	d
	1	1	1	0	d	d	d	d
	1	1	1	1	d	d	d	d

(a) Truth table

$B_3 = A_3 + A_2 A_1 + A_2 A_0$

$B_2 = A_2 \bar{A_1} \bar{A_0} + \bar{A_2} A_1 + \bar{A_2} A_0$

$B_1 = A_1 A_0 + \bar{A_1} \bar{A_0}$

$B_0 = \bar{A_0}$

(b) Karnaugh maps

Figure 3-30 Design of an Excess-3 code converter.

SUMMARY

In this chapter we discussed:

1. The method of expanding functions into standard SOP and POS forms.
2. Plotting the POS and SOP forms on Karnaugh maps.
3. Using Karnaugh maps to obtain the minimal expression for a function.
4. Implementing SOP and POS forms using AND-OR, NAND, and NOR logic.

Finally, several practical examples using these techniques to achieve a minimal expression were presented. The minimal expression is useful because it leads to a minimal hardware implementation.

GLOSSARY

BCD: Binary coded decimal representation.
Domain: The variables for which a function is defined.
Literal: The occurrence of a variable in either its primed or unprimed state.
POS: Product of sum terms.
SOP: Sum of product terms.
Subcube: A group of two or more adjacent 1s on a Karnaugh map.

PROBLEMS

Section 3-3.

3-1. Write each expression below in standard SOP form.
    (a) $f(a,b,c) = a\bar{b} + c$
    (b) $f(w,x,y,z) = w\bar{x} + xyz + \bar{w}\bar{y}z$
    (c) $f(a,b,c,d) = b + \bar{a}\bar{c} + a\bar{b}cd$
    (d) $f(v,w,x,y,z) = \bar{x}\bar{z} + wxy + v\bar{w}yz$

3-2. Write the expressions of Problem 1 in standard POS form.

3-3. Write the numerical expressions for Problem 1 in both SOP and POS form.

3-4. Given:

$$f(W,X,Y,Z)$$

Which numbers do the following expressions represent?
    (a) $\bar{X}\bar{Z}$ **(d)** $\bar{W}XY$
    (b) $W\bar{X}$ **(e)** $(\bar{W} + X + \bar{Y} + Z)$
    (c) $W\bar{X}Z$ **(f)** $(W + \bar{Y})$

3-5. Show by algebraic manipulation that

$$x\bar{y} + w\bar{y}\bar{z} + \bar{w}\bar{y}z + wxz + \bar{w}x\bar{z} = w\bar{y}\bar{z} + \bar{w}\bar{y}z + wxz + \bar{w}x\bar{z}$$

(*Hint:* Multiply $x\bar{y}$ by 1 in the form of $(z + \bar{z})$.)

3-6. Identify each gate in Fig. 3-2b.

3-7. Sketch circuits to implement the following expressions:
    (a) $ABC + DE + F$
    (b) $(A + C + D)(D + B)$
    (c) $ABC + (D + E)B$

3-8. Implement the following expressions:
    (a) Using OR-AND logic.
    (b) Using NOR logic.

        1. $(W + \bar{X})(\bar{W} + \bar{Y} + Z)$
        2. $(W + X + Y + Z)(\bar{X} + \bar{Y})(\bar{W} + X + \bar{Z})$

Section 3-4.

3-9. If $f(x,y,z) = \Sigma(3,4,6)$
    (a) Find the SOP form for the function.
    (b) Find the POS form for the expression.
    (c) Show they are equivalent by algebraic manipulation.

Section 3-5.

3-10. Find the expression for each subcube shown in Fig. P3-10.

3-11. Find the minimum expression for

$$f(W,X,Y,Z) = \Sigma(0,2,3,4,8,9,10,11,15)$$

Is the subcube composed of the four corner squares essential?

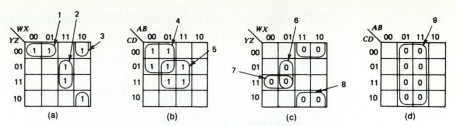

Figure P3-10 Karnaugh maps for Problem 3-5.

3-12. Find the minimum expression for

$$f(W,X,Y,Z) = \Sigma(0,2,3,5,6,7,11,14,15)$$

Find a 4-cell subcube in your map that is not essential and should not be included in the minimal expression.

3-13. Show how

$$\overline{X}\overline{Y}\overline{Z} + \overline{W}X\overline{Y} + WXZ + W\overline{X}Y$$

covers the Karnaugh map of Fig. 3-13c.

3-14. Find a minimal cover for the Karnaugh map of Fig. 3-14 that is different from the result found in Example 3-28.

3-15. Given

$$f(W,X,Y,Z) = \Pi(1,5,6,7,10,12,13,14,15)$$

Draw the Karnaugh map and find the minimal POS expression.

3-16. Find the minimal SOP expression for the function of Fig. 3-15 by placing 1s in the map wherever there are no 0s. Show by algebraic manipulation that your answer is the same as the answer given in Example 3-29.

3-17. Simplify the Karnaugh maps of Fig. P3-17.

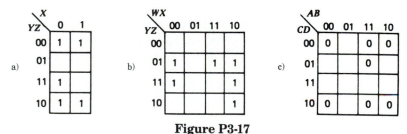

Figure P3-17

3-18. Find the POS form for the functions whose Karnaugh maps are given in Figs. P3-17a and P3-17b. Show that the POS forms are equivalent to the SOP forms by algebraic manipulation.

3-19. Obtain the SOP form for the Karnaugh map of Fig. P3-17c. Show that it is equivalent to the POS form found in Problem 3-17.

Section 3-6.

3-20. Given

$$f_1(a,b,c,d) = \Sigma(3,7,9,10,11,14,15)$$
$$f_2(a,b,c,d) = \Sigma(0,3,4,7,8,9,12,14,15)$$

Minimize f_1 and f_2 together and draw the minimal circuit.

3-21. For the functions given below, draw the Karnaugh maps and express them in POS and SOP form.

$$f(x,y,z) \quad = \Sigma(1,2,3,6) + d(0)$$
$$f(w,x,y,z) = \Sigma(1,10,12,13,14) + d(8)$$
$$f(a,b,c,d) = \Pi(2,3,9,10,12,15) + d(7,11)$$

Section 3-7.

3-22. Given

$$f(V,W,X,Y,Z) = \Sigma(0,1,2,8,9,10,14,15,16,17,22,24,25,28,30,31)$$

Find the minimal SOP expression for the function.

Section 3-8.

3-23. Find the minimal POS and SOP expressions for segments c and f of a 7-segment display.

3-24. Given inputs $A_3A_2A_1A_0$, we are required to produce outputs $B_3B_2B_1B_0$, where $B = A$ for $A < 10$ and $B = A - 4$ for $A \geq 10$. Show your truth tables and Karnaugh maps.

3-25. A hotel has four rooms (W,X,Y,Z) and two bellboys. Each room is capable of ringing for room service and a request will be answered by any one of the two bellboys. Unfortunately, the occupants of rooms W and X dislike the first bellboy and will not accept his services. Show your truth tables and Karnaugh maps for a circuit to sound an alarm whenever a request cannot be answered.

3-26. Design the circuits for the excess-3 code converter of Example 3-39.

3-27. Show the NAND gate implementation for f_1 and f_2 in section 3-5-9.

To be sure you understand this chapter, return to section 3-2 and review the questions. If there is any question you cannot answer, review the appropriate section of the text to find the answers.

4

Logic Families
and Their Characteristics

4-1 Instructional Objectives

This chapter describes the basic logic families in use today. It also describes the various types of TTL and CMOS available. After reading the chapter, the student should be able to:

1. Read and understand IC specifications as published by their manufacturers
2. Decide whether to use TTL or CMOS logic for the particular application.
3. Select the best series of TTL for the application if TTL is the choice.
4. Interface between TTL and CMOS.
5. Calculate the speed-power product of an IC gate.

4-2 Self-Evaluation Questions

Watch for the answers to the following questions as you read the chapter. They should help you understand the material presented. When you have finished the chapter, return to this section and be sure you can answer all of the questions.

1. Why should more than one logical form be investigated before building a logic circuit?

2. Why is power dissipation an important consideration in the design of circuits?

3. What overhead cost is associated with an IC?

4. What precautions must be observed when mixing ICs of different families?

5. What are the speeds of each TTL family?

4-3 Evaluation of IC Families

The simplest way to implement a logic function such as AND or OR is with a *diode gate*, which is the proper connection of several diodes and a resistor. This was tried in the 1960s and was quickly found to be unsatisfactory because voltage levels in diode gates deteriorate and amplification is needed.

Diode gates were replaced by the RTL (Resistor Transistor Logic) and then the DTL (Diode Transistor Logic) series of logic gates. These are now obsolete, and modern designers generally use one of three families of logic: Transistor Transistor Logic (TTL), Complementary Metal-Oxide Semiconductor (CMOS), and Emitter-Coupled Logic (ECL). Each has its advantages, disadvantages, and uses in special applications.

Designers usually select a logic family on the basis of the following criteria:

1. Speed
2. Power dissipation
3. Cost
4. Fanout
5. Availability

4-3-1 Speed

The *speed* of a logic family is measured by the propagation delay time of its basic inverter or NAND gate. A square wave is applied to the input and the output observed. The difference in nanoseconds from when the input has completed 50 percent of its transition to the point where the output has completed 50 percent of its transition is the *propagation delay*. Actually there are two propagation delays, as shown in Fig. 4-1. One, labeled t_{PHL}, is the delay when the input causes the *output* to change from a HIGH to a LOW level, and the other, t_{PLH}, is the delay when the *output* goes from a LOW to a HIGH level. Usually t_{PHL} and t_{PLH} are quite close and the delay of the family is the average of the two.[1]

4-3-2 Power Dissipation

Power dissipation is the amount of power (in milliwatts) that an IC drains from its power supply. Generally each IC contains several circuits. Most typical power measurements are made when half the circuits on the chip are in the 1 state and the others are in the 0 state. Power measurements are also affected by the input frequency—that is, the rate at which the IC changes state. CMOS has almost no power dissipation at low frequencies, but its dissipation increases as the frequency increases (see section 4-7).

[1]Some designers prefer to specify the speed of a family as the maximum clock rate of its flip-flops.

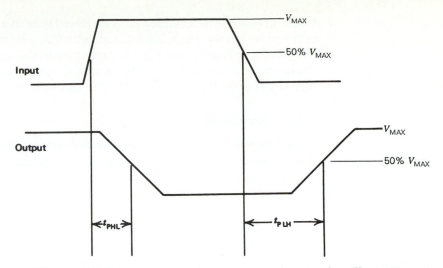

Figure 4-1 Response at an inverter to an input pulse, illustrating t_{PHL} and t_{PLH}.

Power dissipation is important not only because it drives up the cost of the power supplies, but because *it increases the heat in the vicinity of the ICs and within the electronic package. Excess heat is a prime cause of failure in electronic circuits.* If the costs of repairs and field maintenance are considered, we can see why a designer must pay careful attention to the power dissipation in the circuit.

4-3-3 Cost

The cost of ICs should be a relatively minor factor affecting the choice of a family. Generally the cost of an IC depends on the quantity manufactured. Standard TTL has been mass produced in very large quantities and enjoys a price advantage over all other families. A **7400** IC, which contains four NAND gates, can be purchased for less than 20 cents. This reduces the cost per NAND gate to less than 5 cents! More complex ICs cost considerably more, primarily because they are not manufactured in as large quantities as NAND gates.

For inexpensive ICs, the *overhead cost* per IC (overhead cost per IC is obtained by taking the total system cost—the cost of sockets, racks, power supplies, cooling, checkout, repair, and so on—and dividing it by the number of ICs in the system) is far more than the cost of the IC. You can buy an IC for 20 cents but you cannot buy a socket to plug it into for that price. Blakeslee, in a detailed analysis of a system,[2] estimates the overhead cost per IC to be $3.31. Although the overhead for any other system will vary somewhat, if $3.31 is accepted as a "ball-park" figure, it is obvious that the cost of most ICs is low compared to the overhead cost.

4-3-4 Fanout

A single electronic source can never drive an unlimited number of electronic loads. *Fanout* is defined as *the number of loads that can be driven from a single source* or the number of gate inputs that can be driven from a single gate output.

[2]Thomas R. Blakeslee, *Digital Design With Standard MSI & LSI* (see References).

Usually fanout is determined for a standard gate in a family driving other gates in the *same* family. High fanout is an advantage because additional drivers are not needed to supply many loads connected to the same source, but it is a relatively minor factor affecting the choice of logic circuits. CMOS, because of its high input impedance, has the largest fanout capability of any logic family.

4-3-5   Availability

Availability is an important factor in determining the selection of a logic family. If an IC specified in a digital design is scarce or unavailable, progress on digital systems can be delayed for weeks, or even months, awaiting delivery. Availability can be considered in two ways:

1.   The popularity of the series.
2.   The "breadth" of the series.

The popularity of a series among the customers helps availability. If several million ICs of a particular type are manufactured, a supply is generally available.

The breadth of the series refers to the number of different types of chips available. In a wide series, a complex function may be available that would have to be built up from less complex chips in another family. For both popularity and breadth, standard TTL has a distinct advantage over other logic families.

Other factors such as temperature range, noise immunity, and so on, are important for special purpose circuits, but usually the choice of an IC family depends on the factors listed above.

4-4   Transistor Transistor Logic (TTL)

There are three types of digital ICs that have achieved any significant popularity: TTL, CMOS (Complementary Metal-Oxide Semiconductor), and ECL (Emitter Coupled Logic). TTL is often called *bipolar* because its logic gates use bipolar junction transistors. (Ordinary transistors that have a collector, emitter, and base are bipolar junction transistors.) CMOS basically uses MOS transistors, which are more like *field effect transistors* (FETs). ECL uses emitter-follower, current switching circuits.

TTL is still the most popular series, although it is getting serious competition from CMOS. TTL is faster than CMOS, but consumes more power. ECL is faster than TTL, but consumes still more power. ECL is only used where extremely high speed logic is required.

We will define an IC *series* as one where all the ICs in the series with the same basic number have the same function (AND gate, OR gate, etc.). The numbers in the series define the function of the IC. The **7400** series is by far the most popular series ever built. It includes both TTL and CMOS ICs. CMOS was first introduced in a **4000** series of ICs, but most modern CMOS ICs are now part of the **7400** series.

The **7400** series is subdivided into several *families*. The differences between the various families depend on the differences in construction of the basic ICs. The members of the TTL family are discussed in this section. CMOS and ECL are discussed in sections 4-7 and 4-8, respectively.

4-4-1 Transistor Transistor Logic Families

The first TTL family to become popular was the *standard family*, which was introduced around 1970. As time progressed, several other families were added to the TTL line of chips. These are:

1. The **L** family (for low power)
2. The **H** family (for high speed)
3. The **S** family (for Schottky)
4. The **LS** family (for low power Schottky)
5. The **AS** family (for advanced Schottky)
6. The **ALS** family (for advanced low power Schottky)

The **H** and **L** families are currently obsolete, but are often included in tables for completeness. The other TTL families are still being used, but are competing with two popular CMOS families. At present, some engineers and companies prefer TTL for new designs, while others prefer CMOS. CMOS is discussed in section 4-8.

4-4-2 Standard TTL

The standard TTL series dominated digital design for many years. While it is now basically obsolete, due to advances in technology, standard TTL ICs are still inexpensive and readily available. The student will probably encounter them in the laboratory and in existing equipment. The standard TTL gate is, however, the easiest to understand and the gates for the more advanced series are variations on the standard gate. For these reasons, the standard gate is presented first.

The basic TTL 2-input NAND gate is shown in Fig. 4-2. This is the circuit for the type **7400** NAND gate. The inputs, A and B, come into Q_1, which is a *multiple-emitter transistor*. This is an NPN transistor with a *p*-doped silicon

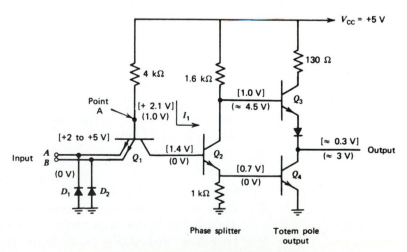

Figure 4-2 A TTL NAND gate.

base, an *n*-doped collector, and *several n*-doped emitters. Each emitter is connected to one of the inputs to the gate.

The TTL output voltage is determined by Q_2, the *phasesplitter* transistor, which drives Q_3 and Q_4, called the *totem-pole* transistors because Q_3 is placed above Q_4 and they look somewhat like heads on a totem pole. They operate as follows:

1. If Q_2 is ON, it supplies current to Q_4, saturating it.
2. At the same time the LOW collector voltage holds Q_3 OFF, and the result is a 0 output.
3. If Q_2 is OFF, no current enters the base of Q_4 and it turns OFF.
4. Now the high collector voltage turns Q_3 ON, which provides a HIGH output voltage (≈ 3.4 V) and a low impedance path to V_{CC}.

There are two sets of numbers indicating the circuit voltages on Fig. 4-2. The upper set applies only when both inputs are HIGH, reverse-biasing both emitter-base junctions. Then the gate operates as follows.

1. Current I_1 flows through the 4-kΩ resistor, the base-to-collector junction of Q_1, and the base-to-emitter junction of Q_2.
2. The voltage at point A is the sum of three forward-biased *pn* junction drops. At 0.7 V per junction this is 2.1 V.
3. The current through Q_2 saturates it and produces a LOW voltage on the base of Q_3, cutting it off. (The function of diode D_3 is to assure that Q_3 is cut off when the phase-splitter is on.)
4. The current through Q_2 enters the base of Q_4, saturating it.
5. Therefore when inputs are HIGH the output is LOW because Q_4, the bottom transistor of the totem-pole pair, is ON and saturated and NAND action has been achieved.

Note that any collector current that flows through Q_4 when it is saturated comes not through Q_3, but from the loads (commonly multiple-emitter transistor inputs to other gates) that are connected to it.

If either *A* or *B* (or both) inputs are LOW, the circuit operates as follows.

1. Current through the 4 kΩ resistor flows through one or both of the multiple-emitters to the low input.
2. The voltage at point A is only 1 V (0.7 V for the base-to-emitter drop and 0.3 V for $V_{CE(sat)}$ of the transistor that is turned ON and causing the input to be LOW).
3. Current I_1 does not flow and Q_2 remains OFF because of the lack of current.
4. Q_4 also receives no base current and remains OFF.
5. There is a high voltage at the collector of Q_2 that turns Q_3 on and effectively connects the output to a high voltage.
6. Therefore, with either input LOW, the output is HIGH and again NAND action has been achieved.

4-4-3 Other TTL Families

Of course, as soon as the standard TTL family appeared, engineers began demanding improvements. As a general rule, one can increase the speed at the cost of increasing the power dissipation. Some engineers demanded higher speed, and the **H** family, which achieved this speed by using more complex circuitry, was an answer, but its power dissipation was high. Other engineers did not need the speed and were more interested in conserving power. The **L** family was brought out for them. The **L** family gate looks like the standard gate except that the resistors are higher. This slows down the circuit, but lowers its power consumption.

The introduction of Schottky TTL made the **H** and **L** families obsolete. Schottky TTL uses Schottky diodes on the base of the transistors to keep them out of saturation. This greatly increases their speed. The Schottky (**S**) family reduced propagation delay from 10 ns to 3 ns. Again, many applications did not require this speed, so the **LS** (low power Schottky) family was introduced. It had higher resistance values, so the speed increased but the power dissipation dropped.

Recently, reductions in device geometry and processing innovations have resulted in the introduction of two new families: Advanced Schottky (**AS**) and Advanced Low Power Schottky (**ALS**). The **AS** is the fastest family, but the **ALS** consumes less power and is a distinct improvement over the older **LS** family.

4-5 Data Books

Most manufacturers of ICs publish data books containing the specifications for their ICs. In the late 1970s the Texas Instrument Corporation (TI) used only a single data book to describe the entire TTL series. Now they must publish a library. TI has recently published data books on the following IC families:[3]

- *TTL Logic: The Basic Families.* This book includes specifications for the standard, Schottky (**S**), and Low-Power Schottky (**LS**) families.
- *TTL Logic: The Advanced Families.* This book supplies information on the advanced Schottky (**AS**) and advanced low-power Schottky (**ALS**) families.
- *High Speed CMOS Logic.* This book provides the specifications for the high-speed (**HC**) and high-speed, TTL compatible (**HCT**) families.
- *Advanced CMOS Logic.* For the user of the Advanced CMOS (**AC**) and Advanced CMOS–TTL compatible (**ACT**) families.

In addition, TI has published data books on memories, bus interface logic, optoelectronics, linear circuits, programmable logic—the list goes on and on. At times it may seem overwhelming, as it does to the engineer in Fig. 4-3, but the reader need not become quite so frustrated. In digital design the engineer should first select the IC family to be used. Then by referencing the data book for that family, he or she can find the specifications for each IC used in the design.

[3]These families are further described in the following sections of this chapter.

Figure 4-3 Using Data Books. (Reprinted by permission of Texas Instruments.)

4-5-1 Description of the 7400 IC

Figure 4-4 consists of three pages from the TI data book on the standard, **S**, and **LS** families of ICs. These pages describe the **7400** NAND gate. The title shows that this page describes the NAND gates in the **5400** and **7400** series, and in three families in each series. The **5400** series is functionally the same as the **7400**, but **5400** series ICs have a wider temperature range and come in a more durable ceramic package. They are also much more expensive and are used pri-

● Package Options Include Plastic "Small Outline" Packages, Ceramic Chip Carriers and Flat Packages, and Plastic and Ceramic DIPs

● Dependable Texas Instruments Quality and Reliability

description

These devices contain four independent 2-input-NAND gates.

The SN5400, SN54LS00, and SN54S00 are characterized for operation over the full military temperature range of −55°C to 125°C. The SN7400, SN74LS00, and SN74S00 are characterized for operation from 0°C to 70°C.

FUNCTION TABLE (each gate)

INPUTS		OUTPUT
A	B	Y
H	H	L
L	X	H
X	L	H

logic symbol†

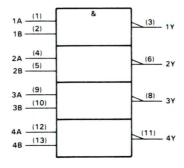

†This symbol is in accordance with ANSI/IEEE Std. 91-1984 and IEC Publication 617-12.
Pin numbers shown are for D, J, and N packages.

SN5400 . . . J PACKAGE
SN54LS00, SN54S00 . . . J OR W PACKAGE
SN7400 . . . N PACKAGE
SN74LS00, SN74S00 . . . D OR N PACKAGE
(TOP VIEW)

```
1A  [1   14]  VCC
1B  [2   13]  4B
1Y  [3   12]  4A
2A  [4   11]  4Y
2B  [5   10]  3B
2Y  [6    9]  3A
GND [7    8]  3Y
```

SN5400 . . . W PACKAGE
(TOP VIEW)

```
1A   [1   14]  4Y
1B   [2   13]  4B
1Y   [3   12]  4A
VCC  [4   11]  GND
2Y   [5   10]  3B
2A   [6    9]  3A
2B   [7    8]  3Y
```

SN54LS00, SN54S00 . . . FK PACKAGE
(TOP VIEW)

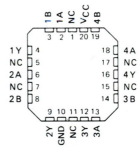

NC - No internal connection

logic diagram (positive logic)

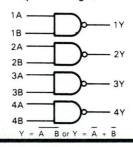

$$Y = \overline{A \cdot B} \text{ or } Y = \overline{A} + \overline{B}$$

TEXAS INSTRUMENTS

POST OFFICE BOX 655012 • DALLAS, TEXAS 75265

2-3

Figure 4-4 The **74x00** NAND gate. (Reprinted by permission of Texas Instruments.)

SN5400, SN54LS00, SN54S00,
SN7400, SN74LS00, SN74S00
QUADRUPLE 2-INPUT POSITIVE-NAND GATES

schematics (each gate)

'00

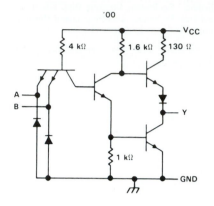

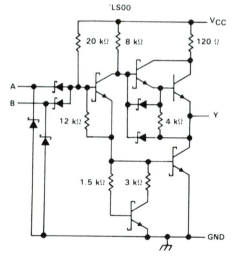

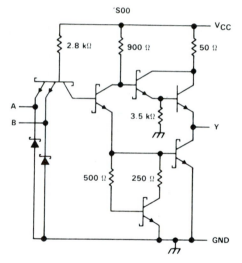

Resistor values shown are nominal.

absolute maximum ratings over operating free-air temperature range (unless otherwise noted)

Supply voltage, V_{CC} (see Note 1) . . . . 7 V
Input voltage: '00, 'S00 . . . . 5.5 V
            'LS00 . . . . 7 V
Operating free-air temperature range: SN54' . . . . −55 °C to 125 °C
                                SN74' . . . . 0 °C to 70 °C
Storage temperature range . . . . −65 °C to 150 °C

NOTE 1: Voltage values are with respect to network ground terminal.

TEXAS
INSTRUMENTS
POST OFFICE BOX 655012 • DALLAS, TEXAS 75265

Figure 4-4 *(continued)*.

recommended operating conditions

		SN5400			SN7400			UNIT
		MIN	NOM	MAX	MIN	NOM	MAX	
V_{CC}	Supply voltage	4.5	5	5.5	4.75	5	5.25	V
V_{IH}	High-level input voltage	2			2			V
V_{IL}	Low-level input voltage			0.8			0.8	V
I_{OH}	High-level output current			− 0.4			− 0.4	mA
I_{OL}	Low-level output current			16			16	mA
T_A	Operating free-air temperature	− 55		125	0		70	°C

electrical characteristics over recommended operating free-air temperature range (unless otherwise noted)

PARAMETER	TEST CONDITIONS†			SN5400			SN7400			UNIT
				MIN	TYP‡	MAX	MIN	TYP‡	MAX	
V_{IK}	V_{CC} = MIN,	I_I = − 12 mA				− 1.5			− 1.5	V
V_{OH}	V_{CC} = MIN,	V_{IL} = 0.8 V,	I_{OH} = − 0.4 mA	2.4	3.4		2.4	3.4		V
V_{OL}	V_{CC} = MIN,	V_{IH} = 2 V,	I_{OL} = 16 mA		0.2	0.4		0.2	0.4	V
I_I	V_{CC} = MAX,	V_I = 5.5 V				1			1	mA
I_{IH}	V_{CC} = MAX,	V_I = 2.4 V				40			40	µA
I_{IL}	V_{CC} = MAX,	V_I = 0.4 V				− 1.6			− 1.6	mA
I_{OS}§	V_{CC} = MAX			− 20		− 55	− 18		− 55	mA
I_{CCH}	V_{CC} = MAX,	V_I = 0 V			4	8		4	8	mA
I_{CCL}	V_{CC} = MAX,	V_I = 4.5 V			12	22		12	22	mA

† For conditions shown as MIN or MAX, use the appropriate value specified under recommended operating conditions.
‡ All typical values are at V_{CC} = 5 V, T_A = 25°C.
§ Not more than one output should be shorted at a time.

switching characteristics, V_{CC} = 5 V, T_A = 25°C (see note 2)

PARAMETER	FROM (INPUT)	TO (OUTPUT)	TEST CONDITIONS		MIN	TYP	MAX	UNIT
t_{PLH}	A or B	Y	R_L = 400 Ω,	C_L = 15 pF		11	22	ns
t_{PHL}						7	15	ns

NOTE 2: Load circuits and voltage waveforms are shown in Section 1.

TEXAS
INSTRUMENTS
POST OFFICE BOX 655012 • DALLAS, TEXAS 75265

2-5

Figure 4-4 *(continued)*.

marily by the military. Almost all commercial work uses the **7400** series, primarily the N (plastic) package.

The left side of page 2-3 (Fig. 4-4) has a description of the chip and its function table. This is the standard function table for a NAND gate. It also contains the newer ANSI/IEEE logic symbols.[4] The right side of the page shows the pinout for the four gates in the IC, in the N (plastic), J (ceramic), and W (flatpack) ICs. The FK package is for a leadless chip carrier. The standard symbols and the logic equations are in the lower right part of the page.

Page 2-4 (Fig. 4-4) gives the schematics for the IC for each family. The schematic for the standard family was discussed in section 4-4-2.

Page 2-5 (Fig. 4-4) gives the specification (V_{OH}, V_{OL}, etc.) for the standard series. There are two additional pages in the manual that contain the specifications for the **S** and **LS** series.

A similar set of pages exists for each of the other ICs in the **7400** series. The data book runs to about 1200 pages.

4-6   Characteristics of TTL Gates

A table of the important characteristics of each of the TTL series is given in Table 4-1. The same basic characteristics apply to all TTL families. They are explained below, with the standard family used as an example. The specifications listed are also shown on page 2-5 of Fig. 4-4.

1. V_{CC}, power supply voltage. The nominal power supply voltage for all series is 5 V. For the **7400** series V_{CC} can vary from 4.75 to 5.25 V.

2. V_{IH}, input voltage when the input is HIGH. The *minimum* voltage *guaranteed* to be recognized as a 1 at the IC input is 2 V. There is also an absolute maximum value of V_{IH} that may be applied to the input of an IC. For the **7400** series this is specified as 5.5 V.

3. V_{IL}, input voltage when the input is LOW. The *maximum* voltage *guaranteed* to be *recognized as a* 0 at the input is 0.8 V.

4. V_{OH}, output voltage when the output is HIGH. The minimum HIGH output voltage for the standard series is 2.4 V. The Schottky and Low Power Schottky series will provide at least 2.7 V output when its output is a 1.

5. V_{OL}, output voltage when the output is LOW. A **7400** IC will produce no more than 0.4 V when its output is a 0. (The bottom transistor of the totem-pole pair must be ON and saturated, producing a $V_{CE(sat)}$ of no more than 0.4 V.)

6. I_{OL}, output current when the output is LOW. This is the minimum amount of current the bottom transistor of a totem-pole pair can "sink" or absorb *when its output is* LOW.

7. I_{IL}, 0 level input current. This is the maximum current an IC will source through the multi-emitter transistor when the input is connected to a 0.

[4]The ANSI/IEEE symbols were created to simplify their use with workstations. They were not popular with many engineers, however, and most workstations do not use them. Therefore, they have been deemphasized in this book. Information is available in *An Overview of IEEE Std 91-1984* published in 1984 by TI.

TABLE 4-1

Characteristics of the TTL Gates

Family	Specific Device	V_{CC} max V	V_{CC} min V	V_{IH} min V	V_{IL} max V	V_{OH} min V	V_{OL} max V	I_{IL} max mA	I_{IH} max mA	I_{OL} min mA	I_{OH} max mA	Propagation Delay ns	Power Dissipation mW	Speed/ Power pJ	Flip-flop frequency MHz
STD	**7400**	5.25	4.75	2.0	0.8	2.4	0.4	−1.6	0.04	16	−0.4	10	10	100	15
LS	**74LS00**	5.25	4.75	2.0	0.8	2.7	0.5	−0.4	0.02	8	−0.4	9.5	2	19	45
S	**74S00**	5.25	4.75	2.0	0.8	2.7	0.5	−2.0	0.05	20	−1.0	3	19	57	125
AS	**74AS00**	5.5	4.5	2.0	0.8	V_{CC}^{-2}	0.5	−0.5	0.02	20	−2.0	1.5	8	12	200
ALS	**74ALS00**	5.5	4.5	2.0	0.8	V_{CC}^{-2}	0.4	−0.1	0.02	8	−0.4	4	1.2	4.8	70

Note: A minus sign indicates that current is flowing *out* of the IC.

131

8. **Propagation delay**. This is the average of T_{PLH} and T_{PHL} for each series under typical operating conditions.

9. **Power dissipation**. This is the average power dissipation for each gate in each series.

10. I_{OH}, the maximum current a gate can supply to other gates when the output is HIGH.

11. I_{IH}, the maximum current into a transistor when the input is HIGH. Because this current is bucking a back-biased diode, it is very small.

12. The maximum flip-flop frequency (f_{max}), the highest frequency at which flip-flops will operate reliably. It is another measure of the speed of a family. Flip-flops are discussed in Chapter 6.

EXAMPLE 4-1

Using the circuit of Fig. 4-2, find the value of I_{IL} if one of the multiple-emitter inputs is connected to ground.

SOLUTION I_{IL} is the current coming out of the multiple-emitter; in this case it is the current flowing to ground. This current travels from the $+5$ V supply through the 4 kΩ input resistor and then the *pn* junction to ground. If the voltage drop across the *pn* junction is assumed to be 0.7 V, the normal drop across a diode, then the current can be calculated as:

$$I_{IL} = \frac{5V - 0.7V}{4 \text{ k}} = 1.075 \text{ mA}$$

The actual specifications for I_{IL} are 1.6 mA. This is the maximum or worst case I_{IL}. The input resistor is *nominally* 4 kΩ, but may vary, resulting in a higher or lower I_{IL}.

4-6-1 Comparison of the TTL Families

Table 4-1 compares the characteristics of the various TTL families. The newer series (**ALS** and **AS**) have the best characteristics, but they may be more expensive and not as readily available as a standard or an **LS** IC. Besides, high speed requires very careful printed circuit board layout, because the runs between the ICs tend to exhibit the characteristics of transmission lines at these speeds. In many circuits, speed and power dissipation are not crucial, and the engineer may be just as well off with an **LS** family IC.

A comparison of the various families can also be shown graphically as in Fig. 4-5. In general, the better families are closer to the origin.

Other characteristics of TTL families, which can be derived from Table 4-1, are discussed in the rest of this section.

4-6-2 Noise Margin

Noise margin is the difference between the voltage produced at the output of a logic circuit and the voltage that will be recognized as that same logic level on its input. This difference, or margin, allows for the rejection of external noise picked up between gates and assures proper operation of the circuit.

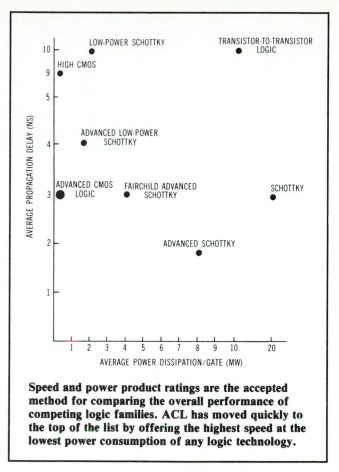

Figure 4-5 Speed and Power for the TTL families. (Reprinted with permission from the January 15, 1987 issue of COMPUTER DESIGN. © 1987, PennWell Publishing Company, Advanced Technology Group.)

For standard TTL, noise margin is illustrated by Fig. 4-6 and can be calculated from the data given in Table 4-1. They show that the minimum HIGH voltage an IC will *generate* is 2.4 V, but the minimum voltage it will *recognize* as a 1 is 2.0 V. This difference, 0.4 V, is the noise margin. It assures us that *any 1 generated by an output will be recognized as a 1 by any input it is connected to.* Similarly, a low level output can be no more than 0.4 V, but any input up to 0.8 V *must be recognized* as a 0. Therefore, TTL logic also has a *low level noise margin* of 0.4 V.

┌ **EXAMPLE 4-2**

Calculate the high level noise margins for Low Power Schottky TTL.

SOLUTION Table 4-1 gives V_{OH}, the minimum output an LSTTL gate will generate, as 2.7 V. Similarly, V_{IH} is found as 2 V, so the high level noise margin for Low Power Schottky TTL is 0.7 V.

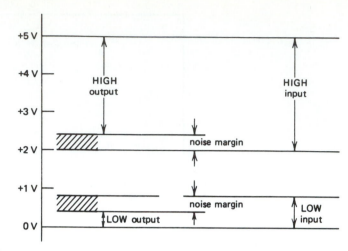

Figure 4-6 Band diagram showing the noise margins of TTL circuits. (From Porat and Barna. *Introduction to Digital Technique.* © John Wiley & Sons, Inc. 1979. Reprinted by permission of John Wiley & Sons, Inc.)

4-6-3 Speed-Power Product

The ideal gate is infinitely fast and consumes no power. The *speed-power product* is the result of multiplying the speed of an IC, as measured by its propagation delay, by the average power consumption per gate. A lower speed-power product means the IC is faster and/or consumes less power. This product is one of the most commonly used criteria for comparing IC families.

EXAMPLE 4-3

Find the speed-power product of:
(a) Standard TTL
(b) Low Power Schottky TTL

SOLUTION   (a) For standard TTL, Table 4-1 gives a power dissipation of 10 mW per gate and a speed of 10 ns. So the speed-power product is 100.
(b) For Low Power Schottky TTL, the product is 9.5 ns × 2 mW = 19.

Low Power Schottky superceded standard TTL because its speed-power product is so much smaller.

4-6-4 Fanout

Fanout is the number of loads that can be driven by a single source. If a source is required to drive more loads than its fanout allows, multiple sources should be used and the loads should be divided.

Fanout must be considered from two perspectives: when the source is LOW and when the source is HIGH. Figure 4-7 shows the situation when the source is LOW. The bottom transistor of the totem-pole pair turns on and absorbs the

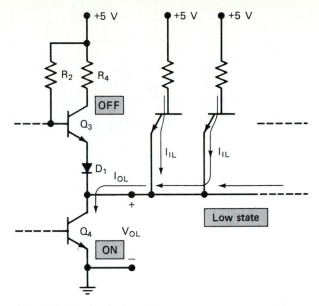

Figure 4-7 TTL loading when the output voltage is LOW. (Ronald J. Tocci, Digital Systems: Principles and Applications, 5ed, © 1991. Reprinted by permission of Prentice Hall, Englewood Cliffs, New Jersey.)

current driven into it by the loads. As the current in the source increases, its voltage V_{OL} increases. I_{OL} is the *maximum current that the transistor can absorb without having its voltage rise above the maximum allowable value of* V_{OL}. The basic rule for low level fanout is:

I_{OL} *must be greater than the sum of all the currents being driven into the source transistor by the loads.*

As Fig. 4-7 shows, the current that flows into the source transistor comes from the I_{IL} of the loads it is connected to. If all gates are from the same family, then the low level fanout can be calculated by Eq. (4-1).

$$\text{Fanout} = \frac{I_{OL}}{I_{IL}} \tag{4-1}$$

EXAMPLE 4-4

What is the low level fanout for the **LS** family?

SOLUTION Table 4-1 shows that I_{OL} is 8 mA and I_{IL} is -0.4 mA for **LS** family ICs. According to Eq. (4-1) the fanout is 20, or an **LS** gate can drive 20 **LS** loads.

Figure 4-8 shows fanout when the output is HIGH. The high level output tries to push current into the diodes on the multiple emitter transistors in the load gates, but these diodes oppose this current flow so it is small. The current that each output gate can source, without its voltage falling below V_{OH}, is I_{OH}, and the current each load can absorb is I_{IH}. Again, the I_{OH} for the source must be greater than all I_{IH}s of the loads it is connected to.

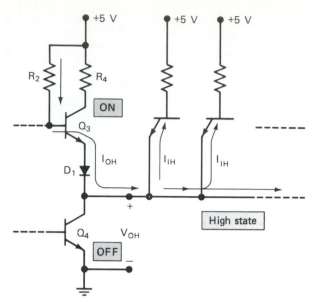

Figure 4-8 TTL loading when the output voltage is HIGH. (Ronald J. Tocci, Digital Systems: Principles and Applications, 5ed, © 1991. Reprinted by permission of Prentice Hall, Englewood Cliffs, New Jersey.)

One of the advantages of **LS**, **AS**, and **ALS** gates is that the fanout on the low side is high because I_{IL} has been reduced. The fanout for **ALS** is often given as 20/80, which means an **ALS** gate can drive 20 other **ALS** gates on the high side and 80 on the low side. Cautious engineers will use the worst-case figure and will not connect more than 20 **ALS** loads to a single source.

If all of the gates in a circuit are of the same family, the fanout is at least 10. If the families are mixed, there is a greater problem. In general, if a gate does not drive many other gates, fanout will not be a problem.

4-6-5 Mixing TTL Families

Because the various TTL families all use the same +5 V power supply, and all have the same logic levels, engineers often substitute one family member for another. For example, if the engineer is using the **LS** family and needs a **74LS00**, but doesn't have one and a **7400** IC is readily available, the **7400** will be used. Although there are some differences in fanout and voltage levels, these will rarely cause a problem. If the ICs are being used to test a circuit on a workbench, the penalty, if there is a problem, is small and we would simply go ahead and do it. We would be very careful about this, however, if the circuit were part of a component that will be manufactured in volume, or if the circuit is to be shipped out to the field where it must work on arrival.

4-7 Emitter-Coupled Logic

Emitter-coupled logic (ECL) is faster than TTL and is used in applications where very high speed is essential. ECL has high switching speeds because the transistors act as difference-amplifier emitter followers, and are never in saturation. Because of this high speed, more attention must be paid to the construction of the boards and the interconnecting wiring, or else noise generated by the fast wavefronts may degrade performance.

The basic ECL gate is shown in Fig. 4-9. It is essentially an OR/NOR gate, and typically both OR and NOR outputs are available. V_{CC1} and V_{CC2} are both connected to ground. Two grounds are provided to eliminate crosstalk within the package (another precaution against high speed noise effects). The supply voltage, V_{EE}, is -5.2 V. A logic 1 output is typically -0.9 V and a logic 0 output is typically -1.75 V.

In the circuit of Fig. 4-9 the input transistors (Q_1 through Q_4) operate as a difference amplifier in conjunction with Q_5. At any time, either Q_5 is ON or one or more of the input transistors is ON. The circuit functions as follows.

1. Resistors R_1 and R_2 and the diodes clamp the base of Q_6 so that its emitter, which is also the base of Q_5, is always at -1.29 V.

2. If all inputs are LOW (-1.75 V), Q_5 has the highest base voltage and turns ON.

3. This clamps the emitters to -2 V, and each input transistor has a base-to-emitter voltage of 0.25 V, which is a forward voltage, but not large enough to turn it ON.

4. The OR output is LOW (≈ -1.75 V). It equals the IR drop across R_5 plus its base-to-emitter voltage drop.

5. The NOR output is HIGH (≈ -0.9 V) because there is only a small voltage drop across R_4 with Q_1 through Q_4 OFF.

6. Consequently, the OR output is a 0 and the NOR output is a 1. This is proper OR gate action when all inputs are LOW.

7. If any input is a 1 (-0.9 V), the base of its transistor is higher than the base of Q_5 and it will turn ON, turning Q_5 OFF.

8. If Q_4 is ON and Q_5 is OFF, the voltage at the base of Q_7 will be close to ground and the OR output will be one base-to-emitter drop below ground or -0.9 V.

9. The current through Q_4 flows through R_4 causing the voltage on the base of Q_8 to be approximately -1 V so the NOR output is approximately -1.75 V.

10. OR action is achieved since with any input HIGH, Q_5 turns OFF and Q_4 turns ON, the OR output is HIGH and the NOR output is LOW.

ECL is currently used only where the very highest speed circuits (less than 1 ns propagation delay) are required. Any slower speed requirements can be satisfied by the **AS** family of TTL logic. ECL has relatively high power requirements.

ECL has been used in some supercomputers, but each IC had to have its own heat sink. It also required fans, and very careful attention to board layout, because transmission line effects are very pronounced at these speeds.

4-8   Complementary Metal-Oxide Semiconductor Gates

Complementary metal-oxide semiconductor gates (CMOS) are currently proving to be popular because of their low power dissipation. A CMOS gate is composed of two *metal-oxide semiconductor* (MOS) gates. There are two types of

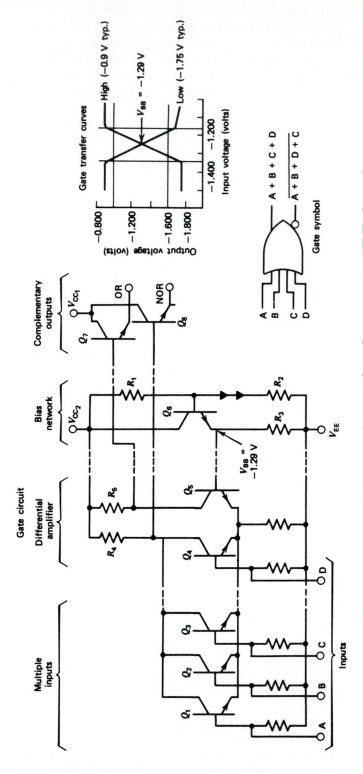

Figure 4-9 A basic ECL gate and its switching characteristics. (MECL, General Information Manual, Motorola, Inc. © 1974. Courtesy of Motorola Integrated Circuits Division.)

MOS gates, n-channel and p-channel. Each gate consists of *a drain, a source, a gate, and a substrate*. For logic circuits, *enhancement* mode gates are used. The symbols for MOS gates and a cross section of an n-channel transistor are shown in Fig. 4-10.

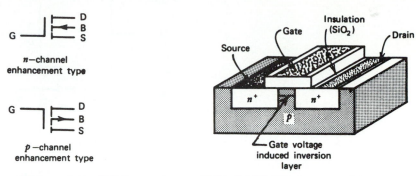

Figure 4-10 MOS transistors. (*RCA COS/MOS Integrated Circuits Manual*, CMS-271, © 1971. Courtesy of Harris Semiconductor.)

An n-channel MOS transistor is similar to a field-effect transistor (FET) and is set up so that the drain is positive with respect to the source. With no enhancement the pn junction between the drain and the substrate is reverse-biased and no current flows. Enhancement occurs if the gate is made positive with respect to the substrate. Note that the gate is separated from the substrate by an oxide insulator, which causes the input impedance of an MOS transistor to be very high. When the gate voltage is positive, electrons in the substrate are attracted toward the gate and the migration of electrons to the gate area effectively changes the p-doped silicon to n silicon. Thus, a low impedance n-channel is formed between source and drain and current flows from drain to source. P-channel MOS transistors work in a similar manner, as the accompanying table shows.

	n-Channel	p-Channel
Substrate material	p	n
Source material	n	p
Drain material	n	p
Gate effects (Enhancement)	Current flows when gate is positive with respect to substrate	Current flows when gate is negative with respect to substrate
Direction of conventional current	Drain to source	Source to drain

MOS transistors are used in large scale memories (see Chapter 15) but are not used as ordinary gates. CMOS, however, does have the advantage of dissipating almost no power. Consequently, it provides an attractive alternative to TTL.

The basic CMOS inverter is shown in Fig. 4-11. As in all CMOS gates it consists of an n-channel transistor whose source and substrate are tied to ground, and a p-channel transistor whose source and substrate are tied to V_{DD}. V_{DD} can be any voltage from $+3$ to $+15$ V. The output voltage is either ground or V_{DD}, denoting a 0 and a 1, respectively.

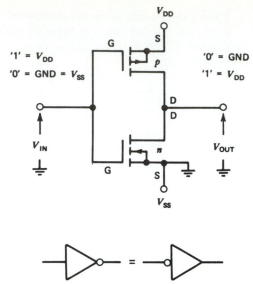

Figure 4-11 COS/MOS inverter circuit. (*RCA COS/MOS Integrated Circuits Manual, CMS-271,* © 1971. Courtesy of Harris Semiconductor.)

The unique feature of a CMOS inverter is that no current flows through it in either the 0 or 1 state. If the input is HIGH, the n-channel transistor is enhanced, but the p-channel is not. The output is disconnected from V_{CC} by the open p-channel transistor. This makes the output LOW and inversion has taken place. Also, since the open p-channel blocks any current flow from V_{DD}, no power is dissipated.

If the input is LOW, the n-channel transistor is effectively open, while the p-channel transistor is enhanced because its gate voltage is negative with respect to its substrate, which is tied to V_{DD}. The output is now shorted to V_{DD} and is HIGH, but there is still no path for current to flow and again the power dissipation is very small. Power is only dissipated in a CMOS gate when it is in the process of *switching states.* Consequently, power dissipation is proportional to the frequency at which the gate is switched, but CMOS power is still much smaller than that of standard TTL. This is very attractive to engineers whose circuits must operate on small amounts of power, or who want to keep their circuits cool. On the other hand, CMOS is slower, more costly, and does not offer all the circuits available in standard TTL.

Logic gates are built from CMOS circuits by adding additional gates. A 2-input NOR gate and a 2-input NAND gate are shown in Fig. 4-12. The tables show how they operate. For the NOR gate, for example, note that if either input is HIGH, at least one of the n-channel transistors is shorted to ground and at least one of the p-channel transistors is open, so that the output is LOW. A HIGH output occurs only if both inputs are LOW, enhancing the two p-channel transistors. In that case, both n-channel transistors are open, isolating the output from ground.

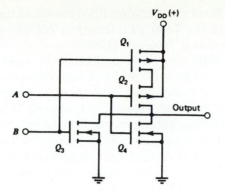

A B	Q_1 Q_2 Q_3 Q_4	Output
0 0	S S O O	H
0 1	O S S O	L
1 0	S O O S	L
1 1	O O S S	L

S = Short
O = Open

(a) A CMOS NOR gate

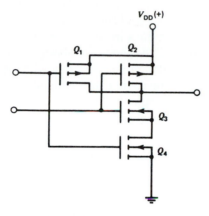

A B	Q_1 Q_2 Q_3 Q_4	Output
0 0	S S O O	H
0 1	S O S O	H
1 0	O S O S	H
1 1	O O S S	L

S = Short
O = Open

(b) A CMOS NAND gate

Figure 4-12 CMOS gates. (From George K. Kostopoulos. *Digital Engineering*. © John Wiley & Sons, Inc., 1975. Reprinted by permission of John Wiley & Sons, Inc.)

4-8-1 CMOS Families

There are six CMOS families that are in common use at the present time:

The **4000** series
The **74C** family
The **74HC** family
The **74HCT** family
The **ACL** series
The **ACT** series

The **4000** series was the first CMOS series. It was developed by RCA.[5] The ICs were very slow, with propagation delays over 100 ns, and it had little drive capability (I_{OL} was 0.4 mA).

[5]The Radio Corporation of America. Although it is now a subsidiary of GE, the label persists in some places.

The **74C** family was essentially a rework of the **4000** series to make it pin-compatible with the **7400** TTL family (a **74C08** was a CMOS AND gate, with the same pinout as a **7408** or **74LS08**).

Both the **4000** series and the **74C** series could operate over a wide range of input voltages ($+3$ to $+18$) and had high noise margins. But, because of their low speed, both of these series are basically obsolete.

The **74HC** (for high speed CMOS) family was a major improvement. The supply voltage for the **74HC** series is limited to between $+2$ V and $+6$ V, but the speed improves as the supply voltage increases. The speed of a **74HC** IC with a $+5$ V power supply is about 9 ns, comparable to a **74LS** IC.

The **74HC** logic levels are not quite compatible with TTL. Engineers started to demand CMOS ICs that could be directly substituted for their TTL counterparts. The **74HCT** series was developed in response. It can only operate with a $+5$ V power supply, but its logic levels are totally TTL compatible. The T in the family designation stands for TTL compatibility. In most cases one can simply pull out a TTL IC and replace it with an HCT IC.

Further improvements in technology have led to the development of the **AC** and **ACT** families. The speeds are about 5 ns for the **AC** and 7 ns for the **ACT** series. These ICs are not pin compatible with other TTL ICs because putting power and ground on the corner pins caused noise problems at the high speeds for which they were designed. Power and ground are brought onto the center pins to help reduce these problems.

4-8-2 Power Considerations Using CMOS

While CMOS absorbs practically no power in the *static state*, if a CMOS gate is switched frequently, its power dissipation increases because it must charge and discharge whatever capacity is connected to its gates.

Figure 4-13 shows the power dissipation of the **4000** series CMOS com-

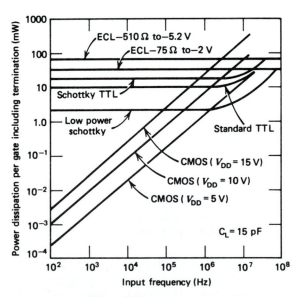

Figure 4-13 Typical power dissipation versus input frequency for several popular logic families. (From the Fairchild CMOS Data Book, 1977. Reprinted courtesy of National Semiconductor.)

pared to **LSTTL**. It shows that a **4000** series gate operating at 1 MHz dissipates as much power as an **LS** gate. Again the **HC** series is superior; it only dissipates about 0.5 mW per gate at 1 MHz, about one quarter of that of an **LS** gate. Furthermore, most gates do not switch at a 1 MHz rate, and therefore dissipate much less power.

4-8-3   TTL Driving CMOS

CMOS gates are separated from the substrate by the metal oxide insulator; therefore they absorb almost no current. This means that the fanout of CMOS gates when driving other CMOS gates is very high.

When TTL drives CMOS as shown in Fig. 4-14, there is no problem with fanout. There could be a problem with voltage level compatibility, however. HCMOS has a V_{IH} of about 3.5 V, with a $+5$ V power supply, but the TTL specification for V_{OH} is only 2.4 V. Therefore, when the TTL gate is putting out as logic 1 its output must be raised. The easiest way to raise this voltage is to add a *pull-up resistor* between the output and the power supply. 10 kΩ is a reasonable value for the pull-up resistor.

There is no problem when TTL drives CMOS and the TTL driver is LOW. Of course, the 10 kΩ pull-up resistor takes an additional .05 mA from the power supply, but this is usually insignificant.

TTL can drive HCT and ACT gates directly.

4-8-4   CMOS Driving TTL

CMOS voltage outputs are generally rail-to-rail. This means that the high output voltage is typically close to the power supply voltage and the low output voltage is close to ground. Because of this, there is no problem with voltage com-

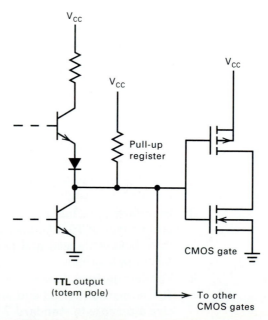

Figure 4-14   TTL driving CMOS.

patibility when CMOS drives TTL, but there could be a fanout problem. The older **74C** family gate had an I_{OL} of 0.4 mA, so it could only drive one LS load. The **74HC** and **74HCT** both have an I_{OL} of 4 mA, so they can drive ten **LS** loads or two standard loads. The newer **ACL** and **ACT** have I_{OL}s of 24 mA, so there is usually no fanout problem with these circuits.

4-9   Decoupling

Decoupling means placing a capacitor between power and ground to minimize noise spikes in a digital circuit. These noise spikes could possibly set a flip-flop or cause other problems.

Both TTL circuits with totem-pole outputs and CMOS have an upper and a lower transistor. One of the pair should always be off. When the circuits switch states, however, there are a few nanoseconds when both transistors are on. This provides a low impedance path to ground and allows for a spike of high current, which can cause noise and negative spikes to appear on the power supply. These spikes may adversely affect other gates driven by the same power supply.

To minimize these spikes, most manufacturers recommend that a 0.01 to 0.1 uF ceramic disk capacitor be placed near each IC in a circuit, and a large capacitor, perhaps 2 uF, be placed between power and ground where the power enters the board. It is now possible to buy sockets with capacitors built in. They assume that power and ground for each IC will be connected to the corner pins. This is generally true except for **ACL** and **ACT** high speed CMOS.

Manufacturers of electronic circuits pay careful attention to decoupling. Small, experimental laboratory circuits, however, will generally work with far less decoupling.

SUMMARY

In this chapter, the three types of digital logic in current usage were described. As a result of these discussions, we make the following suggestions.

1. ECL has the highest speed of any family, but use ECL for high speed circuits only. For low speed circuits there is no need to take on the extra expense, power, and care required to eliminate high speed noise.
2. CMOS is very attractive where low power dissipation or battery operation is required. It is somewhat slower than TTL, but the additional speed is often unnecessary. Fortunately, it is very easy to interface or combine TTL and 5 V CMOS.
3. Standard TTL has the broadest line and can easily be mixed with other types of TTL for special requirements. It offers a good compromise between speed and power dissipation and is a good choice for most applications.
4. For new designs, Low Power Schottky should be seriously considered. Its comparable speed and lower power dissipation make it an attractive alternate to standard TTL.

GLOSSARY

Breadth: The number of different types of gates and circuits available on an IC family.

CMOS: Complementary Metal-Oxide Semiconductor circuit, consisting, at least, of a p-channel and an n-channel MOS transistor.

DTL: Diode Transistor Logic.

ECL: Emitter-Coupled Logic.

Enhancement: Providing the proper voltage at the gate of an MOS transistor to turn it on.

Fanout: The number of loads that can be driven by a single source or input.

LSI: Large scale integration.

MOS: Metal-Oxide Semiconductor Transistor.

Propagation delay: The time required for the output of a gate to respond to a change in the inputs.

RTL: Resistor Transistor Logic.

Saturation: The state of a transistor when it draws maximum current. In saturation V_{ce} is very low (≈ 0).

TTL: Transistor Transistor Logic.

PROBLEMS

Section 4-6-4.

4-1. A hypothetical family has an I_{OL} of 3 mA and an I_{OH} of 0.5 mA. What is the fanout of this family?

4-2. Calculate the high and low fanout of an **LS** IC.

4-3. How many standard gates can an **LS** gate drive?

4-4. How many **ALS** gates can an **HCMOS** output drive?

4-5. What are the high and low level noise margins of an **LS** gate? Of an **ALS** gate?

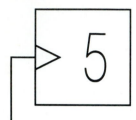

Basic TTL Gates

5-1 Instructional Objectives

In this chapter the basic TTL gates are introduced. After reading this chapter the student should be able to:

1. Use Schmitt triggers to smooth input wave forms.
2. Properly connect, pull-up, or clamp any unused gate inputs.
3. Use open collector and 3-state gates, where necessary.
4. Use strobed gates and AND-OR-INVERT gates.
5. Determine where buffer/drivers are necessary and use them.
6. Build complementers and comparators using EXCLUSIVE-OR circuits.

5-2 Self-Evaluation Questions

Watch for the answers to the following questions as you read the chapter. They should help you understand the material presented. When you have finished the chapter, return to this section and be sure you can answer all of the questions.

1. What is the difference between a Schmitt trigger and an ordinary NAND gate?

2. Describe the three methods of handling unused inputs. State their advantages and disadvantages.

3. Why is wire-ANDing of totem-pole gates prohibited?

4. Why is a large V_{OH} an advantage in an open-collector gate?

5. When several 3-state gates are connected together, why must only one gate be enabled at any time?

6. How does the strobe input affect the operation of a **74x25**?

5-3 Introduction

The most basic IC gates (ANDs, ORs, NANDs, and NORs) were introduced in Chapter 2. We will now begin to consider other ICs and the circuits that can be built from them. Henceforth, all ICs discussed in this book are identified by part number, and can be purchased from the manufacturer, a local distributor, or an electronics discount house. Any circuit described can easily be set up in the laboratory if the reader wants to investigate its behavior.

The ICs considered in this and the following chapters are described in detail in one of the following manuals:

TTL Logic—Standard TTL, Schottky, Low Power Schottky Data Book
Texas Instruments, 1988
ALS/AS Logic Data Book
Texas Instruments, 1986
High-Speed CMOS Logic Data Book
Texas Instruments, 1988
Advanced CMOS Logic Data Book
Texas Instruments, 1988

Data books are also available from other IC manufacturers such as Motorola and National Semiconductor, Inc. These data books are updated approximately every 3 years.

5-4 Schmitt Triggers

The basic NAND gates in the **74x00** series were introduced in section 2-9-1. The Schmitt trigger is a special type of NAND gate that is often used to smooth out or square up noisy or irregular voltage inputs. The most common use of a Schmitt trigger is as a line or *bus* (a bus is a group of lines) receiver, where one device, such as a memory, receives signals generated by another device, such as a microprocessor. The Schmitt trigger is used to smooth out or eliminate any stray noise picked up on the line.

A Schmitt trigger has the following special properties:

1. Assume the input voltage is low, but increasing. Its output will be HIGH. The Schmitt trigger will not turn on (its output will not go LOW) until the input voltage becomes *greater* than a certain voltage called the *positive-going threshold voltage*.

2. If the input is HIGH, but going LOW, the output voltage will not go HIGH until its input voltage becomes *less* than another voltage called the *negative-going threshold voltage*.

The positive-going threshold voltage (V_{T+}) is greater than the negative-going threshold voltage (V_{T-}). The difference between V_{T+} and V_{T-} is called *hysteresis*, and gives the Schmitt trigger the ability to square up slow and jagged wave forms. Sometimes a hysteresis symbol (⎍) is placed within the gate to distinguish a Schmitt trigger from a NAND gate.

The action of a Schmitt trigger is illustrated in Fig. 5-1, where it is compared with the action of an ordinary NAND gate. The ordinary NAND gate (a **74x00** perhaps), is assumed to have a single threshold. The output of the NAND gate is assumed to be LOW if its input is *above* the threshold, and HIGH if its input is *below* the threshold. If the input pulse is slow, noisy, or uneven, spikes appear on the output, as shown in Fig. 5-1b, because the input crosses and recrosses the NAND gate threshold.

The Schmitt trigger output, a clean pulse, is shown in Fig. 5-1c. The Schmitt trigger output does not go LOW until the input wave form crosses the positive-going threshold. Once triggered, however, the Schmitt trigger will not turn off merely because the input recrosses the positive-going threshold. The input must be sufficiently negative to cross the negative-going threshold for the Schmitt trigger to turn OFF.

The **74x14** hex Schmitt trigger inverter is the most commonly used Schmitt trigger IC because it contains six inverters in a single package. Other

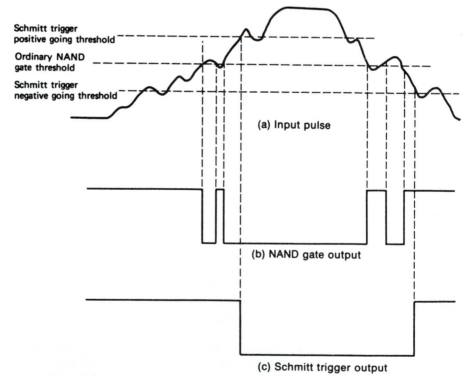

Figure 5-1 Action of a Schmitt trigger compared to the action of a NAND gate.

Schmitt trigger ICs are the **74x132**, a quad 2-input NAND, the **74x13**, a dual 4-input NAND, and the **74LS19** and **74LS24**.

For the **74x13, 74x14**, and **74x132**, the TTL data book contains several pages of specifications and circuits where they are useful. Typical Schmitt triggers have a V_{T+} of 1.7 V and a V_{T-} of 0.9 V.

EXAMPLE 5-1

A 1.8-V, 1-MHz sine wave is applied to a threshold detector that is to produce an output pulse if the input exceeds 1.7 V. How long does the output pulse last if the input is connected to:

(a) A NAND gate whose output changes when the input crosses 1.7 V?
(b) A **74x13** Schmitt trigger whose $V_{T+} = 1.7$ V and $V_{T-} = 0.9$ V?

SOLUTION The solution for both parts is shown in Fig. 5-2.

(a) The NAND gate turns on and off at an angle such that:

$$\theta = \sin^{-1}\frac{17}{18} = 70.8 \text{ degrees (turn ON) and } 109.2 \text{ degrees (turn OFF)}$$

Thus, the output is ON for 38.4/360 or 0.107 of a cycle. Since a cycle takes 1 μs, the output pulse will be 107 ns long.

(b) The Schmitt trigger turns on when $\theta = \sin^{-1} 17/18 = 70.8$ degrees, but does not turn off until:

$$\theta = \sin^{-1}\frac{9}{18} = 150 \text{ degrees}$$

The Schmitt trigger is ON for 79.2 degrees out of 360 degrees or 220 nanoseconds. Clearly the Schmitt trigger has improved the detector by delivering a longer output pulse.

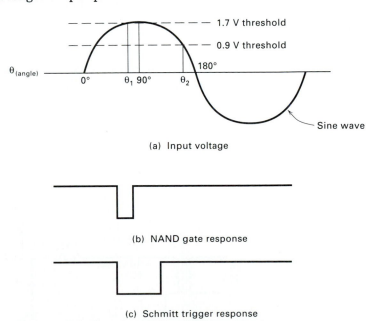

(a) Input voltage

(b) NAND gate response

(c) Schmitt trigger response

Figure 5-2 Response of threshold detectors to a 1.8 V sine wave.

Certain flip-flops and one-shots (multivibrators) in the TTL family have trigger inputs that have a voltage transition instead of a voltage level. They are known as "edge-triggered" devices. Waveform edges are shown in Fig. 5-3. The positive edge is where the waveform rises, or goes from a 0 to a 1. The negative edge is where the waveform falls or goes from a 1 to a 0. Normally the edges must change faster than 1 V per microsecond ($dv/dt \geq 1$ V/μs) to be effective. When a typical TTL gate changes state, it goes from 0.4 to 3.4 V in about 3 ns; therefore, it changes at the rate of 1 V/ns = 1000 V/μs and there is no problem. If the input is too slow, however, Schmitt triggers are usually used to square up and speed up the output, which can then be used as a trigger.

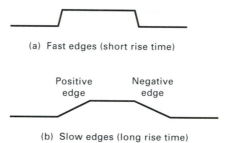

(a) Fast edges (short rise time)

Positive edge   Negative edge

(b) Slow edges (long rise time)   **Figure 5-3**   Waveform edges.

EXAMPLE 5-2

The 120-V, 60-Hz "house lines" are to provide a series of triggers for a TTL circuit. In a practical circuit these triggers could be used to monitor commercial power. The absence of a trigger would then be an early warning that power has failed. The triggers must have edges faster than 1 V/μs. Design a circuit to produce the output pulses.

SOLUTION   A solution is shown in Fig. 5-4. The circuit is designed as follows:

1.   The first problem is to limit the TTL input signals to approximately 0 to 5 V so that the maximum value of V_{IH} will not be exceeded.
2.   The transformer is used to reduce the input ac voltage to a more manageable level. Here a transformer with a peak output voltage of 10 V is selected.
3.   When the transformer voltage is greater than 5 V diode D_1 turns on, clamping the input of the **74x13** to 5 V. The 1-kΩ resistor prevents excessive current through the diode.

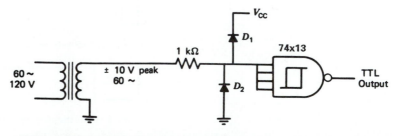

Figure 5-4   A Schmitt trigger used to square up a 60-Hz input.

4. When the transformer voltage is negative, diode D_2 turns on and clamps the input of the **74x13** to ground. The **74x13** is protected because the diodes hold its input voltage between ground and + 5 V.

5. The transformer output is:

$$V = 10 \sin 377t$$

and

$$\frac{dv}{dt} = 3770 \cos 377t$$

6. The maximum rate of change of the input is 0.00377 V/μs. This is too slow to be used as a trigger.

7. The Schmitt trigger **74x13** can accept this slow input and produce a TTL output. The **74x13** output is a square wave at a frequency of 60 Hz. The trigger output switches at TTL speeds, which is much faster than required in this problem.

5-5   Open and Unused Inputs

Occasionally, some of the inputs to a TTL gate are *not* used. This occurs most often in one-shots or flip-flops where the devices have features that are not needed for the particular circuit, but it can also occur with simple gates. Suppose, for example, an engineer needs two 3-input NAND gates and an inverter for his circuit. It is most economical to use a single **74x10** IC for all three circuits, but this means transforming one 3-input NAND gate to an inverter, leaving two unused inputs.

In TTL, *open inputs* almost invariably *act as a logic 1*. It requires a *current* through a multiple-emitter input to ground (or $V_{CE(sat)}$) to produce a 0, and an open input provides no such current path. A **74x10** will function as an inverter if two of its inputs are left open. The output is the complement of the signal on the remaining input.

EXAMPLE 5-3

One of the inputs to a **74x02** 2-input NOR gate (see section 2-9-2) is left open. How does the output behave?

SOLUTION   The open input behaves as a 1. Since the **74x02** sees a 1 on one of its inputs, it produces a low output regardless of the state of the connected input. One must be especially careful *not* to leave OR or NOR gate inputs unconnected.

TTL manufacturers advise against leaving inputs open. Open inputs are susceptible to noise and may occasionally provide an erroneous signal or spike. Also, while open inputs act like a logic 1, they appear as a 0 (or a voltage in the prohibited region) when viewed on an oscilloscope. This can add to the confusion when attempting to debug a circuit.

For MOS or CMOS gates the situation is far worse. If their inputs are floating, the outputs of MOS or CMOS gates drift and are *unpredictable*. Consequently, *unused MOS or CMOS inputs must be pulled up or tied to V_{CC} for a 1 or grounded for a 0*.

5-5-1 Pull-up Resistors

Unused inputs can be "pulled up" by connecting them to V_{CC} via a resistor as in Fig. 5-5a, where a **74x20** with two unused inputs is shown. This decreases the noise susceptibility. Texas Instruments recommends that a 1-kΩ resistor be used, and that no more than 25 unused inputs be connected to the same resistor.

5-5-2 Clamping

Another way to handle unused inputs is to *clamp* them, which effectively ties them to a constant voltage of 3.6 V. A typical clamp circuit, shown in Fig. 5-5b, consists merely of two diodes and a resistor. The voltage drop across the diodes lowers the output voltage to $V_{CC} - 1.4$ V. The advantage of clamping is that it reduces the IC inputs to 1.4 V less than the power supply. Therefore, any voltage spikes that occur on the power supply are unlikely to drive the clamp voltage

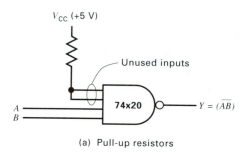

(a) Pull-up resistors

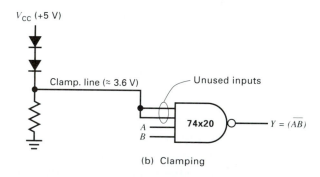

(b) Clamping

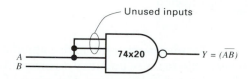

(c) Tying unused inputs to used inputs

Figure 5-5 Methods of handling unused inputs.

above 5 V, where the IC input may be damaged. Because 3.6 V is a logic 1 only for TTL circuits, clamping should not be used for CMOS.

5-5-3   Tying Used Inputs to Unused Inputs

If gate inputs are unused, they can be tied to used inputs on the same gate as shown in Fig. 5-5c. This is the most popular way of handling unused gate inputs, but often leads to mistakes in determining fanout. Engineers usually count the number of inputs connected to an output to determine fanout. If two or more inputs are connected to the same gate of NAND or AND gates, they should only be counted as one load because a single gate cannot supply more than 1.6 mA through its 4-kΩ resistor regardless of how many inputs are connected to the multiple emitters.

TTL OR and NOR gates, however, absorb one standard load for each connected input. In general the input circuit of each load may have to be examined if a precise determination of fanout is required.

Because clamps and pull-up resistors tie unused inputs to a logic 1, they cannot be used on OR or NOR gates, although unused OR and NOR inputs can be tied to ground. Engineers tend to eliminate the whole problem by tying unused gate inputs to used inputs on the same gate. Flip-flops and one-shots, however, contain certain inputs (direct sets and direct clears, etc.) that must be HIGH if unused, and cannot be tied to varying inputs. These unused inputs are usually clamped. Pull-up resistors are used to terminate open collector gates (section 5-6) and cables (section 17-4).

5-6   Wire-ANDing and Open Collector Gates

For gates where the 0 logic level is caused by saturating a transistor, an additional level of logic can be obtained by connecting the outputs or collectors of several gates together. This is called *wire-ANDing*. Figure 5-6 shows the outputs of three gates wire-ANDed together. If any one of the three transistors is saturated, it causes the output to be LOW. Therefore, the output is HIGH only if *all* three transistors are cut off, which corresponds to a 1 output for each individual gate. AND action occurs because the outputs of each of the three individual circuits must be a 1 in order that the wire-AND output be a 1.

If only one of the gates in Fig. 5-6 is LOW, the transistor draws additional current because it is connected to V_{CC} through three load resistors instead of one. This extra current decreases the fanout of the gate.

Unfortunately, TTL circuits with *totem-pole outputs* are *not* amenable to wire-ANDing. The wire-ANDing of totem pole outputs is shown in Fig. 5-7. If the inputs on gate 1 cause its output to be a 1 and the inputs to gate 2 cause its output to be a 0, the top transistor of gate 1's totem-pole pair turns ON and the bottom transistor of gate 2's totem-pole pair turns ON.

Consequently, current I_1 flows. For a standard TTL gate there is only 130 Ω in the current path, and the current I_1 may be excessive and damage one of the gates. Even if the gates remain undamaged, the output voltage may enter the prohibited region because of the heavy current. Therefore, wire-ANDing of TTL gates is very poor design.

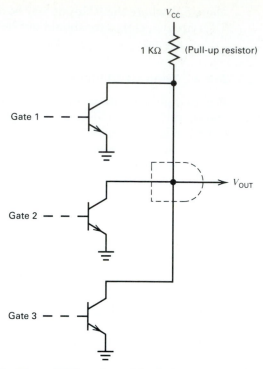

Figure 5-6 Three DTL gates with their outputs wire-ANDed together. Note the dashed AND symbol that is sometimes used to indicate wire-ANDing.

Wire-ANDing of CMOS gates is prohibited for similar reasons. Figure 5-8 shows a pair of CMOS gates wire-ANDed. If the inputs are such that the *p-gate* on the top pair and the *n-gate* on the bottom are both turned on, excessive current will flow through the low impedance path from V_{CC} to ground and the output voltage will be approximately half the drain voltage.

5-6-1 Open-Collector Gates

To allow wire-ANDing, TTL manufacturers have produced a series of *open-collector* gates. The output of an open-collector gate is shown in Fig. 5-9. The collector is tied only to a pin on the IC package and a load must be tied to the collector at this point. If the input conditions at the gate cause base current to flow, the open-collector output transistor saturates and pulls its output voltage to ground. If no base current flows, the output transistor acts like an open circuit.

Figure 5-10 is a table of the open-collector gates available in the standard TTL series. They accomplish the function of wire-ANDing by being tied to each other. To function properly, open-collector gates must be tied to V_{CC} through a pull-up resistor, a resistor connected between the output of the open-collector gate and V_{CC}. Formulas exist for calculating the value of the pull-up resistor, but in almost all cases 1 kΩ is satisfactory.

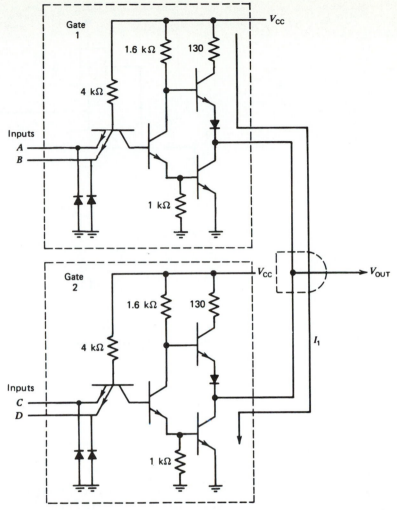

Figure 5-7 Wire-ANDing two **7400**s together. Note that this is *not recommended*. If A OR B is low and C AND D are high, current I_1 will flow, which may damage the transistors.

EXAMPLE 5-4

Find an expression for the output of Fig. 5-11a.

SOLUTION In Fig. 5-11 two **7403** open-collector NAND gates and a **7409** open-collector AND gate are wire-ANDed together, causing V_{OUT} to be LOW unless the output of each gate is HIGH. Therefore,

$$V_{OUT} = (\bar{A} + \bar{B})(\bar{C} + \bar{D})EF$$

The same circuit is shown in Fig. 5-11b, where the alternate gate representation is used for clarity.

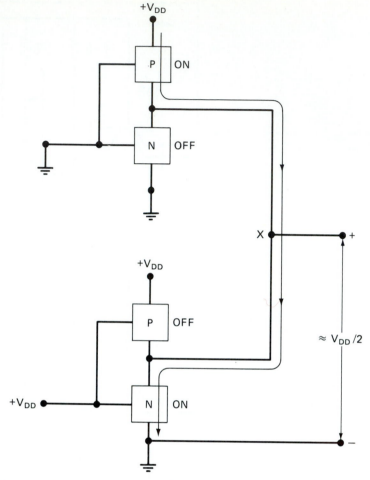

Figure 5-8 Wire-ANDing of CMOS gates. *NOT RECOMMENDED*. (Ronald J. Tocci, *DIGITAL SYSTEMS: Principles and Applications, 5e*, (Reprinted by permission of Prentice Hall, Englewood Cliffs, New Jersey.)

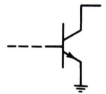

Figure 5-9 Output circuit of an open-collector gate.

Besides the standard series, open collector gates are manufactured in the **LS**, **S**, **ALS**, and **HC** series. Similar circuits are available in CMOS, where they are called *open-drain* gates. Figure 5-12 shows three CMOS open-drain gates wire-ANDed together.

Open-collector circuits are generally slower than TTL or CMOS circuits because any stray capacity on the output line must charge through the pull-up resistor rather than through a transistor.

Open collector IC	Description		Totem-pole Equivalent	Diagram
74x01	Quad NAND	2-input gates	74x00	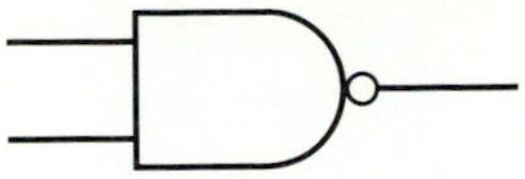
74x03*	Quad NAND	2-input gates	74x00	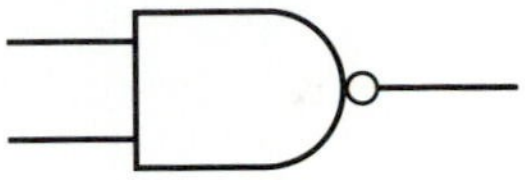
74x05	Hex	Inverter	74x04	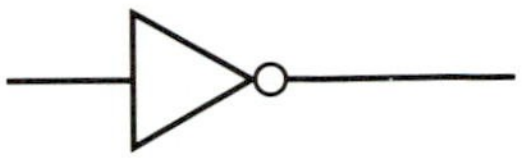
74x09	Quad AND	2-input gates	74x08	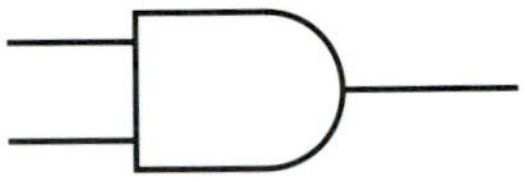
74x12	Triple NAND	3-input gates	74x10	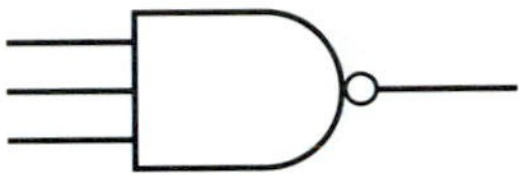
74LS15	Triple AND	3-input gates	74x11	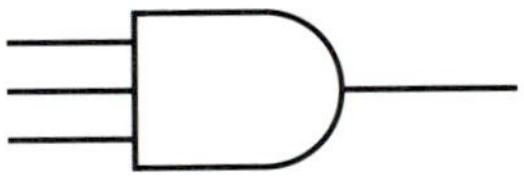
74x22	Dual NAND	4-input gates	74x20	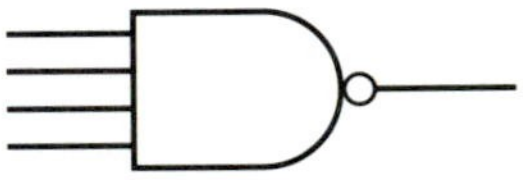
74x33	Quad NOR	2-input gates	74x02	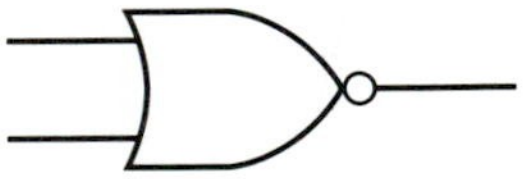

Figure 5-10 A table of the most common open-collector gates.

5-6-2 Open-Collector Buffer/Drivers

Buffer/drivers differ from ordinary gates because they have a larger current sinking capability and a larger fanout. They are used to drive many loads or loads that require high current.

Two of the most popular open-collector buffer/drivers are the **74x06** and **74x07**. The **74x06** is a hex-inverting buffer and the **74x07** is a hex-noninverting buffer/driver. The schematics and circuit diagrams are shown in Fig. 5-13. They can absorb a current of 40 mA (compared to an I_{OL} of 16 mA for an ordinary gate) and have the additional advantage that V_{OH} is 30 V for these ICs. Ordinary open-collector circuits like the **7403** have a V_{OH} of 5 V, which means their collectors may not be connected to a supply voltage greater than 5 V. The **74x06** and **74x07** are often used to interface from TTL to circuits requiring higher voltages or currents than ordinary TTL gates can handle.

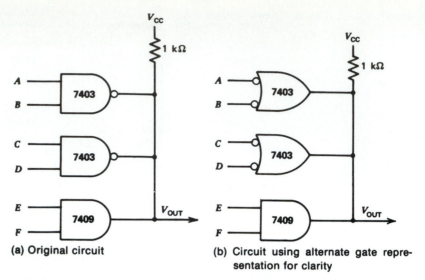

(a) Original circuit

(b) Circuit using alternate gate representation for clarity

Figure 5-11   Circuit for Example 5-4: (a) Original circuit; and (b) Circuit using alternate (or DeMorgan) gate representation for clarity.

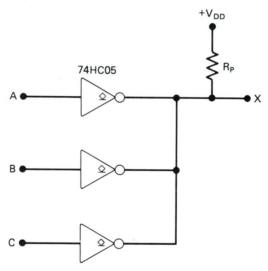

Figure 5-12   Wire-ANDing three CMOS open-drain gates. (Ronald J. Tocci, *DIGITAL SYSTEMS: Principles and Applications, 5e,* (Reprinted by permission of Prentice Hall, Englewood Cliffs, New Jersey.)

Because these ICs are buffer/drivers and must produce high current outputs, they are available only in the standard series.

EXAMPLE 5-5

Small lamps are often used to indicate the state of a digital circuit. If a 10-V, 40-mA incandescent lamp is to indicate the level of a point in a digital circuit, and the lamp is to be lit when the point is a 1, how should the lamp be connected?

(a) **schematic**

'06, '16

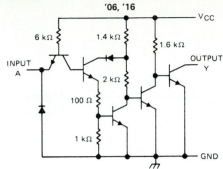

Resistor values shown are nominal.

logic diagram (positive logic)

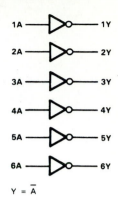

1A ▷○ 1Y
2A ▷○ 2Y
3A ▷○ 3Y
4A ▷○ 4Y
5A ▷○ 5Y
6A ▷○ 6Y

$Y = \overline{A}$

(b) **schematic**

'07, '17

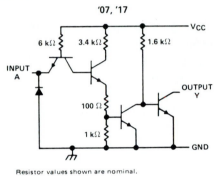

Resistor values shown are nominal.

logic diagram (positive logic)

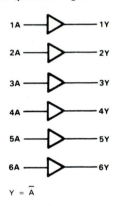

1A ▷ 1Y
2A ▷ 2Y
3A ▷ 3Y
4A ▷ 4Y
5A ▷ 5Y
6A ▷ 6Y

$Y = \overline{A}$

Figure 5-13 **7406** and **7407** circuits and symbols: (a) **7406**; (b) **7407**. (Reprinted by permission of Texas Instruments.)

SOLUTION Because of both the current and voltage requirements, the lamp cannot be connected directly to an ordinary TTL gate. The solution is to connect the lamp to the output of a **74x06**, as shown in Fig. 5-14. When the point in the circuit is HIGH, the output of the **74x06** is LOW, allowing current to flow through the lamp, turning it ON. When the point is LOW, the output of the **74x06** is actually an open-collector transistor that is OFF. This prevents any current flow and keeps the lamp OFF. Note that when the **74x06** gate is OFF, 10 V appear at the open collector, which would damage an ordinary open-collector gate with a V_{OH} of 5 V. However, since V_{OH} is 30 V for the **74x06**, it is well within specifications.

Designers using incandescent indicators usually prefer 5-V lamps to eliminate the need for another power supply. Buffer/drivers are still used, however, to satisfy the current requirements. Pull-up resistors are not required in this circuit, because the lamp itself acts as the load.

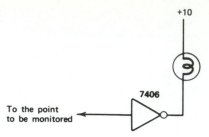

Figure 5-14 Driving an incandescent lamp with a **7406** open-collector inverter.

5-6-3   Other Buffer/Driver Gates

Three buffer/driver gates with totem-pole outputs are available in the **7400** series:

1. The **7428** quad 2-input NOR buffer, which is logically equivalent to the **7402**.
2. The **7437** quad 2-input NAND buffer, which is logically equivalent to the **7400**.
3. The **7440** dual 4-input positive NAND buffer, which is logically equivalent to the **7420**.

The main advantage of these buffer/drivers is that they have an I_{OL} of 40 mA instead of 16 mA for an ordinary gate. This means they have a fanout of 25 instead of 10, or they can drive a 40-mA lamp directly. Usually buffer/drivers are only used where the requirements for large fanout or heavy output current exists.

5-6-4   Transmission Gates

The transmission gate is a newer CMOS gate that has no TTL counterpart. It acts as an open or closed switch in the circuit. The circuit, traction table, and symbol of a transmission gate are shown in Fig. 5-15.

The symbol for a transmission gate is shown in Fig. 5-15c. If the *control* input is LOW, the gate acts as an open switch; there is a very high impedance between points A and B. They are effectively disconnected. If the control input is HIGH, there are only a few hundred ohms between points A and B. The points are effectively short-circuited and current can flow in *either* direction through the gate.

Transmission gates are available as CMOS ICs in DIPs. Figure 5-16 shows the block diagram and logic diagram of the **74HC4016** quad switch, manufactured by Motorola, Inc.

5-7   Three-State Devices

Wire-ANDing is essential in modern digital systems, especially those that involve microprocessors (μPs). Figure 5-17 shows a typical μP system. In such a system, the μP must communicate with other devices such as:

*   **RAM**—The read-write memory.
*   **ROM**—The Read-Only memory.

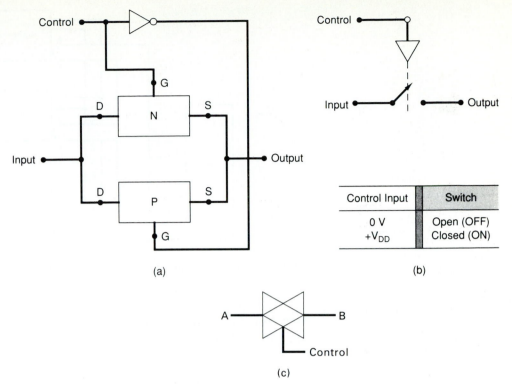

(a)

Control Input	Switch
0 V	Open (OFF)
+V$_{DD}$	Closed (ON)

(b)

(c)

Figure 5-15 CMOS bilateral switch (transmission gate): (a) Circuit; (b) Function table; (c) Control symbol. (Ronald J. Tocci, *DIGITAL SYSTEMS: Principles and Applications, 5e*, (Reprinted by permission of Prentice Hall, Englewood Cliffs, New Jersey.)

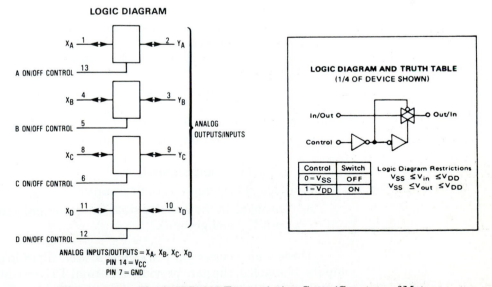

Figure 5-16 The **74HC4016** Transmission Gate. (Courtesy of Motorola, Inc.)

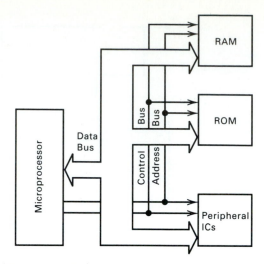

Figure 5-17   A typical microprocessor system.

- **Peripheral ICs**—Printer drivers and disc controllers are typical peripheral ICs. Both the Apple and IBM personal computers have a series of *slots* to accommodate these device drivers.

As shown in Fig. 5-17, the μP communicates with these peripheral ICs by using three buses (a bus is a group of wires). These are the *data bus*, the *address bus*, and the *control bus*. Most 8-bit μPs, such as the Intel **8085**, the Zilog **Z80**, or the Motorola **68HC11**, have an 8-bit data bus and a 16-bit address bus. The address and control buses need not concern us now.

The data bus in Fig. 5-17 is *bidirectional*, which means data can flow either from the μP to the external device or vice versa. As a result any one of the four devices must be capable of driving the same bus. Consequently, their outputs must be connected together by a wire-AND type circuit.

Older computer systems used open-collector circuits to wire-AND their buses together. In newer systems, open-collector gates have been replaced by 3-state devices that have 3 *output states*. They have been developed for both TTL and CMOS.

All modern microprocessors and their peripherals use 3-state outputs, which allow the engineer to directly connect outputs in parallel and to add or remove ICs connected to the output line without affecting circuit operation.

The three possible output states are:

1. Logic 1 (low impedance to V_{CC}).
2. Logic 0 (low impedance to ground).
3. Disabled (in the disabled state the device presents a high impedance to both V_{CC} and ground).

Three-state devices have an *enable/disable* input in addition to the normal inputs. If enabled, the gate provides a normal TTL or CMOS output, but it presents a *very high output impedance if it is disabled*. The principle of 3-state operation is illustrated by the basic inverter of Fig. 5-18. Here, as with most

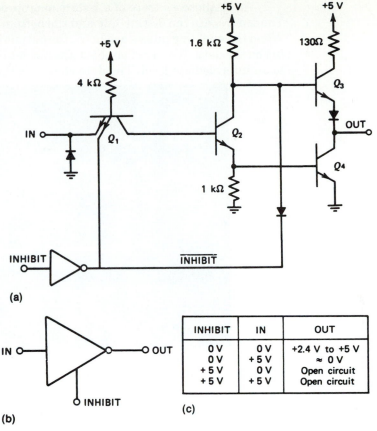

(a)

INHIBIT	IN	OUT
0 V	0 V	+2.4 V to +5 V
0 V	+5 V	≈ 0 V
+5 V	0 V	Open circuit
+5 V	+5 V	Open circuit

(b)

(c)

Figure 5-18 A 3-state TTL inverter circuit. (From Porat and Barna, *Introduction to Digital Techniques*. Copyright John Wiley & Sons, Inc., 1979. (Reprinted by permission of John Wiley & Sons, Inc.)

3-state gates, the gate is enabled if the INHIBIT input is LOW. This causes the output of the internal inverter ($\overline{\text{INHIBIT}}$) to be HIGH, reverse-biasing both its input to the multiple-emitter transistor and the diode to the base of Q_3. Consequently, the circuit functions as though there were no INHIBIT input and the output is the inverse of its input.

If the INHIBIT input is HIGH, $\overline{\text{INHIBIT}}$ is LOW. Now current flows through the base-to-emitter junction of Q_1 and deprives Q_2 of base current, cutting it OFF. The low voltage at $\overline{\text{INHIBIT}}$ drags the base of Q_3 down and cuts Q_3 off. As a result, both Q_3 and Q_4 are cut off, which causes a high output impedance to both V_{CC} and ground. CMOS devices operate similarly; when the device is inhibited, both the p- and the n-channel MOS transistors are open.

Three-state devices are currently being used in more complex devices, such as memories, shift registers, and multiplexers. They are designed for parallel operation and *one and only one of the parallel gates may be enabled at any time*. If no gates are enabled, the output presents a high impedance and its voltage may be in the prohibited region. If more than one gate is enabled, excessive current may flow and damage the ICs.

We should like to emphasize that 3-state gates should *only be used to connect outputs together*. They should not be used for other logic functions.

An oscilloscope trace of a 3-state microprocessor line is shown in Fig. 5-19. The top trace is the 3-state line and the bottom trace is the enabling pulse line. When the enabling pulses are LOW, the output is driven HIGH or LOW depending on the data. When not enabled, the output line is high impedance and floats to an intermediate level. The data shown on this line are 010-01-001.

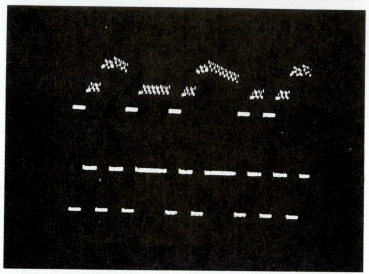

Figure 5-19 Signals on a 3-state bus line. (Courtesy of Ed Pickett and People's Cable TV Company, Rochester, N.Y.)

5-7-1 The 74x125 and 74x126 3-State Gates

Two commonly used 3-state gates are the **74x125** and **74x126**. Their pinouts are shown in Fig. 5-20. Each is a straight-through gate; the output is the same as the input when the IC is enabled. The **74x125** is enabled by a LOW on its enable

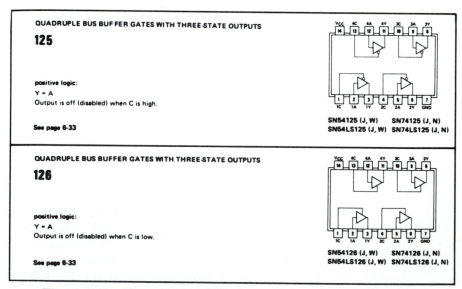

Figure 5-20 The **74125** and **74126** 3-state gates. (From the *TTL Data Book for Design Engineers*, 2nd ed., Texas Instruments, Inc. Courtesy of Texas Instruments, copyright 1976.)

line, while the **74x126** needs a HIGH enable line to function. The **74LS125** is used in the circuitry of the IBM PC (see section 14-5-2).

EXAMPLE 5-6

For the circuit of Fig. 5-21, find the output for each combination of inputs V and W.

SOLUTION   The operation of the circuit is best described by the accompanying table.

Inputs		Points						
V	W	A	B	C	Gate 1	Gate 2	Gate 3	Output
0	0	0	1	1	DISABLED	ENABLED	DISABLED	N
0	1	0	0	0	DISABLED	DISABLED	ENABLED	P
1	0	1	0	1	ENABLED	DISABLED	DISABLED	M
1	1	1	0	1	ENABLED	DISABLED	DISABLED	M

Note that this is a well-designed circuit because no combination of inputs enables more than one gate.

5-7-2   Other 3-State Drivers

A 3-state driver is an IC that puts a set of 3-state gates between an ordinary circuit and a bus. This isolates the circuit from the bus except when the 3-state drivers are enabled. Perhaps the most popular are the octal (8-bit) bus drivers such as the **74LS241** and **74LS244** shown in Fig. 5-22. They have 8 inputs driv-

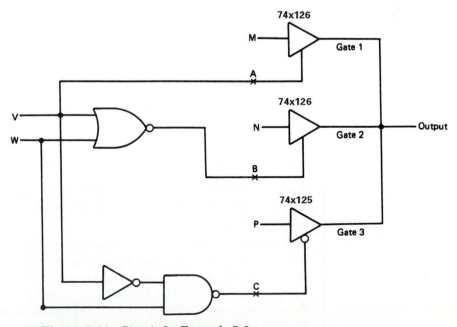

Figure 5-21   Circuit for Example 5-6.

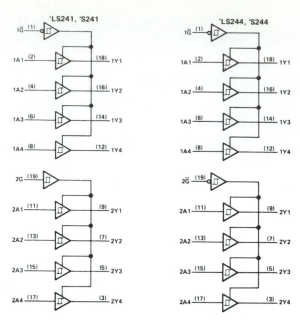

Figure 5-22   The **74LS241** and **74LS244** line drivers. (Reprinted by permission of Texas Instruments.)

ing 8 outputs, and two ENABLE lines. They require a 20-pin package to accommodate the 8 lines. The **74365, 74366, 74367,** and **74368** are other 3-state drivers. They each have six data inputs and six 3-state outputs, but they come in a 16-pin package.

Another advantage of the **74LS241** and **74LS244** is that they are *drivers*; they can drive higher currents than ordinary TTL gates. If an output is high speed or must drive a long line (more than a foot), it tends to produce overshoots and undershoots on the line. These can be minimized by *terminating* the line. This means adding a resistor at the end of the line that is approximately equal to the *characteristic impedance* of the line. The driver must be able to drive this resistor to the minimum voltage required for a logic 1. In a typical digital circuit this impedance is 130 Ω, which can be driven by these ICs.

A third advantage of the **74LS241** and **74LS244** is that they have Schmitt-trigger inputs. Therefore, if they are used as line receivers, they can ignore small voltage perturbations on the line.

EXAMPLE 5-7

A set of 8 switches is to be connected to a μP bus, so the μP can read them. Why can't they be connected directly as shown in Fig. 5-23a? How can they be connected?

SOLUTION   If they were connected directly, any closed switch would short its line to ground, making all signals on that line a logic 0. It could also damage any driver on the bus by shorting its output to ground. The switches should be connected via a **74LS244** as shown in Fig. 5-23b. Here the 3-state gates are only enabled when the μP must read the switches, and it does so by sending out an ENABLE pulse to the **74LS244**.

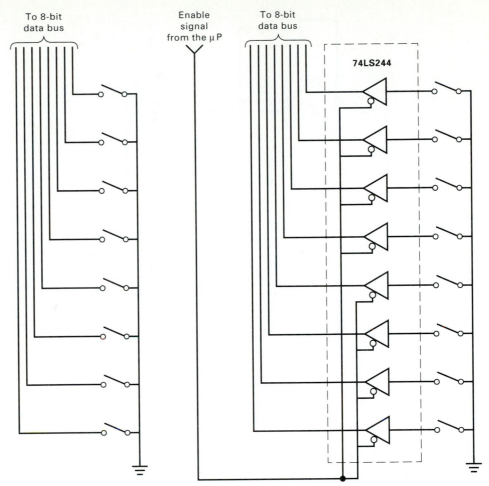

Figure 5-23 Connecting a set of switches to a microprocessor: (a) Direct connection; (b) Connection via a **74LS244**.

5-7-3   3-State Transceivers

A *transceiver* is a circuit that can transmit data in *either* direction. It is used to connect a bidirectional bus to a μP. Memories and peripheral ICs require transceivers at their outputs, but most of these ICs have the transceiver already built in. The circuit for one line of a transceiver is shown in Fig. 5-24.

The ENABLE line must be brought LOW for the chip to transmit data. If it is LOW, the DIRECTION line determines the direction of data flow by enabling one of the 3-state gates. If DIRECTION is LOW, for example, gate 2 is enabled and data flows from *B* to *A*. If ENABLE is LOW and DIRECTION is HIGH, data flows from *A* to *B*.

The **74LS245**, shown in Fig. 5-25, is the most popular 3-state transceiver. It is designed for an 8-bit bus and has DIRECTION and ENABLE signals that apply to all 8 lines. Like the 8-bit drivers, it comes in a 20-pin package.

The **74LS244** and **74LS245** are used extensively in the IBM-PC. Figure 5-26 shows a page from the system board, where the computer is driving the Read-Only Memories. The address bus where data only flows one way, from the

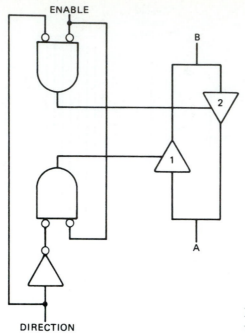

Figure 5-24 One line of a bidirectional transceiver.

microprocessor to the memory, is driven by **74LS244**s. The data bus is bidirectional, however, so this bus is driven by a **74LS245** transceiver.

5-8   Strobed Gates

A strobed gate is simply a gate with a *strobe input*. Typically, if the strobe input is HIGH, the gate will function normally, but if the strobe input is LOW, the output will go to a particular logic level and remain there, regardless of the other inputs. The most common strobed gate is the **74x25** dual 4-input NOR gate, shown in Fig. 5-27. If the strobe input is HIGH, the **74x25** functions as a 4-input NOR gate. But if the strobe is LOW, its output will be HIGH as its output equation indicates.

5-8-1   Expandable Gates and Expanders

Expandable gates have inputs that allow additional logic to be introduced, thus making them more versatile. A second gate, called an *expander*, is used to provide the additional logic. This type of logic is basically obsolete and rarely used. Readers who do encounter expandable gates and expanders may refer to the first or second edition of this book, where they are explained.

5-9   AND-OR-INVERT Gates

SOP expressions can be implemented by AND-OR logic as shown in section 3-3-5. To facilitate this implementation, a series of AND-OR-INVERT gates are

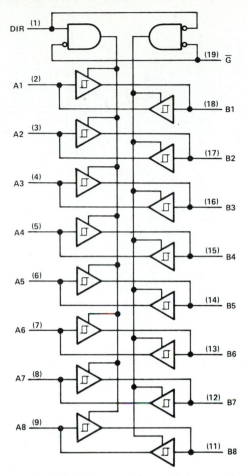

Figure 5-25 The **74LS245 Transceiver**. (Reprinted by permission of Texas Instruments.)

available in the **74x00** line. These consist of a group of AND gates connected to a NOR gate, all on the same chip.

The AND-OR-INVERT (AOI) gates available in the **74x00** series are shown in Fig. 5-28. Note that the word "wide" in each gate refers to the number of inputs to the NOR gate. The **74x51** is a dual 2-wide AOI gate and the **74x54** is a 4-wide AOI gate. Other AOI gates and some gates that are only AND-ORs (no inversion) exist in other series. A typical problem using AOI gates is shown in Example 5-8.

EXAMPLE 5-8

A *register* is a group of logic outputs, usually associated with each other to form a code representing information (a number, a letter, etc.). An n-bit register has n outputs. Given two 4-bit registers, A and B, and a select line, design a circuit so that the output is the complement of register A if the select line is LOW, and the complement of register B if the select line is HIGH.

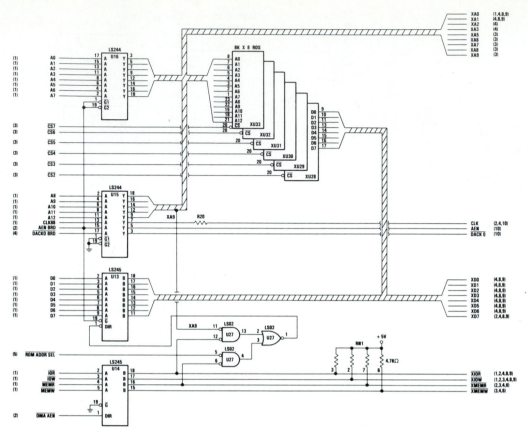

Figure 5-26 Driving the ROM in the IBM-PC. (Courtesy of International Business Machines Corporation.)

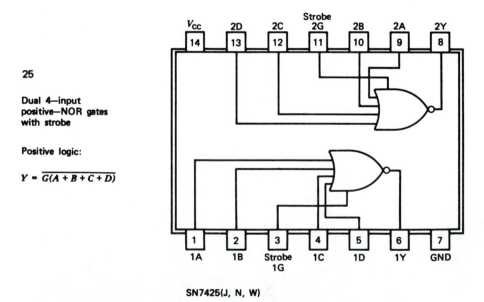

25

Dual 4—input
positive—NOR gates
with strobe

Positive logic:

$$Y = \overline{G(A + B + C + D)}$$

SN7425(J, N, W)

Figure 5-27 The **7425** strobed 4-input NOR gate. (*The TTL Data Book for Design Engineers*, 2nd ed., Texas Instruments, Inc. Courtesy of Texas Instruments, copyright 1976.)

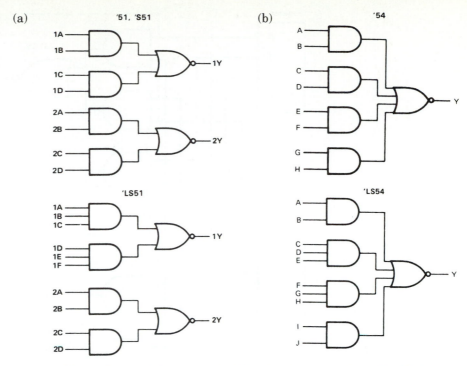

Figure 5-28 AOI gates in the **7400** family: (a) **74x51**; (b) **74x54**. (Reprinted by permission of Texas Instruments.)

SOLUTION In either case, the output is the complement of a 4-bit register; therefore, the output must contain 4 bits. This design can be realized by setting up four, 2-wide, 2-input AOI gates and allowing the select line and its complement to make the choice. Consequently, it can be built with two **74x51**, as shown in Fig. 5-29. When the select line is LOW, all the B inputs to the AND gates are disabled and the A inputs come through the AOI gates. When the select line is HIGH, only the AND gates connected to the B register are enabled, and the 4-bit output is the complement of the B register.

EXAMPLE 5-9[1]

Given: Two 4-bit registers A and B and a SELECT line. Design a circuit to:

(a) Produce a 4-bit output C, such that if SELECT = 1, then $C = A$ (the contents of register C equals the contents of register A), and if SELECT = 0, $C = B$.

(b) Add gating so that if $A = 3$ and SELECT = 1, then $C = 7$.

SOLUTION This problem has two requirements. One approach is to first satisfy requirement a, and then modify the circuit to satisfy the second requirement. The first requirement is similar to Example 5-8 except that the outputs must not be complements of the inputs. This can be satisfied by simply adding an inverter to the outputs of Fig. 5-29.

[1]Examples 5-9 and 5-10 were given as test problems by the author. The examples are presented because the solutions offered by the students were very poor and convoluted.

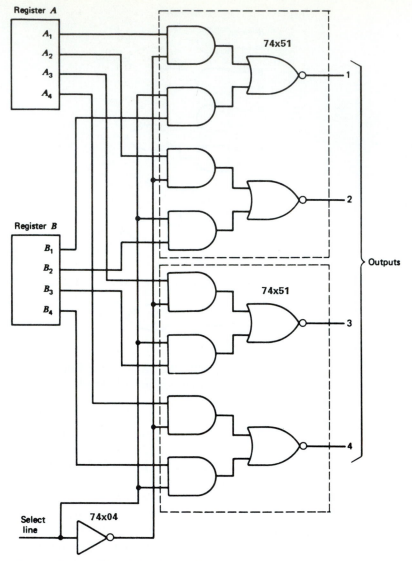

Figure 5-29 Circuit for Example 5-8.

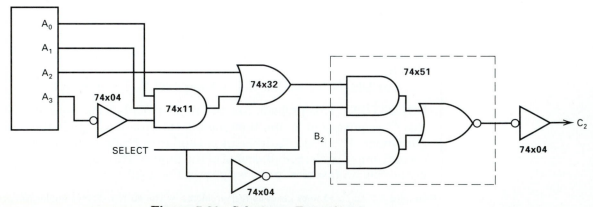

Figure 5-30 Solution to Example 5-9.

Now consider the second requirement. What we are saying is that if $A = 3$ ($A3, A2, A1, A0$ are 0011 respectively), then $C = 7$, which means that $C2$ must change from a 0 to a 1. This means that $C2$ must be a 1 if A is selected and $A2$ is a 1 OR if $A = 3$. A circuit to detect when A is 3 must be added. A little thought, however, should convince the reader that $A2$ is a *don't care* in this circuit. The circuit for generating $C2$ is shown in Fig. 5-30.

EXAMPLE 5-10

Repeat Example 5-9, but replace the AOI gates with 3-state gates.

SOLUTION   The solution is shown in Fig. 5-31. Because the enable levels for the **74x125** and **74x126** are different, no inverter was required for the select lines. It can be seen that:

(a) The outputs of the 3-state gates are connected together, but only one gate of each pair is enabled at any time.

(b) The logic for changing the $C2$ output was placed *before* the 3-state gates. 3-state gates are used for wire-ANDing outputs, not for basic logic functions.

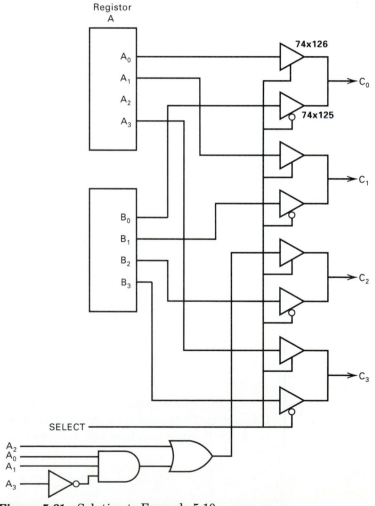

Figure 5-31   Solution to Example 5-10.

An EXCLUSIVE-OR (XOR) gate is a 2-input gate whose output is the same as an OR gate except that it produces a 0 output if *both* inputs are 1. The symbol $\oplus$ is used to indicate the XOR operation. The symbol and truth table for the **74x86** quad 2-input XOR gate are shown in Fig. 5-32.

A	B	Y
0	0	0
0	1	1
1	0	1
1	1	0

(a) Circuit $Y = A \oplus B$ (b) Table truth

Figure 5-32   EXCLUSIVE-OR gates.

The SOP form for the XOR gate, which can be derived from the truth table in Fig. 5-32, is:

$$Y = A\bar{B} + \bar{A}B \tag{5-1}$$

The complement of the XOR gate is the *exclusive-NOR*, a gate whose output is 1 whenever its two inputs are equal. The SOP form for the exclusive-NOR can be obtained by complementing Eq. (5-1) and is:

$$Y = AB + \bar{A}\bar{B}$$

EXAMPLE 5-11[2]

An XOR gate has a signal $X1X2X3$ applied to one input of an XOR gate and $X4$ applied to the other leg as shown in Fig. 5-33. Express its output in SOP form.

SOLUTION

$$Y = A \oplus B = A\bar{B} + A\bar{B}$$

$$\text{Here } A = X1X2X3 \text{ and } B = X4$$

$$Y = X1X2X3\overline{X4} + (\overline{X1X2X3})X4$$

$$= X1X2X3\overline{X4} + \overline{X1}X4 + \overline{X2}X4 + \overline{X3}X4$$

XOR gates and circuits built from them have a large variety of important applications, and are discussed further in Chapter 15. To familiarize the reader with their use, however, two examples are presented here.

[2]This problem also pertains to the material in section 12-6-2.

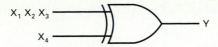

Figure 5-33 The circuit for Example 5-11.

EXAMPLE 5-12

Given two 4-bit registers, A and B, design a circuit to determine whether the numbers in the two registers are the same.

SOLUTION The required circuit is an equality detector; it must do a bit-by-bit comparison of the register outputs. If the output of each pair of corresponding bits are equal ($A_1 = B_1, A_2 = B_2$, etc.), the numbers in the two registers are equal. This comparison can be made by using XOR gates. If the output of an XOR gate is LOW, its two inputs are equal. Four XOR gates, one to compare each bit, are required and the output of each gate must be LOW for equality. Therefore, if the outputs are fed to a 4-input NOR gate, the final output will be HIGH only if all its inputs are LOW, indicating that the two registers contain the same number. The circuit is shown in Fig. 5-34. A **74x25** is used as the 4-input NOR gate. It is shown in its negative-NAND input representation with the strobe clamped. The circuit usually works if the strobe is left unconnected.

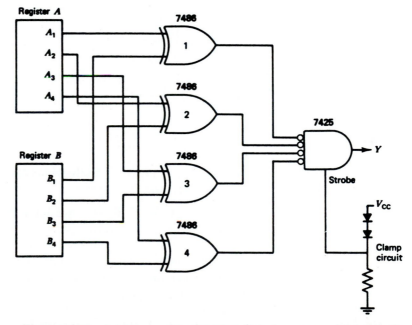

Figure 5-34 A 4-bit equality detector. Gate 1 compares bit 1 of both registers; gate 2 compares bit 2; and so on.

EXAMPLE 5-13

Given a 4-bit register and a select line, design a circuit such that:
    (a) If the select line is HIGH, the circuit output is the same as the register.

(b) If the select line is LOW, the output is the complement of the register outputs.

SOLUTION This circuit can be designed using AOI gates and inverters (see Problem 5-11), but it can be designed more simply by using XOR gates. A little thought reveals that if *one of the inputs to an XOR gate is a 1, the gate inverts the other input. However, if one input to an XOR gate is a 0, the output is the same as the other input*. Therefore, the circuit is designed by tying the select line and the register outputs together in four XOR gates as shown in Fig. 5-35. When the select line is HIGH, one input to each XOR gate is 0 and it does not invert. If the select line is LOW, however, one input to each XOR gate is HIGH and the four outputs are the complements of the four bits of the register.

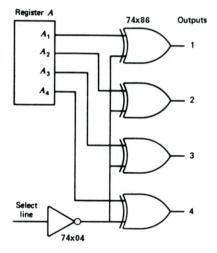

Figure 5-35 Design of the complementation circuit of Example 5-13.

SUMMARY

In this chapter, the basic SSI gates available in the **74x00** series were introduced and their characteristics studied. Some simple circuits using these gates were designed. In addition, open-inputs, wire-ANDing, strobed gates, and EXCLUSIVE-ORs were introduced and examples of their use were presented. These basic gates are used as components in the more elaborate circuits to be covered in succeeding chapters.

GLOSSARY

AOI: AND-OR-INVERT gate.

Clamp: A circuit that ties unused inputs to a voltage in the 1 range, which is less than V_{CC}.

Expandable gate: A gate with inputs to accept additional logic.

Expander: A gate with special outputs to provide additional logic to an expandable gate.

Open-collector gate: A gate whose output is a collector with no internal connection to V_{CC}.

Pull-up resistor: A resistor connected between V_{CC} and a point in the circuit.

Schmitt trigger: A gate that turns on at a different voltage than the voltage that turns it off.

SSI: Small scale integrated circuit.

Strobe: An input signal that can activate or disable a gate.

Three-state gate: A gate having either a 1, 0, or high impedance output.

Transceiver: A bidirectional driver-receiver.

Wire-ANDing: Tying the outputs of several gates together.

XOR: EXCLUSIVE-OR-gate.

PROBLEMS

Section 5-4.

5-1. How long are the responses of the NAND gate and Schmitt trigger of Example 5-2 to a 2-V, 500-kHZ sine wave?

5-2. Design a **74x13** Schmitt trigger detector to determine if the output of a sine wave is greater than 5.1 V. If the input frequency is 1 MHz, what is the minimum width of the output pulse?

5-3. A voltage spike that starts at 0 and rises at the rate of 1 V/μs until it reaches 3 V, after which it falls at the rate of 1 V/μs, as shown in Fig. P5-3, is applied to a **74x13**. Show the output of the **74x13** as a function of time.

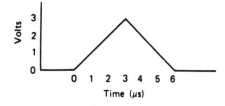

Figure P5-3

Section 5-5.

5-4. An 8-input NOR gate has five unused inputs. Will it function properly if:
  (a) The unused inputs are tied to used inputs.
  (b) The unused inputs are tied to clamp.
  (c) The unused inputs are tied to pull-up resistors.
  (d) The unused inputs are tied to ground.

Section 5-6.

5-5. Why shouldn't an open-collector gate be tied to a totem-pole gate?

5-6. An incandescent lamp is used to monitor various points in a circuit, as shown in Fig. 5-14. Is the lamp ON or OFF if the gate is a **74x06** and the input is connected to:
(a) A logic 1.
(b) A logic 0.
(c) The input is not connected to anything (open).
Repeat this problem if the gate is a **74x07**.

Section 5-7.

5-7. If the output of a disabled 3-state gate is connected to the input of a TTL gate, does the TTL gate see a 1 or a 0? Explain.

5-8. Given three 3-state gates, *A*, *B*, and *C*, and two select inputs, *D* and *E*, design a circuit to function in accordance with the table below:

Select Inputs		Gate that Controls the Output
D	*E*	
0	0	*A*
0	1	*A*
1	0	*B*
1	1	*C*

5-9. For the circuit of Fig. P5-9, find the output for each combination of inputs.

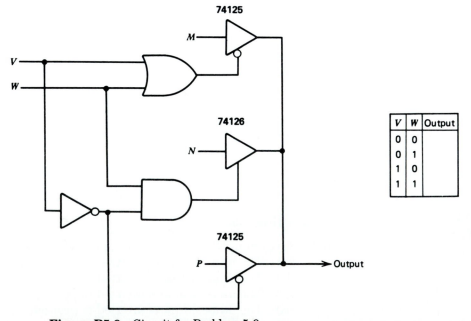

V	W	Output
0	0	
0	1	
1	0	
1	1	

Figure P5-9 Circuit for Problem 5-9.

5-10. Identify the gates and find the output expression for the circuits of:
   (a) Fig. P5-10a.
   (b) Fig. P5-10b.
   What indicates that the gates of Fig. P5-10a are open-collector gates?

Section 5-9.

5-11. Use AOI gates to solve Example 5-13.

Section 5-10.

5-12. Design an EXCLUSIVE-OR gate using:
   (a) Only NAND gates.
   (b) Only NOR gates.

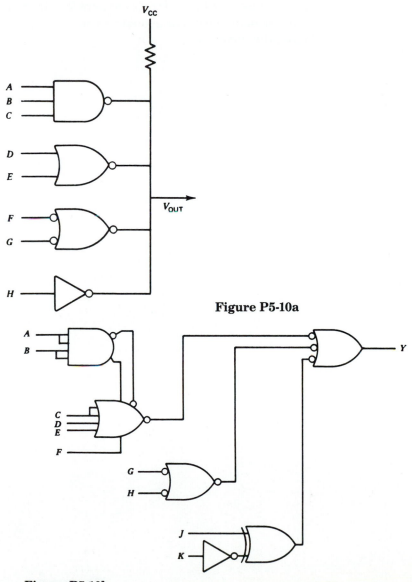

Figure P5-10a

Figure P5-10b

5-13. Implement the function:

$$A \oplus B \oplus C$$

(a) Using **74x86**s.
(b) Express the function in SOP form and implement it without **74x86**s.

5-14. The function $AB + \overline{A}\overline{B}$ can be generated by an XOR gate and an inverter. Show that the function will be generated regardless of whether the inverter is placed on the output of the XOR gate or on one of the inputs.

5-15. Show that $A \oplus (A \oplus B) = B$.

After attempting the problems the student should return to the questions of section 5-2 and be sure their answers are clear. If the student does not understand them all, he or she should reread the appropriate sections of the chapter to find the answers.

6

Flip-Flops

6-1 Instructional Objectives

Except for the gates discussed in Chapter 5, flip-flops (FFs) are the most commonly used digital circuit.

This chapter explains what FFs are, how they work, and why they are used in many circuits. After reading it the student should be able to:

1. Construct a FF from NAND gates and from NOR gates, and explain how they respond to SET and CLEAR pulses.
2. Explain how D and J-K FFs react to pulses on their SET, CLEAR, and CLOCK lines, and how these FFs react to pulse trains that are given by timing charts.
3. Explain how each parameter listed in Section 6-14 affects the action of a FF.
4. Design registers using FFs.
5. Design circuits where FFs monitor and react to sequences of events.

6-2 Self-Evaluation Questions

As the student reads the chapter, he or she should be able to answer the following questions:

1. What are the stable states of a FF? How are they sensed?
2. How do D and latch-type FFs react to inputs on their D and CLOCK/ENABLE lines? What is the difference between a D FF and a latch?
3. What is the difference between master-slave FFs, edge-triggered FFs, and FFs with data lockout?
4. How do direct SETS and direct CLEARS function?
5. What are the limitations on the speed of FFs?
6. How are registers constructed? Why do they employ FFs?
7. How can FFs be used to handle sequence problems?
8. What are races and glitches? How can a glitch be useful?

6-3   Introduction

The circuits considered in previous chapters were *combinatorial* circuits whose outputs depended *solely* on the present inputs. The outputs of such circuits do not depend on the sequence in which those inputs were applied, nor on the state of the circuit before the inputs were applied. In circuits of any size or complexity, however, the *sequence* of events quickly becomes critical and the logic designer must cope with the additional dimension of *time*.

Consider, for example, a problem of the automobile manufacturer, as stated to the logic designer. Before allowing a car to start, he wants the driver to be seated and to buckle the seat belt. To accomplish this, the manufacturer installs appropriate sensors in both the seat and belt. At first it seems that a simple AND gate will suffice. If the driver is seated *and* the belt is buckled, the car can be started without the nasty warning buzzer. But can the manufacturer frustrate the wily driver who dislikes seat belts and, therefore, buckles the belt first and then sits on it? Now the manufacturer specifies a circuit design requiring:

1. The driver is first seated.
2. The belt is then buckled.
3. The driver must have sat down *before* buckling the seat belt.

It is logic problems like these involving sequences of events in time that make logic design more difficult and more challenging.

6-4   The Basic Flip-Flop

To keep track of any sequence of events, a device having the capability of remembering things (memory) is required. The simplest and most widely used *memory cell* is the *bistable multivibrator,* commonly called the *flip-flop* (FF).

The most basic FF is the *SET-RESET* FF shown in Fig. 6-1. This FF produces two outputs. The output labeled Q is also called the SET (or 1) output; the other output, labeled $\bar{Q}$, is called the RESET (or CLEAR or 0) output. When the SET-RESET FF is operating properly, the Q and $\bar{Q}$ outputs are always *complements* of each other. The FF is considered to be in the SET or logic 1 state if the Q output is HIGH and the $\bar{Q}$ output is LOW. Conversely, the FF is considered to be RESET or CLEARED (or to contain a 0) if the Q output is LOW and the $\bar{Q}$ output is HIGH.

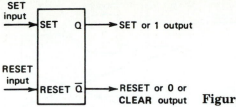

Figure 6-1 The basic SET-RESET flip-flop.

There are two basic inputs to a FF called SET and RESET. Each of these inputs has two levels: *active* and *quiescent*. An active level makes things happen; a quiescent input is passive. When an input is active, it forces the FF output to assume its state (i.e., an active SET input causes a FF to SET), but a quiescent level does not affect the output of the FF.

The FF of Fig. 6-1 operates as follows.

1. An active (1) signal on the SET line causes the FF to SET.
2. Once SET, it remains in its SET state, even after the SET signal has been removed. It remains SET until an active signal is applied to the RESET line.
3. This RESET signal clears the FF and it remains RESET until a SET signal is again applied.
4. For proper operation, the SET and RESET signals should not be applied simultaneously.

A simple FF is often described as a *one-bit memory*. When both inputs are quiescent (neither SET nor RESET is active), the FF "remembers" which input was most recently active. If a SET signal was received last, the FF output will be SET ($Q = 1, \bar{Q} = 0$) and if a RESET signal was received last, the FF output will be RESET.

One problem remains. *The internal circuitry of most FFs is symmetrical*, and when power is applied after the circuit has been turned OFF, there is no way of telling whether the FF will come ON in the SET or RESET state. Many sophisticated systems use a POWER-ON CLEAR signal, generated whenever power is first applied, to clear all critical FFs before operation begins.

In personal computers, the RAM (Random Access Memory, see Chapter 13) is composed of FFs. These FFs lose their information or forget when power is turned off. That is why files must be SAVEd before turning power off, and RETRIEVEd at the start of each computer session.

6-5 NOR Gate Flip-Flops

Perhaps the simplest practical FF is a FF constructed from two **74x02** NOR gates, as shown in Fig. 6-2. It illustrates all the important points of section 6-4. The active signal levels are HIGH and the quiescent (inactive) signal levels are LOW. Note that because of the inversion property of the NOR gates, the Q or SET output is the output of the lower gate, whose input is connected to the RESET line, and the $\bar{Q}$ or RESET output is the output of the upper gate. The FF operates as follows.

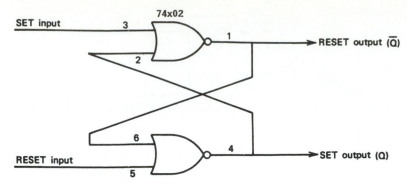

Figure 6-2 The NOR gate flip-flop.

1. If a 1 is applied to the SET input (pin 3) it causes pins 1 and 6 to go LOW, and the $\bar{Q}$ output (pin 1) is a 0.

2. With the RESET input quiescent (0), both pins 6 and 5 are 0, causing pin 4 (the SET or Q output) to be HIGH. This HIGH signal is also applied to pin 2.

3. If the SET input becomes quiescent (0), there is still a 1 input to the top NOR gate at pin 2. Therefore, its output remains LOW and the output of the lower NOR gate, which has two 0 inputs, remains HIGH. The SET output is still 1 and the FF has not changed state; it "remembers" that a SET pulse is the most recent pulse.

4. When the RESET or CLEAR input at pin 5 becomes a 1, pins 4 and 2 go LOW.

5. The LOW inputs at pins 3 and 2 cause pin 1 to go HIGH. The FF is now in its RESET state ($Q = 0$ and $\bar{Q} = 1$).

6. When the RESET input returns to 0, pin 6 is still a 1 and the FF remains in the CLEAR state until the next SET pulse occurs.

To clarify sequential circuits, engineers often construct *timing charts* that show the time relationship of voltages at various points in the circuit. A timing chart for the NOR gate FF is shown in Fig. 6-3. The FF starts in the CLEAR state because the RESET input is initially HIGH. After the RESET signal goes LOW, the FF remains RESET until the leading edge of the SET pulse occurs. The FF then SETS (the SET output goes to 1 and the RESET output goes to 0) and remains SET until the RESET input again goes HIGH.

When a short pulse occurs, such as the SET pulse in Fig. 6-3, the *leading edge* is the edge which occurs first and the *trailing edge* is the pulse edge that follows. For positive pulses, the leading edge is also the positive-going edge and the trailing edge is the negative-going edge. For negative-going pulses, however, the leading edge is the negative-going edge.[1]

---

[1]The term *Leading-Edge* has been used by a personal computer manufacturer to name their computers.

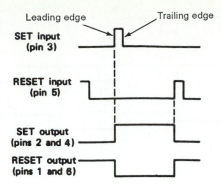

Leading edge   Trailing edge

SET input
(pin 3)

RESET input
(pin 5)

SET output
(pins 2 and 4)

RESET output
(pins 1 and 6)

Figure 6-3   Timing chart for the NOR gate flip-flop (Fig. 6-2).

6-6   NAND Gate Flip-Flops

The most common SET-RESET FF is built with two **74x00** NAND gates as shown in Fig. 6-4. For clarity, the NAND gates are shown as NORs with negative inputs because *the active level of the input signals is negative*. Negative active signals are often written with a bar above the signal name, as shown in the figure. The FF operates as follows.

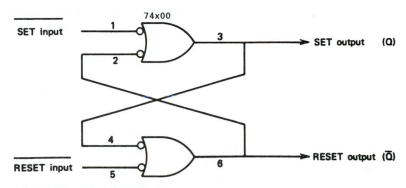

Figure 6-4   The NAND gate flip-flop.

1. Assume the FF is being SET. This means the $\overline{SET}$ input to pin 1 is LOW, which causes the SET output at pin 3 to go HIGH.
2. Since the RESET input is quiescently HIGH, both pins 4 and 5 are HIGH, causing pins 6 and 2 (the RESET output) to be LOW.
3. If the SET input now goes HIGH, the FF remains in its SET state because the level at pin 2 remains LOW and both inputs to the lower NAND gate remain HIGH.
4. This condition continues until a RESET input causes pin 5 to go LOW, which forces the RESET output HIGH. Now pins 1 and 2 are both HIGH, causing the SET output to go LOW.
5. The FF remains in this reset condition until the next SET pulse causes pin 1 to go LOW again.

The timing chart of Fig. 6-5 was constructed to help explain the NAND gate FF:

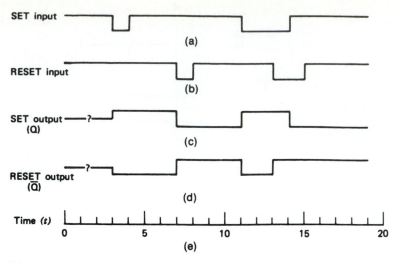

Figure 6-5 Timing chart for the **74x00** NAND gate flop of Fig. 6-4.

1. It is assumed that the circuit is turned on at $t = 0$.
2. With both the SET and RESET inputs quiescent (HIGH), we cannot tell in which state the FF came up; therefore, the outputs are represented with a question mark.
3. At $t = 3$, the SET input goes LOW, setting the FF, as indicated by the fact that the SET output goes HIGH and the RESET output goes LOW.
4. Despite the disappearance of the SET input at $t = 4$, the FF remains SET until the leading edge of the RESET pulse occurs at $t = 7$.
5. This resets the FF and it remains RESET until the next SET pulse at $t = 11$.
6. At $t = 13$ both the SET and RESET become active (LOW) simultaneously. This forces both the Q and $\bar{Q}$ outputs HIGH. Thus, the *penalty* paid for having both inputs active is the *loss of* FF *action* because Q and $\bar{Q}$ are no longer complementary.
7. When the SET input goes HIGH at $t = 14$, the FF assumes the RESET state since the RESET is still active.
8. For t greater than 15 neither input is active, but the FF remains RESET. It "remembers" that the RESET input was the last active input.

EXAMPLE 6-1

The circuit of Fig. 6-6 uses three **74x20** NAND gates to form a 3-state FF. This is a circuit with 3 inputs—$\bar{A}$, $\bar{B}$, and $\bar{C}$—and 3 outputs—A, B, and C. Only one of the inputs may be LOW (active) at any one time and only the output corresponding to the LOW input should be LOW. This output must remain LOW until the next active input occurs.

    This circuit might be used in a digital computer that is required to remember whether the result of the last arithmetic operation was greater than, less than, or equal to 0.

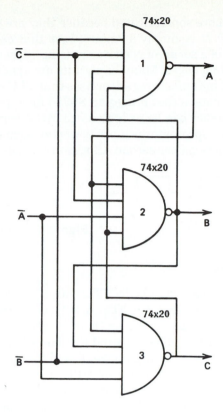

Figure 6-6 A **74x20** 3-state flip-flop.

Explain how the circuit operates as:

1. $\bar{A}$ goes LOW.
2. $\bar{A}$ returns to its quiescent HIGH state.
3. $\bar{C}$ goes LOW.

SOLUTION

1. When $\bar{A}$ goes LOW, it forces outputs B and C HIGH. Since $\bar{B}$ and $\bar{C}$ are HIGH (quiescent), all four inputs to gate 1 are HIGH and output A goes LOW, holding the inputs to gates 2 and 3 LOW.
2. Output A remains LOW after $\bar{A}$ returns to its HIGH state and holds B and C HIGH.
3. When $\bar{C}$ goes LOW, it forces the outputs of gates 1 and 2 HIGH. This causes four HIGH inputs to be present at gate 3 and output C now goes LOW.

6-7   D-Type Flip-Flops

The NAND and NOR gate FFs described in the previous paragraphs are very simple, unclocked FFs. Instead of coupling two NAND or NOR gates to form a FF, most engineers use ICs that are specifically designed to be FFs.

The IC manufacturers have produced a large variety of flip-flop ICs. Most of these ICs are more sophisticated because they are *clocked*. A digital clock is shown in Fig. 6-7. It is simply a square wave that oscillates between 0 V and +5 V. It consists of a series of positive or rising edges alternating with a series of negative or falling edges. Many IC circuits are synchronized with these *clock edges*. When a PC manufacturer advertises that a PC contains a 25 MHz **80386**, for example, this means that it uses an **80386** microprocessor that is driven by clock whose frequency can be as high as 25 MHz. Higher clock rates generally mean that the microprocessor is more powerful because it can run faster or execute more instructions per second. Clocks are discussed further in Chapter 7.

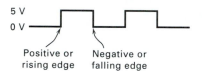

Figure 6-7   A digital clock.

6-7-1   The 74x74

ICs designed specifically as FFs by TTL manufacturers fall primarily into two categories: D-type FFs and J-K FFs. A D-type FF contains a D input, whose function is described later in this section. The most commonly used D FF is the **74x74,** dual D, positive-edge-triggered FF. The **74x74** has two identical FFs in a 14-pin package. The circuit, pin layout, and function table are shown in Fig. 6-8. A *Function Table* shows how a FF will respond to the edges or levels on its various inputs. The Function Table for the **74x74** is given in Fig. 6-8b.

Note first the direct CLEAR and direct SET inputs that enter the first FF on pins 1 and 4 (Fig. 6-8a). These are called the *asynchronous* inputs because their effect is *independent* of the clock (CK). If only these two inputs are used, the FF will function as a simple SET-RESET FF. The bubbles shown on the asynchronous inputs (pins 1, 4, 10, 13) indicate that the active level is LOW.

The direct SET or PRESET input, pin 4, is generally drawn at the top of the FF, as shown in Fig. 6-8a. A negative pulse on this input immediately SETS the FF regardless of the clock or the D input, as shown on line 1 of Fig. 6-8b. Conversely, the direct CLEAR or RESET input is generally drawn at the bottom of the FF and a negative pulse at this point clears the FF. If the SET and CLEAR inputs are both active (LOW) simultaneously, both the Q and $\bar{Q}$ outputs go to a 1, as shown on line 3 of Fig. 6-8b, and remain there until either one of the

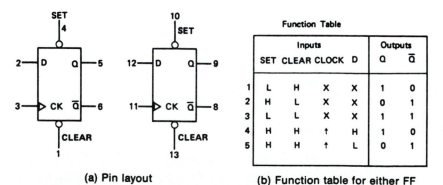

(a) Pin layout

Function Table

	Inputs				Outputs	
	SET	CLEAR	CLOCK	D	Q	$\bar{Q}$
1	L	H	X	X	1	0
2	H	L	X	X	0	1
3	L	L	X	X	1	1
4	H	H	↑	H	1	0
5	H	H	↑	L	0	1

(b) Function table for either FF

Figure 6-8   The **74x74** dual-D positive-edge-triggered flip-flop and its function table. In (b), X = irrelevant.

active inputs is removed. This mode of operation is not recommended (and is avoided by all but the most adventuresome engineers). If the SET or CLEAR inputs are not used, they may be left open, or preferably tied to a clamp (see section 5-5-2) or some other HIGH level.

When the direct inputs are inactive, the synchronous inputs may be used. The clock inputs are used to control the FF. Note that there is no bubble on the clock inputs (pins 3 and 11) of Fig. 6-8a. The absence of a bubble indicates that the FF reacts to *positive edges* of the clock. On each *positive edge,* the level on the D input (pin 2 on FF1 and pin 12 on FF2) is set into the FF. Thus, if the D input is HIGH (logic 1) when the clock goes HIGH, the Q output goes to 1 and the FF remains SET until the next clock pulse. Conversely, if the D input is LOW, a clock pulse causes the Q output to go LOW. The inputs are called *synchronous* because the outputs are *synchronized to the clock* and *only change when it goes positive.* This is also shown on lines 4 and 5 of the Function Table of Fig. 6-8b, where the up-arrow symbol (↑) indicates that changes occur on positive transitions only.

EXAMPLE 6-2

Devise a circuit to toggle a D FF. Toggling means causing the FF to reverse its state on each clock pulse.

SOLUTION In order to force the D FF to toggle, a 0 must be clocked in whenever the FF is SET and a 1 must be clocked in whenever the FF is CLEAR. But the $\bar{Q}$ output is LOW whenever the FF is SET, and HIGH whenever it is CLEAR. Therefore, if the $\bar{Q}$ output is connected to the D input of the same FF, it will toggle on each clock pulse. The circuit and timing chart solutions are shown in Fig. 6-9.

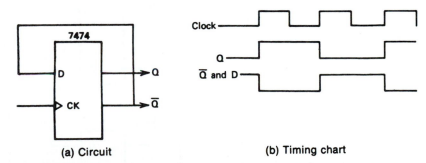

(a) Circuit   (b) Timing chart

Figure 6-9 Circuit and timing chart for Example 6-2. *Note:* The direct SET and direct CLEAR are not shown and are assumed to be HIGH.

6-7-2 The 74x174

Another D-type FF often used because of its packaging density is the **74x174** hex D shown in Fig. 6-10. Six D-type FFs have been placed inside one 16-pin DIP. To achieve such packaging density, the following price was paid.

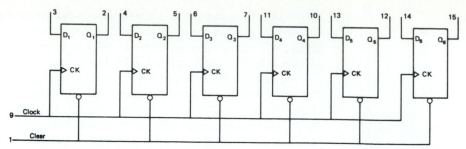

Figure 6-10 The **74x174** HEX D flip-flop. V_{CC} is on pin 16; GND is on pin 8.

1. A positive transition (↑) at pin 9 clocks *all* FFs simultaneously.
2. A negative pulse or level at pin 1 CLEARS *all* six FFs.
3. The $\bar{Q}$ outputs are not available.

The $\bar{Q}$ outputs, if required, can be obtained by connecting an inverter to the Q outputs.

6-7-3 Bistable Latches

Bistable latches are a variation of the D-type FF. They have a D input and an ENABLE input. If the ENABLE input is HIGH, the output follows the D signal. But if the ENABLE signal goes LOW, the output remains where it was at the last instant the ENABLE was HIGH, and is not affected by any changes on the D input. The output is therefore latched. The action of a bistable latch is illustrated by the timing chart of Fig. 6-16 (see also Example 6-6).

The most popular bistable latch is the **74x75** QUAD LATCH shown in Fig. 6-11. Note that four FFs are contained in the 16-pin DIP, but there are only two

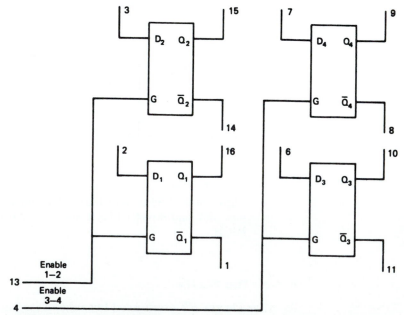

Figure 6-11 The **74x75** quad latch flip-flop. V_{CC} is on pin 5; GND is on pin 12.

ENABLE gates (labeled G), each connected to a pair of latches, and there are no direct CLEARS or SETS.

6-7-4 The 74x373 and 74x374

There are two objections to the **74x174**:

- The **74x174** accommodates only 6 FFs. Most microprocessors transmit data a byte (8 bits) at a time, so engineers would prefer an IC that contains 8 FFs.
- The CLEAR is not used very often. A better use for this pin might be found.

In response to these objections, the IC manufacturers have produced the **74x373** and **74x374**s shown in Fig. 6-12. They both contain 8 FFs, and have replaced the CLEAR input with a 3-STATE ENABLE. This allows the outputs to be disabled, or disconnected from a 3-state bus. These ICs are very often used on a 3-state bus. The drawback to these ICs is that they come in a 20-pin package and require a 20-pin socket.

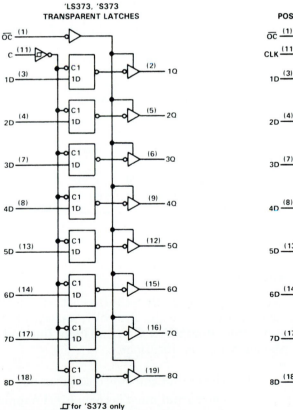

Figure 6-12 The **74x373** and **74x374** ICs. (Reprinted by Permission of Texas Instruments.)

The **74x373** is a latch; it responds whenever the latch input is high. The **74x374** contains D FFs. Its output only changes when a positive edge occurs on the CLOCK input.

6-8 J-K Flip-Flops

J-K FFs are very versatile and widely used. These FFs are so named because they have a J and a K input as well as a clock. Typically, the outputs of a J-K FF change on the *negative-going* edge of the clock, although positive-edge triggered J-K FFs are also available.

J-K FFs conform to the function table of Fig. 6-13, where Q_N is the state of the FF before the clock transition and Q_{N+1} is the state of the FF after the transition. A J-K FF operates as follows.

1. If both J and K inputs are 0 (LOW), the FF will not change state. The table indicates that $Q_{N+1} = Q_N$.
2. If J = 1 and K = 0, the FF SETS on the next negative clock transition. In this case the Q_{N+1} output is 1.
3. Conversely, if K = 1 and J = 0, the FF CLEARS.
4. If both J and K are 1, the FF toggles, or changes state on each clock transition, as indicated in the function table, which shows that $Q_{N+1} = \bar{Q}_N$.

Inputs		Output
J	K	Q_{N+1}
0	0	Q_N
1	0	1
0	1	0
1	1	$\bar{Q}_N$

Figure 6-13   Function table for a J-K flip-flop.

EXAMPLE 6-3

A clock line is a line that alternates between HIGH and LOW levels at a predetermined frequency. Typically, a clock line can be produced by a square wave generator (SWG). Given an input line and a clock line, design a circuit to the following specifications:

1. The output can only change on negative clock transitions.
2. The output must become 1 if the input is 1 and the present state of the output is 0.
3. If the input is 0, the output must be 0 after the next clock pulse.
4. The output must not remain a 1 for two consecutive clock pulses.

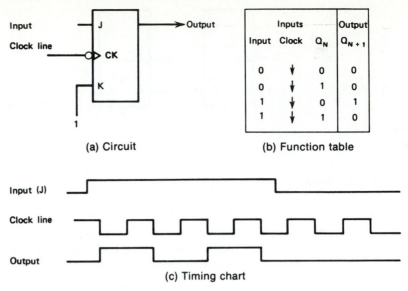

(a) Circuit

| Inputs | | | Output |
Input	Clock	Q_N	Q_{N+1}
0	↓	0	0
0	↓	1	0
1	↓	0	1
1	↓	1	0

(b) Function table

(c) Timing chart

Figure 6-14   Solution for Example 6-3.

The input-output relationships are also defined by the function table of Fig. 6-14, where the down arrow (↓) indicates a negative transition of the clock. A negative transition is also indicated by the bubble shown on the clock input.

SOLUTION   The solution and timing chart are also shown in Fig. 6-14. The input line is connected to the J input of a J-K FF, while the K input is clamped HIGH. Therefore, if J is 0, the FF CLEARS, while if J is a 1, as it is for the first four negative transitions, the FF toggles.

6-8-1   Master-Slave and Data Lockout FFs

In earlier years, manufacturers built two varieties of J-K FFs with peculiar features. These are the Master-Slave FFs, the **7473** and **74x107**, and a FF with data lockout, the **74111**. Both are more complex, and the master-slave FFs can cause problems for an unwary engineer. Because these FFs are basically obsolete, we will not discuss them further here. Should the engineer encounter them, a thorough discussion of their behavior can be found in the second edition of this book.

6-8-2   Edge-Triggered Flip-Flops

An edge-triggered FF changes on the negative edge of the clock, depending on the levels on its J and K inputs *at the time of the clock edge,* as shown in Fig. 6-14b. The **7473** and **74x107** were both master-slave FFs with the problems alluded to in the previous section. The newer **LS** and **HC** versions of these ICs were changed to edge-triggered FFs to eliminate these problems. The pinout of the standard and **LS** versions of the **74x107** is shown in Fig. 6-15. The **74x109,** **74x113,** and **74x114** are other examples of edge-triggered, J-K FFs.

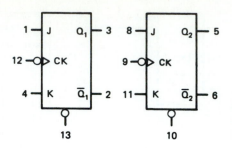

Figure 6-15 Pin layout for the **74x107** dual J-K master-slave flip-flop. V_{CC} is on pin 14; GND is on pin 7.

6-9   Timing Charts

Timing charts are helpful in clarifying the operation of the various FFs as well as in showing how the voltages throughout a circuit vary as a function of time.

Example 6-4 shows how the outputs of the most commonly used FFs behave in response to their inputs.

EXAMPLE 6-4

Given the input and clock shown on lines 1 and 2 of Fig. 6-16, find the output:

1. If the input and clock are applied to the D and CLOCK inputs of a **74x74**.
2. If the input and clock are applied to the D and ENABLE of a latch-type FF (the **74x75**, for example).
3. If the input is applied to the J input of an edge-triggered J-K FF with the K input tied HIGH.

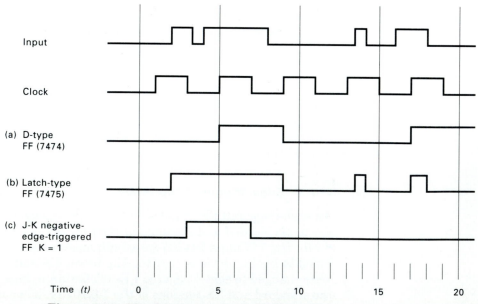

Figure 6-16   Flip-flop timing charts for Example 6-4.

SOLUTION

1. (Figure 6-16a) The D-type FF changes state at positive clock edges only. These occur at $t = 1, 5, 9, 13$, and 17. But since the input (D) is only HIGH at $t = 5$ and 17, two positive pulses will occur at these times.

2. (Figure 6-16b) The latch output is the same as the input whenever the clock is HIGH (times between 1 and 3, 5 and 7, 9 and 11, etc.). Therefore, the short pulse at $t = 4$ is missed because the clock is LOW, but the short pulse at $t = 14$ is transferred to the output because the clock is HIGH. Note that the latch retains the last output level whenever the clock goes LOW.

3. (Figure 6-16c) The J-K negative-edge-triggered FF behaves as the master-slave FF, except that it does not respond to the pulse at $t = 14$ because the J and K inputs are only effective at the negative edge (at $t = 15$). At $t = 15, J = 0$, and $K = 1$; therefore, the FF does not change state and it does not produce the second output pulse.

6-10   Direct SETS and Direct CLEARS

Direct CLEARS and direct SETS (the asynchronous inputs) override *any* clock inputs. They are called *asynchronous* because they affect the FF whenever they are applied, and do not depend on a clock edge to clock them in. They are internally wired to prevent any clock action from affecting the FF when they are active. For example, a clock cannot SET a 1 into a FF while the direct CLEAR is LOW. The timing charts of Fig. 6-16 were drawn on the assumption that the direct CLEARS and SETS were inactive, allowing the FFs to respond to clock pulses only. A CLEAR pulse at any time, however, would have RESET any of the FFs.

The asynchronous outputs are active only on LOW inputs. This means FFs should respond only to their clock inputs if the direct inputs are left unwired. As a precaution, many engineers wire unused direct inputs to clamp. Usually both the direct and clocked inputs are used to make optimum use of a FF.

EXAMPLE 6-5

An engineer must CLEAR a **74x74** when a pulse line (a line carrying a train of pulses of varying durations) is negative, and clock in the level on the D input when the pulse goes positive. He builds the circuit of Fig. 6-17a, reasoning that a LOW level on the pulse line CLEARS the FF and the following positive edge both removes the direct CLEAR and clocks in the D level. Is his reasoning valid? Explain reasons for your answer.

SOLUTION   The direct CLEAR will remain active anywhere from 10 to 30 ns after the CLEAR input has gone HIGH. In the circuit of Fig. 6-17a, the clock occurs simultaneously with the removal of the direct CLEAR.

The CLEAR does not have sufficient time to release and still overrides the clock. The **74x74** will remain in the CLEAR state regardless of the level on the D input. The circuit does not meet the specifications.

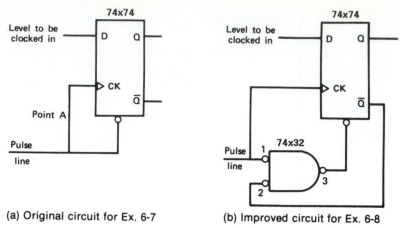

(a) Original circuit for Ex. 6-7　　　(b) Improved circuit for Ex. 6-8

Figure 6-17　Circuits for Examples 6-5 and 6-6.

One of the most common errors made by designers is their attempt to clock a FF too soon after the removal of a direct input, although it is rarely as obvious as it is in Example 6-5. The time required between the removal of a direct input and the successful application of a clock is not specified by the manufacturer, but experience has shown it to be between 10 and 30 ns. Inserting two inverters at point A of Fig. 6-17 may delay the clock long enough to allow the circuit to work, but this also depends on the transition times of the inverters as well as the speed of the direct CLEAR.

6-11　Race Conditions

A race condition exists if a circuit output depends on which of two *nearly* simultaneous inputs arrive at a point in the circuit first, or where there may not be enough time between the removal of one input and the arrival of another. Inserting inverters at point A of Fig. 6-17a may create the latter type of race condition.

Circuits that contain race conditions are unreliable. If the same circuit is built several times or mass-produced, the output of each circuit is liable to be different due to the varying speeds of the circuit components. What is worse, the output of the same circuit may be different at different times if this output depends on the results of a *very* close race, because those results may not always be the same. Consequently, engineers devote considerable time and effort toward eliminating race conditions. A thorough examination of the specific circuit usually reveals a satisfactory way of controlling pulses so that races are eliminated. Only when all else fails would we resort to brute force techniques, such as using cascaded inverters, to delay inputs. Superior logic design often eliminates race conditions, as Example 6-6 demonstrates.

EXAMPLE 6-6

Design a circuit to meet the specifications of Example 6-5 without race problems.

SOLUTION   If the CLEAR input is removed as soon as the FF is RESET, it will no longer override the clock! The solution is shown in Fig. 6-17b. If the pulse line goes LOW while the FF is SET, a negative pulse comes through the **74x32** to CLEAR the FF. This releases the direct CLEAR by causing pins 2 and 3 of the **74x32** to go HIGH. Since the direct CLEAR is now released at the beginning of the negative input pulse, the clock will be effective when it goes HIGH.

The circuit of Fig. 6-17b was set up in the laboratory with the D input held HIGH, and the resultant wave forms are shown on the oscilloscope (CRO) traces in Fig. 6-18. The top trace is the clock, the second trace is the FF output, and the bottom trace is the direct CLEAR or the output of the **74x32**. It can be seen that the FF output goes HIGH on the positive edge of the clock, and when the clock goes negative, the spike on the CLEAR input RESETS the FF, eliminating any and all race difficulties.

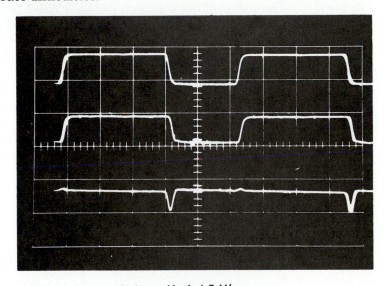

Scales: Vertical 5 V/cm

Horizontal 200 ns/cm

Figure 6-18   Oscilloscope traces for the circuit of Fig. 6-17b.

6-11-1   Glitches

The short negative-going spike on the lower trace of Fig. 6-18 is often called a *glitch*. Glitches may occur when several inputs to the same circuit are occurring nearly simultaneously. They depend on gate delays and are very short. The glitch in Fig. 6-18 only lasted about 20 ns. Glitches are often hard to see on an oscilloscope trace. They showed up clearly in Fig. 6-18 only because the sweep speed of the oscilloscope was very high (200 ns per cm). If the frequency of the applied wave were slower, the pulses on the two upper traces would be longer, but the time of the glitch would remain constant. The glitch would be harder to see because it would occupy a smaller part of the trace at lower speeds.

The circuit of Fig. 6-17 depended on the glitch, but glitches can often be troublesome. A little glitch in the wrong place can often cause a large problem, and engineers often go to great pains to eliminate glitches. Logic analyzers (see section 19-8) often employ a circuit called a *glitch-latch* to detect glitches that may occur between clock edges.

6-11-2 Clock Skew

Many circuits have several FFs that should operate *synchronously*. This means they should all change on a positive or negative clock edge, and the clock should be applied to all the FFs *at the same time*. The word synchronous means controlled or synchronized by the same clock.

Unfortunately, due to propagation delays or other circuit problems, the clock may not arrive at all FFs at precisely the same time. The time difference is called *clock skew* and can cause glitches in the output.

As an example of clock skew, consider the circuit of Fig. 6-19a. Both the D and J-K FFs are set up to toggle. Assume they start with both Q1 and Q2 at 0. The output of the XOR gate will be 0. When the edge occurs on the INPUT line, Q1 and Q2 will both become 1 and the output will still be 0. Note, however, that due to the presence of the inverter, there has been clock skew, and FF2 toggled after FF1 toggled. This time difference produces a glitch on the output of the XOR gate. The timing chart of Fig. 6-19b shows the relationship of the various pulses.

The clock skew problem of Fig. 6-19 might be reduced by replacing the J-K FF with a second D FF and eliminating the inverter. Now both D FFs would toggle in response to the same clock pulse. If one of the D FFs, however, is faster than the other (its propagation delay is less), there still might be a glitch on the output.

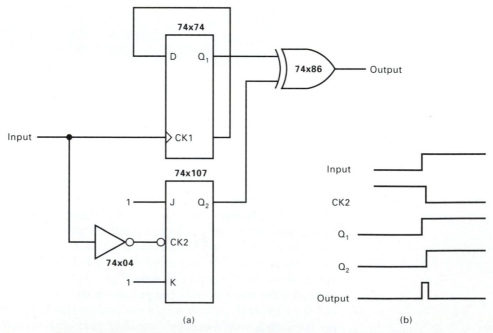

Figure 6-19 A circuit illustrating clock skew: (a) circuit; (b) timing.

6-11-3 Synchronous versus Asynchronous Inputs

Some engineers advocate using only the synchronous inputs to a FF. This means all FFs will change state only in response to a clock edge, or at approximately the same time. This tends to minimize glitches and race conditions.

If a FF or other circuit uses both synchronous and asynchronous conditions, the timing between the two types of inputs often causes races, glitches, and other unforeseen events. The asynchronous outputs, however, give the engineer additional capability, especially if a CLEAR pulse is required. We feel that the asynchronous outputs can be used in some circuits where race conditions or glitches will not disturb the operation.

6-12   Flip-Flop Parameters

One of the most important characteristics of an IC family is the *speed* or *propagation delay* of its FFs. The speed of the various families is determined by several parameters, as shown in Table 6-1. This table was constructed using the **74x74** D-type FF, because it is available in almost all families. The times for J-K FFs in the various families to react are similar. The parameters listed in Table 6-1 are *worst-case*; most FFs will react faster. If a FF is slower, it should be rejected by the manufacturer as being out-of-specification.

The parameters listed in Table 6-1 are:

1. f_{max}. This is the highest frequency at which clock pulses may be applied to a FF and still maintain proper and stable clocking. For a **7474** or a **74107**, the minimum value of f_{max} is 15 MHz. The manufacturers state that typical **7474**s and **74107**s will toggle at 25- and 20-MHz rates, but they do not guarantee the FFs will toggle faster than 15 MHz.

2. t_{PD}. This is the propagation delay, in ns, of the FF. It is the time from the start of a clock or clear pulse until the FF has reacted to the pulse (been totally clocked or cleared).

3. t_W. This is the minimum pulse width that can be applied to the SET or CLEAR inputs to a FF, with a guarantee that it will be effective. FFs may respond properly to shorter pulses than those specified. However, the manufacturer does not guarantee this; therefore, operating with short pulses is risky and should be avoided. On the other hand, the manufacturer does not guarantee that a very short, sharp pulse, often called a *glitch,* will not affect the output. For the **7474**, the CLEAR line must be held LOW for at least 30 ns to *guarantee* that the FF will CLEAR, but a 5-ns glitch may also CLEAR the FF. System failures are often traced to glitches on critical inputs.

4. t_{setup}. This is the time a signal must be present on one terminal before an active transition occurs at another terminal. For a **7474**, t_{setup} is 20 ns (listed in some catalogs as 20 ↑). This means that the D input must be held constant for at least 20 ns before a positive clock edge to assure a reliable output.

5. t_{hold}. This is the time a signal must remain at a terminal *after* an active transition occurs. For a **7474**, this is 5 ns. The signal at the D input

TABLE 6-1

Times for a 74x74 in the various families.

Family IC	Standard 7474	Low Power Schottky 74LS74	Schottky 74S74	Advanced Schottky 74AS74	Advanced Low Power Schottky 74ALS74	High Speed CMOS 744C74	High Speed TTL compatible CMOS 74HCT74	Advanced CMOS 74ACT11074	
f_{max}	15	25	75	105	34	31	23	100	MHz
t_{PD}	40	40	18	10	18	46	42	12.5	ns
t_w	30	25	15	4	15	25	20	5	ns
t_{setup}	20	20	15	5	20	20	15	2	ns
t_{hold}	5	5	0	0	0	0	0	0	ns

should be removed no sooner than 5 ns after the positive edge of the clock. There is zero hold time for most J-K FFs.

The setup and hold times are shown in Fig. 6-20.

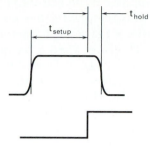

Figure 6-20 Setup and hold times.

EXAMPLE 6-7

How fast can a **7474** be clocked?

SOLUTION   The maximum clock frequency is 15 MHz. This indicates that positive transitions on the clock cannot come at a rate faster than 15 MHz or they must be separated by a time of $\frac{1}{15}$ MHz or 66.7 ns. A consideration of the clock LOW and clock HIGH times indicates that the pulses must be no closer than 67 ns apart. (The clock must be HIGH for 30 ns and LOW for 37 ns.) If only a symmetric (50 percent duty cycle) clock is used, the clocks can be no closer than 74 ns or 13.5 MHz.

The manufacturers' specifications on FFs must also be checked for *loading*. This is done by checking the unit load or I_{IL} specifications. In the standard series, one standard load is -1.6 mA. For the **7474**, I_{IL} is -1.6 mA, or one standard load, for the D and PRESET inputs. But it is -3.2 mA, or two standard loads for the CLOCK and CLEAR inputs. Therefore, only five CLEAR or CLOCK inputs can be driven from an ordinary driver. Buffer drivers are often used when the loading becomes very heavy.

6-13   Uses of Flip-Flops

Flip-flops are the logic designer's most versatile tool and are used in many industrial circuits. To acquaint the reader with their use, some of the more common and more interesting circuits containing FFs are described in this section.

6-13-1   Registers

Registers were defined in Chapter 1 as a repository for a group of bits. Registers are usually composed of a group of FFs; that is, an N-bit register consists of N FFs. Very often no logic or arithmetic operations are performed on the bits in the register; the function of the register is simply to retain (store) the word for a period of time.

There are many uses for registers. The bits may be used to hold a number, an ASCII character, or they can form a *status register* that informs the computer of the status of conditions in the external world. As an example of a status register, consider a printer connected to a computer. One bit of the register can be used to tell the computer if the printer is "off-line"; another bit can be used to tell the computer that the printer has no paper in it. In either case, the computer should not attempt to send characters to the printer. Computers use registers extensively. The Intel **8085** 8-bit microprocessor, for example, uses six 8-bit registers called B, C, D, E, H, and L, where it can store six bytes of data. This data can be transferred to another special 8-bit register called an *accumulator* very quickly. The Intel **8088** microprocessor, which is used in the IBM PC (and compatibles), has fourteen 16-bit registers within it.

An example of a 4-bit register consisting of two **74x74** FFs is shown in Fig. 6-21. The register may be cleared, if necessary, by a single pulse on the CLEAR line. Data on the D line are clocked in on the positive edge of each clock, and replace the previous contents of the register.

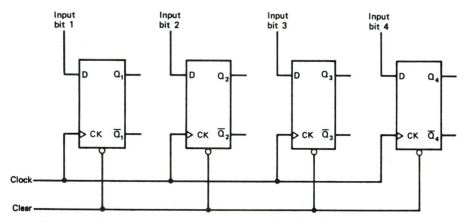

Figure 6-21 A 4-bit register composed of **74x74** FFs.

Where all stages of a register are to be cleared and loaded simultaneously, it is often wiser to build the register with **74x174** chips because they contain 6 FFs in a single package. They also save on loading. The CLOCK and CLEAR inputs of a **74x174** take only one load each, whereas it requires a drive capability of 12 loads to CLOCK or CLEAR six individual **7474** FFs (see section 6-12).

Registers built using **74x373** and **74x374** ICs are very popular because each IC can hold an entire byte. Figure 6-22 shows an example using **74x374**s. These can be the B, C, and D registers on the Intel **8085**. Each register can hold one byte of data and must be capable of sending its byte to the accumulator on command. The outputs of each register are connected together on an 8-bit bus. Normally the ENABLE inputs to each **74x374** are HIGH, so the 3-state output of each register is off.

Suppose that the microprocessor receives an instruction MOVE THE CONTENTS OF REGISTER C TO THE ACCUMULATOR which is sometimes called the A register. The timing for the execution of this instruction is shown in Fig. 6-22b. First ENABLE REG C goes low, which allows register to put its data out on the bus. While this data is on the bus, there must be a positive edge on the REG A CLOCK input, so register A can receive the data on the bus.

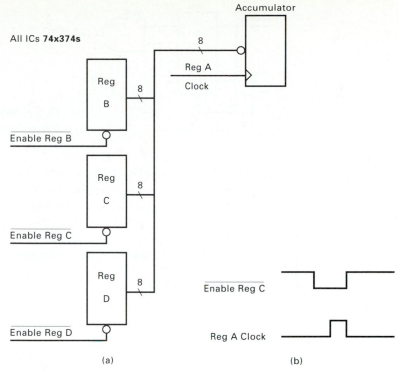

All ICs **74x374s**

Accumulator

Reg
B

Enable Reg B

Reg
C

Enable Reg C

Reg
D

Enable Reg D

Reg A
Clock

8

8

8

8

Enable Reg C

Reg A Clock

(a)

(b)

Figure 6-22 A register example: (a) circuit; (b) timing.

6-13-2 Sequence Problems

As stated at the beginning of this chapter, FFs can be used to keep track of sequences of events. The auto seat belt problem is now restated in more formal terms and solved in Example 6-8.

EXAMPLE 6-8

Given two input lines, A and B, design a circuit to produce a HIGH output only if both A and B are both HIGH and A becomes HIGH before B becomes HIGH. (For the auto problem, the A line corresponds to the event "the passenger is seated," and the B line corresponds to the event "the seat belt is buckled.")

SOLUTION The FF and gate of Fig. 6-23 determine whether the three conditions necessary for proper operation are satisfied:

1. A is HIGH (the driver is in the car).
2. B is HIGH (the seat belt is buckled).
3. A went HIGH before B went HIGH (the driver sat down before he buckled his seat belt).

If both inputs to the AND gate (**74x08**) are HIGH, the circuit is acting properly. A LOW output from the **74x08** indicates an alarm condition.

The FF sets only if A goes HIGH (the driver sits down), placing a 1 on the D input and removing the direct CLEAR before B goes HIGH (the belt is

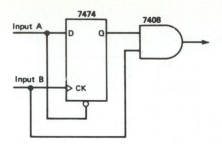

Figure 6-23 Circuit for solution of Example 6-8.

buckled). Now the positive-going edge of the B input sets the FF. This causes two HIGH inputs to appear at the AND gate, indicating proper operation. Note that we have taken the precaution of clearing the FF to cause an alarm condition whenever A goes LOW (the driver is not seated).

Unfortunately, the FF output cannot be directly connected to the alarm because we must still guard against the remaining possibility that:

1. A goes HIGH.
2. B goes HIGH setting the FF.
3. B goes LOW (this will not reset the FF).

This corresponds to the driver sitting down, buckling the seat belt, and then unbuckling it.

By connecting the B input (the belt sensor) to the AND gate, we prevent this possibility. Now the alarm will sound whenever B is LOW (the belt is unbuckled).

6-13-3   Counters

Digital counters are often needed to count pulses or events. Counters are usually composed of FFs forming a register, and the pulses to be counted are clocked into the least significant stage of the register.

A 3-bit up-counter consisting of three **74x107** FFs is shown in Fig. 6-24. The J and K inputs to all FFs are clamped HIGH so that each FF will toggle on each negative edge of its clock. The pulses to be counted are applied to the CLOCK input of the least significant stage. The *Q* output of each stage is connected to the CLOCK input of the next (more significant) stage.

To build an up-counter, which increases its count by 1 (increments) as each clock pulse occurs, the following procedure can be applied.

1. Examine the least significant stage; if it is a 0, change it to a 1 and stop.
2. If it is a 1, change it to a 0, go to the second least significant stage, and repeat the foregoing procedure.
3. Continue until a stage that contains a 0 is found. Change this stage to a 1 and stop.

The circuit and timing chart of Fig. 6-24 implement this procedure in hardware. The circuit is built by cascading toggle FFs. If any stage changes

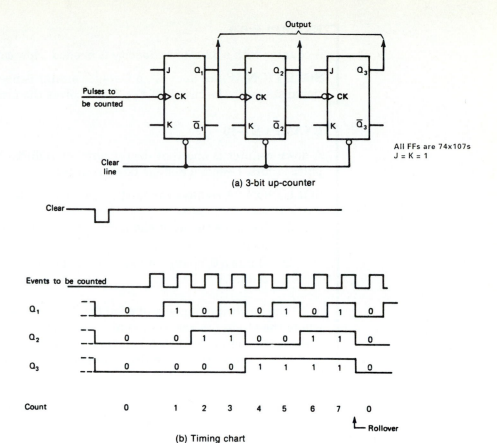

(a) 3-bit up-counter

All FFs are 74x107s
J = K = 1

(b) Timing chart

Figure 6-24 A 3-stage up-counter and timing chart.

from a 0 to a 1, the Q output goes from LOW to HIGH. This provides a positive transition that does *not* clock the next stage and, therefore, does not affect the higher order stages. However, if a stage goes from a 1 to a 0, its Q output goes LOW and toggles the following stage.

Refer to the timing chart (Fig. 6-24b); the counter output is initially set to all 0s by the clear pulse. The first clock pulse toggles Q_1 HIGH, but the LOW to HIGH transition does not affect Q_2 or Q_3. Now reading the output as $Q_3Q_2Q_1 = 001$, the output is the binary number 1 indicating that one pulse has passed. On the second pulse, Q_1 returns to its 0 state. Since Q_1 changes from a 1 to a 0, the negative-going edge clocks Q_2 into the 1 state. The LOW to HIGH transition on Q_2 does not affect Q_3, and the output now reads $Q_3Q_2Q_1 = 010$ or the binary number 2.

The counter continues to increment on successive clock pulses until the count reaches 111 (binary 7). On the next clock pulse, the counter "rolls over" to 000. A three-stage counter has a capacity of 2^3 or 8 counts before rolling over (as shown in Fig. 6-24b). If additional stages are added in the same manner, each additional stage doubles the capacity of the counter. Also, although the clocks are drawn at equal intervals, this is not a requirement; the counter counts irregularly as well as regularly spaced pulses.

EXAMPLE 6-9

A counter with a 32-count capacity is needed. How can it be built?

SOLUTION Since $32 = 2^5$, a 5-stage counter is needed. Adding two additional stages to the circuit of Fig. 6-24a solves the problem.

EXAMPLE 6-10

A down-counter is one that decrements, or reduces its count by 1, as each pulse occurs. Design a 3-stage down-counter.

SOLUTION A counter can be decremented by the following procedure:

1. Examine the least significant stage; if it is a 1, change it to a 0 and stop.
2. If it is a 0, change it to a 1, go to the next stage and repeat the procedure.

The hardware implementation for the solution is shown in Fig. 6-25a, where the $\bar{Q}$ outputs are connected to the CLOCK of the succeeding stages. The counter output is still taken from the Q outputs. As the timing chart (Fig. 6-25b) shows, the first pulse causes the output to change from 000 to 111.

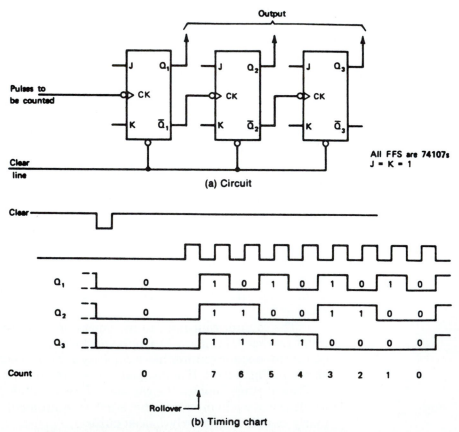

Figure 6-25 A 3-stage down-counter and timing chart.

This is the rollover situation for a down-counter.[2] Each additional pulse causes the output of the counter to be reduced by 1.

6-13-4 Counters Made From D-Type FFs

Counters can also be built from D FFs such as the **74x74**. To do so, the FFs must be made to toggle by connecting their D inputs to $\bar{Q}$. An up-counter can then be built by connecting $\bar{Q}_1$ to CLK_2, $\bar{Q}_2$ to CLK_3, and so on. Connecting the Q outputs to the succeeding CLK inputs produces a down counter (see Problems 6-16 and 6-17).

6-13-5 Presettable Counters

A counter is presettable if a number can be loaded into it. It can then start counting from that number instead of from 0.

Figure 6-26 shows two stages of a D-type up-counter which is presettable. There is a *mode switch* in the upper left corner of the figure. This sets the mode of the counter. If the switch is closed, one input to each **74x00** gate will be 0. All the outputs will be HIGH. Thus, the PRESET and CLEAR of each FF will be HIGH, or inactive, and the circuit can count.

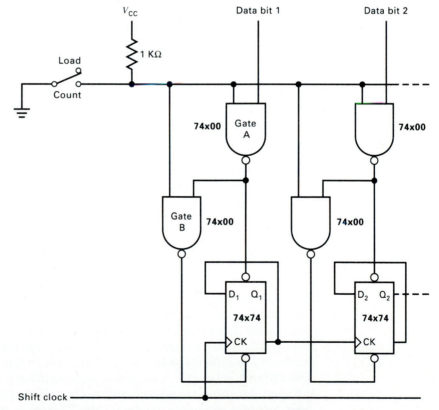

Figure 6-26 A D-type counter that is presettable.

[2]Note that the number 111 is also the number -1 in 2s complement arithmetic. We could therefore consider that the counter has counted down from 0 to -1.

If the switch is open, the circuit will be in LOAD mode. Data bit 1 will be jammed into Q_1 and data bit 2 will be jammed into Q_2. If data bit 1 is HIGH, for example, there will be a LOW on the PRESET of Q_1 and a HIGH on its CLEAR. This will force Q_1 to SET.

EXAMPLE 6-11

In Fig. 6-26, show that Q_1 will clear if data bit 1 is 0 in LOAD mode.

SOLUTION   In LOAD mode the LOAD input to gate A will be high, but the data input will be 0. Therefore its output, and the direct set of Q_1, will be HIGH. Now, however, gate B will receive two HIGH inputs and produce a 0 output. This 0 will appear on the CLEAR input of Q_1 and clear the FF.

Figure 6-26 shows a counter built of **74x74** type FFs. If the counter is to be built of J-K FFs, a **74x76** is preferable to a **74x107** because the **74x76** has both PRESET and CLEAR inputs.

6-13-6   Synchronizing Flip-Flops

The operation of many complex digital circuits is controlled by a fixed frequency *clock,* often derived from a crystal oscillator. These are *synchronous circuits* because most or all *state changes are synchronized with the internal clock.* A problem arises when these circuits must also respond to external events, such as signal changes on a switch or other input line, and these events *cannot* be synchronized with the external clock.

There are two ways of handling these events.

1. **Asynchronous.** The external event or transition is used to direct SET or CLEAR a FF when it occurs.
2. **Synchronous.** The external event is used to produce a pulse that is synchronized with the internal clock.

The asynchronous method is slightly simpler to implement in hardware, but dangerous, as the author unfortunately learned when he was in industry. In a complex system there are many state changes on each clock edge, and some ICs respond faster than others. If an external event is allowed to affect a system whenever it occurs, a small percentage of the time it will occur near a clock transition, after some ICs have responded, but before others have. This intermingling of transitions often causes erroneous operation. What is worse, these problems are difficult to test for and difficult to foresee because of the random time difference between the events. In the author's case, the problems became apparent only after the equipment was installed at the customer's site, and fixing it meant a 1500-mile airplane ride. Therefore, it is usually wise to prevent transition from occurring at random times.

Transitions can be synchronized with the internal clock by the use of synchronizing FFs, as Example 6-12 shows.

EXAMPLE 6-12

A debounced switch[3] is connected to a circuit driven by a clock. Typically the clock is much faster than the switch, which is operated by humans, so the switch is up or down for many clock cycles. Design a circuit so that a switch thrown at any time within cycle 1 will produce a single output pulse during cycle 2, as shown in Fig. 6-27. Notice that the output pulse is synchronized with the clock as required.

SOLUTION   This type of circuit usually requires two FFs. The first FF is SET asynchronously by the switch. The second FF can then be set by the clock provided that the first FF is set. The circuit is shown in Fig. 6-28.

The switch sets FF1 and FF2 sets on the next positive clock edge. When FF2 sets, it clears FF1 and FF2 then clears on the following positive clock edge.

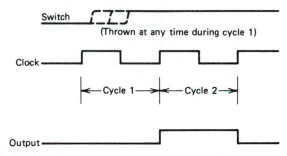

Figure 6-27   Synchronizing an output with a clock.

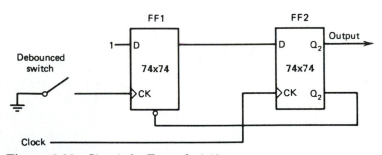

Figure 6-28   Circuit for Example 6-12.

6-13-7   Glitches

Glitches are short, sharp spikes that occur on signal lines. If they are unexpected, they can be very troublesome, especially if they cause FFs to CLEAR or SET. On the other hand, *controlled glitches* can be useful in certain circuits.

A controlled glitch is one that has been *deliberately* designed into a circuit to produce certain events. Controlled glitches are designed in the following manner.

1.   Event 1 initiates a glitch.
2.   The glitch causes event 2.
3.   The occurrence of event 2 terminates the glitch.

[3]Switch debouncing is discussed in section 7-8.

Here the glitch must be very short and sharp, since the event caused by the glitch terminates it.

EXAMPLE 6-13

Given two input lines, A and B, design a circuit to the following specifications:
(a) The output goes to 1 on every positive transition of the A line.
(b) The output goes to 0 on every positive transition of the B line.

The output must be capable of being SET by another A line transition immediately after being CLEARED, regardless of the level of the B line.

SOLUTION   A controlled glitch should occur whenever the B line makes a positive transition. A circuit to solve this problem is shown in Fig. 6-29. Its operation is as follows.

1. Q_1 is SET by every positive transition of line A.
2. If Q_1 is SET, FF2 is enabled and SETS on a positive transition of line B.
3. This causes FF1 to CLEAR, which now clears FF2 via its CLEAR input.
4. Thus FF2 is only SET for the time it takes itself to SET, plus the time it takes for FF1 to CLEAR. Since this is about two propagation delay times, FF2 creates a glitch of perhaps 40 ns. Here event 1 is the positive transition of the B line.
5. This causes the glitch, which is the setting of FF2.
6. When FF2 is SET, it causes event 2, the clearing of FF1, and when FF1 CLEARS, it CLEARS FF2 terminating the glitch.

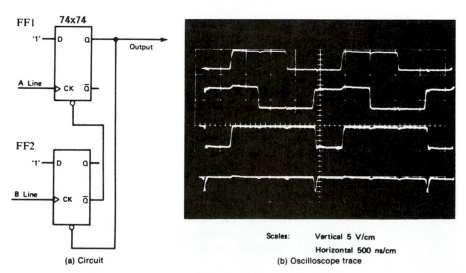

(a) Circuit

Scales: Vertical 5 V/cm
          Horizontal 500 ns/cm

(b) Oscilloscope trace

Figure 6-29   Solution of Example 6-13.

This circuit (Fig. 6-29) was set up in the laboratory, and the important CRO wave forms are shown in Fig. 6-29b. The top trace is the A line and the second trace is the B line. The output is shown on the third trace and the glitch is seen on the bottom trace. Clearly, the positive edge of A causes the output to rise, and the positive edge of B causes the glitch, which causes the output to fall, as noted in the specification of Example 6-13.

6-13-8 Serial Data Transformation

As a final, more difficult problem, consider the two forms of data transmission shown in Fig. 6-30. These are the NRZ format and the Bi-Phase format. The problem is to design a circuit to convert NRZ data to bi-phase. This problem requires some thought, but it can be solved using the ICs we have considered thus far.

The first step is to clearly determine the nature of NRZ and bi-phase data. The output clock can be divided into *bit cells,* as shown by the dashed lines in Fig. 6-30. The bit cells are between the negative edges of the clock and each cell

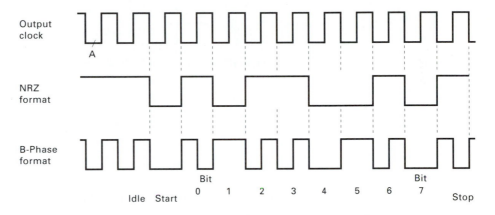

Data 01001101 (S4D)

Figure 6-30 NRZ and Bi-Phase formats.

contains one data bit. The NRZ data assumes a constant level during each bit cell. The bi-phase data is more complex. It makes a transition on each negative edge of the clock. If the bit is a 1 it will also make a transition on the positive edge of the clock, or in the middle of the bit cell. If the bit is a 0, there will be no transition at this time. Note that both clock and data transitions can be either positive or negative.

After a little thought, the foregoing suggests that there ought to be one FF that toggles on every negative transition of the clock, and one FF that toggles on the positive transition, providing the input data is positive. This suggests putting the clock line through an inverter so that negative edges appear at both FFs at opposite transitions of the clock.

The first FF is strictly a toggle FF. The design of the second FF is also simple if J-K FFs are used. The J and K inputs can be tied to the NRZ data. If it is a 1, the second FF will toggle in the middle of the bit cell; if the NRZ data is a 0 it will not. Finally, the two FFs must be connected so that there is a change in the bi-phase output whenever either one of them changes. This suggests an XOR gate, especially since both FFs cannot change at the same time. The final circuit is shown in Fig. 6-31.

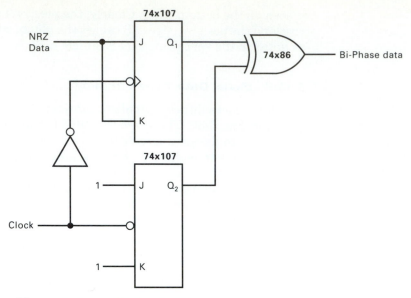

Figure 6-31 An NRZ-to-Bi-Phase converter.

SUMMARY

The basic FFs available in TTL were introduced in this chapter, and some examples of circuits using these FFs were presented. Other FFs exist (consult the manufacturers' catalogs for a complete list of available FFs) and may be useful in special cases, but the vast majority of circuits are designed using D and J-K FFs. NAND and NOR gate FFs are often used where only SET-RESET FFs are required, but D and J-K FFs are generally preferred; they are more versatile because they can be CLEARED or SET directly and information can also be clocked into them.

There is no formula we can give to enable one to design a circuit that will satisfy a given specification. To be successful, the engineer must first be very familiar with the characteristics of the available FFs and gates; then a review of the design examples presented in this chapter will (hopefully) lead to ideas necessary for the design.

GLOSSARY

Active signals: Signal levels that cause a FF to change state.

Clock: A continuous square wave of a constant frequency.

Down-counter: A counter that decrements on every pulse.

Glitch: A short, sharp pulse on a signal line.

Quiescent level: The signal level that does not cause a FF to change state.

RESET: The state of a FF when $Q = 0$ and $\bar{Q} = 1$.

SET: The state of a FF when $Q = 1$ and $\bar{Q} = 0$.

Toggling: Causing a FF to change or reverse its state.

Up-counter: A counter that increments on every pulse.

PROBLEMS

Section 6-5.

6-1. How would the circuit of Fig. 6-2 operate if the SET input (pin 3) of the NOR gate FF became disconnected?

6-2. What levels would the outputs of the NOR gate FF assume if both the SET and RESET inputs are HIGH?

Section 6-6.

6-3. How would the circuit of Fig. 6-4 operate if the SET input (pin 1) of the NAND gate FF became disconnected?

6-4. Design a 3-state FF using **7425** ICs.

6-5. Design a 3-state FF using 2-input open collector ICs. The design should use **7403** chips. The input should come in through **7405** gates.

Section 6-7.

6-6. Design a latch FF using only two input logic gates (**7400s, 7402s, 7408,** etc.) and inverters.

6-7. Solve the problem of Ex. 6-3 using only a **7474** and a **7408** chip.

Section 6-8.

6-8. On a logic laboratory training board a variable is provided from a switch input as shown in Fig. P6-8. A **74107** and a source of clock pulses are also available. To obtain the variable and its complement the rest of the circuit of Fig. P6-13 was constructed. Explain how it works and which outputs contain the variable and its complement.

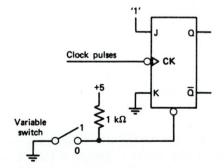

Figure P6-8

Section 6-9.

6-9. For Fig. P6-9, sketch the output if the input is:
   (a) The D input to a **7474**.
   (b) The input to a latch.

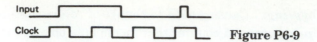

Figure P6-9

(c) The J input to a master-slave FF **74107** and the K input is tied HIGH.

(d) The J input to a negative edge-triggered FF and the K input is tied HIGH.

(e) The J input to a J-K master-slave FF with data lockout and the K input is tied HIGH.

6-10. Explain the operation of the circuit of Fig. P6-10 as it is clocked. Draw a timing chart for Q_1 and Q_2 as the clocks proceed.

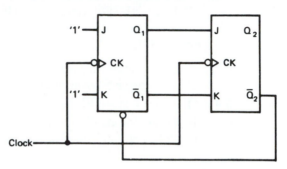

Figure P6-10

Section 6-12.

6-11. Considering HIGH and LOW clock widths only, find the maximum frequency at which you can toggle a **74107**. Use the TTL Data Book.

Section 6-13-1.

6-12. Draw a 6-stage register made up of **74174**s. What must you add if the $\bar{Q}$ outputs are also required?

6-13. Design a 16-bit register:
(a) Using **74x74**s.
(b) Using **74174**s.
Assume the clocks and clears are originally driven by **7400** NAND gates.

Section 6-13-2.

6-14. Design a circuit to produce a 1 output if $ABC = 1$ and only if A precedes B, and B precedes C.

Section 6-13-3.

6-15. Using two FFs, Q_1 and Q_2, which form a 2-stage counter whose outputs are shown in Fig. P6-15a, design a circuit whose output is HIGH for counts 1 and 2 and LOW for counts 0 and 3.
(a) Use a third FF. Assume the $\bar{Q}$ outputs are also available.
(b) Use another gate but no additional FFs.
Note: The solution of Problem 6-15b was used to generate the waveforms shown in Fig. P6-15b. Observe and explain the notches in the fourth trace.

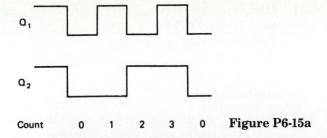

Count   0   1   2   3   0   **Figure P6-15a**

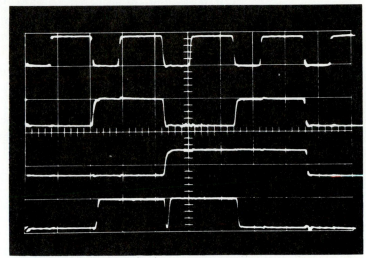

Figure P6-15b   Waveforms for Problems 6-19. Scales: Vertical 5 V/cm; horizontal 500 ns/cm.

Section 6-13-4.

6-16. Design a 3-stage up-counter using **74x74**s.

6-17. Design a 3-stage down-counter using **74x74**s.

6-18. Design a 3-stage up-down counter. An up-down line is provided as well as a clock line. If the up-down line is HIGH, the counter is to increment on each clock pulse, and if the up-down line is LOW, the counter is to decrement.

Section 6-13-5.

6-19. Design a 4-bit register with the following inputs:
    (a)   LOAD 1—on the leading edge of the LOAD 1 pulse four bits of input data are loaded into the FF.
    (b)   CLEAR—whenever the CLEAR is LOW the register is cleared.
    (c)   LOAD 2—whenever the LOAD 2 input is LOW the register must contain a binary 13.
    Design it on the assumption that the CLEAR and LOAD 2 inputs cannot occur simultaneously.

6-20. Repeat Problem 6-19 with the additional specification that the CLEAR and LOAD 2 pulse may occur simultaneously, and when they do, the LOAD 2 pulse predominates.

6-21. Design a 5-stage counter that can have a binary 13 loaded into it:
  (a) Using **74x74**s.
  (b) Repeat using **74107**s.

Section 6-13-6.

6-22. Given a constant clock and an input line, design a circuit so that if there is a LOW on the input line at any time during cycle 1, the output will be HIGH throughout cycle 2. Note that the output can be HIGH for many consecutive cycles if the input is constantly LOW. This circuit can be considered a *glitch-latch* since it transforms a glitch into a pulse whose duration is one clock cycle.

6-23. The circuit of Fig. P6-23a was designed to produce a single pulse on the cycle following the switch throw only when the clock is HIGH. The output waveform is shown in Fig. P6-23b.
  (a) Explain why the trailing glitch occurred.
  (b) Redesign the circuit to eliminate the glitch.

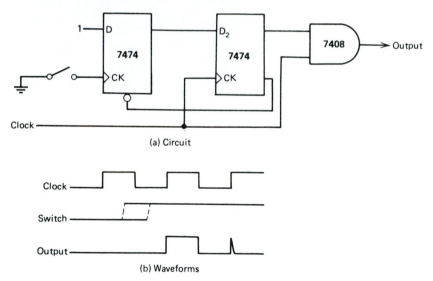

Figure P6-23 (a) Circuit. (b) Waveforms.

6-24. Repeat Problem 6-22 if the output is to be HIGH only if the input was LOW during the previous cycle and the clock is HIGH.

6-25. A student designed the circuit of Fig. P6-25 to solve the single pulse problem. Like the student, the circuit worked half the time. Explain what happened by drawing timing charts to show how the circuit operates:
  (a) If the switch is opened when the clock is LOW.
  (b) If the switch is opened when the clock is HIGH.

6-26. The input to the circuit of Fig. P6-26 is a 1 MHz clock. Sketch Q_1 and Q_2. Assume they start at 0.

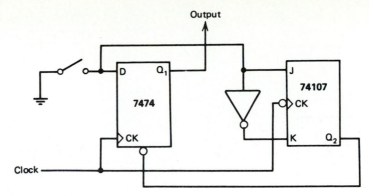

Figure P6-25

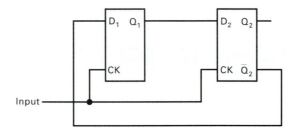

Figure P6-26

Section 6-13-8.

6-27. Design a bi-phase to NRZ converter.

6-28. The circuit of Fig. P6-28 is an NRZ to bi-phase converter using D FFs. Plot the levels at points B, C, D, E, and F for the NRZ input of Fig. P6-28. Is the output bi-phase?

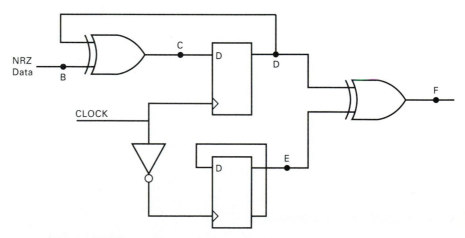

Figure P6-28

After attempting the problems, return to Sec. 6-2 and be sure you can answer the questions. If any of them still seem difficult, review the appropriate sections of the chapter to find the answers.

7

One-Shots
and Oscillators

7-1 Instructional Objectives

This chapter introduces the student to the uses of *one-shots* (monostable multi-vibrators) in digital circuits. It also covers the construction of clock generators (oscillators) and methods of switch debouncing. After reading it, the student should be able to:

1. Use a one-shot to produce a pulse of predetermined duration.
2. Design triggering circuits for one-shots.
3. Use one-shots to control the timing or sequence of events of a circuit.
4. Design one-shot, **555**, and Schmitt-trigger oscillators to run at a pre-determined frequency.
5. Debounce a switch.
6. Design a one-shot discriminator.

7-2 Self-Evaluation Questions

Watch for the answers to the following questions as you read the chapter. They should help you understand the material presented.

1. What is the function of the resistor and capacitor connected to a one-shot?

2. Why is the maximum duty cycle a limitation on the maximum frequency of triggers to a one-shot?

3. Explain the conditions necessary to trigger a **74x221**.

4. What are the differences between a *retriggerable* one-shot and an ordinary one-shot?

5. Explain how to design one-shot timing chains.

6. What is switch bounce? What problems does it cause? How can switch bounce problems be eliminated?

7-3   Introduction to One-Shots

The purpose of a one-shot (also called a monostable multivibrator) is to produce an output pulse having a duration or pulse length determined by the designer. One-shots are used to set timing and control the sequence of events in a digital system.

A one-shot is essentially a FF with only one stable state. Like a FF, it has two outputs, Q and $\bar{Q}$. Normally it is in the CLEAR state ($Q = 0$ and $\bar{Q} = 1$). When a trigger (a pulse edge for **74x00** ICs) is applied, the one-shot turns on or flips to the 1 or SET state. It remains there for a time period determined by a resistor and capacitor that are used to set the duration of the pulse, and then switches back to the 0 state.

The action of a one-shot is shown in Fig. 7-1, where the incoming trigger is a positive edge on the waveform of Fig. 7-1a and the response of the one-shot is shown in Fig. 7-1b. The time period of the one-shot has been set to 1 ms, so a trigger will cause it to go from the 0 state to the 1 state, remain there for 1 ms, and then flip back to the 0 state.

The 1 state of a one-shot is called *quasistable* because it must return to the 0 state when the output pulse time expires. *If a one-shot is continually in the 1 state, it indicates either the IC or the circuit is faulty.*

There are two types of one-shots: *non-retriggerable* and *retriggerable*. For either type, a trigger initiates the one-shot action. If additional triggers are applied to a non-retriggerable one-shot when it is in the 1 state, they are ineffec-

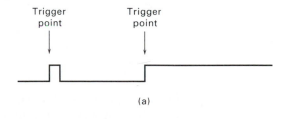

(a)

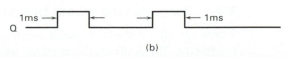

(b)

Figure 7-1   Response of a one-shot to a trigger: (a) trigger (positive-edge); (b) one-shot output.

tive; the one-shot will return to the stable state when its time expires. Retriggerable one-shots restart their timing whenever a trigger is applied. Retriggerable one-shots are discussed in Section 7-5.

7-4   The 74x221 One-Shot

The most popular and commonly used *non-retriggerable* one-shot in the TTL series is the **74x221**. It comes in the standard and LS series. It contains two one-shots in a single 16-pin package. It functions the same way as the older **74121**, that was a single one-shot in a 14-pin package.

The NAND gate (shown as a NOR gate with negative inputs) and the Schmitt-trigger AND gate shown in Fig. 7-2. are part of the one-shot, and are *not* external gates. In order to trigger the one-shot, which causes it to produce an output pulse, there must be a *rising pulse edge* at point Z (Fig. 7-2) that is inside the **74x221** and *cannot be examined by the engineer.*

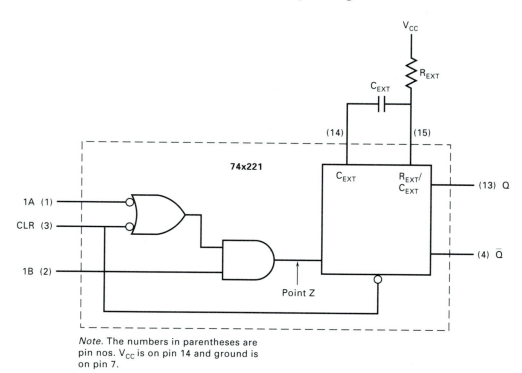

Note. The numbers in parentheses are pin nos. V_{CC} is on pin 14 and ground is on pin 7.

Figure 7-2   This pertains to Fig. 2.2 (Reprinted by Permission of Texas Instruments.) (a) Pinout (b) Function Table (c) Circuit Diagram

7-4-1   Triggering the 74x221 One-Shot

There are three inputs to each section of the **74x221** that are concerned with triggering: the A input, the B input, and the CLR (Clear) input. Whenever CLR is LOW, it will clear the one-shot. The one-shot cannot be triggered when CLR is LOW. The one-shot can be triggered under the following conditions:

- **Triggering from the A input**—If both B and CLR are high, a *negative* edge at A will cause a positive edge at point Z of Fig. 7-2c and a trigger.
- **Triggering from the B input**—If CLR is HIGH and A is LOW, a *positive* edge on the B input will cause the one-shot to trigger.
- **Triggering from the CLR input**—If CLR goes low, it will clear the one-shot. If B is HIGH and A is LOW when the CLR goes back to its high state, the positive edge on the CLR input will cause a trigger.

The triggering is also shown in the three lower lines of the function table for the **74x221** (Fig. 7-2b).

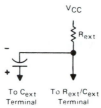

Figure 7-3 Timing component connections for the **74x221**.

7-4-2 Timing

The timing of the one-shot, or the duration of its output pulse, is determined by the timing resistor and capacitor[1] connected to it. The timing resistor and capacitor must be connected between the C_{EXT} and R_{EXT}/C_{EXT} pins as shown in Fig. 7-3. For long pulses, large value capacitors are required. If electrolytic capacitors are used, the positive terminal must be connected to the C_{EXT} input.

The duration of the output pulse is given by

$$t_W = 0.7 C_T R_T \qquad (7\text{-}1)$$

where C_T and R_T are the values of timing capacitor and resistor, respectively. Manufacturers also supply curves giving the output pulse width as a function of the timing resistors and capacitors, but for the **74x221**, Eq. (7-1) is simpler and easier to use.

Because of variations between ICs and the tolerances on commercial resistors and capacitors, measured pulse durations may differ by as much as 20 percent from the values calculated by the above equations. We recommend that the circuit be breadboarded by using the values obtained from the equations (or the closest standard values of resistors and capacitors). Then, using the oscilloscope, apply whatever tweaking (adjusting) is necessary to precisely time the pulses.

The IC manufacturers place limits on both the external resistor and external capacitor that can be used with a one-shot. These limits are listed in the specifications. For the **74x221** the *minimum* output pulse width is obtained using no external capacitor (there is always some stray capacity between pins 14 and 15) and a 1.5 kΩ resistor. Typically, this results in an output pulse width of

---

[1]Most commercially available IC mounting panels make provision for mounting some discrete components. These external components consist of pull-up resistors and decoupling capacitors, as well as timing resistors and capacitors.

30 to 35 ns. The maximum allowable value for an external capacitor is 1000 μF (see the manufacturer's specifications). The longest pulse width results from using both the maximum external capacitor and resistor:

$$t_W(\text{MAX}) = 0.7 \times (1000 \times 10^{-6}) \times (100 \times 10^3) \qquad (7\text{-}2)$$
$$= \textbf{70 seconds}$$

By the proper choice of external resistors and capacitors, a pulse width anywhere between 30 ns and 70 seconds can be obtained. Longer pulse widths can be generated by setting up an oscillator followed by a divide-by-N circuit (see Problem 7-17).

EXAMPLE 7-1

Use a **74x221** to produce an output pulse of 1 μs. Find the required capacitance using a timing resistor of 10 kΩ.

SOLUTION   Using an external resistor of 10 kΩ, we have

$$C = \frac{T}{0.7R} = \frac{10^{-6}}{7000} = \textbf{143 pF}$$

7-4-3   Duty Cycle Limitations

For a repetitive waveform, *duty cycle* is defined by Eq. (7-1) as the ratio of the ON time to the total time (ON time plus OFF time), as shown in Fig. 7-4.

$$\text{Percent duty cycle} = \frac{T_{\text{ON}}}{T_{\text{ON}} + T_{\text{OFF}}} \times 100 \qquad (7\text{-}3)$$

After the pulse of a non-retriggerable one-shot has expired, it must remain OFF for a period of time to *recover*. The ratio of the pulse time to the time the one-shot must remain OFF is the maximum duty cycle of the one-shot. If this duty cycle is exceeded, its output pulse will *jitter*. This means the width of each pulse will not be constant.

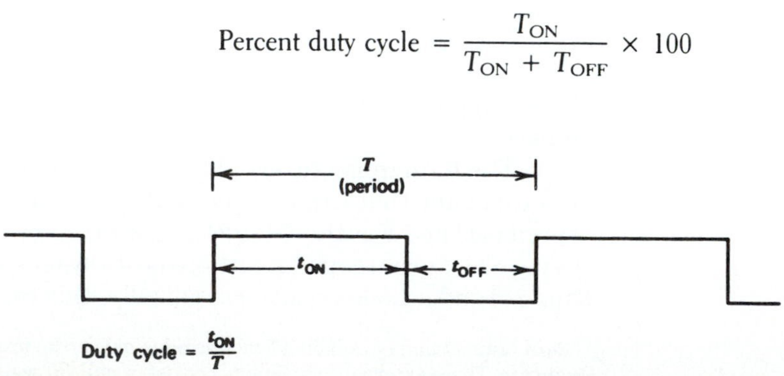

$$\text{Percent duty cycle} = \frac{T_{\text{ON}}}{T_{\text{ON}} + T_{\text{OFF}}} \times 100$$

Duty cycle = $\frac{t_{\text{ON}}}{T}$

Figure 7-4   Duty cycle of a repetitive "square" wave.

EXAMPLE 7-2

A one-shot is designed to produce a 100 μs pulse and is specified to have a maximum duty cycle of 75 percent. Calculate the minimum time that should be allowed between triggers.

SOLUTION Since the maximum duty cycle is 75 percent, the minimum time between triggers can be found from Eq. (7-3).

$$T_{\text{OFF}} + T_{\text{ON}} = \frac{T_{\text{ON}} \times 100}{\text{percent duty cycle}} = \frac{100 \times 100}{75} = 133.3 \text{ μs}$$

Example 7-2 shows the one-shot has 33.3 μs between pulses to recover. More frequent triggers could cause the output pulse to jitter; its duration would not be 100 μs consistently.

7-4-4 One-Shot Operating Conditions

Table 7-1 lists the operating conditions for a **54LS221** and **74LS221** one-shot. The table was taken from the Texas Instruments TTL Data Book, and similar tables exist for other TTL one-shots. Those lines in the table that pertain particularly to one-shots will be discussed in this section.

Line 6 of the table refers to the *rise time* of the trigger pulse. The B input to the **74x221** is a Schmitt-trigger input (note the hysteresis symbol on the gate in Fig. 7-2c) that responds to very slowly changing inputs. The A inputs are normal TTL inputs and should change faster than 1 V/μs, but the B inputs can respond to pulse edges as slow as 1 V/s. The A inputs will respond to TTL or CMOS outputs, but if slower wave forms are required to trigger a one-shot, they must be applied to the B input.

Lines 9 and 10 give the limitations of the timing resistor and capacitor. For the **74LS221** the capacitor values can range between 0 and 1000 μF, but the resistor must be between 1.4 kΩ and 100 kΩ. Other one shots have different limitations on these parameters.

Line 11 considers the duty cycle of the one-shot. The recovery time required for one-shots imposes a duty cycle limitation on them, as explained in section 7-3. The duty cycle depends on the value of the timing resistor. For the

TABLE 7-1   Operating conditions for the 74LS221

		SN54LS221			SN74LS221			UNIT
		MIN	NOM	MAX	MIN	NOM	MAX	
Supply voltage, V_{CC}		4.5	5	5.5	4.75	5	5.25	V
High-level input voltage at A input, V_{IH}		2			2			V
Low-level input voltage at B input, V_{IL}				0.7			0.8	V
High-level output current, I_{OH}				−400			−400	μA
Low-level output current, I_{OL}				4			8	mA
Rate of rise or fall of input pulse, dv/dt	Schmitt, B	1			1			V/s
	Logic input, A	1			1			V/μs
Input pulse width	A or B, $t_{w(in)}$	50			50			ns
	Clear, $t_{w(clear)}$	40			40			
Clear-inactive-state setup time, t_{su}		15			15			ns
External timing resistance, R_{ext}		1.4		70	1.4		100	kΩ
External timing capacitance, C_{ext}		0		1000	0		1000	μF
Output duty cycle	$R_T = 2$ kΩ			50			50	%
	$R_T = $ MAX R_{ext}			90			90	
Operating free-air temperature, T_A		−55		125	0		70	°C

(Reprinted by permission of Texas Instruments.)

74x221 the maximum allowable duty cycle is 50 percent for a 2-kΩ timing resistor, but rises to 90 percent if an external timing resistor of 100 kΩ is used (see Problem 7-2).

If the maximum duty cycle is exceeded, the one-shot still triggers, but its pulse duration is no longer stable. The variation of pulse length is called *jitter*. Generally jitter is undesirable because one-shot output pulses are usually required to have a specific duration.

7-4-5 The 74121 One-Shot

The **74121** was the original one-shot in the **7400** standard series. It has two A inputs and no CLR input. There is only one one-shot in a **74121** IC, so most users prefer the dual **74x221**. The **74121**, however, has the advantage of having an internal 2 kΩ timing resistor. If this value is satisfactory, the user does not have to provide an external timing resistor. Figure 7-5 shows a **74121** connected to produce a 1 μs pulse, using both the internal 2 kΩ and an external 10 kΩ resistor. The operating conditions for the **74121** are somewhat different than the **74LS221**, so the user should consult the data book.

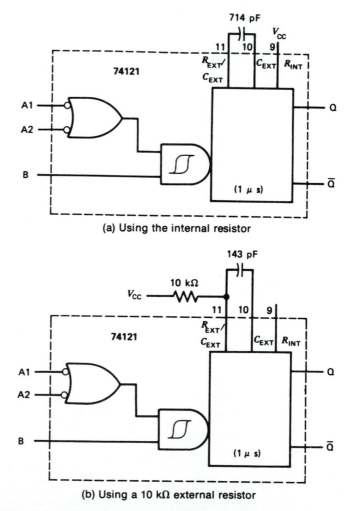

(a) Using the internal resistor

(b) Using a 10 kΩ external resistor

Figure 7-5 **74121**s designed to produce a 1-μs pulse.

7-4-6   Logic Probes

A logic probe was shown in Fig. 2-15. The basic circuit for a logic probe uses a one-shot and is shown in Fig. 7-6. It operates as follows:

a) When the voltage on the probe is a logic 0, the output of the first **74x04** is HIGH and drives current into the **LOGIC 0** Light-Emitting Diode (LED) and turns it on.

b) When the probe voltage is a logic 1, the output of the first **74x04** is LOW, but the output of the second **74x04** is HIGH and the **LOGIC 1** LED turns on.

c) Whenever there are transitions on the line, the one-shot is triggered. In response to each trigger, it produces a 100 ms. pulse, which lights the **PULSE** LED. Thus the engineer can determine if voltage transitions are occurring at point under test. This is similar to the trigger light on an oscilloscope, which illuminates whenever a trigger pulse occurs.

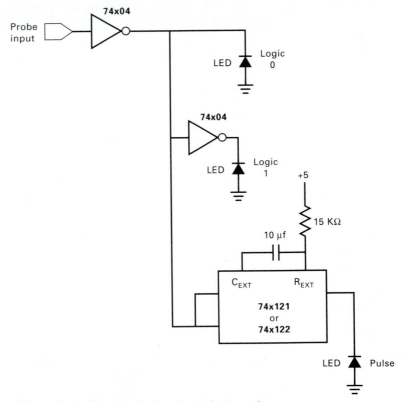

Figure 7-6   The circuit for a basic logic probe.

7-5   Retriggerable One-Shots

A *retriggerable* one-shot responds to a trigger when it is ON as well as when it is quiescent. If a trigger occurs when the one-shot is ON, it resets the timing. The one-shot does not turn OFF until one pulse duration after the last trigger. Retriggerable one-shots have unlimited duty cycles: they can be ON continuously if triggered by a pulse train whose frequency is shorter than the pulse time of the one-shot.

Retriggerable one-shots also have a CLEAR input. The presence of a LOW signal on the CLEAR input immediately *resets* the one-shot.

EXAMPLE 7-3

A retriggerable one-shot is timed to produce a 10-μs pulse. Find its output if triggers occur at times of 1, 7, 20, and 23 μs:
    (a)   If no CLEAR pulses are applied.
    (b)   If CLEAR pulses are applied at times of 12, 22, and 32 μs. Assume the trigger and CLEAR pulses are of very short duration.

SOLUTION   The solution is shown in Fig. 7-7.
    (a)   Without CLEAR pulses the one-shot is ON from $t = 1$ until $t = 17$ μs (10 μs after the trigger at $t = 7$) and from $t = 20$ to $t = 33$ μs, as shown in Fig. 7-7a.
    (b)   The output is SET by the triggers as in part A, but it is RESET whenever a CLEAR pulse occurs, as Fig. 7-7b shows.

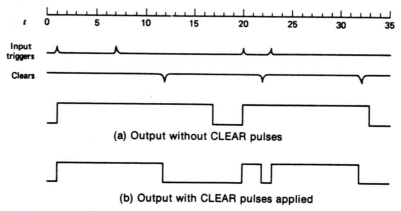

Figure 7-7   Response of a 10-μs retriggerable one-shot to a series of triggers and clears.

7-5-1   The 74LS123

There are basically four retriggerable one-shots available in the TTL families. Two are in the standard series, the **74122** and **74123**, and two are in the LS series, the **74LS122** and **74LS123**.[2] Of these, the **74LS123** is probably the most popular and we will concentrate on it.

The **74LS123** is a *dual* retriggerable one-shot in a 16-pin DIP package. The circuit and function table are shown in Fig. 7-8. The A, B, and CLR inputs function the same way as on the **74x221**. The function table (Fig. 7-8b) indicates that the **74LS123** triggers in the same manner as a **74121** or **74LS221**, by a positive-going edge at the output of the internal AND gate.

The timing components for the **74LS123** are shown in Fig. 7-9. The C_{EXT} pin is connected to ground internally, but an external ground connected to the

[2]As of this writing, one-shots are not produced in the advanced TTL or CMOS series.

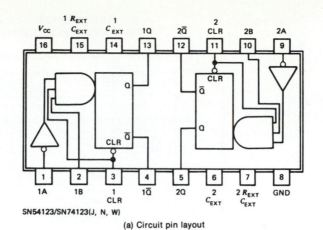

(a) Circuit pin layout

SN54123/SN74123(J, N, W)

FUNCTION TABLE

INPUTS			OUTPUTS	
CLEAR	A	B	Q	$\overline{Q}$
L	X	X	L	H
X	H	X	L	H
X	X	L	L	H
H	L	↑	⊓	⊔
H	↓	H	⊓	⊔
↑	L	H	⊓	⊔

(b) Function table

Figure 7-8 The **74123** and **74LS123** dual retriggerable one-shot. (From *The TTL Data Book for Design Engineers*, 2nd ed., Texas Instruments, Inc. Courtesy of Texas Instruments, copyright 1976.)

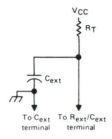

Figure 7-9 Timing component connections. (Reprinted by Permission of Texas Instruments.)

pin is also recommended. The positive side of an electrolytic capacitor should be connected to the R_{EXT}/C_{EXT} pin.

The timing of the **74LS123** is given by the charts of Fig. 7-10 for short pulses and by Eq. (7-4) for longer pulses.

$$t_W = K \cdot R_T \cdot C_{EXT} \tag{7-4}$$

Warning: Both the equation and the chart do not use standard units. R_T is in kΩ, C_{EXT} is in pF, and t_W is in ns.

To calculate the pulse time of the one-shot, we recommend:

a) If the time of the pulse or the value of the capacitor and resistor are short enough to be on the chart, Fig. 7-10a, use the chart. C_{EXT} must be less than 1000 pF or .001 μF. The chart also implies that R_T must be between 5 kΩ and 260 kΩ.

b) For longer pulse widths or higher capacitors, use Eq. (7-4). The value of K in the equation depends on the value of C_{EXT} and is given by Fig. 7-10b. If C_{EXT} is greater than 1 μF, the value of K can be taken to be 0.33.

(a)

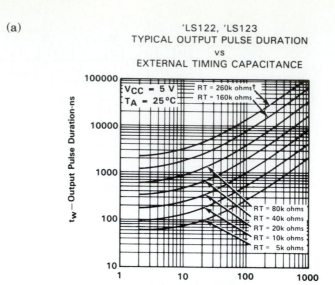

'LS122, 'LS123
TYPICAL OUTPUT PULSE DURATION
vs
EXTERNAL TIMING CAPACITANCE

†This value of resistance exceeds the maximum recommended for use
over the full temperature range of the SN54LS circuits.

(b)

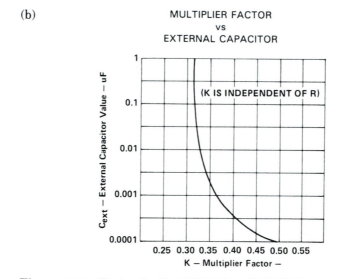

MULTIPLIER FACTOR
vs
EXTERNAL CAPACITOR

Figure 7-10 Timing for the **74LS122** and **74LS123**: (a) timing chart;
(b) multiplier factor. (Reprinted by Permission of Texas Instruments.)

EXAMPLE 7-4

Find the pulse width for a **74LS123** if R_T is 10 kΩ and C_{EXT} is

   a)   10 pF
   b)   1000 pF.
   c)   10 μF.

SOLUTION
   a)   This value is on the chart, Fig. 7-10a, at the intersection of the 10 kΩ
        line and the 10 pF line. The time is 120 ns.

b) 1000 pF is on the edge of the chart. From Fig. 7-10a, we read about 4000 ns or 4 μs. This value can also be calculated from Eq. (7-3). Figure 7-10b shows that the value of K when $C = 1000$ pF, or .001 μF, is 0.37. Eq. (7-4) becomes:

$$t_W = 0.37 \cdot 1000 \text{ pF} \cdot 10 = 3700 \text{ ns}$$

The two values are in reasonable agreement.

c) For 10 μF, Eq. (7-4) is simply:

$$t_W = 0.33 \cdot 10^7 \text{ pF} \cdot 10 = 33 \cdot 10^6 \text{ ns or 33 ms}$$

EXAMPLE 7-5

A system obtains its basic power from the 120V, 60-Hz commercial power lines. Design a circuit to monitor these lines. If the commercial ac power ever misses a cycle, a buzzer is to sound continuously until a reset push button is manually depressed. This circuit could be used to warn a computer or its operator that an emergency exists, that power may be lost. It is similar to the circuits used for heart pacemakers.

SOLUTION The design is shown in Fig. 7-11 and proceeds as follows:

1. The circuit to reduce the 120 V power to TTL levels and speeds has already been found in Example 5-2 and shown in Fig. 5-4. The output of the Schmitt trigger is a 60-Hz square wave, which means it produces pulse edges every 16.7 ms.

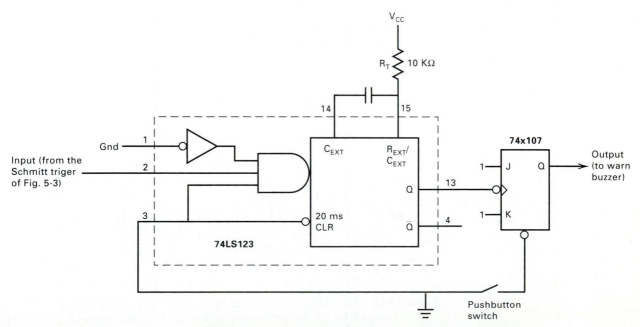

Figure 7-11 Circuit for commercial power dropout detection.

2. If these pulse edges are fed to a **74LS123** retriggerable one-shot, whose time is set slightly longer than 16.7 ms, say 20 ms, the one-shot never is reset (its 20-ms output pulse never expires) unless a cycle is missed.

3. If we assume an R_T of 10 kΩ, the capacitor required is given by Eq. (7-4).

$$C = \frac{20 \times 10^6 \text{ ns}}{.33 \times 10 \text{ k}\Omega} = 6 \times 10^6 \text{ pF} = 6 \text{ } \mu\text{F}$$

To use standard capacitors, a 6 μF can be made up of a 5 μF in parallel with a 1 μF capacitor. Note that the pulse time of the one-shot is written within the one-shot. This makes it much easier for an engineer to understand the circuit.

4. When the **74LS123** output goes LOW, it sets the **74x107**. This drives the alarm buzzer. If its voltage and current requirements are high, the alarm buzzer may be coupled to the **74x107** through a buffer/driver. If an ac cycle is missed, the **74LS123** output goes LOW, setting the **74x107**.

5. When the push button is depressed, it clears the FF and also puts a low voltage on the CLR input of the **74LS123**.

6. When the push button is released, it triggers the **74LS123** and the circuit resumes normal operation. Note that it will sound the alarm again in 20 ms if the ac power is *not* working properly.

7-5-2   The Other Retriggerable One-Shots

The pin connection for the **74123** is the same as for the **74LS123**. The **74LS122** and **74122** are single retriggerable one-shots in a 14-pin DIP package. Their pinouts are shown in Fig. 7-12. They contain more logic and an internal resistor, but only have a single one-shot in the IC. The internal resistor is 10 kΩ and can

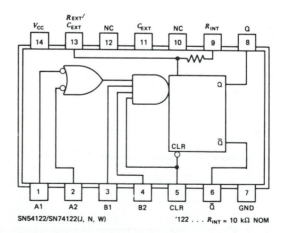

Figure 7-12   The **74122** and **74LS122** retriggerable one-shots. (From *The TTL Data Book for Design Engineers*, 2nd ed., Texas Instruments, Inc. Courtesy of Texas Instruments, copyright 1976.)

be used by connecting pin 9 to V_{CC}. The timing capacitor is connected between pins 11 and 13 (the two C_{EXT} connections). An external resistor, if used, is connected between pin 13 (R_{EXT} C_{EXT}) and V_{CC}, with pin 9 left open. The external timing resistor must be between 5 and 50 kΩ for the **74122**.

The timing for the standard series one-shots is slightly different than for the LS series. Details are given in the TTL data book.

When the CLEAR input or electrolytic capacitors are used in the **74122** or **74123**, the diode circuit of Fig. 7-13 should be used to prevent reverse voltage across C_{EXT}. Electrolytic capacitors should be connected with their positive terminal to R_{EXT}/C_{EXT}. They will generally be used when the pulse time is greater than 3 ms.

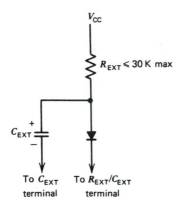

Figure 7-13 Timing component connection when CLEAR or electrolytic capacitors are used.

7-6   Integrated Circuit Oscillators

Oscillators are circuits that produce a repetitive output waveform. In analog circuits, this waveform is usually a sine wave. An AM radio station transmitting at 1 MHz, for example, needs a 1 MHz oscillator to produce its basic waveform.

In digital circuits, oscillators are used to produce *clocks* or *square waves*. If the frequency of these oscillators must be precise, such as the clock driving a microprocessor, highly stable crystal-controlled oscillators are available in DIP packages and produce TTL compatible outputs. If the tolerance on the required frequency is not extremely critical, however, serviceable oscillators can be built from a variety of digital circuits.

7-6-1   The 74LS123 Oscillator

Two one-shots, or two sections of a dual one-shot such as the **74LS123**, can be coupled together to produce an oscillator, as Fig. 7-14a shows. The waveforms, Fig. 7-14b, show that when a pulse on one of the one-shots expires, the resulting negative-going edge triggers the other one-shot. If this is allowed to continue indefinitely, the circuit functions as an oscillator or clock generator.

EXAMPLE 7-6

Design an oscillator, using two halves of a **74LS123**, to produce output pulses of 1 μs ON, 2 μs OFF, and so forth. Include a START switch.

SOLUTION   The circuit is the same as shown in Fig. 7-14. The timing components are selected as follows:

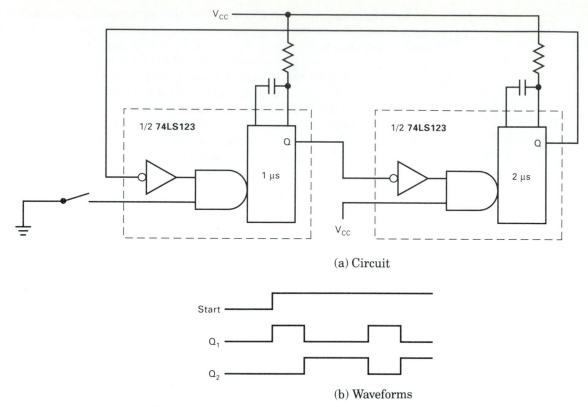

(a) Circuit

Start

Q_1

Q_2

(b) Waveforms

Figure 7-14 A one-shot oscillator composed of two halves of a **74LS123**: (a) circuit; (b) waveforms. *Note:* For clarity, the CLR inputs are not shown. They should be tied HIGH.

1. The timing of the two one-shots is set for 1 μs and 2 μs, respectively. 10 kΩ timing resistors were selected. These times are small enough to be on the chart, Fig. 7-10a, and the capacitors were found to be 220 and 450 pF, respectively.

2. The START switch is connected to the B input of the first one-shot. For simplicity, we assume it is free of bounces (see section 7-8). While the START switch is LOW, both one-shots are OFF. When the START switch is turned ON, it fires the first one-shot for a period of 1 μs. When the first one-shot turns OFF, the negative edge at Q1 triggers the second one-shot, which fires for a time of 2 μs.

3. When the pulse output of the second one-shot expires, its negative-going edge retriggers the first one-shot. Thus, the circuit continues to oscillate and produce these clock pulses as long as the START switch is HIGH.

EXAMPLE 7-7

Design circuitry to add to Fig. 7-14, so that a single-cycle/continuous switch can be used. When the switch is in the SINGLE-CYCLE position, and the START switch is flipped ON, each one-shot should pulse once. But if the switch is in CONTINUOUS mode, a continuous train of pulses should result when the START switch is turned ON.

SOLUTION   The solution is shown in Fig. 7-15. It operates much like the circuit of Fig. 7-14, except that a **74x08** AND gate (shown as a negative OR gate) has been added between the single-cycle/continuous switch and the A input of the first one-shot.

At the instant the START switch is opened, the level at A1 will be low. Therefore, the positive transition on the B input will trigger the first one-shot. When its pulse expires, the negative transition on Q1, which also appears on A2, will trigger the second one-shot.

To trigger the first one-shot a second time, there must be a *negative transition on A1*. If the single-cycle/continuous switch is open, the negative transition when the output at Q2 expires will come through the **74x08** and retrigger Q1. This will result in continuous oscillations, as in Example 7-6. If, however, the switch is closed, the output of the **74x08** will always be low. Then A1 will not see the negative transition at Q2, and will not trigger. In this situation Q1 will pulse, Q2 will pulse, and then the pulses will stop. Circuits like this are often designed to pulse in response to a signal from another circuit. It would operate in single-cycle mode, but the continuous mode might be used during test and debugging.

The circuit of Fig. 7-15 could also be built using two **74121**s. Because the **74121** has an internal AND gate, the **74x08** would not be needed, and its internal resistance could be used. If one pulse is much shorter than the other, however, duty cycle problems could occur when using a **74121**.

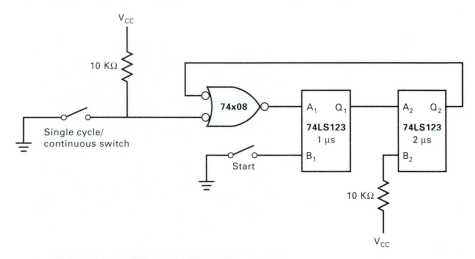

Figure 7-15   Circuit for Example 7-7.

7-6-2   The Single 74123 Oscillator

An oscillator can also be built from a single **74123** or **74LS123** by tying its Q output to its A input. When the pulse expires, Q goes LOW, retriggering the one-shot. The outputs of this oscillator are *glitches*, but the time between the glitches is controlled by the RC of the **74x123**.

A variation of this circuit that provides for external control is shown in Fig. 7-16a.[3] When the CONTROL input is LOW, point A is always HIGH and the oscillator is quiescent. Oscillations start when the CONTROL input goes HIGH. By using the chart (Fig. 7-10), the time of this oscillator was set for 0.5 μs and the oscillator's output is shown in Fig. 7-16b. The glitches last for about 40 ns. This oscillator can also be controlled by using its CLEAR input.

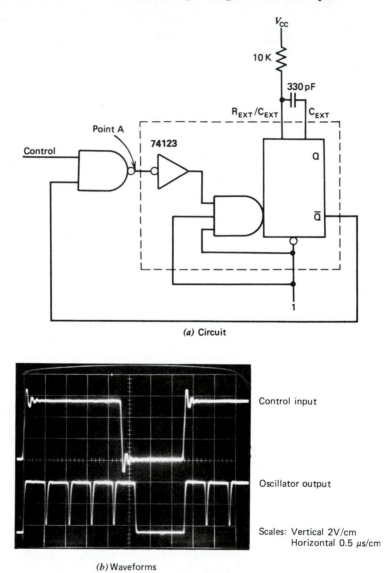

(a) Circuit

(b) Waveforms

Figure 7-16 The single one-shot oscillator.

[3]A circuit like this has been successfully used to drive the high frequency clock and shift register in a video display driver kit.

7-6-3 Schmitt-Trigger Oscillators

A very simple oscillator can be made from a Schmitt trigger as shown in Fig. 7-17. The external resistor must be kept small or the circuit will fail to oscillate (see Problem 7-14). A 330-Ω resistor is a good choice for this oscillator. Schmitt-trigger oscillators can be designed to produce outputs from 0.1 Hz to 10 MHz.

The oscillator of Fig. 7-17 operates as follows.

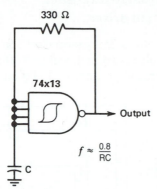

$f \approx \dfrac{0.8}{RC}$

Figure 7-17 A Schmitt-trigger oscillator. *Note:* This circuit can also be built using a **74x14**.

1. The **74x13** inverts. If its output is HIGH the capacitor charges. When the capacitor voltage reaches V_{T+} (the positive-going threshold voltage of 1.7 V), the **74x13** output goes LOW.

2. Now the capacitor discharges toward V_{OL} (0.2 V, typical). When it reaches V_{T-} (0.9 V), the Schmitt-trigger output goes HIGH again and the capacitor starts to recharge again.

3. The frequency of oscillation depends on the resistor and capacitor. It is given by Eq. 7-5.

$$f \approx \frac{0.8}{RC} \qquad (7\text{-}5)$$

EXAMPLE 7-8

Design a 100-kHz Schmitt-trigger oscillator.

SOLUTION From (7-5) we obtain

$$C = \frac{0.8}{R \cdot f} = \frac{0.8}{330 \times 10^5} = 0.0242 \ \mu F$$

The circuit should be breadboarded as designed above and its output observed on a CRO. Any tweaking necessary to produce a precise pulse width is done by changing values of C.

7-6-4 The 555 Timer

The **555** timer is a very popular and commonly used oscillator.[4] It comes in an 8-pin DIP package (about half the size of an ordinary IC) and can be set up either as a one-shot or an astable circuit (clock).

[4]Available as a 555 from National Semi-Conductor, Signetics and Fairchild or as an MC 1455 from Motorola. A 556, which consists of two 555s in a 14-pin DIP, is also available.

A simplified equivalent circuit for the **555** set up as a one-shot is shown in Fig. 7-18. Internally, the **555** have five basic parts:

1. The lower comparator.
2. The upper comparator.
3. The internal FF.
4. The discharge transistor.
5. The output driver.

The lower comparator compares[5] the voltage on its input (pin 2) with $\frac{1}{3}V_{CC}$. V_{CC} can be any voltage from 4.5 to 18 V. If the voltage on pin 2 is less than $\frac{1}{3}V_{CC}$, the lower comparator produces a LOW output voltage, which CLEARS the FF.

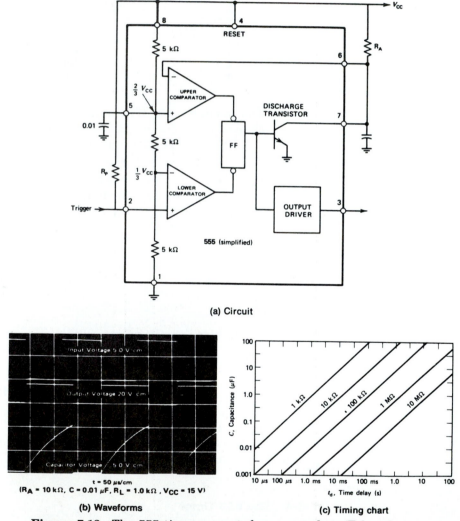

(a) Circuit

(b) Waveforms

(c) Timing chart

Figure 7-18 The **555** timer connected as a one-shot. (Taken from Motorola MC 1455/1555 data sheets and Motorola Semiconductor Data Library, Vol. 7, pp. 2–25. Courtesy of Motorola Integrated Circuits Division.)

[5]Comparators are discussed further in Sec. 18-5.

The upper comparator compares the voltage on its input (pin 6) with $\frac{2}{3}V_{CC}$. If its input is greater than $\frac{2}{3}V_{CC}$, it SETS the FF. Pin 5 of the **555** can be used to control or change the threadhold voltage of the upper comparator. Normally this feature is not used and pin 5 is connected to ground through a 0.01 μF capacitor.

If the internal FF is SET, it saturates the discharge transistor and also causes the output driver to produce a 1 output. Conversely, if the FF is CLEAR, the discharge transistor is cut off and the output driver produces a 0 output.

For monostable (one-shot) operation, an external resistor and capacitor are connected as shown in Fig. 7-18. In the quiescent state the FF is SET, and the discharge transistor is ON. It effectively shorts the capacitor to ground. The voltage at pin 6 is therefore $V_{CE(sat)} \approx 0.2$ V. Pin 2 is used as the trigger input. Normally the voltage at pin 2 is HIGH because of the resistor R_p. A trigger causes one-shot operation as follows.

1. An input trigger drives pin 2 to ground.
2. This fires the lower comparator, resetting the FF, which causes the discharge transistor to turn OFF.
3. The timing capacitor now charges toward V_{CC}, through R_A and raises the voltages on pin 6.
4. When the voltage on pin 6 reaches $\frac{2}{3}V_{CC}$, the upper comparator fires.
5. This SETS the FF, which turns on the discharge transistor and shorts out the capacitor. The waveforms are shown in Fig. 7-13b.
6. The time of the pulse is given by the timing chart of Fig. 7-18c, or:

$$t_W \approx 1.1\, R_A C \qquad\qquad (7\text{-}6)$$

7. If pin 4 (RESET) is connected to ground, it turns ON the discharge transistor. This immediately discharges the capacitor and prevents it from charging. If it is not used, it is connected to V_{CC}. For clarity, the reset logic is not shown in the simplified schematic.
8. The trigger pulse must be shorter than the output pulse, and additional triggers, while the output is LOW, do not affect the pulse width.

EXAMPLE 7-9

Design a 10-ms one-shot using a **555**.

SOLUTION   We arbitrarily choose a 1-μF capacitor. From the chart or Eq. (7-6), we find

$$R_A = \textbf{9.1 k}\Omega$$

The **555** can also be used as an astable circuit or clock generator, by connecting it as shown in Fig. 7-19a. Here, both comparator inputs (pins 2 and 6) are connected to the capacitor. The circuit operates as follows:

1. Assume the discharge transistor is off. Then the capacitor charges toward V_{CC} through R_A and R_B.

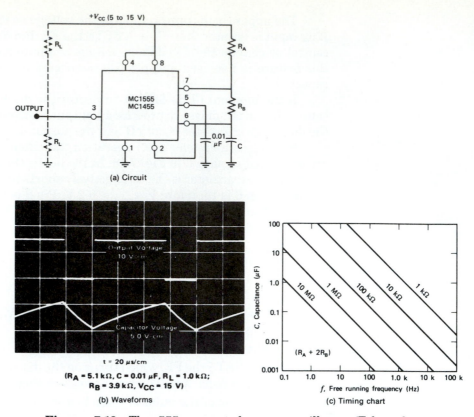

Figure 7-19 The **555** connected as an oscillator. (Taken from Motorola MC 1455/1555 data sheets and Motorola Semiconductor Data Library, Vol. 7, pp. 2–25. Courtesy of Motorola Integrated Circuits Division.)

2. When the capacitor voltage reaches $\frac{2}{3}V_{CC}$ the upper comparator SETS the FF. This turns on the discharge transistors and the capacitor discharges toward ground through R_B.

3. When the capacitor voltage becomes as low as $\frac{1}{3}V_{CC}$, the lower comparator CLEARS the FF. The discharge transistor turns OFF and the capacitor starts to charge again. This process continues indefinitely, resulting in an oscillator or clock generator.

4. The output frequency can be obtained from the timing chart (Fig. 7-19c) or

$$f = \frac{1.44}{(R_A + 2R_B)C}$$

5. The output and capacitor voltage waveforms for an 11-kHz oscillator are shown in Fig. 7-19b. Note that $R_A + 2R_B \approx 13$ kΩ, and a frequency of 11 kHz results using either Eq. (7-7) or the chart.

EXAMPLE 7-10

Design a 100-kHz oscillator using a **555**.

SOLUTION Using Eq. (7-7) we find

$$f(R_A + 2R_B)C = 1.44$$

If we choose $C = 10^{-9}$ (0.001 μF), then $R_A + 2R_B = 14,400\ \Omega$. This is a good choice as it keeps R_A and R_B in the kΩ range. There are many ways to select R_A and R_B. One choice, which makes them approximately equal, is to set $R_A = \mathbf{5k\Omega}$ and $R_B = \mathbf{4.7\ k\Omega}$.

　　To check, we use the timing chart with $R_A + 2R_B = \mathbf{14,400}\ \Omega$ and $C = .001$ μF. This combination does yield a frequency of approximately 100 kHz.

7-6-5 Crystal-Controlled Oscillators

A crystal is a small piece of quartz. When an alternating current is applied, it oscillates at a resonant frequency that depends on its physical dimensions. Thus, crystals can be sized or cut to resonate at any required frequency. The frequency of oscillation of crystals is highly stable; therefore, they are used in oscillators whose frequency must be very precise. The clock generators that drive microprocessors and determine their timing are all crystal-controlled.

　　Consider the circuit of Fig. 7-20. It is simply an inverter with its output tied back to its input. If the input is HIGH, the output will be LOW, but that makes the input LOW, which makes the output HIGH, and around we go. This circuit oscillates at a high frequency determined by the gate delay of the inverter. If, however, a crystal is placed in the loop, it will oscillate at the resonant frequency of the crystal, and its frequency will be highly stable.

　　Figure 7-21a shows the crystal oscillator circuit used for the **68HC11** microprocessor.[6] The oscillations enter the **68HC11** at the EXTAL and XTAL pins. Typical values of R_f, C1, and C2 are:

R_f = 1 MΩ–20 MΩ Higher values are sensitive to humidity; lower values reduce gain and could prevent startup.
C1 = 5 pF–25 pF　Value is usually fixed.
C2 = 5 pF–25 pF　Value may be varied to trim frequency.

　　The inverter has been replaced by a NAND gate within the **68HC11**. The STOP signal will halt the oscillations and the microprocessor if it is LOW. This will cause the microprocessor to idle, but will save energy. The crystal and its physical connections to the **68HC11** are shown in Fig. 7-21b.

74x04

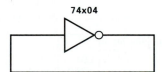

Figure 7-20　An inverter oscillator.

[6]This circuit is discussed more thoroughly in the **M68HC11** Reference manual, available from Motorola or Prentice Hall.

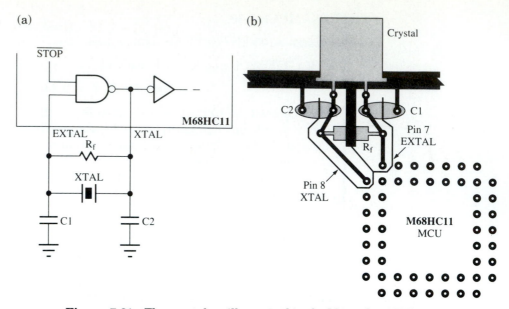

Figure 7-21 The crystal oscillator used in the Motorola **68HC11** microprocessor: (a) crystal oscillator circuit; (b) crystal layout example. (Courtesy of Motorola, Inc.)

The circuit for the crystal oscillator on the IBM PC is shown in Fig. 7-22. The crystal is connected to the **8284** clock generator manufactured by Intel, Inc. The crystal frequency is 14.31818 MHz. This is exactly four times the *color-burst* frequency used in a TV. The Apple computer uses the same frequency crystal. Figure 7-22 shows that the **8284** divides the frequency by three to provide the 4.77 MHz clock to drive the **8088** microprocessor in the IBM PC. It also divides the frequency down to 1.1931817 MHz to drive the **8253** interval timer IC.

Crystal oscillators are available in 8- and 14-pin DIP packages from commercial manufacturers. They come in a variety of the frequencies from 1 MHz to 80 MHz. One such supplier is the Connor-Winfield Corporation of Aurora, Illinois.

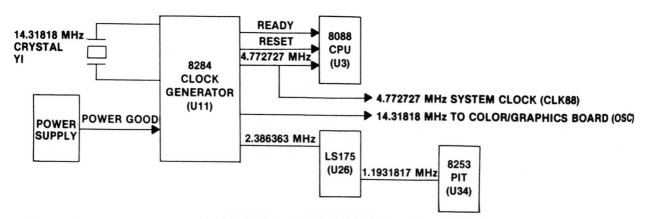

Figure 7-22 The IBM PC clock circuitry. (Courtesy of International Business Machines Corporation.)

7-6-6 Other Oscillators

There are two other types of oscillator circuits in common use. These are voltage controlled oscillators (VCOs) and phase-locked loops. A VCO is an oscillator whose frequency varies as an input voltage to it changes. The **74x124** is an example. A phase-locked loop synchronizes two different frequencies. Neither of these oscillators is, strictly speaking, digital, and they are mentioned to make the reader aware of their existence. Further details can be found in other books.

7-7 Timing Generation Problems

Engineers are very frequently required to design circuits that produce a sequence of pulses to control the operation of a larger circuit or system. Generally these pulses must conform to a timing chart. There are several ways to design such a timing circuit. Designs using one-shots are considered below, and designs using shift registers are considered in Chapter 10.

7-7-1 The Design of Timing Circuits Using One-Shots

Circuits to generate pulse sequences can be built from one-shots by having the *negative-going* edge (generated when the one-shot pulse expires) trigger the next succeeding one-shot. Simple examples of this were presented in Examples 7-5 and 7-6. SINGLE-CYCLE or CONTINUOUS modes of operation can be easily incorporated into this design, as shown below.

> ## EXAMPLE 7-11
>
> Design a circuit using **74x123**s to generate a series of pulses conforming to the timing chart given in Fig. 7-23. Include a START switch and a SINGLE-CYCLE/CONTINUOUS mode switch.
>
> **SOLUTION** This circuit requires five one-shots as follows.

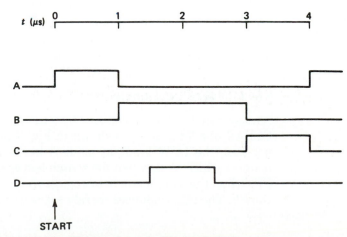

Figure 7-23 Timing chart for Example 7-11.

1. A one-shot to generate pulse A.
2. A one-shot to generate pulse B (triggered by the expiration of pulse A).
3. A one-shot to generate pulse C (triggered by the expiration of pulse B).
4. A one-shot to generate pulse D.
5. A one-shot to generate the time required between the expiration of pulse A and the start of pulse D. This one-shot fires at the end of pulse A, and when it expires, it triggers pulse D.

The final circuit is shown in Fig. 7-24. The START switch (assumed debounced) is connected to the B input of the A one-shot. The output of the C one-shot is fed back to the A one-shot, so that in CONTINUOUS mode the expiration of the C one-shot triggers the A one-shot and restarts the cycle. The switch on the **74x08** controls the mode of operation.

7-8   Switch Bounce

Switches are often used to start or control a circuit's action, but whenever a switch is designed into a circuit, we must be very careful to avoid problems that could be caused by switch "bounce." The action of a switch is shown in Fig. 7-25a. Most toggle switches in current use are the BREAK-BEFORE-MAKE type. This means that, in the process of switching, the switch blade first leaves one contact (BREAK), touches neither contact for a few milliseconds during its traverse, and then mates with the other contact (MAKE).

Unfortunately, both the MAKE and BREAK processes consist of a series of connections and disconnects called *bounces*. The action of a switch as it is toggled is shown in Fig. 7-25b and the bounces are plainly seen. Some switches bounce for a duration of up to 50 ms.[7]

Many digital circuits require a single transition from a switch. The oscillator of Fig. 7-14, for example, when operated in SINGLE-CYCLE mode produces pulses that last for only 3 μs. If, however, the START switch bounced for several milliseconds, it would generate many START pulses and cause many cycles to occur before the switch stopped bouncing. To convert the bouncy output of a switch to a clean pulse, *switch debouncing* circuits must be used. Some simple debouncing circuits are described below.

7-8-1   74x74 Debouncing

A noisy switch can be debounced by connecting its outputs to the direct SET and CLEAR of a **74x74** FF as shown in Fig. 7-26a. When the switch is in the UP position, the LOW input at point A keeps the FF in the SET position. The bounces which occur when the switch leaves the *upper* position do not affect the FF. When the switch starts to engage the *lower* contact, the first bounce clears the FF. The other bounces merely serve as additional CLEAR pulses and have

---

[7]Some switch manufacturers include a maximum bounce time as part of their specifications.

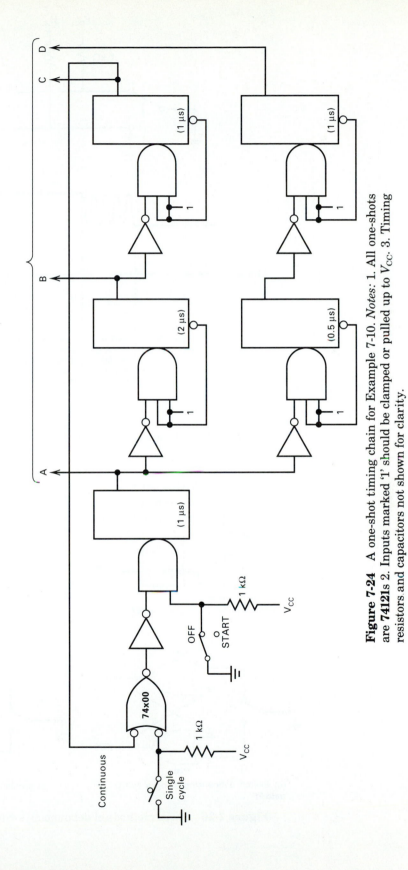

Figure 7-24 A one-shot timing chain for Example 7-10. *Notes:* 1. All one-shots are **74121**s 2. Inputs marked '1' should be clamped or pulled up to V_{CC}. 3. Timing resistors and capacitors not shown for clarity.

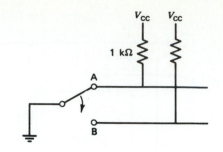

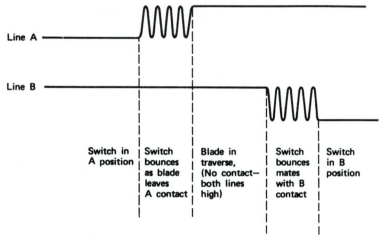

Figure 7-25 The action of a toggle switch as it is thrown from A to B.

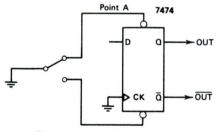

(a) Flip-flop debouncing with a **7474** FF

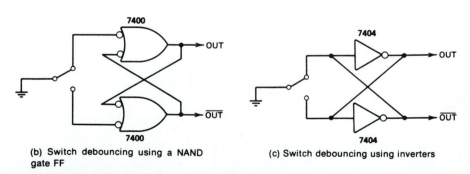

(b) Switch debouncing using a NAND gate FF

(c) Switch debouncing using inverters

Figure 7-26 Three methods of debouncing switches.

no further effect on the FF. Therefore, the FF produces a single output change, as required, when the switch is thrown from UP to DOWN.

The situation is similar when the switch is thrown to the UP position. The FF does not set until the first negative bounce occurs when the blade starts to engage the upper contact. The following bounces merely serve as additional SET pulses. The FF of Fig. 7-26a is a successful debouncing circuit because only a single transition occurs for each throw of the switch.

7-8-2   74x00 Debouncing

NAND gate FFs built from **74x00**s are also often used to debounce switches. The circuit for a NAND gate debouncer is shown in Fig. 7-26b and its operation is the same as the **74x74** debouncer. The **74x00** FF is also SET (or CLEARED) by the first negative bounce after the switch has been thrown, and further bounces have no effect on it.

7-8-3   74x04 Debouncing

A clever debouncing circuit can be built using two **74x04** inverters, as shown in Fig. 7-26c. When the switch is in the UP position, the output of the upper inverter is HIGH and the output of the lower inverter, which is shorted to ground, is LOW. When the switch enters the break portion of its downward traverse, the 0 output of the lower inverter keeps the upper output HIGH, until the first negative bounce when the switch blade touches the lower contact. This negative spike causes the output of the lower inverter to go HIGH and the output of the upper inverter to go LOW. The LOW output of the upper inverter holds the lower inverter HIGH during the bounces and a clean pulse is produced.

EXAMPLE 7-12

In the circuit of Fig. 7-24 (see Example 7-11), which switches should be debounced?

SOLUTION   Certainly the START switch should be debounced. The first pulse is only 1 μs long, but the first one-shot would be erratically triggered for several milliseconds if the START switch were not debounced.

Generally mode control switches like the SINGLE-CYCLE/CONTINUOUS switch do not need debouncing. They are set before the START switch is thrown. Since setting the two switches are human actions, whose time durations are long compared to switch bounce times, it is safe to assume the mode switch stops bouncing long before a person can throw the START switch.

7-8-4   Other TTL Debouncing Circuits

Some special TTL circuits can act effectively as switch debouncers. One such circuit is the **74265** quadruple complementary output element, shown in Fig. 7-27. For each input, the **74265** provides both the same output and its complement, with minimum time skew between them as shown in Fig. 7-27c and d.

By using feedback, the **74265** can be an effective switch debouncer, as shown in Fig. 7-27e. This would give the user four debouncers in a single IC.

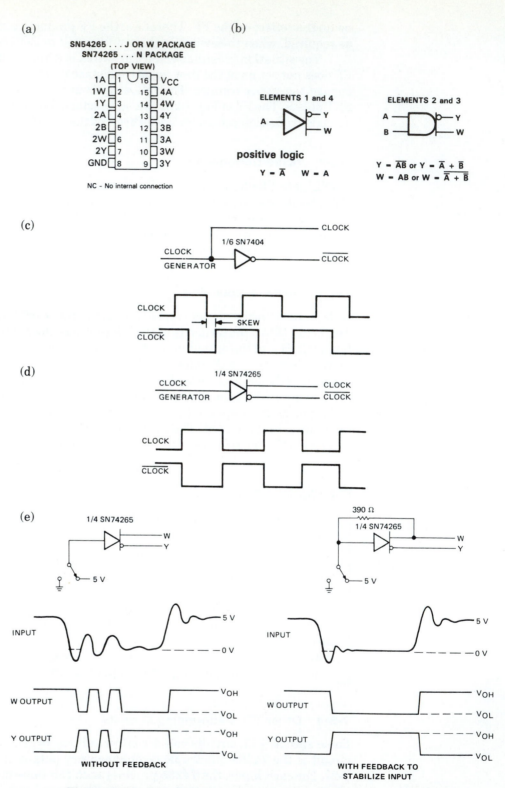

Figure 7-27 The **74265**: (a) pinout; (b) logic diagram; (c) typical clock/clock generator circuit; (d) skewless clock/clock generator circuit; (e) switch debouncer. (Reprinted by Permission of Texas Instruments.)

Class A or single-pole, single-throw switches, have only two contacts (they are either open or closed) and cannot be debounced by the methods of section 7-8.

The simplest method of debouncing class A switches is to connect them to the D input of a FF and clock it with a constant clock whose period is longer than the bounce time of the switch. This is shown in Fig. 7-28.

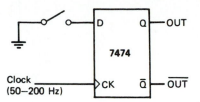

Figure 7-28   Debouncing a class A switch using a FF.

7-9-1   Schmitt-Trigger Debouncing

Class A switches can generally be debounced by connecting them to an *RC* network and then a Schmitt trigger, as shown in Fig. 7-29. The time constant of the *RC* network should be longer than the switch bounce time so that the capacitor charges relatively slowly.

The circuit of Fig. 7-29 operates as follows.

1. When the switch is opened, the bounces discharge the capacitor and it does not have time to charge to V_{T+} until the bounces stop completely. Only then does the Schmitt-trigger output go LOW.

2. When the switch closes, the first bounce discharges the capacitor, causing the Schmitt-trigger output to go HIGH. Since the capacitor does not have sufficient time to charge up to V_{T+} between bounces, the Schmitt-trigger output remains HIGH.

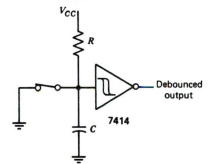

Figure 7-29   Debouncing a single-pole, single-throw switch.

7-9-2   One-Shot Debouncing

One-shots can also be used to debounce switches as shown in Fig. 7-30. The first one-shot is set for a long pulse, compared to the bounce time of the switch (8.8 ms in Fig. 7-30). When the switch stops bouncing and retriggering the first one-shot, it times out and triggers the second one-shot (for 1 µs in Fig. 7-30). The function of the NOR gate is to produce an output pulse when the switch is thrown from HIGH to LOW (because the switch line will be LOW when the second one-shot fires) but not when the switch is thrown from LOW to HIGH. This

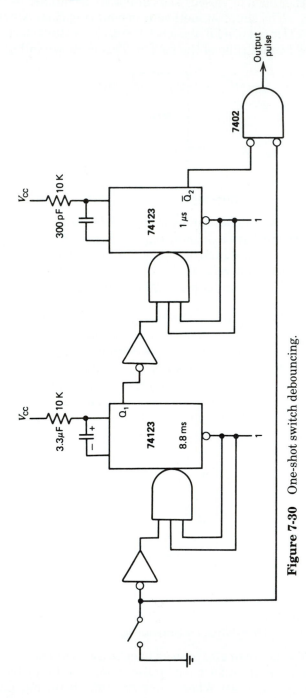

Figure 7-30 One-shot switch debouncing.

circuit produces a single output pulse for each switch throw. The length of that pulse is variable and determined by the second one-shot.

7-10 The One-Shot Discriminator

A single pair of one-shots (a single **74x123**) can be used as an FM *discriminator* that determines which one of two signal frequencies is being received. This can be used in Frequency Shift Keying (FSK, see section 17-10-3) circuits. The time of the first one-shot should be set between the periods of the two input frequencies, and the first one-shot should trigger the second one-shot. If the lower frequency (longer pulse width) signal is applied to the first one-shot, it will constantly time out and provide a series of triggers to the second one-shot, which will always be SET. If the higher frequency is being received, the first one-shot will be constantly SET; the second one-shot will not see any trigger pulses and will remain CLEAR. This discriminator is illustrated in Example 7-13 (also see Problem 7-18).

EXAMPLE 7-13

The "Kansas City Standard" used by some engineers for transmitting digital data to tape cassettes uses a 2400 Hz signal for a 1 and a 1200 Hz signal for a 0. Design a one-shot discriminator to produce a 1 output when the input is 2400 Hz and a 0 output when the input is 1200 Hz.

SOLUTION   The solution is shown in Fig. 7-31. The periods of the input waves are approximately 400 μs (for a 2400 Hz input) and 800 μs (for a 1200 Hz input). The time of the first one-shot is set between them or at approximately 600 μs. If the 1200-Hz signal is coming in, the first one-shot resets every 800 μs, as shown in the waveforms of Fig. 7-31b. The second one-shot is set for a pulse width of 1 ms and it is always triggered. If the input frequency is 2400 Hz, the first one-shot is always SET; the second one-shot gets no trigger pulses and is always CLEAR. Thus, $\overline{Q}_2$ will be a 1 if the input frequency is 2400 Hz and a 0 if the input frequency is 1200 Hz.

SUMMARY

In this chapter the design of one-shots to generate pulses of predetermined lengths and the design of TTL oscillators of any desired frequency were studied. Switch debouncing circuits were also covered. These circuits are widely used to generate and control the timing or sequence of events in a digital system. Consequently, the engineer should know how to design them.

GLOSSARY

Astable circuit: A square wave oscillator or clock generator.
Debouncing circuit: A circuit designed to produce a clean output in response to a switch closure.

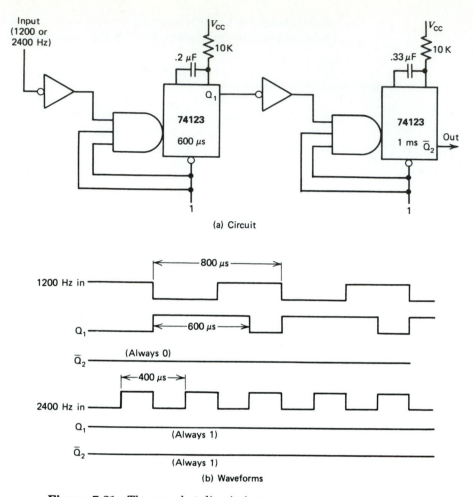

(a) Circuit

(b) Waveforms

Figure 7-31 The one-shot discriminator.

Discriminator: A circuit whose output depends on the frequency of the input wave.

Duty cycle: The ratio of ON time to total pulse period for a repetitive square wave.

Jitter: One-shots that produce output pulses of different lengths are said to jitter.

Monostable circuit: See one-shot.

One-shot: A circuit that produces an output pulse for a fixed period of time in response to a trigger, and then returns to its quiescent state.

Recovery time: The time required by a one-shot to recover (to recharge its internal capacitor) after having been fired.

Retriggerable one-shot: A one-shot that can be restarted by triggers when it is in its 1 state.

Switch bounce: Fluctuations in output levels when a switch MAKES or BREAKS with a contact.

Trigger: A short pulse or edge that initiates a one-shot pulse.

Tweaking: A small change made in resistance or capacitance to time a circuit precisely.

PROBLEMS

Section 7-4.

7-1. Design a **74LS221** to produce a 5-ms pulse using:

(a) A 2-kΩ resistor.

(b) A 40-kΩ external resistor.

What is the minimum time that should be allowed between pulses in each case?

7-2. A **74LS221** has a timing capacitor of 1 μF. The timing resistor consists of a 5-kΩ fixed resistor in series with a 20-kΩ potentiometer. Calculate:

(a) The minimum pulse width of the output.

(b) The maximum pulse width of the output.

(c) Why is it unwise to use a potentiometer without the fixed resistor?

7-3. A traffic light is red for 25 seconds and green for 45 seconds. Calculate its duty cycle.

Section 7-5.

7-4. Design a one-shot to produce pulses of 0.5 μs using:

(a) A **74LS221**.

(b) A **74LS123**.

In each case above calculate the required timing resistors and capacitors and draw the circuit showing how they are connected to the one-shot.

7-5. Repeat Problem 7-4 if the required pulse is 0.5 ms.

7-6. A one-shot has a timing resistor of 10 kΩ and a timing capacitor of 0.5 μF. Calculate its output pulse width if the one-shot is:

(a) A **74LS221**.

(b) A **74LS123**.

7-7. Add an ENABLE-DISABLE toggle switch to the circuit of Fig. 7-11. If the switch is in the DISABLE position the circuit should never alarm, but if the switch is in the ENABLE position the circuit should function normally.

7-8. A circuit has two 100-kHz clocks, A and B, and two output points, C and D. If *both* clocks are operational, A must be connected to C and B to D. If either clock fails, the working clock must drive both points. Design a circuit to detect a clock failure and produce the required outputs. (A clock failure means the clock stops making transitions.) (*Hint:* Retriggerable one-shots are excellent devices for this type of problem.)

7-9. Design a clock that is ON for 19 ms and OFF for 1 ms using:

(a) **74LS221s**.

(b) **74LS123s**.

Section 7-6.

7-10. Design a one-shot to produce a pulse width of 5 ms using:
    (a) A **74LS221**.
    (b) A **74LS123**.
    (c) A **555**.

7-11. Design an oscillator to produce a 50-kHz square wave using:
    (a) A **74LS221**.
    (b) A **74LS123**.
    (c) A Schmitt trigger.
    (d) A **555**.

7-12. Explain with diagrams, how you would build a circuit to produce the waveform of Fig. P7-12. Include a SINGLE-CYCLE/CONTINUOUS option. Debounce switches where necessary. Show specifically where you would take the output to obtain the continuous waveform shown.
    (a) Use a **74LS221s**.
    (b) Use a **74LS123**.

⎡‾2 µs‾⎤___⎡3 µs⎤‾‾‾⎡2 µs⎤___⎡3 µs⎤‾‾‾⎣   **Figure P7-12**

7-13. A **555** timer has RA = RB = 100 kΩ and C = 1 µF. Find its frequency of oscillation from the curves (Fig. 7-19) and verify it by the equation.

7-14. An attempt was made to build a Schmitt-trigger oscillator using a 1-kΩ resistor. It would not oscillate, but it was found that the Schmitt-trigger output voltage was LOW continuously and its input voltage was 1 V. Using these clues, explain what happened and why the Schmitt trigger would not oscillate.

7-15. Design a 200-kHz oscillator made up of a **74LS123s** with the following stipulations:
    (a) When an external CLEAR line goes LOW both sections of the oscillator shall CLEAR immediately.
    (b) When the CLEAR line goes HIGH only the first section of the oscillator shall trigger.

Section 7-7.

7-16. Design a circuit to generate the timing pulses shown in Fig. P7-16.

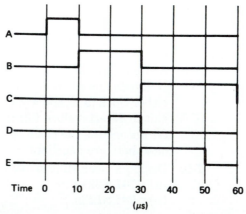

Figure P7-16

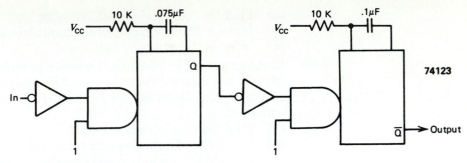

Figure P7-19

7-17. Design a 10-minute clock by building a 1-Hz oscillator and using IC counters to divide the frequency down.

Section 7-8.

7-18. A one-shot is to be triggered from a switch. Is it necessary to debounce it if the pulse time is:
  (a) 5 μs?
  (b) 500 μs?
  (c) 500 ms?
  Explain the reasons for your answer in each case.

Section 7-10.

7-19. The circuit of Fig. P7-19 is a discriminator. The output is different if a 4 kHz or a 5 kHz wave is applied.

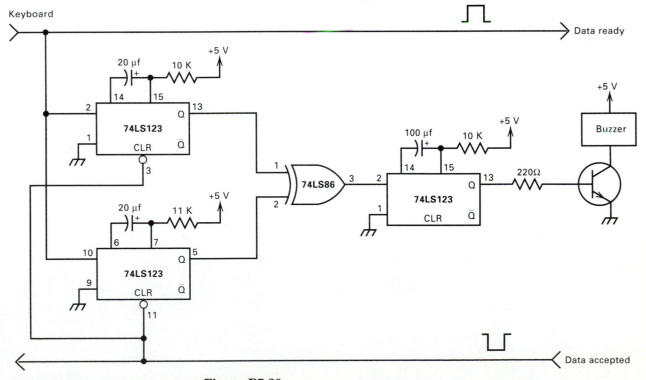

Figure P7-20

(a) Find the "on time" of each one-shot.

(b) Make a timing chart for each input frequency.

(c) What is the output in each case?

7-20. In the circuit of Fig. P7-20, the keyboard produces a DATA READY signal when it has data for the computer. The computer must respond with a negative-going DATA ACCEPTED signal. If it waits too long the buzzer will sound.

(a) Find the time of each one-shot.

(b) How fast must the computer respond if the buzzer is not to sound?

(c) Explain in your own words how the circuit works.

After attempting the problems, return to Sec. 7-2 and review the self-evaluation questions. If you cannot answer certain questions, return to the appropriate sections of the chapter to find the answers.

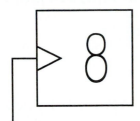

Counters

8-1 Instructional Objectives

Counters are digital circuits that increment on each input or clock pulse that is applied. They have many uses. The display of time on a digital watch, for example, is controlled by a counter. Counters also control the timing of events, such as the cycles in a washing machine.

This chapter explains the operation of digital counters. After reading it the student should be able to:

1. Design a counter to count to N, and recycle, where N is any given number.
2. Design a circuit to divide down a clock frequency to a desired sub-multiple.
3. Design a counter to count in an irregular or shortened sequence.
4. Explain the principle underlying the design of a synchronous counter.
5. Design counters using the **7490** series of TTL counters.

8-2 Self-Evaluation Questions

Watch for the answers to these questions as you read this chapter. They should help you understand the material presented. When you have finished the chapter, return to this section and be sure you can answer all of the questions.

1. Why are crystal oscillators used? Why must their frequency be as high as the highest frequency clock required in the circuit?
2. How is a specific count decoded from an N-bit counter?
3. What is the difference between synchronous and ripple counters? What are the advantages and disadvantages of each?
4. How are the CARRY and BORROW outputs of a **74x193** used to construct a cascaded up-counter or down-counter?
5. How does the LOAD input function on a **74x193**?

8-3   Introduction

Counters are special purpose digital circuits, composed of FFs or ICs designed specifically as counters, whose function is to count any number of events required by the system specifications. A count-by-N circuit, which is capable of counting up to a specified number N, will require at least K FFs, where $2^K > N$. We have already considered a count-by-8 or 3-bit counter in section 6-13-3. In this chapter various methods of designing counters are examined.

8-4   Divide-by-N Circuits

A divide-by-N circuit can count to a given number, N. It is continuously clocked, and is allowed to recycle, or rollover to 0, after reaching the Nth count. If the output is taken from the most significant stage of the counter, it provides a pulse that is N times as long as the original clock pulse. The frequency of this output is equal to the frequency of the original clock divided by N: hence the name divide-by-N. The circuit of Fig. 6-24 is a 3-bit counter, or a divide-by-8 circuit, composed of three FFs. The frequency of the Q_3 output is one-eighth the frequency of the clock.

Complex systems that require several clocks of various frequencies are designed in accordance with these principles. If the timing for the system is critical, it is usually provided by a *crystal controlled oscillator* (see section 7-6-5), which is highly stable and jitter-free. The frequency of the crystal is chosen to be as high or higher than the highest frequency clock required by the circuit. The various clocks needed throughout the circuit are then obtained by counting down the clock pulses of the crystal oscillator. Crystal oscillators in DIP packages, which operate from a 5-V supply and have TTL compatible outputs, are commercially available. For inexpensive oscillators, typical crystal frequencies range from 4 to 20 MHz.

> **EXAMPLE 8-1**
>
> The basic timing for a system comes from a 1 MHz crystal controlled oscillator. At a point in the circuit a 500 Hz clock is required. What is the capacity of the counter and how many FF stages will be required to build it?
>
> **SOLUTION**   To reduce 1 MHz to 500 Hz a divide-by-2000 counter is required (1 MHz ÷ 500 = 2000). Since 11 is the smallest number such that $2^{11} \geq 2000$, at least 11 stages will be required for a divide-by-2000 counter. It is now specified and can be designed by using the principles developed later in this chapter.

Ripple counters are the simplest counters to build; consequently, they are very commonly used. The circuit of Fig. 6-24 is a 3-bit ripple counter. It is called a ripple counter because the clock ripples through the counter stage by stage. The clock can cause the first (least significant) stage to toggle or change. If the first stage changes, its changing output could cause the next stage to toggle, and so on. Whether stage two changes depends on the output of stage one, and whether the stage three changes depends on the output of stage two. (This limits the speed of the counter or the frequency at which the counts to be counted may occur, and is the main disadvantage of a ripple counter.) For a **K**-bit counter, there are some counts for which all **K** stages change. There must be sufficient time for the clock to ripple through the **K** stages, plus any additional gating and decoding time, before the next clock pulse can occur.

EXAMPLE 8-2

How fast can an 11-stage ripple counter built of **74107**s be clocked? Allow an additional 60 ns for decoding or gate delays.

SOLUTION   We must first determine the propagation delay for a single **74107**. This can be found from the manufacturer's specifications. From page 2-322 of the *TTL Data Book* (1988 edition) the worst case propagation delay from the clock to the Q or $\bar{Q}$ output is seen to be 40 ns. To produce a reliable output the circuit will require

$$11 \text{ stage delays} + \text{gate delays} = \text{total delay}$$
$$11 \text{ stages} \times 40 \text{ ns/stage} + 60 \text{ ns} = 500 \text{ ns}$$

Thus, the events to be counted may not be closer than 500 ns apart and the clock frequency is limited to 2 MHz.

The counter of Example 8-2 was slow because it used a standard series FF, which has a relatively long propagation delay. Typical worst case propagation delays for J-K FFs in other series are given in Table 8-1.

TABLE 8-1

Propagation Delays for JK FFs

FF	Series	Propagation Delay
74LS107	Low power Schottky	20 ns
74ALS113	Advanced low power Schottky	19 ns
74AS109	Advanced Schottky	10 ns
74HC107	High speed CMOS	39 ns
74ACT11109	Advanced CMOS	11 ns

8-5-1   Decoding Specific Counts

Digital systems are often required to generate a pulse when a counter reaches a specified number. A *decoder* is a circuit that can produce this pulse. The simplest way to design a decoder is to AND together the Q outputs of all stages that are a 1 and the Q outputs of all stages that are a 0 for the required count. An example of a 4-stage counter decoding a count of 6 is shown in Fig. 8-1. For the moment ignore the dashed line. Since 6 is binary 0110, the $\bar{Q}$ outputs of stages 0 and 3 and the Q outputs of stages 1 and 2 are ANDed together. The output of the NAND gate is low only when the counter contains a binary 6.

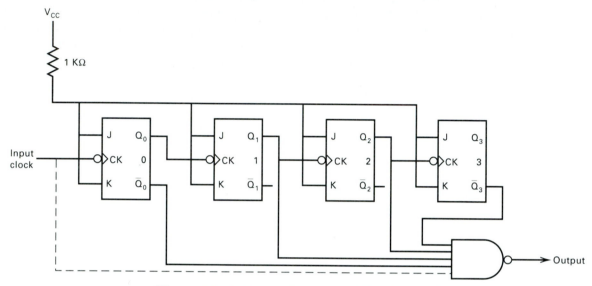

Figure 8-1   A 4-bit ripple counter with a decoder for a count of 6.

Counters are most often decoded at counts of 0 or when they reach their maximum count (all 1s). When the count is 0, the $\bar{Q}$ outputs are all HIGH. Therefore, to decode a count of 0, the $\bar{Q}$ outputs of each stage are connected to a NAND gate decoder, while to decode the maximum count the Q outputs of each stage are connected to a NAND gate decoder.

Counters that are connected to decoders often produce glitches at the outputs of the decoders. This occurs because the clock edge causes the FFs to change state, and not all FFs change at precisely the same time. Therefore, some false states may exist for a few nanoseconds before the transition is complete.

The circuit of Fig. 8-1 was connected in the laboratory and the results are shown in Fig. 8-2. The top CRO trace is the clock input. The most significant stage of the counter is shown on trace 2 (it is a 1 for counts 8 through 15); the second most significant stage is shown in trace 3. Trace 4 shows the decoder output. It is LOW at a count of 6, but a *glitch* can be seen clearly when the count changes from 7 to 8. Whether these unwanted glitches cause a real problem depends on the rest of the circuit. If the glitch is applied to the CLOCK, PRESET, or CLEAR of a FF, it could cause the FF to change state, which might be very troublesome. The glitch of Fig. 8-2 was tested and found strong enough to direct clear a FF.

Counters can be made glitch-free by tying the clock to the input of the decoder as shown by the dashed line of Fig. 8-1. If the dashed line is connected, the negative edge of the clock cuts off the decoder before the glitches occur. Now the

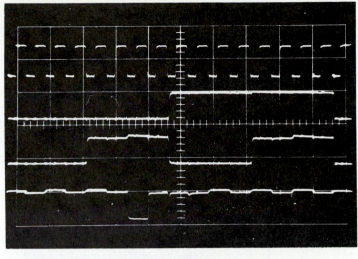

Scales: Vertical 5 V/cm

Horizontal 1 μs/cm

Figure 8-2 CRO traces for a 4-bit counter and the output of a decoder for a count of 6. Vertical scale 5 V/cm; horizontal scale 1 μs/cm.

decoder cannot function until the clock goes HIGH, but if this delay in the output is acceptable, a glitch-free output is obtained.

8-6 Synchronous Counters

Synchronous counters are often used to reduce the time delay and glitch problems associated with ripple counters. A synchronous counter *has one common clock that is connected to all FFs.* This clock causes all FFs to change at the same time (synchronously). Consequently, all FFs change within one propagation delay time regardless of the number of stages. Also, with all stages changing together, glitches are greatly reduced because of the shorter transition time interval. Glitches are not completely eliminated because one stage may respond to the same clock faster than another stage.

8-6-1 K-Bit Synchronous Counters

As in ripple counters, the simplest K-bit synchronous counter counts to N, where $N = 2^K$. A synchronous counter is designed in accordance with the following principle:

    The Mth stage toggles only on clock pulses that occur when all less significant stages are one.

    Consider, for example, counting numbers in binary. The least significant bit changes on each count. The second least significant bit counts 00110011. . . . It changes on counts that occur after the least significant bit is a 1. The third least significant stage counts four 0s, four 1s, four 0s, and so on. It changes at counts of 4, 8, 12, . . . or on the pulses *following* counts of 3, 7, 11,. . . . But 3, 7, 11, . . . are exactly those counts where the two least significant bits are both 1. This conforms to the principle stated above.

EXAMPLE 8-3

Design a divide-by-16 synchronous counter using 5-K FFs.

SOLUTION   The counter is designed as follows.

1. Since there are 16 states and $16 = 2^4$, at least 4 FFs are required.
2. Because this is a synchronous counter, the clocks of all stages must be connected together. The circuit and timing chart for the 4-stage synchronous counter is shown in Fig. 8-3. To develop this circuit, first the four FFs were drawn, then the common clock was sketched in.
3. The first stage of the counter toggles on each clock pulse, so J_1 and K_1 are tied high.
4. According to the synchronous principle, the second stage toggles only when the first stage is a 1. This is accomplished by tying Q_1 to the J and K inputs of stage 2.
5. Stage 3 toggles only when Q_1 and Q_2 are both HIGH. Q_1 and Q_2 are ANDed together at point A, which is wired to J_3 and K_3.
6. FF4 cannot toggle unless Q_1, Q_2, and Q_3 are all HIGH. The second AND gate is used to produce this signal at point B (J_4 and K_4).

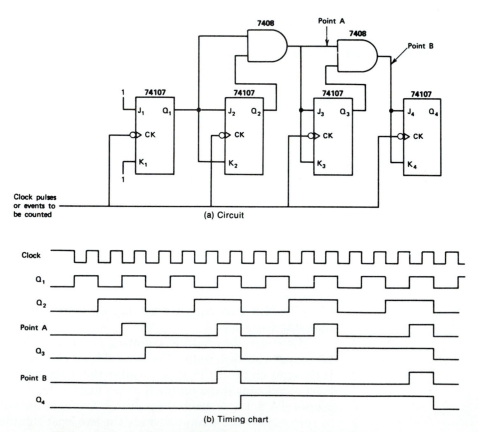

Figure 8-3   A 4-bit synchronous counter and timing chart.

8-6-2 Synchronous D-Type Counters

The algorithm for building an $n + 1$ stage synchronous counter using J-K FFs was explained in the previous paragraph, and is very simple:

AND together the Q outputs of stages 0, 1, . . . n − 1.

Apply the output of the AND gate to the J and K inputs of stage N.

Synchronous counters using D-type FFs have become more important recently, because these FFs are used within Programmable Array Logic (PAL) ICs (see Chapter 12). The algorithm for constructing these counters is slightly more complex. To construct an n stage synchronous counter, do the following:

1. AND together the Q outputs of stages 0, 1, . . . n − 1.
2. Connect the output of the AND gate to the input of an XOR gate. The other input to the XOR gate must be connected to Q_n.
3. The output of the XOR gate must be connected to D_n.

EXAMPLE 8-4

Design a 4-stage, synchronous counter using D FFs.

SOLUTION The solution is shown in Fig. 8-4. It follows the algorithm, ANDing the lower stages, and XORing the output with the output of the upper stage. To get a feel for how it works, the reader should concentrate on stages 0 and 1. Stage 0 simply toggles in response to each clock pulse. Now consider stage 1, when Q_0 is 0, the XOR gate does not invert Q_1. Therefore, Q_1 and D_1 are at the same level, and Q_1 will not change in response to a clock pulse. If Q_0 is a 1, however, the XOR gate inverts its other input. This effectively connects $\overline{Q_1}$ to D_1 and the FF will toggle at the next positive edge of the clock. This is correct counter action. The higher stages will also toggle only when the output of their AND gates is 1.

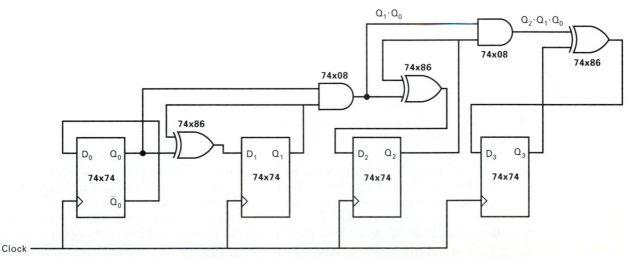

Figure 8-4 A 4-stage synchronous counter using D FFs.

EXAMPLE 8-5

Suppose the counter of Fig. 8-4 was built in the laboratory and did not work. Explain how to debug it.

SOLUTION   Probably the best way to debug a circuit is to reduce it to simpler components and debug each one. Here, we would disconnect stage 0 from stage 1, and use an oscilloscope to determine if stage 0 was properly toggling in response to the clock. If stage 0 were working, we would connect it to stage 1 and test this portion of the circuit. We would continue until we found the stage causing the problem.

8-6-3   Timing Synchronous Counters

In a synchronous counter, all FFs toggle simultaneously, but sufficient time must be allowed for the gates to establish the proper J and K levels on all FFs before the next pulse can occur.

The maximum propagation delay time for the synchronous counter of Fig. 8-3 may be calculated by adding the two delays caused by the **7408**s to the propagation time for the FF, plus the decoding time. The worst-case propagation delay time for a **7408** is 27 ns (from the manufacturer's specifications), and if the decoding time is again assumed to be 60 ns, the propagation delay for the counter is:

One FF propagation delay + two **7408** delays + decoder delay =
total delay 40 ns + 54 ns + 60 ns = 154 ns

This compares to a worst-case delay of 220 ns for a 4-stage ripple counter.

For minimum time delay, the J-K synchronous counter can have a propagation delay of one FF and one AND gate. The minimum delay time for the D-type counter of Fig. 8-4 is the sum of one FF plus one AND gate plus one XOR delay time.

8-7   The 3s Counter

If a certain clock frequency is needed in a circuit, it is necessary to divide the master oscillator by the number required to obtain the desired clock frequency. Therefore, we must be able to design circuits that count to, or divide by, any given number.

Divide-by-N circuits where $N = 2^K$ have already been discussed. When N is not equal to 2^K the problem is more complex. The smallest such number is 3. Fortunately, a 3s counter or divide-by-3 circuit is easy to build and requires no additional gating. The circuit and timing chart of a 3s counter are shown in Fig. 8-5. By connecting $\bar{Q}_2$ to J_1, FF1 cannot toggle when $Q_2 = 1$, which causes the circuit to divide by 3. Note that the K_1 input could either be connected to J_1 or clamped HIGH. This circuit also gives us the capability of building a divided-N counter if N can be expressed as 3×2^K.

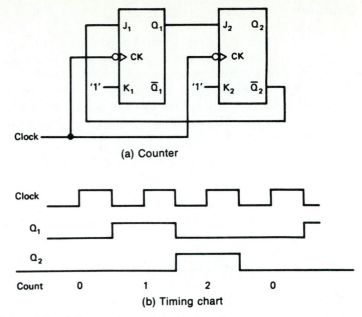

(a) Counter

(b) Timing chart

Figure 8-5 A 3s counter and its timing chart.

EXAMPLE 8-6

Design a divide-by-12 counter using the minimum number of FFs and no external gates.

SOLUTION The solution proceeds as follows.

1. Since 4 is the smallest number such that $2^4 > 12$, four FFs are required.
2. The fact that $12 = 3 \times 2^2$ suggests that a 2-stage counter and a 3s counter solve the problem.
3. If the count is to go from 0 to 11, the 3s counter must be placed at the most significant stages.
4. The circuit is shown in Fig. 8-6 where a divide-by-4 counter precedes a divide-by-3. The circuit is a hybrid because the count ripples through the divide-by-4 counter but the output of the divide-by-4 acts as a synchronous input to the divide-by-3 stages.

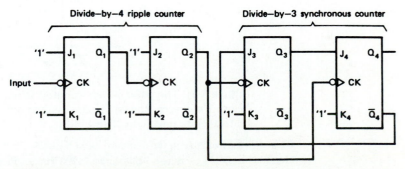

Figure 8-6 A divide-by-12 counter.

Because of the general need for counters, many are already packaged in TTL ICs. The three simplest counters available are the **7490, 7492,** and **7493.**[1] They are a decade counter, a divide-by-12, and a divide-by-16 (binary) counter, respectively. Unfortunately, these counters have power on pin 5 and ground on pin 10,[2] so care must be taken when wiring them into a circuit. Also, to some extent they are all ripple counters, and may produce glitches on decoder outputs.

8-8-1   The 7490 Decade Counter

The **7490** is a decade counter that actually consists of a single J-K FF and a divide-by-5 circuit. The IC's pin configuration, functional block diagram, and count sequence are shown in Fig. 8-7. In the **7490,** FF A functions as a divide-

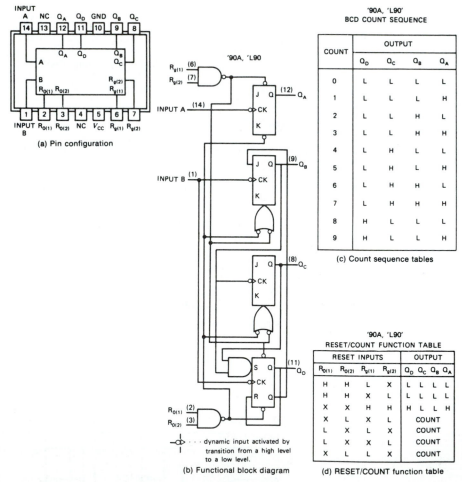

(a) Pin configuration

(b) Functional block diagram

'90A', 'L90'
BCD COUNT SEQUENCE

COUNT	OUTPUT			
	Q_D	Q_C	Q_B	Q_A
0	L	L	L	L
1	L	L	L	H
2	L	L	H	L
3	L	L	H	H
4	L	H	L	L
5	L	H	L	H
6	L	H	H	L
7	L	H	H	H
8	H	L	L	L
9	H	L	L	H

(c) Count sequence tables

'90A', 'L90'
RESET/COUNT FUNCTION TABLE

RESET INPUTS				OUTPUT			
$R_{0(1)}$	$R_{0(2)}$	$R_{9(1)}$	$R_{9(2)}$	Q_D	Q_C	Q_B	Q_A
H	H	L	X	L	L	L	L
H	H	X	L	L	L	L	L
X	X	H	H	H	L	L	H
X	L	X	L	COUNT			
L	X	L	X	COUNT			
L	X	X	L	COUNT			
X	L	L	X	COUNT			

⊲ ⋯ dynamic input activated by transition from a high level to a low level.

(d) RESET/COUNT function table

Figure 8-7   The **7490** decade counter. (From the *TTL Data Book for Design Engineers*, 2nd ed., Texas Instruments, Inc. Courtesy of Texas Instruments, copyright 1976.)

[1]These ICs are available in the standard and LS series.

[2]Manufacturers now produce a **74290** and a **74293** that are identical to the **7490** and **7493** except for the pinout. Power and ground are on pins 14 and 7.

by-2 counter, and the other three FFs function as a divide-by-5 counter. The divide-by-5 circuit counts 0, 1, 2, 3, 4, 0, ... If the input signal is connected to input A, and the Q_A output is connected to input B, a decade counter results and the output is given by the BCD count table of Fig. 8-7c.

The $R_{0(1)}$ and $R_{0(2)}$ inputs form a direct clear. The Q outputs of the **7490** are all LOW whenever both $R_{0(1)}$ and $R_{0(2)}$ are HIGH, as shown by the RESET/COUNT function table of Fig. 8-7d. The counter can also be direct SET to a count of 9 if both the $R_{9(1)}$ and $R_{9(2)}$ inputs are HIGH, which is a useful feature when performing decimal arithmetic. If the counter is to count normally, at least one of the R_0 inputs and one of the R_9 inputs must be held LOW.

EXAMPLE 8-7

Design a divide-by-25 counter using **7490**s.

SOLUTION   Two divide-by-5 circuits in two **7490**s can be cascaded, and the A FFs left unconnected, to build a divide-by-25 counter. The circuit is shown in Fig. 8-8. Note that the count sequence does not progress from 0 to 24. Note also that one of the R_0 and R_9 inputs on each IC are tied to ground.

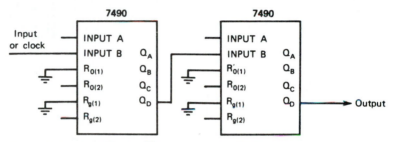

Figure 8-8   A divide-by-25 counter using **7490**s.

8-8-2   The 7492 Divide-by-12 Counter

The **7492** is a divide-by-12 circuit that consists of a J-K FF followed by a divide-by-6 counter. The pinout, block diagram, and count sequence are shown in Fig. 8-9. FF A is simply a divide-by-2 circuit; FFs B and C form a divide-by-3 counter, exactly as shown in Fig. 8-5. FF D is another divide-by-2 circuit. A divide-by-12 counter is made by putting the clock into input A and connecting output Q_A to input B. As in the **7490**, $R_{0(1)}$ and $R_{0(2)}$ are ANDed together to form a direct clear. At least one of these inputs must be LOW in order for the IC to count.

An examination of the count sequence table for the **7492** connected as a divide-by-12 counter (Fig. 8-9c) reveals that it does not count in a 0 to 11 sequence. This is because the 3s counter (Q_B and Q_C) is before the final J-K FF (Q_D). Actually it counts the binary equivalent of 0 through 5 followed by 8 through 13. However, the Q_D output is symmetric when a continuous pulse train is applied, which is an advantage in some circuits.

EXAMPLE 8-8

The basic oscillator for a system runs at 1.2 MHz. Design a circuit that will produce a symmetric 2000 Hz square wave from the basic system clock.

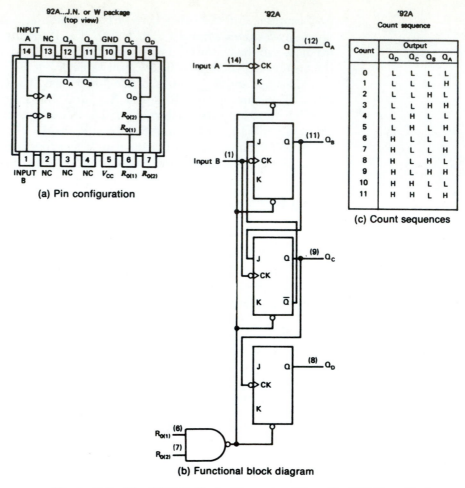

Figure 8-9 The **7492** divide-by 12 counter. (From the *TTL Data Book for Design Engineers*, Texas Instruments, Inc. Courtesy of Texas Instruments, copyright 1976.)

SOLUTION Since 1.2 MHz ÷ 2000 = 600, a divide-by-600 counter is needed. It can be built by using two divide-by-10 counters followed by a divide-by-6 counter, which suggests two **7490**s and the divide-by-6 portion of a **7492**. The circuit is shown in Fig. 8-10. There are also other ways to design this circuit (see Problem 8-19).

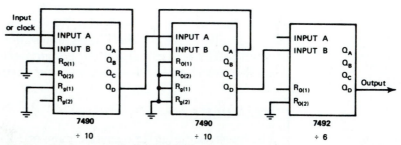

Figure 8-10 A divide-by-600 counter.

8-8-3 The 7493 4-Bit Binary Counter

The **7493** is a simple 4-bit binary counter, consisting of a single FF, Q_A, followed by three cascaded FFs that form a divide-by-8 counter. The pin layout, block diagram, and count sequence are shown in Fig. 8-11. If Q_A is connected to input B, the circuit forms a divide-by-16 ripple counter. It counts in a straightforward binary sequence, as the count sequence table (Fig. 8-11c) indicates. Again the $R_{0(1)}$ and $R_{0(2)}$ inputs are ANDed together to form a direct CLEAR for all four stages of the counter.

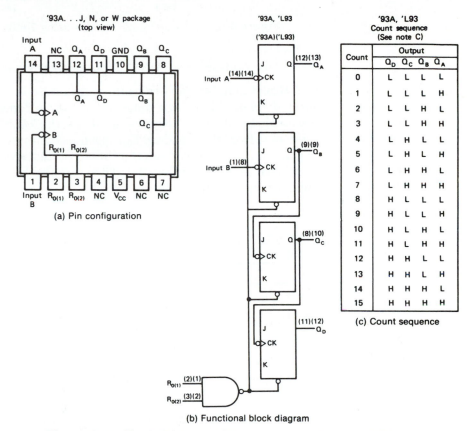

(a) Pin configuration

(b) Functional block diagram

(c) Count sequence

Figure 8-11 The **7493** 4-stage binary counter. (From the *TTL Data Book for Design Engineers*, Texas Instruments, Inc. Courtesy of Texas Instruments, copyright 1976.)

EXAMPLE 8-9

Design a divide-by-128 counter.

SOLUTION This counter could be built using seven stages from four **74107**s. Similar counters were designed in Chapter 6. Fewer ICs are needed, however, if **7493**s are used. This approach is shown in Fig. 8-12, where the first **7493** is used as a 4-stage counter and cascaded to the 3-stage counter of the second **7493**.

An alternate solution is to connect both **7493**s as 4-stage counters and take the output from Q_C of the second **7493**.

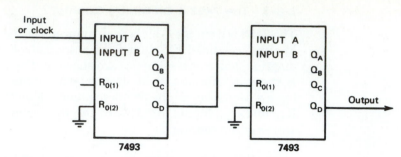

Figure 8-12 A divide-by-128 counter using **7493**s.

8-8-4 The 74LS56 and 74LS57 Frequency Dividers

The **74LS56** and **74LS57** are counters in an 8-pin DIP that are used as frequency dividers. Their logic diagram and wave forms are shown in Fig. 8-13. They divide by 50 and 60 respectively. Thus, the **74LS56** is ideal for generating a 1 Hz wave from the standard European power lines (50 Hz), and the **74LS57** can generate a 1 Hz wave from the United States power lines if it is driven by a circuit such as shown in Fig. 5-4. They can also be cascaded to generate any combination of 50 Hz and 60 Hz signals.

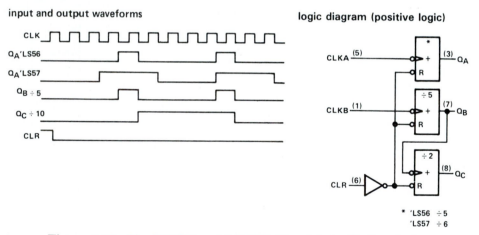

Figure 8-13 The **74LS56** and **74LS57**. (Reprinted with Permission of Texas Instruments.)

EXAMPLE 8-10

Using the waveforms of Fig. 8-13, explain how the **'LS56** and **'LS57** divide the input clock by 50 and 60, respectively.

SOLUTION The waveforms show that Q_A divides by 5 for the **'LS56** (it produces one output pulse for every 5 input pulses) and divides by 6 for the **'LS57**. For both ICs, FF B divides its input by 5 and FF C divides the output of FF B by 2. If the Q_A output of the **'LS56** is connected to the CLKB input, it will provide a divide-by-five circuit followed by a divide-by-ten, or a divide-by-fifty. The frequency of the pulse output at Q_C will be one-fiftieth of the input frequency. The **'LS57** functions similarly, but divides by sixty.

Up-down counters are counters that can either count up or count down depending on the mode of the input. Two of the most popular 4-bit up-down counters available are the **74x192**, which is a decade counter, and the **74x193**, which is a binary counter. These counters are available in the standard, **LS, ALS**, and **HC** series.

8-9-1   The 74x193

The **74x193** is a synchronous, 4-bit, up-down counter with direct CLEAR and direct LOAD capabilities. This IC contains the following useful features:

- It can be cleared
- It can be preset to any number between 0 and 15.
- It can count-up.
- It can count-down.

The block diagram of the **74x193** is shown in Fig. 8-14 with the pin numbers for each input and output signal in parentheses. The T on each FF stands for toggle and the bubble indicates the FF toggles on the negative-going edge of a pulse. This is equivalent to a J-K FF where J = K = 1.

There are six outputs available from a **74x193**:

1. The CARRY output.
2. The BORROW output.
3. The four FF outputs—Q_A, Q_B, Q_C, and Q_D—which retain the count. Q_D is the most significant bit of the count.

The inputs to the **74x193** consist of:

1. A CLEAR line.
2. A LOAD input.
3. Four data inputs (A,B,C,D).
4. A COUNT-UP input.
5. A COUNT-DOWN input.

The function of each input and output is explained below and is best understood by referring to Fig. 8-14.

- *Clear:* The counter can be cleared (all FF outputs 0) by placing a HIGH level on the CLEAR input. This input should be tied to ground if it is not to be used.
- *Load:* The counter can be LOADed, or PRESET. Whenever the LOAD input is taken LOW, the number on inputs A, B, C, and D will be jammed into the four FFs.
- *Count-Up:* The count-up mode is entered by holding CLEAR low (the counter is not being cleared), LOAD high (the counter is not being loaded), and the count-down input HIGH. Then the counter

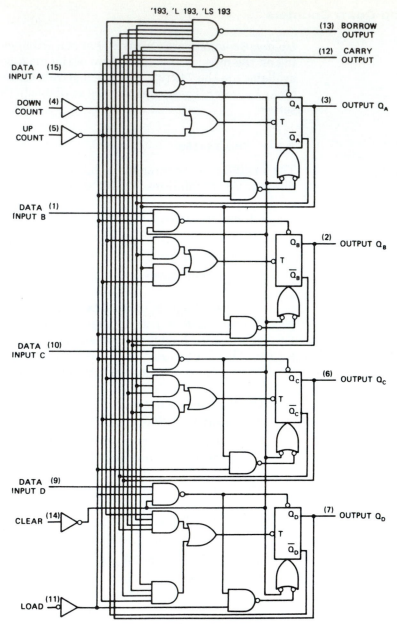

Figure 8-14 Block diagram of the **74x193**. (From the *TTL Data Book for Design Engineers*, 2nd ed., Texas Instruments, Inc. Courtesy of Texas Instruments, copyright 1976.)

will increment whenever a positive edge is applied to the count-up input.

- *Count-Down:* The count-down mode is similar to count-up. In the count-down mode, the count-up input must be HIGH, and the counter will decrement on each positive edge applied to the count-down input.
- *Carry-Out:* The carry-out pin goes LOW only if the counter is in count-up mode, the count in the FFs is 15, and the level on the count-up input is LOW.

- *Borrow-Out:* The borrow-out pin goes LOW only if the counter is in count-down mode, the count in the FFs is 0, and the level on the count-down input is LOW.

8-9-2 Cascading 74x193s

The CARRY and BORROW outputs are used when cascading several **74x193**s to accommodate a larger count than a single IC can handle. To cascade stages of an UP-COUNTER, the CARRY output of the least significant **74x193** must be connected to the COUNT-UP input of the next **74x193**, as shown in Fig. 8-15a. If, for example, the count in IC1 is 15, the counter operates as follows.

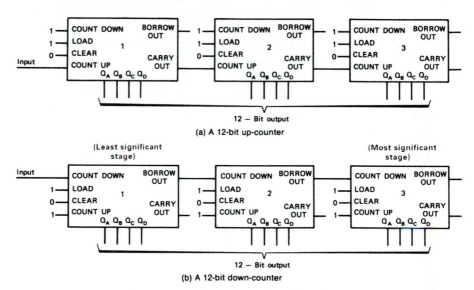

Figure 8-15 12-bit counters made by cascading **74x193**s.

1. When the COUNT-UP line goes LOW, the CARRY-OUT line, which is connected to the COUNT-UP line of IC2, also goes LOW.
2. The next positive transition of the COUNT-UP line causes IC1 to roll over from all 1s to all 0s.
3. This, in turn, causes the CARRY-OUT line to go HIGH, applying a COUNT-UP pulse to IC2.
4. Therefore, after the count pulse, IC1 is 0 and IC2 contains a count of 1. The output of the counter is now:

$$0000 \quad 0001 \quad 0000$$

$$IC3 \qquad IC2 \qquad IC1$$

This is binary 16.

5. IC2 does not increment again until IC1 rolls over once again.

IC3 receives a COUNT-UP pulse only when both IC1 and IC2 have outputs which are all 1s. Then the counter functions as follows.

1. When the input goes LOW, the CARRY output of IC1 goes low.
2. IC2 now has a low input on its COUNT-UP line and, because it contains all 1s, it produces a low level at its CARRY output. This is connected to the COUNT-UP line of IC3.
3. When the input again goes HIGH, IC1 and IC2 both receive a COUNT-UP pulse, which rolls them over to all 0s.
4. IC3 also receives a COUNT-UP pulse. Therefore, IC3 increments once for every 256 input pulses.

When it is counting down, a **74x193** rolls over from all 0s to all 1s if it receives a DOWN-COUNT pulse when its count is 0. DOWN-COUNTERS may be cascaded by connecting the BORROW-OUT line to the COUNT-DOWN line of the next IC, as shown in Fig. 8-15b. Assume, for example, that the count in the DOWN-COUNTER is 48, so that IC1 contains a 0 and IC2 contains a 3. This makes the count:

$$0000 \quad 0011 \quad 0000$$

$$IC3 \qquad IC2 \qquad IC1$$

Where the four least significant bits are in IC1, the next four bits are in IC2, and the most significant bits are in IC3. The circuit now operates as follows.

1. When the DOWN-COUNT line goes LOW, the BORROW output of IC1 and the DOWN-COUNT input to IC2 also go LOW.
2. When the DOWN-COUNT line goes HIGH again, IC1 rolls over to all 1s.
3. The BORROW output of IC1 now goes HIGH, causing IC2 to receive a DOWN-COUNT pulse and decrement to 2.
4. The count now reads:

$$0000 \quad 0010 \quad 1111$$

$$IC3 \qquad IC2 \qquad IC1$$

This is the binary equivalent of 47.

If all stages of a DOWN-COUNTER contain 0s, the next downcount will roll the entire counter over to all 1s, but a BORROW-OUT pulse will appear at the output of the most significant IC to flag the event.[3]

The timing chart of Fig. 8-16 has been prepared to illustrate all the features of the **74x193**.

---

[3]In 2s complement arithmetic (see section 1-9), an array of all 1s represents the number -1. Therefore, if the counter is at 0 and it receives a DOWN-COUNT pulse, it will roll over to all 1s for the number -1. This is very convenient for many applications using 2s complement arithmetic.

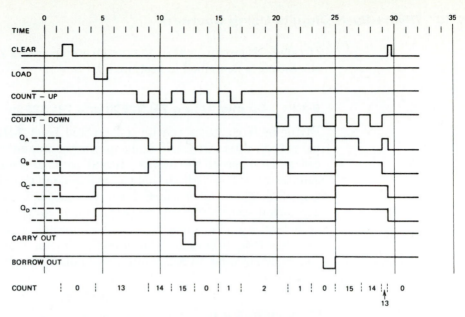

Figure 8-16 Timing chart for a **74x193**.

1. The count is undetermined until $t = 1.5$, when the leading edge of the CLEAR pulse resets all the Q outputs to 0.

2. At $t = 4.5$ a LOAD pulse occurs. The inputs (not shown) are assumed to be 13 (input A = 1, B = 0, C = 1, D = 1) and are transferred to the outputs, making the count 13. The A, B, C, and D inputs have no effect unless the LOAD input is LOW.

3. The LOAD pulse is followed by a series of COUNT-UP pulses. The positive transitions of these pulses occur at $t = 9, 11, 13, 15,$ and 17 and increment the counter to 14, 15, 0 (rollover), 1, and 2.

4. For t between 12 and 13 the count is 15 and the COUNT-UP line is LOW, so that a CARRY-OUT pulse is generated.

5. A series of COUNT-DOWN pulses follows, with positive transitions at $t = 21, 23, 25, 27,$ and 29. This causes the count to decrement from 2 to 1, 0, 15 (rollover), 14, and then 13.

6. A BORROW-OUT pulse is generated when t is between 24 and 25, where the count is 0 and the COUNT-DOWN level is LOW.

7. A positive pulse or spike occurs on the CLEAR line at $t = 29.5$, and RESETS all the outputs.

8-9-2 The 74x192 Decade Counter

The **74x192** is a decade counter, which is identical to the **74x193** in all other respects. All input and output signals are the same, and they are placed on the same pins. Since the **74x192** is a decade counter, it rolls over from 9 to 0 when counting up and from 0 to 9 when counting down. CARRY-OUT and BORROW-OUT signals are generated when the count is 9 or 0 and the proper count line goes LOW. The **74x192** is a useful counter for circuits that work on a decimal base.

Divide-by-N circuits where N is not an even power of 2, can be constructed using the counters described in section 8-9.

8-10-1 Counter Using 74LS290s and 74LS293s

Divide-by-N counters can be constructed from **74LS290**s and **74LS293**s by using their direct CLEARS and a minimum amount of gating. The general procedure is to decode the number N and use it to clear the counter. This type of counter generates a glitch that clears itself and the count is N only for the duration of the glitch, after which it is 0. The output is usually taken from the most significant stage of the counter that changes, and is usually not symmetric. Nevertheless, it produces one output pulse for every N input pulses and satisfies the specifications. If N is an even number, a symmetric output can always be obtained by building an N/2 counter and following it by a divide-by-2 (J-K) FF.

EXAMPLE 8-11

(a) Design a divide-by-87 counter using **74LS290**s. What will the output look like?

(b) If the input is a constant clock of 200 kHz, calculate the time the output is 1 and the time the output is 0 for each output cycle.

SOLUTION (a) Since the count is less than 100, the counter can be built by cascading two **74LS290**s. The circuit is shown in Fig. 8-17a. The most significant IC of the counter, which is connected to $R_{0(2)}$, goes HIGH at a count of 80. The AND gates decode a count of 7 from the least significant stage and are connected to $R_{0(1)}$. At a count of 87 both $R_{0(1)}$ and $R_{0(2)}$ are HIGH and the counter resets.

The waveforms are shown in the CRO traces of Fig. 8-17b. The top trace is Q_{D2}, which is also used to trigger the oscilloscope. In circuits of this type, the oscilloscope must be synchronized by the *lowest frequency generated*; otherwise the oscilloscope won't trigger repetitively (see section 8-10-2).

The bottom trace is Q_{A1} and the glitch is clearly seen. When Q_{A1} is LOW, the count is even because Q_{A1} is the least significant bit (LSB) of the count. The waveforms show that when Q_{D2} goes HIGH, Q_{A1} goes LOW; this is a count of 80. The count then progresses to a count of 87, which is the glitch that resets the counter to 0.

If the input frequency were decreased, all of the pulses in Fig. 8-17b would get longer except the glitch. Thus, the glitch would occupy a smaller portion of the trace and would be more difficult to see.

(b) The output should be taken at the point that changes least often, Q_{D2} in this case. If the input is 200 kHz, then each input point is 5 μs (the input to Fig. 8-17b is much faster to bring out the glitch). Q_{D2} is up for 7 counts (80−86) plus the glitch time, and down for 80 counts (0−79) minus the glitch time. If the glitch time is ignored, Q_{D2} is up for 35 μs and down for 400 μs. Thus, a 5-μs input produces a 435-μs output and the circuit divides the input frequency by 87.

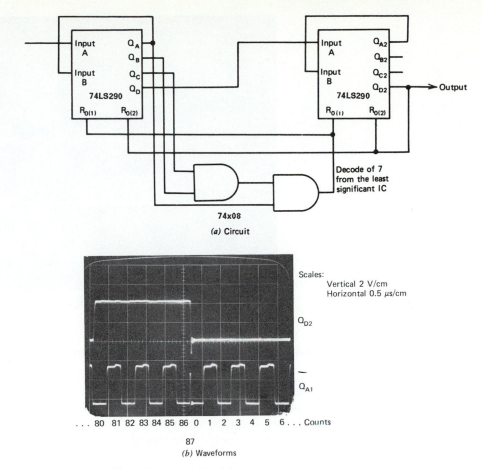

(a) Circuit

Scales:
Vertical 2 V/cm
Horizontal 0.5 μs/cm

Q_{D2}

Q_{A1}

. . . 80 81 82 83 84 85 86 0 1 2 3 4 5 6 . . . Counts

87

(b) Waveforms

Figure 8-17 A divide-by-87 counter.

8-10-2 Oscilloscope Shadows[4]

If the waveforms of complex circuits are to be observed on an oscilloscope (CRO), the CRO should always be triggered on the stage that changes least often (Q_{D2} in Fig. 8-17). Figure 8-18 shows what happens if we attempt to observe Q_{A1} while also triggering from it; a common mistake. Note that the trace is clear on the left side, but shadows start to come in and get progressively deeper as we move to the right.

The reason for the shadows is that the CRO is being triggered by an *almost repetitive waveform*. Q_{A1} is an identical pulse 86 times, but a glitch the 87th time. If the glitch is within the trace, it inverts the waveform and causes the shadow. As we move farther to the right, there are more pulses following the trigger, and a higher probability that the glitch is within them; hence the deepening shadows.

---

[4]This section contains advanced material and may be omitted on first reading.

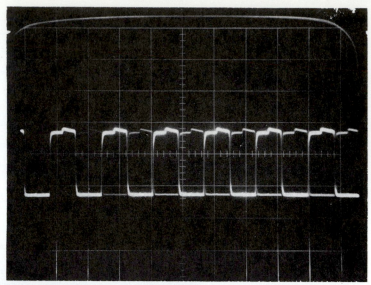

Figure 8-18 Oscilloscope shadows caused by triggering on an almost repetitive waveform.

8-10-3 Divide-by-N Counters Using 74x193s and 74x192s

The simplest way to design divide-by-N counters using **74x192**s and **74x193**s is to decode the desired count and connect the output of the decoder to the CLEAR input of the IC. When the count reaches N, the decoder output clears the counter, and it starts counting again from 0. The IC is cleared by a glitch. This is similar to designing a counter made up of **74290**s (section 8-10-1).

EXAMPLE 8-12

Design a divide-by-147 counter using **74x192**s or **74x193**s. Glitch clears are acceptable.

SOLUTION   Since 147 is a number requiring three decimal digits, three **74x192**s would be required, but 147 = 10010011, which is an 8-bit binary number and can be contained in two **74x193**s; therefore, the **74x193** is selected for this design. The circuit is shown in Fig. 8-19. The binary equivalent of 147 is decoded by the **74x20** and applied to the CLEAR inputs of the **74x193**s. The **74x04** inverts the decoded pulse to make the CLEAR pulse positive.

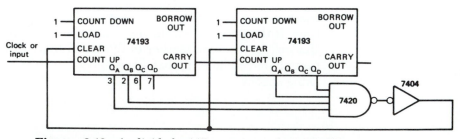

Figure 8-19 A divide-by-147 counter using **74x193**s as an up-counter.

Down counters provide a more elegant solution to the problem of Example 8-12, a solution that requires no external gates. The number N can be placed on the inputs to the **74x193**s as shown in Fig. 8-20a, and the BORROW-OUT signal of the most significant **74x193** connected to the LOAD input of both **74x193**s. The input pulses are applied to the COUNT-DOWN line. When the count reaches 0 and the clock goes LOW, the BORROW-OUT pulse from the most significant **74x193** loads the number N back into the counter. Figure 8-20a is a divide-by-147 counter using this circuit.

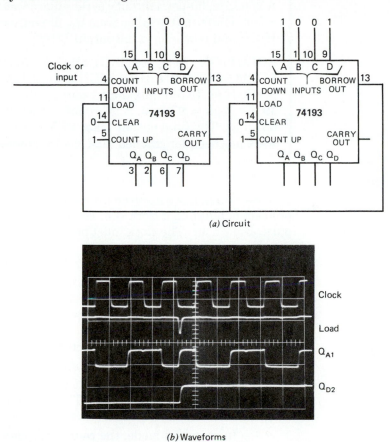

(a) Circuit

(b) Waveforms

Figure 8-20 A divide-by-147 counter using **74x193**s as a down-counter.

The waveforms are shown in Fig. 8-20b. The top trace is the input.[5] The second trace is the BORROW-OUT of the most significant stage. It is a glitch that only occurs when the count is 0 and the input goes LOW. It loads the number 147 into the counter. The third trace is Q_{A1}, the least significant bit of the count. When the counter rolls over, Q_A is 0 for half an input pulse (when the count is 0, but before the input goes LOW), and 1 for half a count (when 147 is loaded into the counter). Otherwise Q_A changes only on positive edges of the input clock. The bottom trace is Q_{D2}, the output and MSB of the counter. It is 0 when the count is LOW, but goes HIGH when 147 is loaded into the counter. Note that the count changes on the positive edge of the input but the glitch oc-

[5]Delayed triggering was used to obtain the CRO traces.

curs on the negative edge because the BORROW-OUT goes LOW at this time. During a complete output cycle there are 146 full counts (counts 146–1), and the counts 0 and 147 each last for half of the input pulse. There are no glitches on Q_A in this circuit; the only glitch is on the BORROW-OUT line.

EXAMPLE 8-13

A 200 kHz, 50-percent duty cycle square wave is applied to the input of Fig. 8-20a. The output is taken from Q_{D2} during each cycle. How long is the output HIGH and how long is the output LOW?

SOLUTION Q_{D2} goes HIGH when the **74x193**s are loaded by the glitch on BORROW-OUT-2. It stays HIGH until the count decrements to 127. Thus, it is HIGH for 19.5 counts (half a count at 147 plus counts 146 through 128), or 19.5×5 μs = 97.5 μs. It is low for 127 counts plus half the count at 0 or for 637.5 μs. The final output takes $97.5 + 637.5$, or 735 μs, so the 5-μs input wave has been lengthened (and the frequency divided) by 147.

As a final, more involved example using these counters, let us consider the problem of building a clock that counts hours. The clock is assumed to get one pulse every hour. The clock must be able to count in one of two modes:

a)   *Standard:* the clock counts hours from 1 to 12.
b)   *Military:* the clock counts hours from 0 to 23.

After a little thought, the following points present themselves:

1.   A switch is required to select the standard or military mode of operation.
2.   Because the maximum count is 23, two cascaded counters are needed.
3.   The output should be in BCD rather than binary. This indicates that **74x192**s should be used. Then the output will be in BCD and it can drive displays using a **74x47** (see section 11-10-3) directly.
4.   In standard mode, the count must start at 1. This suggests that the counter must be LOADed, and should not be CLEARed.

In accordance with this thinking, the counter of Fig. 8-21 was designed. The counter itself consists of two cascaded **74x192**s. The output of the counter is the hours in BCD. All direct inputs except the LSB are wired to ground.

In standard mode, the switch is open and the LSB input to the counter is a 1. AND gates 1, 2, and 3 form a 4-input AND gate. Their output is 1 if the switch is in standard mode and the counter output is 13 (Q_{A1}, Q_{B1}, and Q_{A2} are all HIGH). As soon as the counter reaches 13, the positive output of AND gate 3 is inverted by the **74x02** and LOADs the counter with a 1. Thus, the count progresses from 12 to 13 to 1, but the count of 13 is just a glitch.

In military mode, all the direct inputs to the **74x192**s are 0. AND gates 1, 2, and 3 will never produce a HIGH output because the mode line is now 0. This line is inverted, however, and fed into AND gate 4. In this mode, when the count reaches 24 (Q_{B2} and Q_{C1} both HIGH), gate 5 will produce a HIGH output, which will come through the **74x02** and load a 0 into the counter.

One final observation: The inverter was not identified. It could be a **74x04** of course, but it might also be another gate on the **74x02**. This could save an IC when the circuit is built.

8-11 Testing Counters

This section describes methods of testing IC counters in the laboratory. Perhaps the simplest way is to connect the output of each stage of the counter to a light or a light-emitting diode (LED) and connect the clock input to a switch. Now every switch throw should increment the counter, and the count will be displayed in the lights. If the counter is operating properly, it will progress as specified.

An alternate method of testing counters is to connect the input to a high-frequency clock. Then the counter will cycle around quickly, and its performance can be monitored on an oscilloscope. To test the hours clock of Fig. 8-21, for example, we would replace the one-hour clock by a 1 μs clock and let it run.

This method of testing counters is very good for *debugging* faulty counters. Every point in the circuit can be monitored and this will help locate the fault. The oscilloscope must always be triggered from the most significant stage, the stage that changes least often. This is the only way to obtain a set of reliable traces. The traces of Fig. 8-17 were obtained by triggering the oscilloscope on the positive edge of Q_{D2}. Sometimes it is necessary to delay the scope trace because the part of the waveform of interest occurs both before and after the trigger. Figure 8-20 was obtained that way.

When circuits become larger and more complex, *logic analyzers* might help in testing them. Logic analyzers are discussed in Chapter 19.

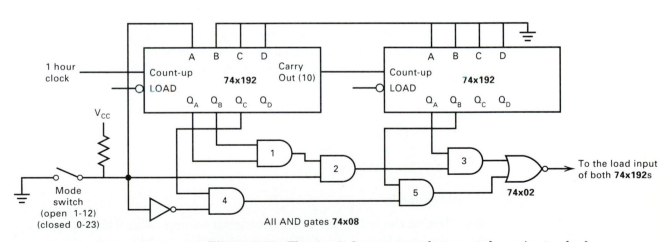

Figure 8-21 The circuit for a counter that counts hours in standard or military time.

SUMMARY

In this chapter we considered the design of counters to count to any arbitrary number N. First ripple and synchronous counters were discussed; then, the use of existing IC counters was explained. Simple methods were developed for making these IC counters count to any desired number.

GLOSSARY

Binary counter: A K stage counter that recycles after 2^K counts.

BORROW-OUT: A pulse indicating that a counter is about to underflow.

CARRY-OUT: A pulse indicating that a counter is about to overflow.

Decade counter: A counter that recycles after 10 pulses.

Decoder: A circuit or gate that detects when a counter reaches a specific count.

LOAD: An input that causes a counter to be directly set.

Ripple counter: A counter built so that the stages change sequentially. A particular stage will change only if the previous stage changes.

Synchronous counter: A counter built so that all stages change simultaneously in response to a clock pulse.

UP-DOWN counter: A counter that will count either up or down, depending on its inputs.

PROBLEMS

Section 8-5.

8-1.   Design a circuit to decode a count of 23 from a 5-stage ripple counter.

8-2.   Design a counter to decode a count of 0 from a 16-stage ripple counter using ICs that have been previously discussed. Find the worst case delay through your counter and decoder. Use standard series ICs.

8-3.   Find the time delay of a 10-stage ripple counter in
   (a)   The standard family.
   (b)   The **LS** family.
   (c)   The **ALS** family.
   (d)   The **HC** family.

Section 8-6.

8-4.   Design the fifth stage of a synchronous counter using **74x74**s.

8-5.   Design a 5-stage synchronous counter using **74x107**s under the following assumptions:
   (a)   Minimum time delay is the most important consideration.
   (b)   Minimum hardware is the most important consideration.

Section 8-7.

8-6. Design a synchronous divide-by-12 counter.

8-7. Design a divide-by-48 counter that counts from 0 to 47.

8-8. Design a divide-by-18 counter and construct your count table.

8-9. **(a)** Design a divide-by-27 counter by cascading three divide-by-3 counters and construct your count table.

  (b) Design a divide-by-27 counter using five FFs and a decoder.

8-10. **(a)** Draw the timing chart and find the count sequence for the counter of Fig. 8-6.

  (b) Repeat this problem if the 3s counter precedes the divide-by-4 counter.

Section 8-8.

8-11. Design a divide-by-7 counter:
  (a) Use **74x107**s.
  (b) Use **74x74**s.
  Hint: Set up a circuit to detect a count of 6 and cause the FF to roll over on the next count.

8-12. Draw a timing chart for the four outputs of a **74x90** as the input is clocked. Notice the Q_D output is not symmetrical. How can the highest output be made symmetrical?

8-13. Given a 1 MHz clock, design a circuit to produce a 25 kHz clock.

8-14. Example 8-8 could have been solved by using a divide-by-12 followed by a divide-by-50, or a divide-by-6 followed by a divide-by-100. Draw these circuits. Which have symmetrical outputs?

8-15. Given a 1.2 MHz clock, design a circuit to produce a 100-Hz clock.

8-16. Given a 1 MHz clock, design a circuit to produce a 12.5 kHz clock. Use no more than two chips.

8-17. Design a divide-by-2048 counter.

Section 8-10.

8-18. Design a circuit to count from 5 to 15 and then repeat. Use only one chip.

8-19. Build a divide-by-75 counter using:
  (a) **74LS290**s.
  (b) **74LS293**s.
  In each case show specifically where you take the outputs from and describe the outputs.

8-20. What is the count of the circuit of Fig. P8-20 if the counters are:
  (a) **74LS290**s
  (b) **74LS293**s

8-21. Design a divide-by-138 counter using two **74LS290**s and a **74x107**. Produce a symmetrical output.

8-22. Design a divide-by-205 counter:
  (a) Using **74x193**s as UP-COUNTERS.
  (b) Using **74x193**s as DOWN-COUNTERS.

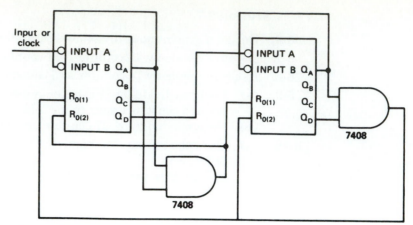

Figure P8-20

8-24. Figure P8-24 shows a timing chart for a **74LS290** counter. What are the various counts shown? Where does the counter roll over?

8-25. Figure P8-25 shows the count sequence for a **74x193** counter. What are the various counts? Expand the figure and show the action of the two BORROW-OUT lines.

8-26. Design a divide-by-166 counter. Show where you take your outputs.
   (a) Use two **74LS290**s and a J-K FF.
   (b) Use two **74x193**s in the countdown mode.
   (c) If the input is a 100 kHz symmetric square wave, how long is your output HIGH and how long is your output LOW in each case?

8-27. An input line is supposed to produce a microsecond pulse every 9 milliseconds. Occasionally, it misses a pulse, but it never misses two in a row.

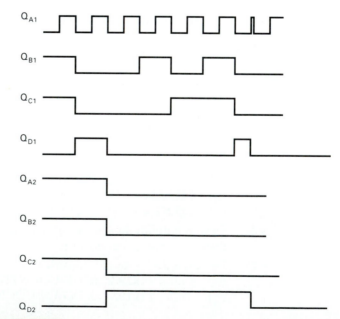

Figure P8-24

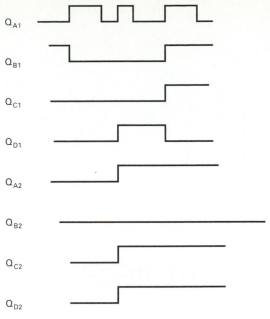

Q_{A1}

Q_{B1}

Q_{C1}

Q_{D1}

Q_{A2}

Q_{B2}

Q_{C2}

Q_{D2}

Figure P8-25

Design a circuit to detect when it has missed 73 pulses and sound an alarm. The alarm is to sound until a reset toggle switch is flipped to the off position. When the toggle switch goes to the on position the circuit restarts. (*Hint:* Retriggerable one-shots do this job nicely.)

8-28. Given: An input line and a pushbutton switch.
  (a) Build a 100 kHz clock.
  (b) Input data is to be sampled on the positive edge of the clock. The third consecutive 1 and subsequent 1s are to be counted. When the count reaches 213, a lamp is to light and the clock must stop. Depressing the pushbutton clears everything. When the pushbutton is again released, the circuit restarts. Ignore switch bounce problems.

8-29. You are an oscilloscope manufacturer and want to include a delay feature on your oscilloscope. You have three BCD thumbwheel switches. Design the circuit to delay the scope by the number of μs set into your thumbwheel switches. Include a delay/no delay option. (*Hint:* **74x192s** might be good ICs to use.)

After attempting the problems the student should return to the questions of section 8-2 and be sure all can be answered. If the student does not understand them all, he or she should reread the appropriate sections of the chapter to find the answers.

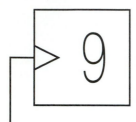

State Tables

9-1   Instructional Objectives

State tables are a tool that allow the digital engineer to analyze and design many complex circuits. After reading this chapter the student should be able to:

1. Determine the next state of a circuit given its present state and the state of any external variables.
2. Construct a bubble diagram for a given circuit.
3. Construct a state checkout table for a given circuit.
4. Construct a state table and use it to design a circuit.
5. Design circuits that function as up-down counters, input sequence detectors, and traffic controllers.

9-2   Self-Evaluation Questions

Watch for the answers to the following questions as you read the chapter. They should help you understand the material presented.

1. How is the numerical value of a state related to the condition of the FFs in the circuit?
2. Does the inclusion of an external variable increase the number of states? Does it increase the complexity of analysis?

3. What are extra states? What are minor loops?

4. What are the advantages of using D-type FFs in state table design? What are the advantages of J-K FFs?

9-3 Introduction

Many digital circuits can be described as existing in one of a limited number of *states* or *conditions*. As an example, consider a traffic light controlling a North–South (N–S) road intersecting with an East–West (E–W) road. In its simplest form the traffic light may assume one of four states:

State	Condition
1	N–S green, E–W red.
2	N–S yellow, E–W red.
3	N–S red, E–W green.
4	N–S red, E–W yellow.

The traffic light will progress from state 1 to 2 to 3 to 4 and back to 1. The light may be controlled by a set of Flip-Flops (FFs) that change synchronously with a clock. State tables and state diagrams are used to design or analyze such a system.

Many digital circuits consist of a set of FFs and gates. The state of these circuits can best be described by a *number*, which is determined by the conditions (SET or CLEAR) of the FFs in the circuit. The circuit of Fig. 9-1 can be used as an example. Note that the FFs in Fig. 9-1 are *synchronous*; they all change at the negative edge of the clock. Circuits that are designed using state tables are usually synchronous.

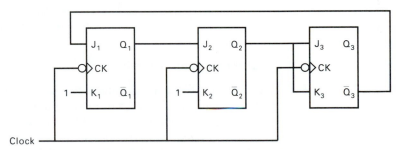

Figure 9-1   The circuit for Examples 9-1 and 9-2.

EXAMPLE 9-1

What state is the circuit of Fig. 9-1 if $Q_1 = 1$, $Q_2 = 0$, and $Q_3 = 1$?

SOLUTION   The FFs can be read as 101 or 5. This circuit is now in state 5.

EXAMPLE 9-2

Given that the circuit is in state 5, what state will it go to after the next negative clock pulse?

SOLUTION The next state can be determined by finding the values on the J and K inputs to each FF. The initial state of the circuit is 5, which means that Q_1 and Q_3 are SET and Q_2 is CLEAR. In this state $\bar{Q}_3$, which is connected to J_1, is 0. Therefore, FF1 has $J_1 = 0$ and $K_1 = 1$. It will clear on the next clock pulse. Proceeding to FF2, J_2 is a 1 before the clock because the FF1 is SET. It will toggle. Because it is presently in the 0 state, it will set. FF3 sees 0s on both J_3 and K_3. It will not change state; it will remain set. Therefore, the next state is $Q_3Q_2Q_1 = 110$ or 6.

9-3-1 Bubble Diagrams

A state or bubble diagram is a group of circles, or bubbles, with the number of the state within the bubble, and a set of arrows showing how the system progresses from state to state. Figure 9-2 is a bubble diagram segment for Example 9-2. It shows that the circuit progresses from state 5 to state 6. Using the methods of Example 9-2, we can determine that after state 6 the circuit goes to state 0, and so on (see Problem 9-1). A bubble diagram for the main loop of this circuit is shown in Fig. 9-3. It shows that the states progress 0,1,2,5,6,0 . . . .

⑤ ⟶ ⑥   **Figure 9-2**   A bubble diagram segment.

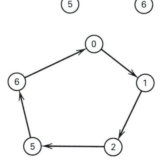

Figure 9-3   A bubble diagram for the main loop of the circuit of Figure 9-1.

9-3-2 State Checkout Tables

A state checkout table is a systematic way to determine how the states progress in a given system; it is a tool for *analyzing* a given circuit. Table 9-1 is the state checkout table for the circuit of Fig. 9-1. It was started when the circuit was in state 5, the given state. Next the Q values (101) were written. From the Q values, the J and K values on each FF were listed; this allows the engineer to determine which state will be entered after the next clock pulse.

The top line of Table 9-1 shows that the circuit progresses from state 5 to state 6. State 6 becomes the present state on line 2. The states progress until it returns to state 5. After that it will recycle, as shown in Fig. 9-3.

9-3-3 Extra States and Hang-up States

There are three FFs in Fig. 9-1. This implies eight possible states. The bubble diagram of Fig. 9-3 contains only 5 states; states 3, 4, and 7 are not in the main loop. If the circuit ever gets into a main loop state, it will continue to circulate through the main loop and never enter one of these extra states. If the circuit

TABLE 9-1

A State Checkout Table for the Circuit of Fig. 9-1.

Present State	$Q_2 Q_1 Q_0$	$J_2 K_2$	$J_1 K_1$	$J_0 K_0$	Next State $Q_2 Q_1 Q_0$	Numerical Next State
5	1 0 1	0 0	1 1	0 1	1 1 0	6
6	1 1 0	1 1	0 1	0 1	0 0 0	0
0	0 0 0	0 0	0 1	1 1	0 0 1	1
1	0 0 1	0 0	1 1	1 1	0 1 0	2
2	0 1 0	1 1	0 1	1 1	1 0 1	5

were to come up in one of the extra states, as it might when power is first applied, it could cause the circuit to malfunction. Therefore, the behavior of the circuit in these states should be checked.

To check the progress of the circuit if it is indeed in an extra state, we can assume it is in such a state and analyze it using the method of Example 9-2, or by the State Checkout Table. If the circuit of Fig. 9-1 comes up in state 3, it will progress to state 4. If it is in state 4, however, it will remain in state 4 (J_1, J_2, J_3, and K_3 will all be 0, thus FFs 1 and 2 will CLEAR, while FF3, which is SET, will not change). This is called a *hang-up situation*; if the circuit enters state 4 it never leaves. Another possible problem, which does not exist in this circuit, is a *minor loop*. The circuit could go from state 4 to state 3 and then back to state 4, for example. If it entered the minor loop it could never get to the major loop.

If the circuit comes up in state 7 it will progress to state 0, and then around the main loop. It will never go back to state 7.

A complete bubble diagram, which takes all possible states into account, is shown in Figure 9-4. It shows that state 7 progresses to state 0 and then around the loop. It also shows that state 3 progresses to state 4. Notice that the output of state 4 loops back on itself. This shows the hang-up situation.

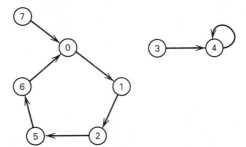

Figure 9-4 A complete bubble diagram for the circuit of Figure 9-1.

There are two possible ways to prevent the circuit from getting stuck in the hang-up state:

1. Add additional logic to exit the hang-up state. In this circuit, logic could be added so that the next state after state 4 is a state in the main loop. This is discussed in section 9-5-3.

2. Connect a CLEAR pulse to the direct CLEARs of all FFs. Here, the CLEAR pulse would set the circuit into state 0, which is in the main loop.

Method 1 requires more thought and more logic but is preferable because the CLEAR pulse is asynchronous and, more importantly, if the circuit ever gets to the hang-up state (perhaps caused by a glitch), it will not exit until another CLEAR pulse occurs.

9-3-4   A Simple Example

Some problems can be solved with a little thought, and do not need the more formal methods presented later in this chapter. As an example, consider an elementary traffic light controller. In the simplest case (a more complex case is presented in section 9-6), where the yellow light is allowed to stay on as long as the green light, it may appear that six states are needed:

1. N–S green
2. N–S yellow
3. N–S red
4. E–W green
5. E–W yellow
6. E–W red

But N–S green or N–S yellow implies that E–W must be red and vice versa. Therefore, only four states are required. This implies only two FFs. We could assign the states as:

State	Q_1	Q_0	Condition
0	0	0	N–S green
1	0	1	N–S yellow
2	1	0	E–W green
3	1	1	E–W yellow

All that is needed to implement this circuit is a synchronous counter (see section 8-6) and some gating. The circuit is shown in Fig. 9-5, with the 6 outputs and their corresponding states clearly marked.

9-4   State Tables

Checkout tables are used whenever the circuit and its starting state are given. They allow the engineer to *analyze* an existing circuit. State tables are used when *designing* a particular circuit. The design typically starts by specifying a particular sequence of states that the final circuit must assume. Then the circuit that will progress through those states must be designed.

Perhaps the simplest example of the use of a state table design is the design of a *truncated counter*. A truncated counter does not progress through all possible states, but truncates or starts over after a specific count is reached. We will start by designing a counter that counts 0,1,2,3,4,0 as shown in the bubble diagram of Fig. 9-6. It can be said that this counter truncates after the count of 4.

The first step in the design is to determine how many states there are and how many FFs will be required. In this example there are five states and three

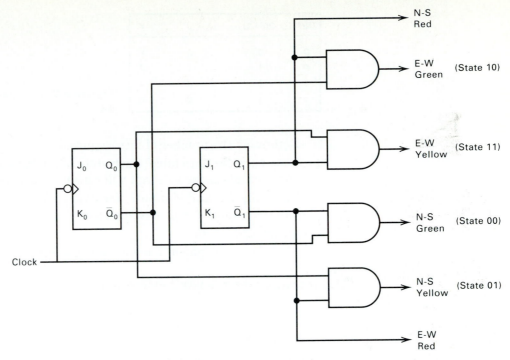

Figure 9-5 A very simple traffic controller.

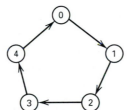

Figure 9-6 A bubble diagram for a truncated counter.

FFs will be needed because three is the smallest number such that $2^N \geq 5$. Next the designer has to decide whether to use D or J-K FFs. J-K FFs are simpler, because the design will lead to Karnaugh maps, and the J-K FFs will have more *don't cares*; this will simplify the maps and lead to circuits with fewer gates. D-type FFs, however, are important because they are contained in PALs (see chapter 11). Examples of both designs will be presented, but this first example will use J-K FFs.

9-4-1 Transition Tables

Before a counter can be designed, the basic transition table for the type of FF to be used must be constructed. The basic transition table for the J-K FF is shown in Fig. 9-7, where $Q(t)$ is the state of a FF before the clock pulse and $Q(t + 1)$ is the desired state of the FF after the transition. The states of the J and K inputs required to cause the transition are also given in the table. For example, if the present state of a FF is a 0, and the next state is to be a 0, the table shows that J must be a 0 but K is a don't care. If J is a 0, the FF will remain cleared regardless of the K input. Once the desired states of a counter have been specified, the counter can be designed by the following steps:

Q(t)	Q(t + 1)	J	K
0	0	0	d
0	1	1	d
1	0	d	1
1	1	d	0

Figure 9-7 The basic transition table for J-K FFs.

1. Determine the number of FFs required.
2. Construct the state table. Each state of the table should include the J and K inputs to the FFs that will cause it to progress to the next state.
3. Using these J and K inputs, construct a Karnaugh map so that the required inputs can be designed into the circuit.
4. Sketch the circuit.
5. Check it out on paper using a state checkout table.

This procedure is best illustrated by examples.

9-4-2 Design of the Truncated Counter

The steps listed in the preceding paragraph may be used to design the truncated counter that counts 0,1,2,3,4,0 . . . .

- **Step 1.** Determine the number of states and FFs required. This has already been done. Here we need five states and three FFs.
- **Step 2.** Construct the state table. The state table for this counter is shown in Fig. 9-8a. It was constructed in accordance with the following substeps:

1. The three FFs are designated as Q_3, Q_2, and Q_1, where Q_1 is the least significant bit. All possible states of the three FFs are listed in column 1, and their corresponding Q outputs are listed in columns 2, 3, and 4 just as they would be written in a truth table.
2. The desired next state, which is the state that must follow the present state, is listed in column 5, and the corresponding FF outputs are then listed in columns 6, 7, and 8. For example, if the present state of the counter is 0 ($Q_3Q_2Q_1 = 000$), the next state must be 1 ($Q_3Q_2Q_1 = 001$) as specified by the given count sequence, or if the present state is 4, the next state must be 0.
3. Since the count sequence goes from 0 through 4, the counter should never be in states 5, 6, or 7. The next state for these present states is arbitrary. To simplify the design, the next states are listed as don't cares.
4. Columns 9 through 14, containing the required J and K conditions to cause each transition, can now be filled in. Here, use is made of the J-K FF transition table, Fig. 9-7. When the counter goes from 0 to 1, Q_2 and Q_3 do not change, but Q_1 changes from 0 to 1. Therefore, J_3 and K_3, and J_2 and K_2, both have values of 0 and d, respectively, since the J-K transition table indicates these are the values J and K must have for a FF

in the 0 state to remain in the 0 state. But Q_1 SETS, so that J_1 must be a 1 while K_1 can be either 1 or 0.

5. To go from state 1 to state 2 (the second line in the table), Q_1 must CLEAR, Q_2 must SET, and Q_3 must remain CLEARED. The J and K inputs required to cause these transitions are listed in columns 9 through 14 of line 2. The rest of the J and K inputs were also obtained by going from the present state to the next state in accordance with the J-K transition table. Where the next state is a don't care, all the J and K inputs are written as don't cares.

- **Step 3.** Draw the Karnaugh maps. Now that the state table is finished, six Karnaugh maps must be drawn (see Fig. 9-8b), one for each J and K input. The Q variables in the Karnaugh maps correspond to the present state table. Since Q_3 is the most significant variable, it is written across the top of the map, and the $Q_2 Q_1$ variables are written down the side in order of significance. For example, J_3 is a 0 for present states 0, 1, and 2, a 1 for state 3, and a don't care otherwise. These values are placed in the J_3 Karnaugh map and the minimal simplification results in the equation: $J_3 = Q_1 Q_2$.

  Fortunately, many don't cares simplify the Karnaugh maps and the circuitry. The minimal expression for each J and K input is obtained and the six equations are written below each Karnaugh map. Where the Karnaugh map contains no 0s (K_3 and K_1 in this example), the J or K input should be permanently HIGH or 1.

- **Step 4.** Sketch the circuit. With the Karnaugh maps finished, the circuit can be constructed. It is drawn as shown in Fig. 9-8c. It is started by drawing the three FFs and the clock. Then the J and K connections are made in accordance with the six equations obtained from the Karnaugh maps, and the circuit is complete. For this counter only J_3 required a gate. All other Js and Ks were either connected to 1s or to another Q or $\bar{Q}$ output.

- **Step 5.** Check the circuit. After the circuit is drawn it should be checked out on paper by constructing the state checkout table as shown in Fig. 9-8d. Since 0 is in the count sequence, it is convenient to start by assuming a present state of 0. (Any other number in the count sequence would also work.) Therefore, Q_1, Q_2, and Q_3 are all 0. Looking at the circuit diagram (Fig. 9-8c), we see that J_2, K_2, and J_3 are also 0 but that J_1, K_1, and K_3 are all 1s. The J and K values are now written into the state checkout table.

  The Q outputs for the next state can now be determined. Since J_1 and K_1 are both 1, FF1 will toggle to the 1 state. Since J_2 and K_2 are both 0, FF2 will not change state, and FF3 will also remain a 0 because J_3 is a 0 and K_3 is a 1. The next state Q outputs are found to be 001 so that the next state is 1.

  The second line of the checkout table is written using the next state from the line above. Thus, the present state on the second line is 1, the next state result from line one. The J and K inputs for this state are found and a new next state is determined. This procedure continues until the circuit returns to its original state. The count found using the table should agree with the specification of the problem.

1	2	3	4	5	6	7	8	9	10	11	12	13	14
Present State	Q_3	Q_2	Q_1	Next State	Q_3	Q_2	Q_1	J_3	K_3	J_2	K_2	J_1	K_1
0	0	0	0	1	0	0	1	0	d	0	d	1	d
1	0	0	1	2	0	1	0	0	d	1	d	d	1
2	0	1	0	3	0	1	1	0	d	d	0	1	d
3	0	1	1	4	1	0	0	1	d	d	1	d	1
4	1	0	0	0	0	0	0	d	1	0	d	0	d
5	1	0	1	?	d	d	d	d	d	d	d	d	d
6	1	1	0	?	d	d	d	d	d	d	d	d	d
7	1	1	1	?	d	d	d	d	d	d	d	d	d

(a) State table

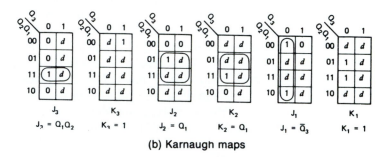

(b) Karnaugh maps

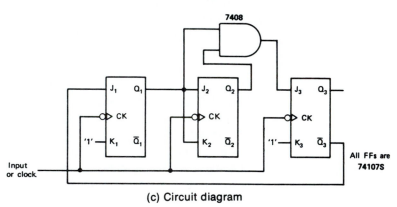

(c) Circuit diagram

Line No.	Present State	Q_3 Q_2 Q_1	J_3 K_3	J_2 K_2	J_1 K_1	Next State Q_3 Q_2 Q_1	Numerical Next State
1	0	0 0 0	0 1	0 0	1 1	0 0 1	1
2	1	0 0 1	0 1	1 1	1 1	0 1 0	2
3	2	0 1 0	0 1	0 0	1 1	0 1 1	3
4	3	0 1 1	1 1	1 1	1 1	1 0 0	4
5	4	1 0 0	0 1	0 0	0 1	0 0 0	0

(d) State checkout table

Figure 9-8   Design of a divide-by-5 counter.

9-4-3 Design of Counters Having Irregular Count Sequences

The methods of section 9-4-2 may also be used to design counters that must have highly irregular count sequences, as Example 9-3 illustrates.

EXAMPLE 9-3

Design a counter to produce the following sequence of outputs: 0, 2, 4, 3, 6, 7, 0, ...

SOLUTION   The solution proceeds as follows.

1. Because there are six distinct counts in this sequence, three FFs are required.
2. As before, the outputs of these FFs are labeled Q_3, Q_2, and Q_1 and the state table is constructed as shown in Fig. 9-9a.
3. All present state possibilities are listed in order, but note the irregular next state listings caused by the specifications.
4. The corresponding next state values of Q_3, Q_2, and Q_1 are then entered in the table and the Js and Ks calculated. Since states 1 and 5 are not in the count sequence, the Js and Ks for these states are don't cares.
5. The Karnaugh maps are drawn as shown in Fig. 9-9b, and the minimal equations obtained. Here K_3 requires an OR gate with Q_1 and $\bar{Q}_2$ as inputs, and K_2 appears to require two AND gates and an OR gate. For K_2, however, a little inspiration simplifies the circuit:

$$K_2 = \bar{Q}_1\bar{Q}_3 + Q_1Q_3 = \overline{(Q_1 \oplus Q_3)} = Q_1 \oplus \bar{Q}_3$$

Hence K_2 is the output of a **74x86** gate whose inputs are Q_1 and $\bar{Q}_3$.

6. The circuit is then drawn as shown in Fig. 9-9c.
7. Finally the state checkout table is carefully constructed as shown in Fig. 9-9d. It verifies that the count sequence conforms to the specifications of the problem.

9-5   Design of Synchronous Counters Using D-Type FFs

Because synchronous D-type FFs are used extensively in Programmable Array Logic (PALs—see Chapter 11), their design should be studied. The secret of designing synchronous counters using D-type FFs is to realize that the *Q outputs for the next state must be the same as the present state D inputs*. The transition table for a D-type FF is given in Fig. 9-10. Knowing the present state of the counter, and its desired next state, is enough to construct the Karnaugh maps and design the circuitry for the D inputs.

The algorithm for using D-type FFs to build counters that progress normally has been given in Section 8-6-2. Counters having irregular count sequences may also be built using D-type FFs as illustrated in Example 9-4.

Present State	Q_3 Q_2 Q_1	Next State	Q_3 Q_2 Q_1	J_3 K_3	J_2 K_2	J_1 K_1
0	0 0 0	2	0 1 0	0 d	1 d	0 d
1	0 0 1	?	d d d	d d	d d	d d
2	0 1 0	4	1 0 0	1 d	d 1	0 d
3	0 1 1	6	1 1 0	1 d	d 0	d 1
4	1 0 0	3	0 1 1	d 1	1 d	1 d
5	1 0 1	?	d d d	d d	d d	d d
6	1 1 0	7	1 1 1	d 0	d 0	1 d
7	1 1 1	0	0 0 0	d 1	d 1	d 1

(a) State table

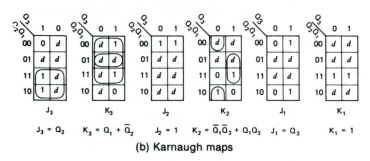

$J_3 = Q_2$ $K_3 = Q_1 + \bar{Q}_2$ $J_2 = 1$ $K_2 = \bar{Q}_1\bar{Q}_3 + Q_1Q_3$ $J_1 = Q_3$ $K_1 = 1$

(b) Karnaugh maps

(c) Circuit diagram

Present State	Q_3 Q_2 Q_1	J_3 K_3	J_2 K_2	J_1 K_1	Next State Q_3 Q_2 Q_1	Numerical Next State
0	0 0 0	0 1	1 1	0 1	0 1 0	2
2	0 1 0	1 0	1 1	0 1	1 0 0	4
4	1 0 0	0 1	1 0	1 1	0 1 1	3
3	0 1 1	1 1	1 0	0 1	1 1 0	6
6	1 1 0	1 0	1 0	1 1	1 1 1	7
7	1 1 1	1 1	1 1	1 1	0 0 0	0

(d) State checkout table

Figure 9-9   Design of the counter of Example 9-3.

Transition table

Q(t)	Q(t+1)	D
0	0	0
0	1	1
1	0	0
1	1	1

Figure 9-10 The transition table for a D-type FF.

EXAMPLE 9-4

Redesign the counter of Example 9-3 using **74x74**s.

SOLUTION   The steps in the solution of this problem are shown in Fig. 9-11:

1. The state table is drawn up as shown in Fig. 9-11a. The present state and next state entries are identical to Fig. 9-9a because the count sequence is the same.

2. The Karnaugh maps for each D input are constructed. Since the D inputs are the same as the next state Q outputs [$D_3 = Q_3(t + 1)$, etc.] the next state Q outputs become the entries in the Karnaugh maps.

3. The minimal equations are taken from the Karnaugh maps and used to implement the circuit of Fig. 9-11c.

4. Finally the state checkout table of Fig. 9-11d is constructed. Again the first state is assumed to be 0. D_3, D_2, and D_1 are determined from the circuit with Q_1, Q_2, and $Q_3 = 000$. The D inputs are calculated for each present state and become the Q outputs of the next state. Thus the D inputs on line 1 become the Q outputs on line 2, and so on. This procedure is continued until the counter rolls around and the next state is 0.

9-5-7 Circuits with Extra States

As explained in section 9-3-3, circuits with extra states can result in a hang-up condition. The circuit of Fig. 9-1 was an example; if it entered state 4, it would remain there indefinitely.

If a state table is constructed for the circuit of Fig. 9-1, it will progress as shown in the bubble diagram of Fig. 9-4. The hang-up state can be eliminated if the next state following state 4 is declared to be a state in the main loop, state 0, for example. This changes the logic slightly and modifies the state table. The state table for the original circuit is given in Fig. 9-12a, and clearly shows that state 4 goes to state 4, clearly a hang-up condition. Figure 9-12b shows how the table is modified to cause state 0 to follow state 4. This changes the K_3 Karnaugh map. The new logic indicates that if K_3 is connected to $\overline{Q_1}$, state 0 will follow state 4 and there will be no hang-up conditions.

9-6   Circuits with External Inputs

Many circuits depend on both their states and the condition of *external inputs*. The external input might be the setting of a switch or the logic level of the output of a gate. An external input can be used to determine the *mode* of a state; it

Present State	Q_3 Q_2 Q_1	Next State	Q_3 Q_2 Q_1
0	0 0 0	2	0 1 0
1	0 0 1	?	d d d
2	0 1 0	4	1 0 0
3	0 1 1	6	1 1 0
4	1 0 0	3	0 1 1
5	1 0 1	?	d d d
6	1 1 0	7	1 1 1
7	1 1 1	0	0 0 0

(a) State table

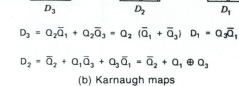

$$D_3 = Q_2\bar{Q}_1 + Q_2\bar{Q}_3 = Q_2(\bar{Q}_1 + \bar{Q}_3) \quad D_1 = Q_3\bar{Q}_1$$

$$D_2 = \bar{Q}_2 + Q_1\bar{Q}_3 + Q_3\bar{Q}_1 = \bar{Q}_2 + Q_1 \oplus Q_3$$

(b) Karnaugh maps

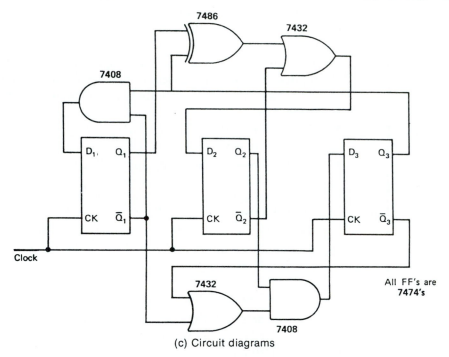

(c) Circuit diagrams

Present State	Q_3 Q_2 Q_1	D_3 D_2 D_1	Next State
0	0 0 0	0 1 0	2
2	0 1 0	1 0 0	4
4	1 0 0	0 1 1	3
3	0 1 1	1 1 0	6
6	1 1 0	1 1 1	7
7	1 1 1	0 0 0	0

(d) State checkout table

Figure 9-11   Circuit for Example 9-4.

Present state			Next state			$J_3 K_3$		$J_2 K_2$		$J_1 K_1$	
Q_3	Q_2	Q_1	Q_3	Q_2	Q_1						
		0									
0	0	1	0	0	1	0	d	0	d	1	d
0	0	0	0	1	0	0	d	1	d	d	1
0	1	1	1	0	1	1	d	d	1	1	d
0	1	0	1	0	0	1	d	d	1	d	1
1	0	1	1	0	0	d	0	0	d	0	d
1	0	0	1	1	0	d	0	1	d	d	1
1	1	1	0	0	0	1	d	d	1	0	d
1	1		0	0	0	1	d	d	1	d	1

$$(K_3 = Q_2)$$

(a)

1	0	0	0	0	0	d	1	0	d	0	d

$$(K_3 = \overline{Q}_1)$$

(b)

Figure 9-12 State tables for Figure 9-1: (a) original circuit; (b) modification of line 4 so that state 0 follows state 4.

does not create additional states, but it does determine how the states will progress. Each external input does, however, become an additional variable that must be considered in the Karnaugh maps as shown in Example 9-5.

EXAMPLE 9-5

How many variables would the Karnaugh map for the circuit of Fig. 9-13 have?

SOLUTION An inspection of Fig. 9-13 shows that the circuit has three FFs (it has, at most, eight states) and two *external* inputs. Thus, the Karnaugh maps for this circuit would have one variable for each FF, and one variable for each external state. Five-variable Karnaugh maps would be required.

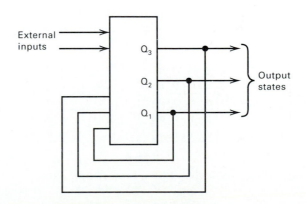

Figure 9-13 A circuit with external inputs.

In this section, three circuits that depend on external inputs will be considered.

9-6-1 Design of an Up-Down Counter

An up-down counter can operate in one of two modes; counting up or counting down. The mode is often selected by a switch. The setting of this switch is the external variable. The design is best illustrated by an example.

EXAMPLE 9-6

Design an up-down counter to count 0, 1, 7, 3, 4, 5, 0. Use J-K FFs. A mode switch can be used to determine if the counter is counting up or down.

SOLUTION The bubble diagram for this circuit is shown in Fig. 9-14. There are two paths between each state; the count-up path when the external input is 0, and the count-down path when the external input is 1. The 0 or 1 next to the path line indicates the condition of the external inputs.

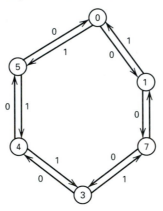

Figure 9-14 The bubble diagram for Example 9-6.

Because the circuit contains six states, it will require three FFs. The three Q outputs plus the external input comprise the four variables that determine the operation of the circuit. The state table is shown in Table 9-2, where the circuit counts up if X is 0 and down if X is 1. The J and K values for each FF can be determined from the state table and are also shown in Table 9-2.

The six Karnaugh maps for the three J and three K inputs are shown in Fig. 9-15. Actually, only five K-maps are required because an inspection of

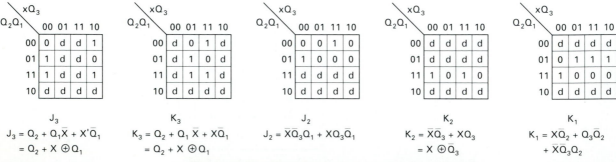

$$J_3 = Q_2 + Q_1\bar{X} + X'\bar{Q}_1$$
$$= Q_2 + X \oplus Q_1$$

$$K_3 = Q_2 + Q_1\bar{X} + X\bar{Q}_1$$
$$= Q_2 + X \oplus Q_1$$

$$J_2 = \bar{X}\bar{Q}_3Q_1 + XQ_3\bar{Q}_1$$

$$K_2 = \bar{X}\bar{Q}_3 + XQ_3$$
$$= X \oplus \bar{Q}_3$$

$$K_1 = X\bar{Q}_2 + Q_3\bar{Q}_2$$
$$+ \bar{X}\bar{Q}_3Q_2$$

Figure 9-15 The Karnaugh maps for Example 9-6.

TABLE 9-2

The State Table for Example 9-6

Numeric Present State	$Q_3\,Q_2\,Q_1$	$x = 0$ Numeric Next State	$Q_3\,Q_2\,Q_1$	$x = 1$ Numeric Next State	$Q_3\,Q_2\,Q_1$
0	0 0 0	1	0 0 1	5	1 0 1
1	0 0 1	7	1 1 1	0	0 0 0
2	0 1 0	d	d d d	d	d d d
3	0 1 1	4	1 0 0	7	1 1 1
4	1 0 0	5	1 0 1	3	0 1 1
5	1 0 1	0	0 0 0	4	1 0 0
6	1 1 0	d	d d d	d	d d d
7	1 1 1	3	0 1 1	1	0 0 1

$x = 0$ $J_3\,K_3$	$x = 1$ $J_3\,K_3$	$x = 0$ $J_2\,K_2$	$x = 1$ $J_2\,K_2$	$x = 0$ $J_1\,K_1$	$x = 1$ $J_1\,K_1$
0 d	1 d	0 d	0 d	1 d	1 d
1 d	0 d	1 d	0 d	d 0	d 1
d d	d d	d d	d d	d d	d d
1 d	1 d	d 1	d 0	d 1	d 0
d 0	d 1	0 d	1 d	1 d	1 d
d 1	d 0	0 d	0 d	d 1	d 1
d d	d d	d d	d d	d d	d d
d 1	d 1	d 0	d 1	d 0	d 0

the logic for J_1 reveals that it can be connected to logic 1. The circuit simplifications are also shown. We do get a break. Although the K-maps for J_3 and K_3 are different, the 1s can both be covered by the same subcubes. Therefore, their logic is the same. The circuit is shown in Fig. 9-16; it is quite complex.

9-6-2 Input Sequences

An input sequence is typically a series of bits received from a serial transmission line. Data received from a MODEM is the most common example. This data is commonly sent using the ASCII code,[1] and each sequence of bits represents a character.

Certain input sequences may have, or may be assigned, special meanings. These sequences can be detected by state table logic, as Example 9-7 illustrates.

---

[1]The ASCII code for data transmission is given in Appendix B and is discussed in section 17-8-1.

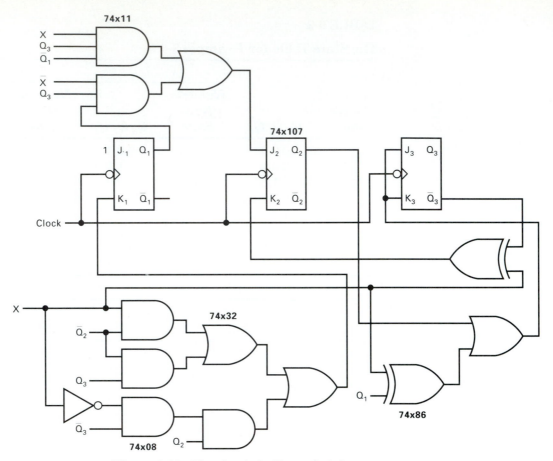

Figure 9-16 The circuit for Example 9-6.

EXAMPLE 9-7

Design a circuit with a clock and an input, so that any time four consecutive 1s appear on the input a counter will count them. A typical input is shown in Fig. 9-17. The output is a 1 any time the input is a 1 and three or more consecutive 1s have been received.

SOLUTION   The circuit can be designed and constructed using state table methods. The first problem is to decide on the number of states. In this case we can get by with four states, defined as follows:

- S0—The last input received was a 0.
- S1—The last input received was a 1.
- S2—The last two inputs were 1s.
- S3—The last three inputs were ones.

In any state if the input is a 0, the system goes back to S0. There will be an output if the system is in S3 (three ones received), and it receives a fourth 1.

A bubble diagram for the circuit is shown in Fig. 9-18. A pair of numbers, separated by a slash, is shown beside each path arrow. The upper number is the input and the lower number is the output. The diagram shows that if the

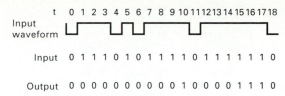

Figure 9-17 Input for Example 9-7.

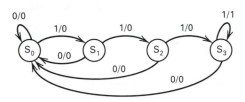

Figure 9-18 A bubble diagram for Example 9-7.

circuit is in state 0 and it receives a 0, it simply loops back or remains in S0. If it receives a 1, it progresses to S1, and then to S2 and S3 if subsequent 1s are received. If a 1 is received in S3, the diagram shows that the output is a 1, and the circuit remains in S3.

Because this circuit only has four states, it only requires two FFs. The state table requires three variables, Q_1, Q_0, and the input, called X. D FFs were selected for this circuit, so the D inputs are the same as the next state. The 3-variable Karnaugh maps were plotted from the state table and the circuit's equations were derived from them. Finally, the circuit was drawn as shown in Fig. 9-19.

To count the number of 1s, the output of the circuit must be connected to a counter, as shown in Fig. 9-20. If the circuit is in S3 (Q_1 and Q_0 both 1s) AND the input, X, is a 1, FF Q_2 will toggle. This toggling FF can then be the least significant stage of a counter (See Chapter 8).

9-6-3 A More Sophisticated Traffic Controller

A very simple traffic light controller, where the yellow light stayed on as long as the green light, has already been discussed in Section 9-3-4. Figure 9-21 shows a more sophisticated traffic controller.[2] When a light goes green, it stays green for several states. The controller responds to two sensors in the road, SENA and SENB, which detect whether traffic is coming in the two directions.

The bubble diagram is shown in Fig. 9-22. Space precludes going over it entirely, but consider the S0 state, where the light is green in the A direction and red in the B direction. There are three possibilities:

1. Traffic is coming in both directions or traffic is coming in neither direction. In either case the light should cycle normally. Observe that the circuit proceeds to S1 in these cases. The $\overline{SENA} \cdot \overline{SENB}$ arrow is when traffic is coming in both directions and the $\overline{SENA} \cdot \overline{SENB}$ arrow is valid when no traffic is coming.

[2]This figure and some of the material were taken from "Designing with Programmable Array Logic" by the technical staff of Monolithic Memories, Inc. The book was published in 1981. Monolithic Memories has since been taken over by Advanced Micro Devices, Sunnyvale, CA.

(a) State table, (b) Karnaugh maps and equations

	X = 0	X = 1
Present state	Next state	Next state
Q_1 Q_0	Q_1 Q_0	Q_1 Q_0
0 0	0 0	0 1
0 1	0 0	1 0
1 0	0 0	1 1
1 1	0 0	1 1

(a) State table

D_1

Q_1Q_0 \ X	0	1
00	0	0
01	0	1
11	0	1
10	0	1

D_0

Q_1Q_0 \ X	0	1
00	0	1
01	0	0
11	0	1
10	0	1

$D_1 = XQ_1 + XQ_0$

$D_0 = X\overline{Q}_0 + XQ_1$

(b) Karnaugh maps and equations

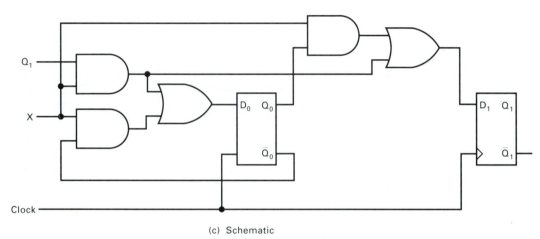

(c) Schematic

Figure 9-19   State tables and the schematic for Example 9-7: (a) state table; (b) Karnaugh maps and equations; (c) schematic.

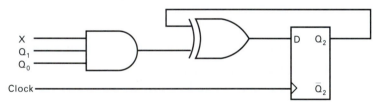

Figure 9-20   The first stage of a counter for Example 9-7.

2.   Traffic is coming in the A direction, but no traffic is coming in the B direction. The light should remain green as long as this condition persists. The condition is indicated when SENA·$\overline{\text{SENB}}$ is 1. Note that under this condition the circuit remains in S0.

3.   Traffic is coming in the B direction, but not in the A direction. This is valid when $\overline{\text{SENA}}$·SENB is 1. In this case, we want to turn off the green light in the A direction as soon as reasonably possible. Here the circuit immediately proceeds to S4, which will be followed by a yellow light in the A direction and then a red light, allowing the B direction light to turn green.

This circuit requires 11 states, which implies four FFs. These four FFs, together with the two external variables, SENA and SENB, give the system 6

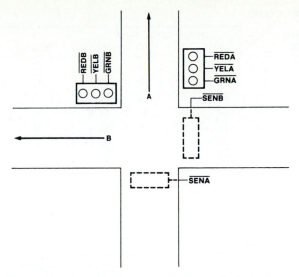

Figure 9-21 Traffic intersection. (Reprinted Courtesy of AMD Corporation.)

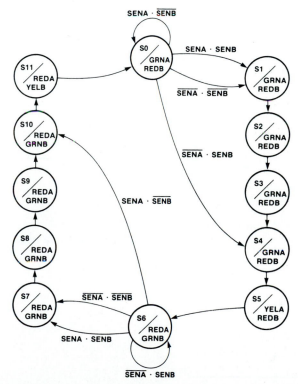

Figure 9-22 State graph—traffic signal controller. (Reprinted Courtesy of AMD Corporation.)

variables. The state table for D-type FFs is shown in Fig. 9-23. It results in four 6-variable Karnaugh maps, one for the D input to each FF. The logic for each D input is then found from the K-map simplification. The reader who requires all the details should consult the original text.

		INPUT SENA, SENB				OUTPUT					
CURRENT STATE	NEXT STATE	00	01	10	11	REDA	YELA	GRNA	REDB	YELB	GRNB
	S0	S1	S4	S0	S1	0	0	1	1	0	0
	S1	S2	S2	S2	S2	0	0	1	1	0	0
	S2	S3	S3	S3	S3	0	0	1	1	0	0
	S3	S4	S4	S4	S4	0	0	1	1	0	0
	S4	S5	S5	S5	S5	0	0	1	1	0	0
	S5	S6	S6	S6	S6	0	1	0	1	0	0
	S6	S7	S6	S10	S7	1	0	0	0	0	1
	S7	S8	S8	S8	S8	1	0	0	0	0	1
	S8	S9	S9	S9	S9	1	0	0	0	0	1
	S9	S10	S10	S10	S10	1	0	0	0	0	1
	S10	S11	S11	S11	S11	1	0	0	0	0	1
	S11	S0	S0	S0	S0	1	0	0	0	1	0

CURRENT STATE	INPUT/NEXT STATE				OUTPUTS
	00	01	10	11	
0 0 0 0	0 0 0 1	0 1 0 0	0 0 0 0	0 0 0 1	0 0 1 1 0 0
0 0 0 1	0 0 1 0	0 0 1 0	0 0 1 0	0 0 1 0	0 0 1 1 0 0
0 0 1 0	0 0 1 1	0 0 1 1	0 0 1 1	0 0 1 1	0 0 1 1 0 0
0 0 1 1	0 1 0 0	0 1 0 0	0 1 0 0	0 1 0 0	0 0 1 1 0 0
0 1 0 0	0 1 0 1	0 1 0 1	0 1 0 1	0 1 0 1	0 0 1 1 0 0
0 1 0 1	1 0 0 0	1 0 0 0	1 0 0 0	1 0 0 0	0 1 0 1 0 0
1 0 0 0	1 0 0 1	1 0 0 0	1 1 0 0	1 0 0 1	1 0 0 0 0 1
1 0 0 1	1 0 1 0	1 0 1 0	1 0 1 0	1 0 1 0	1 0 0 0 0 1
1 0 1 0	1 0 1 1	1 0 1 1	1 0 1 1	1 0 1 1	1 0 0 0 0 1
1 0 1 1	1 1 0 0	1 1 0 0	1 1 0 0	1 1 0 0	1 0 0 0 0 1
1 1 0 0	1 1 0 1	1 1 0 1	1 1 0 1	1 1 0 1	1 0 0 0 0 1
1 1 0 1	0 0 0 0	0 0 0 0	0 0 0 0	0 0 0 0	1 0 0 0 1 0

STATE	Q3 Q2 Q1 Q0
S0	0 0 0 0
S1	0 0 0 1
S2	0 0 1 0
S3	0 0 1 1
S4	0 1 0 0
S5	0 1 0 1
S6	1 0 0 0
S7	1 0 0 1
S8	1 0 1 0
S9	1 0 1 1
S10	1 1 0 0
S11	1 1 0 1

Input = SENA, SENB

Current/Next State = Q3, Q2, Q1, Q0/Q3+, Q2+, Q1+, Q0+

Output = REDA, YELA, GRNA, REDB, YELB, GRNB

Figure 9-23 State tables for the traffic controller. (Reprinted Courtesy of AMD Corporation.)

9-7 Advantages and Disadvantages of State Tables

When faced with the design of a project, the engineer must decide whether to use state tables. The advantages of state tables are:

1. They are a *systematic* approach to a design problem.
2. They *minimize* the gating required.
3. They result in *synchronous circuits*. Some engineers feel it is dangerous to allow circuits to change states asynchronously.
4. There are some problems where it is not easy to see any reasonable alternative. Counters with irregular count sequences are an example. When the count sequence is irregular, it is difficult to decide how to proceed from one count to the next. The systematic use of state tables will solve this problem.

There are also some disadvantages of using state tables:

1. They require considerable time and effort to construct.
2. The user cannot take advantage of the asynchronous inputs on the chips. The direct SETS and CLEARS on a **74x74** are an example.
3. Gate minimization may be unimportant. If the circuit is to be burned into a PAL, extra gates may be available on the PAL.
4. The number of variables is limited. Karnaugh maps become unwieldy when the number of variables is large as Example 9-8 shows.

EXAMPLE 9-8

How many states are involved in the hours counter of Fig. 8-21?

SOLUTION   At first glance there are 8 counter outputs and the switch input. The total is 9 variables or 512 states. After a little thought we may realize that the first counter can only progress through 10 states $(0-9)$, and the second counter can only progress through 3 states $(0, 1,$ and $2)$ because the count cannot exceed 23. With the switch considered, this reduces the number of states to 60, which is still formidable.

The engineer should consider the pros and cons of state tables when approaching a design.

9-7-1   An Example

Suppose we want a counter to count 14 states (S0–S13). This counter requires four FFs, Q_4–Q_1. The simplest solution is to AND together Q_4, Q_3, and Q_2, and clear the counter when they are all 1s. This method was discussed in Chapter 8. Suppose, however, that glitch clears are ruled out, as they would be in a PAL. Then there are two approaches.

1. Use state tables.
2. Decode a count of 13 (Q_4 AND Q_3 AND Q_1 are all 1). If J-K FFS are being used, OR this with the K input of each FF. This will cause all stages of the counter to CLEAR. If D-type FFs are being used, invert this output and AND it with the D inputs to each FF. This will cause all FFs to be cleared on the next clock pulse.

The latter method requires more gates, but is simpler to understand and probably easier to debug. We have presented the alternatives; the choice is up to the reader.

SUMMARY

This chapter began with the definition of the state of a system, which was determined by the settings of the FFs within the system. The use of the state checkout table to determine the sequence of states the circuit followed was explained. Bubble diagrams were introduced, and their uses in both analysis and design were discussed.

The design of these types of circuits was discussed next. State tables were introduced as a means of systematically designing such a circuit. Up-down counters, traffic controllers, and sequence detectors were presented as examples of this type of design. Finally, the advantages and disadvantages of state table design was discussed.

GLOSSARY

Bubble diagram: A diagram consisting of a set of circles, or bubbles. Each bubble contains a number of a state within it. The bubble diagram shows how a system progresses from state to state.

Hang-up state: A state which "hangs-up" the system. If the system enters a hang-up state it can never leave or progress to another state.

Irregular counter: A counter where the numerical progression is not regular or in sequence.

Main loop: The major progression of states in a system.

Minor loop: A complete loop of states which is smaller than the main loop.

State: The condition of a system as determined by the settings of the FFs within it.

State checkout table: A table to determine how the states progress in a given system.

State table: A table that allows the engineer to specify the progress of states in a system, and then design the system accordingly.

Synchronous: All FF changes are synchronized with an event, usually a clock edge.

Truncated counter: A counter that counts sequentially, but does not step through all possible states; it truncates or returns to 0 after a particular state.

PROBLEMS

Section 9-3.

9-1.   For the circuit of Fig. P9-1, draw a state checkout table. Assume the count starts at 0. Use the table to determine the count sequence and draw the bubble diagram.

9-2. Repeat Problem 9-1 for the circuit of Fig. P9-2.

9-3. What is wrong with the circuit segment of Fig. P9-3?

9-4. Repeat Problem 9-1 for the circuit of Fig. P9-4.

Section 9-4.

9-5. Design a traffic controller where the green light stays on twice as long as the yellow light.

9-6. Design a truncated counter to count 0,1,2,3,4,5,0 . . . .
    (a) Use J-K FFs.
    (b) Use D-Type FFs.

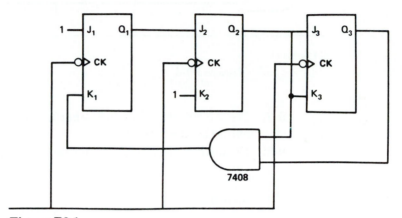

Figure P9-1

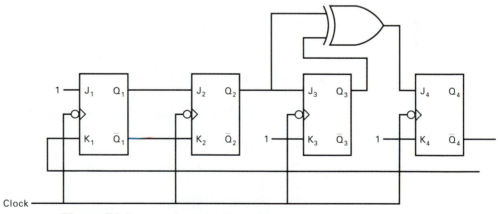

Figure P9-2

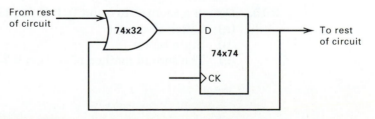

Figure P9-3

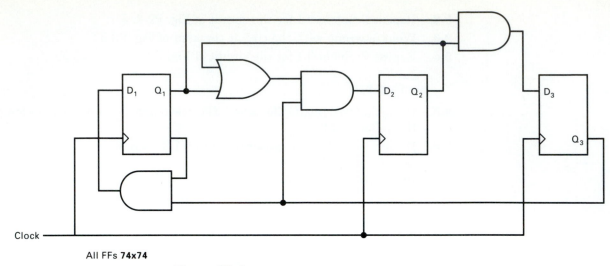

Clock

All FFs **74x74**

Figure P9-4

9-7. For the circuit of Fig. 9-8, what is the next state if the circuit somehow enters states 5, 6, or 7?

9-8. Draw a complete bubble diagram (showing all possible states) for the circuit of Fig. 9-10.

Section 9-5.

9-9. Design a counter to count the following sequence:

$$0,1,4,5,6,0, \ldots$$

    (a) Use **74x107**s.
    (b) Use **74x74**s.

9-10. Modify the circuit of Fig. 9-1 so that state 2 follows state 4.

9-11. Modify the circuit of Fig. 9-1 to create a minor loop so that it progresses 3,4,7,3 . . . .

9-12. Repeat Problem 9-9, but design the counter as an up-down counter.

Section 9-6.

9-13. Design a circuit to detect if the code 1010 has been received on a serial input line.
    (a) Use state tables.
    (b) Use a shift register.

9-14. In Fig. 9-22, explain how the circuit exits in state S6.

9-15. Design a counter to count 0, 1, 2, . . . 13 using
    (a) Glitch clear.
    (b) State tables.
    (c) The second method of section 9-7-1.

Shift Registers

10-1 Instructional Objectives

A shift register is used in digital and computer circuits for multiplication, division, timing, and parallel-to-serial data conversion. This chapter introduces the student to the design and application of shift registers. After reading it, the student should be able to:

1. Design shift registers for both left and right shifting.
2. Include such features as parallel loading and control of serial inputs in shift registers.
3. Utilize currently existing TTL shift registers.
4. Design timing chains using shift registers.
5. Design parallel-to-serial and serial-to-parallel converters using shift registers.
6. Decode the count of a pseudo-random sequence generator.

10-2 Self-Evaluation Questions

Watch for the answers to the following questions as you read this chapter. They should help you to understand the material presented.

1. When a register shifts, which bits are shifted out of the register? Which stages are vacated?
2. How can multiplication and division of binary numbers by 2 be accomplished using shift registers? Is this analogous to a procedure using decimal arithmetic?
3. What is circular shifting?
4. Explain why a **74x164** is used to perform a serial-to-parallel conversion and a **74x165** or **74x166** to perform a parallel-to-serial conversion.
5. In a shift register timing chain, why is only one pulse allowed in the chain at any one time?
6. How are pseudo-random sequence generators constructed?

10-3   The Basic Shift Register

A *shift register* consists of a group of FFs connected so that *each FF transfers its bit of information to the next most significant FF of the register when a clock occurs*. The action of a shift register is illustrated in Fig. 10-1, where each bit shifts one place to the right after each clock. In Fig. 10-1 it is assumed, for clarity, that bits shifting out of the most significant stage (the rightmost stage) are lost, and that 0s shift into the least significant stage (the leftmost stage).

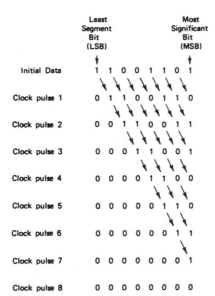

Figure 10-1   The action of a shift register.

10-3-1   Shift Registers Built from D-Type FFs

Shift registers can be built from D-type FFs, as shown in Fig. 10-2a, by connecting the Q output of each stage to the D input of the next succeeding stage. Since the D input to each stage is the same as the Q output of the previous stage, data will be shifted one bit to the right on the positive edge of each clock pulse.

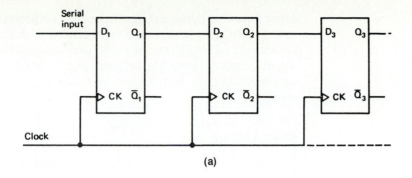

(a)

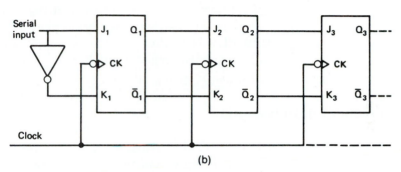

(b)

Figure 10-2 Shift registers built from D and J-K FFs: (a) D-type shift register using **74x74** FFs; (b) Shift register built from J-K FFs.

10-3-2   Shift Registers Built from J-K FFs

Shift registers can also be built from J-K FFs by connecting the Q and $\bar{Q}$ outputs of each stage to the J and K inputs of the succeeding stage, as Fig. 10-2b shows. The information in the register shifts one stage to the right on each negative transition of the clock. Since the data are constant for an entire clock period, master-slave type problems are not encountered in a J-K shift register.

Note that shift registers are generally shown as shifting to the right,[1] as in Fig. 10-2. Therefore, if the contents of a shift register are to be considered as a

---

[1]Authors face a problem when discussing shift registers: whether to show them as shifting left-to-right or right-to-left. There are two advantages to placing the LSB on the right and having the register shift from right-to-left:

1. If the bits of a shift register are interpreted as a binary number, it appears as it normally would rather than inverted.
2. This method is consistent with the microprocessor literature, where the LSB is always on the right, and a left shift moves a bit to the next higher position in the word.

There are also two advantages, however, to placing the LSB on the left and shifting left-to-right:

1. The normal flow of a person's eye when scanning a page is from left to right. This is now consistent with the data movement.
2. Digital literature, when showing shift registers and timing charts, has always shown a left-to-right flow.

Because this is a digital textbook, we have selected the latter presentation.

binary number, the *least significant bit* (LSB) must be on the *left*. Consequently, in this chapter, numbers are represented with the LSB on the *left* and the *most significant bit* (MSB) on the *right*, to correspond to the normal direction of shifting in a shift register. Unfortunately, this is a *reversal* of the usual way of representing numbers, but it is clearer when the numbers are contained in shift registers.

EXAMPLE 10-1

A 10-bit shift register contains the number 1011011101. What does it hold after two shift clocks if 0s are shifted into the vacated positions?

SOLUTION   The clock pulses shift the word two places to the right and shift 0s into the two least significant places. The resulting output is therefore **0010110111**. Note that the two most significant (rightmost) bits have been shifted out of the register and lost.

10-4   Left-Right Shift Registers

A LEFT-RIGHT shift register can shift information either left or right. The direction of shifting is controlled by a toggle switch or a control FF. Computers require some form of a LEFT-RIGHT shift register to execute shift instructions.

EXAMPLE 10-2

Design a LEFT-RIGHT shift register using **74x74** D-type FFs. The mode of shifting is to be controlled by a toggle switch.

SOLUTION

1. The toggle switch positions are labeled LEFT and RIGHT to indicate the direction of shifting. If the toggle switch is in the RIGHT position, the D input to a stage must be connected to the Q output of the next stage to the left.

2. If the toggle switch is in the LEFT position, the D input to a stage must be connected to the next stage to the right.

3. An AND-OR-INVERT (AOI) gate can be used to implement this logic (see section 5-9). Each contact of the toggle switch is connected to one of the AND gates so that only one AND gate is enabled at any time.

4. The AOI gate could be followed by an inverter to compensate for the AOI inversion. A more elegant solution, however, is to move the inversion to the front of the AOI gate by using the $\bar{Q}$ output of the FF instead of the Q output. This eliminates an inverter at each stage.

5. Figure 10-3 shows three stages of a LEFT-RIGHT shift register with the mode switch and interconnections required to make it function properly.

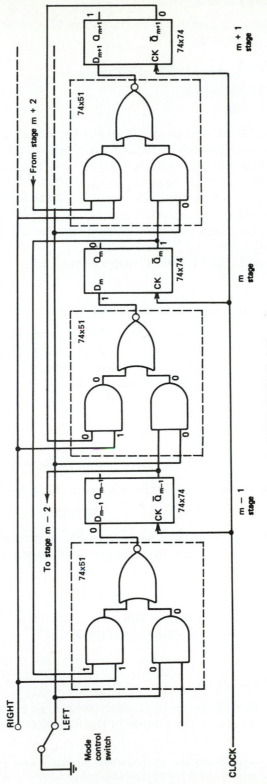

Figure 10-3 Three stages of a LEFT-RIGHT shift register.

By placing the correct 1s and 0s on Fig. 10-3, show that the shift register functions properly if:

1. It is in SHIFT-LEFT mode.
2. $Q_{m+1} = 1$
3. $Q_m = 0$

SOLUTION The proper 1s and 0s are already placed on Fig. 10-3. With the mode control switch in the LEFT position, all the lower AND gates in the AOI ICs are connected to ground (0). The mode inputs to the upper AND gates are open and act as a 1. Since $Q_{m+1} = 1$, the 0 at $\bar{Q}_{m+1}$ is connected to the AOI gate of stage m, which causes the input, D_m, to be HIGH. Similarly, since $Q_m = 0$, the 1 on $\bar{Q}_m$ is fed to the leftmost AOI gate and causes D_{m-1} to be 0. Therefore, at the next clock pulse, the D inputs are entered into the FFs; Q_m becomes 1, and Q_{m-1} becomes 0. The 1 and 0 in Q_{m+1} and Q_m are transferred to Q_m and Q_{m-1} and a LEFT SHIFT occurs as specified.

10-4-1 Multiplication and Division by 2

A binary number can be multiplied by 2 simply by placing it in a shift register and shifting it one stage to the right (providing a 1 is not shifted out of the most significant stage of the shift register). Similarly, a binary number can be divided by 2 by shifting it one stage to the left. In division any remainder is lost unless additional circuitry is used to preserve it.[2]

┌ **EXAMPLE 10-4**

The number 25 is in an 8-stage shift register. Multiply the number by 8.

SOLUTION Initially the number 25 in an 8-stage register looks like:

$$10011000$$

Note that the leftmost bit is the LSB. Multiplication by 8 is equivalent to three multiplications by 2, or three shifts to the right. After three shifts, the register contains

$$00010011$$

This is $128 + 64 + 8 = 200$, or 8×25. This example worked properly because 1s were not shifted out of the MSB.

┌ **EXAMPLE 10-5**

Given four 10-bit registers, each containing a binary number, describe how one could obtain the average of the four numbers.

---

[2]In microprocessor literature, where the shift directions are reversed, a left shift is equivalent to a multiplication and an *arithmetic right shift* is equivalent to a division.

SOLUTION   Since the average is the sum of the four numbers divided by 4, an adder circuit is needed to sum the numbers. The output of the adder could be transferred into a shift register. Two left shifts would then divide the output by four and leave the average in the register. Unfortunately, any remainder would be lost (i.e., the fractional part of the average would be discarded. See Problems 10-4 and 10-5).

10-5 Serial Inputs and Parallel Loading of Shift Registers

When shifting right, the least significant stage of the shift register is vacated. The designer has the option of entering any data into this stage. The input to the least significant stage is generally called the *serial input*. In Fig. 10-2a, for example, the *serial input* is connected to D_1, the D input of the first stage.

There are four common ways of using the serial input to a shift register.

1. **Pulling in 0s.** This places 0s in all the vacated stages of a shift register. It is accomplished by tying the serial input (D_1) to ground.
2. **Pulling in 1s.** This places 1s in all the vacated slots of a shift register and is accomplished by tying D_1 HIGH.
3. **Circular shifting.** This connects the output of the most significant stage to D_1. Consequently, the MSB before the shift becomes the LSB after the shift. Circular shifting can also be accomplished in SHIFT-LEFT mode by connecting the output of the LSB to the input of the MSB. Most computers use these circuits to execute circular shift instructions. In microprocessors, circular shift instructions are called ROTATES.
4. **Serial data input.** Serial input from an external source can be shifted into the register by tying the output of the external source to D_1. Care must be taken to synchronize the external data with the shift clocks.

EXAMPLE 10-6

Use a 4-position rotary switch to implement the four shift register options listed above.

SOLUTION   The solution is shown in Fig. 10-4. The rotary switch selects one gate of a 4-wide AOI gate (**74x54**) and the proper data are connected to the other input. The output is then inverted and fed to the D input of the least significant stage of the shift register. The **74x04** inverters cause the inputs to all AND gates except the selected gate of the AOI circuit to be LOW.

10-5-1 Microprocessor Shifting

Shifting in microprocessors (μPs) generally makes use of the CARRY FF. This is a FF within the μP chip whose prime function is to retain the state of the carry after an arithmetic operation. In SHIFT or ROTATE instructions it is often

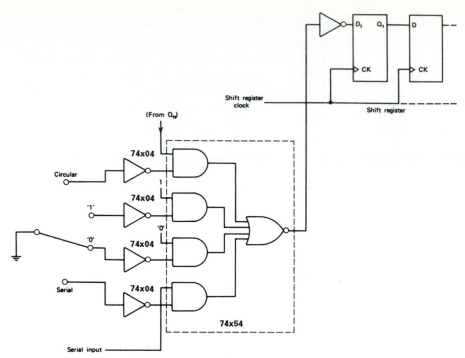

Figure 10-4 Circuit for Example 10-6.

used both to provide a place to retain the bit shifted out and to supply the bit being shifted in.

The operation of a microprocessor ROTATE instruction is shown in Fig. 10-5. Note that in a microprocessor B0 is the LSB and B7 the MSB. In a ROTATE LEFT instruction the MSB of the 8-bit (1-byte) μP word is shifted into the CARRY FF, while it supplies the input to the LSB. This is really a 9-bit ROTATE using the 8 bits of the register plus the CARRY FF. RIGHT ROTATES are accomplished similarly by reversing the arrows as shown in Fig. 10-5b. Since the contents of the CARRY FF can be retained and the CARRY FF manipulated in other ways, the μP user can readily control shift operations using the CARRY FF.

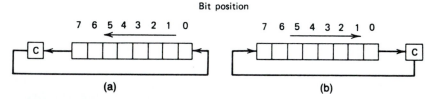

Figure 10-5 A microprocessor ROTATE instruction: (a) rotate left (ROL); and (b) rotate right (ROR).

10-5-2 Parallel Loading

Many shift registers must be capable of accepting data from an external source as well as shifting bits. These shift registers operate in SHIFT mode or LOAD mode. In SHIFT mode they operate normally, but in LOAD mode the external data are jammed into the register and replace any previous data.

EXAMPLE 10-7

Design a shift register controlled by a switch. When the switch output is LOW the register shifts RIGHT, but when the switch output is HIGH, external data are loaded into the register.

SOLUTION The first two stages of the shift register are shown in Fig. 10-6. When the switch is down, or in the SHIFT position, the output of each of the **74x00** gates is HIGH, and these outputs do not affect the direct SETS and CLEARS of the FFs. If the switch is in the LOAD position, however, the external data bits are jammed into the shift register FFs through the NAND gates. If data bit 1 is a 1, for example, the direct set to stage 1 is LOW, setting it; while if data bit 1 is a 0, the lower NAND gate has two HIGH inputs, and its LOW output clears FF1.

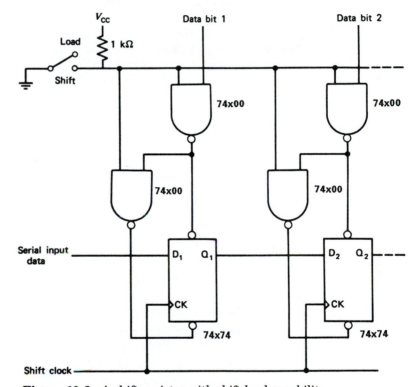

Figure 10-6 A shift register with shift-load capability.

There are also two submodes for loading a shift register: asynchronous and synchronous. The asynchronous mode, where data on the external lines is immediately jammed into the shift register whenever it is in LOAD mode, was shown in Example 10-7. In the synchronous mode, data on the external lines is only loaded when a clock edge occurs.

10-6 Parallel Load and Parallel Output Shift Registers

Many shift registers are available in TTL ICs. The important features of these shift registers are:

1. **The number of bits in the shift register.** If longer shift registers are required, they can be built up by cascading ICs.
2. **Parallel load.** The ability to parallel load data simultaneously into the shift register.
3. **Serial input.** The ability to serially shift data into the shift register.
4. **Parallel output.** The ability to obtain all output bits of a shift register simultaneously in parallel.
5. **Serial output.** The ability to obtain serial output from a shift register.
6. **Shift frequency.** The maximum permissable frequency of the shift clock.
7. **LEFT-RIGHT shift register.** The ability of the shift register to shift data in either direction.

To acquaint the student with shift registers, the **74x164, 74x165,** and **74x166** shift registers are described in detail. These are 8-bit shift registers in 14- and 16-pin packages. Other shift registers, such as the **74x194** and **74x195**, are also available. In general these have more capabilities, but they are only 4-bit shift registers.

10-6-1   The 74x164 Serial-in Parallel-Output Shift Register

The **74x164** is an 8-bit shift register in a 14-pin package with all 8 outputs available. The pin configuration and block diagram are shown in Fig. 10-7. A parallel load is not available on this IC and the data are loaded serially via an input AND gate. Serial inputs A and B must both be HIGH to load a 1 into the shift register. Data is shifted right on every positive edge of the clock. The **74x164** also has an asynchronous CLEAR, which clears all stages of the shift register whenever it goes LOW.

The operation of the **74x164** is illustrated by the timing chart of Fig. 10-8.

1. The clear pulses at $t = 1$ and $t = 24$ clear all outputs
2. The positive clock transitions, which clock in the input data, occur at $t = 3, 5, \ldots$ The only transition times when inputs A and B are both HIGH are at $t = 9, 11,$ and 15. Q_A goes HIGH for one clock period following each of these times.
3. Since the Q_A output is shifted to Q_B one clock period later, Q_B is HIGH at $t = 11, 13,$ and 17.
4. The data continue to shift right through the shift register on succeeding clock pulses.
5. Note that the output pulses on Q_E, Q_F, Q_G, and Q_H are truncated by the clear pulse at $t = 24$.

10-6-2   The 74x165 and 74x166 8-bit Parallel-Load Shift Registers

The **74x165** and **74x166** are two very similar 8-bit parallel-in, serial-out shift registers. The **74x165** is first described in detail, then the differences between the **74x165** and the **74x166** are explained.

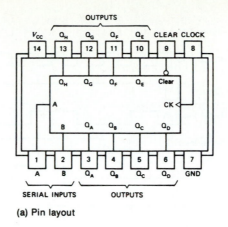

(a) Pin layout

(b) Block diagram

Dynamic input activated by transition from a high level to a low level

Figure 10-7 The **74x164** shift register. (From the *TTL Logic Data Book*, 1988, Texas Instruments, Inc. Courtesy of Texas Instruments.)

The pin layout, block diagram, and function table of the **74x165** are shown in Fig. 10-9. From it we see that there are 8 parallel inputs and a SHIFT/LOAD input (pin 1). When the SHIFT/LOAD input is LOW (LOAD mode), the clocks are inhibited and the input data are jammed into the 8 stages of the shift register as it was in Example 10-7. When the SHIFT/LOAD input is HIGH, loading is inhibited, and data are shifted one bit to the right on each positive transition of the clock provided the CLOCK INHIBIT input is LOW. An examination of Fig. 10-9b reveals that the CLOCK input is ineffective if the CLOCK INHIBIT input is HIGH. Note also that the CLOCK and CLOCK INHIBIT inputs can be interchanged. The CLOCK INHIBIT line should be changed from LOW to HIGH only while the clock line is HIGH. A positive transition of the CLOCK INHIBIT line when the clock line is LOW causes an extra clock to occur.

In Shift mode data is shifted on each positive clock edge. A serial input line is also provided on pin 10. The serial data are shifted into the A FF on each clock pulse.

It was impossible to make the parallel outputs available and package the **74x165** in a 16-pin DIP. Therefore, only the outputs of the most significant stage are available (Q_H and $\overline{Q}_H$). To study the operation of the **74x165** in detail, consult

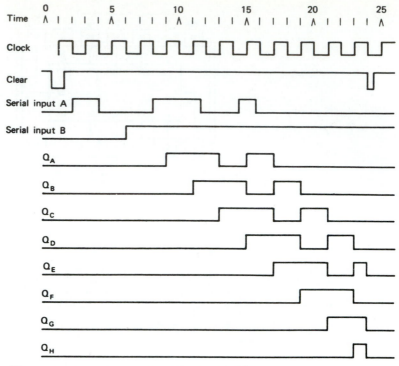

Figure 10-8 A timing chart for the **74x164**.

the timing chart of Fig. 10-10. For clarity, the output of each *internal* FF is shown, but only the Q_H output is *actually available* at an external connection. The **74x165** operates as follows.

1. Positive edges or clocks occur at $t = 1, 3, 5, \ldots$
2. The first pulse on the serial input is clocked into Q_A at $t = 1$.
3. This pulse shifts to the right and appears at the output, Q_H, seven clock pulses later ($t = 15$).
4. The short pulse at $t = 8$ does not enter the shift register because no clock occurs while it is HIGH.
5. The pulse on the serial input from $t = 16.5$ to $t = 19.5$ starts a double-width pulse down the shift register, because clocks occur at $t = 17$ and $t = 19$ while it is HIGH. This double-width pulse, however, is truncated at $t = 20$.
6. From $t = 20$ to $t = 22$, the shift register is in the LOAD mode. The data loaded in are assumed to be 11011001 (input H to input A, respectively).
7. At $t = 22$ the shift register returns to shift mode. The next shift clock OCCURS at $t = 23$, and the word starts to form at the output.
8. From $t = 25.5$ to $t = 28.5$, the CLOCK INHIBIT input goes high. Note that it makes a LOW-to-HIGH transition when the clock is HIGH. The negative transition of the CLOCK INHIBIT input can be made at any time. This pulse inhibits the clock at $t = 27$, and the shift register stages cannot change at this time.

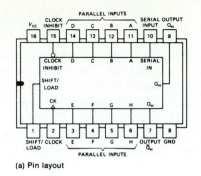

(a) Pin layout

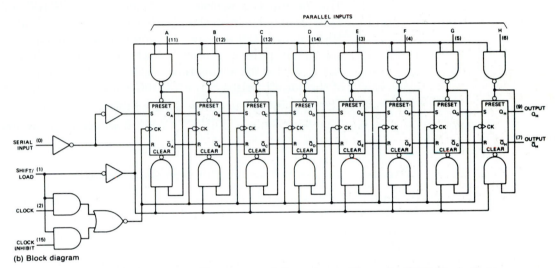

(b) Block diagram

Figure 10-9 The **74x165** shift register. (From the *TTL Logic Data Book*, 1988, Texas Instruments, Inc. Courtesy of Texas Instruments.)

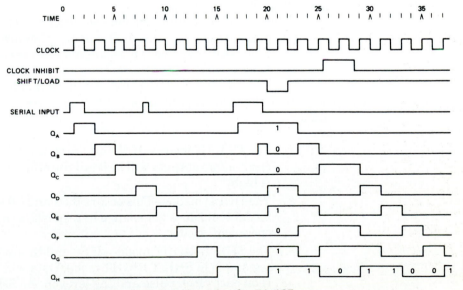

Figure 10-10 A timing chart for the **74x165**.

9. At $t = 29$, shift clocks resume and the data loaded in are shifted out at Q_H, one bit at a time.

10. The data, which appear serially at Q_H, are the parallel-loaded data, 11011001. The first 0, however, is twice as long as the other pulses because of the inhibited clock at $t = 27$. This shows that one must be careful when using the CLOCK INHIBIT line.

The **74x166** is also an 8-bit PARALLEL-LOAD, SERIAL-OUTPUT shift register. The differences between the **74x165** and the **74x166** are as follows.

1. The **74x166** is *synchronously* loaded. If the SHIFT/LOAD input is LOW, it only loads on the positive edge of the clock. Referring to Fig. 10-10, if a **74x166** were used it would have loaded at $t = 21$, instead of $t = 20$. The pulse at this time would be only two time units long ($t = 21$ to $t = 23$) for the asynchronously loaded **74x165**. Therefore, the time of the output pulses of the **74x166** would be more uniform.

2. The **74x166** does not have $\overline{Q}_H$ available.

3. The **74x166** has a DIRECT CLEAR, which asynchronously clears all stages.

10-6-3   Parallel-In/Parallel-Out 8-bit Shift Register

An 8-bit shift register that allows for both parallel input and parallel output requires more than 16 pins. The **74198** is an 8-bit shift register in a 24-pin package that has left or right shifting, parallel outputs, and parallel load. It is only available in the standard TTL series.

The **74x299** and **74x323** are two more modern shift registers. The pinout and function table of the **74x299** are shown in Fig. 10-11. The pinout shows that the **'299** comes in a 20-pin DIP. There are eight input/output pins labelled A/Q_A . . . H/Q_H. These pins have a dual function: they are the lines to be connected to the external inputs when the **'299** is being loaded, otherwise they hold the output of the shift register.

There are two auxiliary outputs at either end of the shift register: Q_A' and Q_H'. Their outputs are the same as Q_A and Q_H; they are used for cascading ICs to make a longer shift register. There are also two serial inputs: SL (Shift Left) and SR (Shift Right).

The function table shows that the operation of the shift register is controlled by the mode inputs, S1 and S0. There are five sections in the function table:

1. The CLEAR mode—If the input to the $\overline{CLR}$ pin is low, the IC will clear; all its outputs will be 0. All the other modes function only if $\overline{CLR}$ is high.

2. The HOLD mode—This occurs if S1 and S0 are both 0s. The IC holds its data and does not change in response to a clock pulse. This is a "no operation" mode.

3. The SHIFT-RIGHT mode—If S0 and S1 are 1 and 0, respectively, the register is in SHIFT RIGHT mode. Data will be shifted from A toward H on each clock pulse and the level on the SHIFT RIGHT serial input (SR) will be entered into the Q_A FF.

SN54LS299, SN54S299 . . . J OR W PACKAGE
SN74LS299, SN74S299 . . . DW OR N PACKAGE
(TOP VIEW)

```
         S0  [ 1   U   20 ] Vcc
         G1‾ [ 2       19 ] S1
         G2‾ [ 3       18 ] SL
      G/Q_G  [ 4       17 ] Q_H'
      E/Q_E  [ 5       16 ] H/Q_H
      C/Q_C  [ 6       15 ] F/Q_F
      A/Q_A  [ 7       14 ] D/Q_D
       Q_A'  [ 8       13 ] B/Q_B
        CLR‾ [ 9       12 ] CLK
        GND  [ 10      11 ] SR
```

MODE	INPUTS								INPUTS/OUTPUTS								OUTPUTS	
	CLR‾	S1	S0	G1‾†	G2‾†	CLK	SL	SR	A/Q_A	B/Q_B	C/Q_C	D/Q_D	E/Q_E	F/Q_F	G/Q_G	H/Q_H	Q_A'	Q_H'
Clear	L	X	L	L	L	X	X	X	L	L	L	L	L	L	L	L	L	L
	L	L	X	L	L	X	X	X	L	L	L	L	L	L	L	L	L	L
	L	H	H	X	X	X	X	X	X	X	X	X	X	X	X	X	L	L
Hold	H	L	L	L	L	X	X	X	Q_{A0}	Q_{B0}	Q_{C0}	Q_{D0}	Q_{E0}	Q_{F0}	Q_{G0}	Q_{H0}	Q_{A0}	Q_{H0}
	H	X	X	L	L	L	X	X	Q_{A0}	Q_{B0}	Q_{C0}	Q_{D0}	Q_{E0}	Q_{F0}	Q_{G0}	Q_{H0}	Q_{A0}	Q_{H0}
Shift Right	H	L	H	L	L	↑	X	H	H	Q_{An}	Q_{Bn}	Q_{Cn}	Q_{Dn}	Q_{En}	Q_{Fn}	Q_{Gn}	H	Q_{Gn}
	H	L	H	L	L	↑	X	L	L	Q_{An}	Q_{Bn}	Q_{Cn}	Q_{Dn}	Q_{En}	Q_{Fn}	Q_{Gn}	L	Q_{Gn}
Shift Left	H	H	L	L	L	↑	H	X	Q_{Bn}	Q_{Cn}	Q_{Dn}	Q_{En}	Q_{Fn}	Q_{Gn}	Q_{Hn}	H	Q_{Bn}	H
	H	H	L	L	L	↑	L	X	Q_{Bn}	Q_{Cn}	Q_{Dn}	Q_{En}	Q_{Fn}	Q_{Gn}	Q_{Hn}	L	Q_{Bn}	L
Load	H	H	H	X	X	↑	X	X	a	b	c	d	e	f	g	h	a	h

†When one or both output controls are high the eight input/output terminals are disabled to the high-impedance state; however, sequential operation or clearing of the register is not affected.

a . . . h = the level of the steady-state input at inputs A through H, respectively. These data are loaded into the flip-flops while the flip-flop outputs are isolated from the input/output terminals.

Figure 10-11 The **74x299** shift register: (a) pinout; (b) function table. (From the *TTL Logic Data Book*, 1988, Texas Instruments, Inc. Courtesy of Texas Instruments.)

4. The SHIFT-LEFT mode—If S1 and S0 are 1 and 0 respectively, the register will shift left on each clock pulse. Data on the SL input will be entered into the rightmost FF: Q_H.

5. The LOAD mode—If S0 and S1 are both 1s, the IC is in LOAD mode. The output drivers on the shift register are 3-state, and are disabled in LOAD mode. This allows external data to be placed on the input pins. It will be loaded into the register on the positive edge (↑) of the clock. Like the **74x166**, the **74x299** is synchronously loaded.

10-7 Applications of Shift Registers

There are many applications of shift registers. Several examples are presented in this section.

10-7-1 Timing Circuits

Sequences of timing pulses can be generated and controlled by using shift registers. Perhaps the simplest way to do this is to use a **74x164** as shown in the circuit of Fig. 10-12a. First the **74x164** is cleared by a LOW pulse on the CLR input. Then 1s are shifted in. Consequently, Q_A goes HIGH one clock pulse later, Q_B goes HIGH on the following clock pulse, and so on as the timing chart of Fig. 10-12b shows.

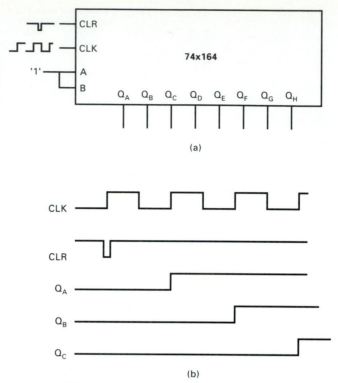

Figure 10-12 Shift register timing using a **74x164**: (a) circuit; (b) timing chart.

A variation on this circuit can be used to generate timing sequences, as Example 10-8 shows. The basic idea is to shift 1s into the shift register until the timing sequence is complete. The pulses then cause the shift register to be cleared. EXCLUSIVE-OR gates are used to select the proper timing pulses.

EXAMPLE 10-8

A circuit is required to produce a series of four pulses repetitively every 70 μs. The pulses are to be HIGH during the following times:

A	0–20 μs
B	15–45 μs
C	35–60 μs
D	55–70 μs

as shown in Fig. 10-13a. Design the circuit.

SOLUTION The solution is shown in Fig. 10-13b. It was designed as follows.

1. An examination of the pulses shows that the shortest time required is 5 μs. Therefore, the shift register must be driven by a 200 kHz clock to shift in 1s at 5 μs intervals. The SERIAL INPUTS are tied HIGH so that 1s are constantly shifted into the register until it clears.

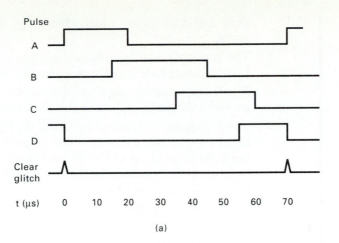

(a)

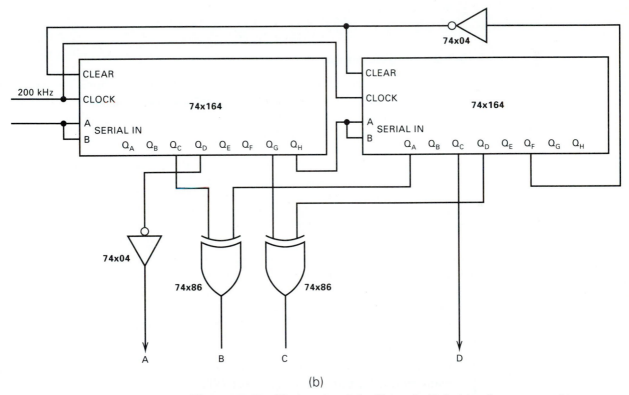

(b)

Figure 10-13 Timing circuit for Example 10-8: (a) pulse outputs; (b) circuit.

2. Because the total cycle time is 70 μs, 14 outputs are needed and two **74x164** shift registers must be cascaded together by tying Q_H of the first **74x164** to Q_A of the second **74x164**.

3. Assuming the shift register is cleared at $t = 0$, Q_{A1} goes HIGH at 5 μs, Q_{B1} goes HIGH at 10 μs, and so on. Q_{F2} goes HIGH at 70 μs. This is inverted and clears both shift registers, thus causing a 70 μs cycle. Note that the output at Q_{F2} is a *glitch*; it is HIGH only until the

shift registers clear, which clears Q_{F2} and removes the CLEAR pulse.

4. The A output can be obtained simply by inverting Q_{D1}. Q_{D1} goes LOW when the shift register is cleared and goes HIGH at 20 μs.

5. The D output can be obtained directly from Q_{C2}. It goes HIGH at 55 μs and LOW at 70 μs, when the shift registers CLEAR.

6. The B output is obtained by using an EXCLUSIVE-OR gate with inputs Q_{C1} and Q_{A2}. From $t = 0$ to 15 μs, both inputs and the output are 0. At 15 μs Q_{C1} goes HIGH, and the output goes HIGH. At 45 μs Q_{A2} goes HIGH; so both inputs are HIGH and the output of the EXCLUSIVE-OR goes LOW.

7. The C output is obtained similarly by tying the EXCLUSIVE-OR inputs to Q_{G1} (which goes HIGH at 35 μs) and Q_{D2} (which goes HIGH at 60 μs).

10-7-2   Ring Counters

A ring counter is a *circulating* shift register, where one stage has a 1 output and all the other stages have a 0 output. The 1 output advances through the shift register at each clock pulse, and recirculates to the first stage after it reaches the last stage. The single HIGH pulse can be used as a timing marker in many circuits.

A simple, 4-bit ring counter and its timing are shown in Fig. 10-14. More versatile ring counters, of up to 8 bits, can be designed using **74x164**. If a 7-bit counter (six 0s and a 1) are required, for example, the Q_G output can be tied to the serial input. Ring counters longer than 8 bits can be built by cascading **74x164**s (See Problem 10-20).

Ring counters must be started properly. When power is first applied, the shift register will probably not come up with only a single 1 in it. A circuit to properly initialize the ring counter may have to be designed.

EXAMPLE 10-9

Design a circuit to properly initialize an 8-bit ring counter using a **74x164** whenever a LOW pulse occurs on an INITIALIZE line.

SOLUTION   First consider the thought process involved. Here, as in any design problem, the problem must be defined precisely. This will often point the way to a correct solution. In this problem the circuit must clear the shift register, place a single 1 in one of the bits, and then have that 1 recirculate.

One solution is shown in Fig. 10-15. A LOW pulse on the INITIALIZE line will clear the shift register and set the INITIALIZE FF. After the INITIALIZE line goes HIGH, the next clock pulse will clear the INITIALIZE FF and clock a 1 into QA. This 1 will then circulate through the shift register, advancing on each clock pulse. The ring counter will continue to recirculate the pulse indefinitely.

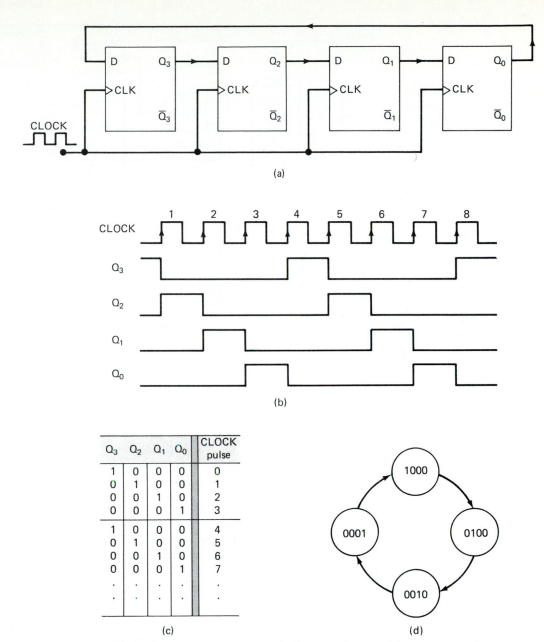

Figure 10-14 (a) Four-stage ring counter; (b) waveforms; (c) sequence table; (d) bubble diagram. (Ronald J. Tocci, DIGITAL SYSTEMS: Principles and Applications, 5e. Reprinted by permission of REGENTS/PRENTICE HALL, Englewood Cliffs, New Jersey.)

10-7-3 Low-Going Ring Counters

A low-going ring counter circulates a single negative pulse through a shift register. Figure 10-16 shows a circuit for an 8-bit, low-going ring counter. If any of the A through G outputs of the **74x164** are LOW, the SERIAL INPUT is HIGH and 1s are shifted into the **74x164**. When the A through G outputs are all HIGH, the SERIAL INPUT to the **74x164** goes LOW. The next clock pulse clocks the LOW

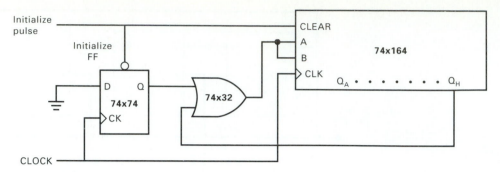

Figure 10-15 An initialization circuit for a ring counter.

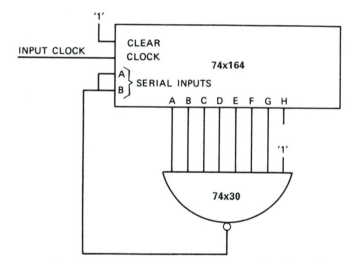

Figure 10-16 Circulating a single negative pulse through a shift register.

input to output A, which now causes the output of the **74x30** to be 1. At any one time only one output is LOW,[3] and the LOW output progresses down the line with each clock pulse.

10-7-4 Johnson Counters

A Johnson or "twisted-tail" counter is a ring counter where the output of the last stage is inverted and fed back to the first stage. A simple, 3-stage Johnson counter is shown in Fig. 10-17a, and its count sequence is given in Fig. 10-17b. The count sequence has six states. In general, an n-stage Johnson counter will produce 2n states.

10-7-5 Serial-to-Parallel and Parallel-to-Serial Conversion

It is often necessary to convert serial data to parallel or vice versa. Computers, for example, transmit and receive data in parallel form, but many of the devices

---

[3]This may not be true for the first seven clock pulses following power turn on because of residual LOWS that may appear in the shift register at that time.

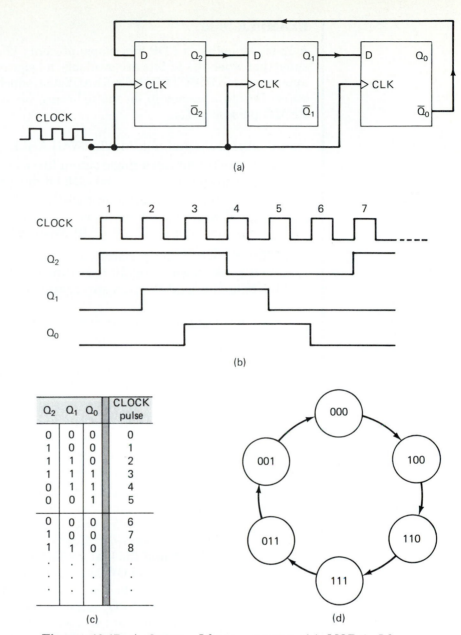

Figure 10-17 A 3-stage Johnson counter: (a) MOD-6 Johnson counter; (b) waveform; (c) sequence table; (d) bubble diagram. (Ronald J. Tocci, DIGITAL SYSTEMS: Principles and Applications, 5e. Reprinted by permission of REGENTS/PRENTICE HALL, Englewood Cliffs, New Jersey.)

they communicate with, such as video terminals or modems, require the data in serial form. Therefore, a parallel-to-serial conversion must be made when going from a computer to the display section of a terminal, and a serial-to-parallel conversion must be made when going from the keyboard section of the terminal to a computer. Shift registers are generally used to make these conversions.

EXAMPLE 10-10

A 12-bit computer (a PDP-8 perhaps) presents 12 data bits in parallel. Assume that when the 12 bits are available, a negative-going pulse, 2 μs long, appears on a COMPUTER DATA READY line, which is controlled by the computer. Design a circuit to do the following when the COMPUTER DATA READY pulse expires:

1. Raise a signal called DATA AVAILABLE.
2. Place the bits on a single output line called the DATA line. The bits are to appear serially and each bit must remain for 1 μs.
3. After the 12 bits have appeared, the DATA AVAILABLE line must go LOW and 0s must appear on the DATA line. The DATA line must remain 0 until the next DATA AVAILABLE pulse.

SOLUTION When the specifications get this complex, it is best to draw a timing chart as shown in Fig. 10-18a. If we refer to the timing chart, we see that the following solution steps are required for the designer.

1. A parallel-to-serial conversion is required. This suggests **74x165**s.
2. Since 12 bits are involved in the conversion, two **74x165**s are needed. They can be cascaded to form a 16-bit shift register by tying Q_H of the first shift register to the SERIAL INPUT of the second shift register.
3. Because 0s are required on the DATA line after the 12 data bits have passed, the four unused inputs and the unused SERIAL INPUT to the least significant shift register are connected to ground.
4. The COMPUTER DATA READY signal can be connected to the SHIFT/LOAD input and used to load the data.
5. To shift out the data, as required, the shift register must be clocked by a 1 μs clock, which starts at the trailing edge of the COMPUTER DATA READY pulse. Here we can use a 1 μs one-shot oscillator (see section 7-6-1). Connecting COMPUTER DATA READY to the B input of the first oscillator stage shuts the oscillator off while the pulse is LOW and synchronizes the clocks with it.
6. The **74x165** is clocked on a positive edge, so the shift register clocks are tied to the $\bar{Q}_2$ output of the one-shot oscillator. The first positive-going edge at this point occurs exactly 1 μs after the expiration of COMPUTER DATA READY.
7. To generate DATA AVAILABLE, 12 clocks must be counted. Therefore, a **74x92** is used. The DATA AVAILABLE FF is SET by COMPUTER DATA READY, but it is masked from the output by the AND gate until COMPUTER DATA READY expires. The DATA AVAILABLE FF is cleared by the **74x92** after 12 counts. Note that the Q_D output of the **74x92** must be inverted to produce a positive edge at this time. Note also that the **74x92** is cleared by COMPUTER DATA READY (inverted and applied to $R_{0(1)}$ and $R_{0(2)}$) so that it always starts at a count of 0.
8. The shift register continues to shift and the **74x92** to count after the 12 data pulses expire. However, since the shift register is con-

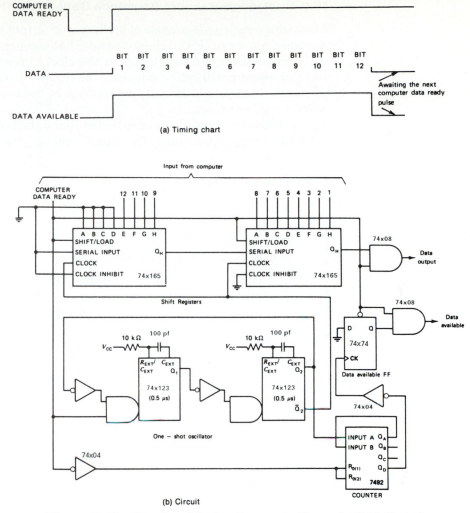

Figure 10-18 Circuit and timing diagram for Example 10-10. Capacitor values obtained from Fig. 7-10c.

tinually shifting out 0s and the DATA AVAILABLE FF remains clear, the output is unaffected by this activity.

9. The next COMPUTER DATA READY pulse stops the clock and clears the **74x92**. The circuit is ready to restart when the pulse expires.

10. The final circuit, designed with these considerations in mind, is shown in Fig. 10-18b.

The circuit for Example 10-10 is a typical, but simple, illustration of digital design. No formulas exist to solve these types of design problems. The solution depended on thought, imagination, familiarity with ICs, and the use of circuits considered here and in previous chapters.

10-7-6 Pseudo-Random Sequence Generators

Any n-stage register is capable of producing 2^n output states. A pseudo-random sequence generator (PRSG) has one state that is constant. If it is in that state it will remain there and not change as the clocks progress. If it is in any other state, however, it will cycle through all the other 2^n-1 states in a random fashion. Of course, the count is actually determined by the circuitry, which is why it is called pseudo-random.

PSRGs are generally constructed by using shift registers with the outputs fed back through XOR gates. The count sequence depends on the circuitry.

EXAMPLE 10-11

The circuit of Fig. 10-19 is a 3-stage PSRG. Determine the count sequence if initially:
   (a) Q_A, Q_B, and Q_C are each 0.
   (b) Q_A, Q_B, and Q_C are each 1.

SOLUTION   (a) With all the outputs 0, the serial input will also be 0. After the clock the situation will still be the same. This is the constant state.

(b) With all the outputs 1, the serial input will be 0. The clock will cause the 1 in Q_A to shift to Q_B and the 1 in Q_B to shift to Q_C, but the 0 on the serial input will be shifted into Q_A. Thus, the count will be 011 or 6 (remember that the rightmost bit is the MSB). The count following 6 is 4. The count progresses 7, 6, 4, 1, 2, 5, 7. Note that there is one constant state and 7 states in the pseudo-random sequence.

PSRGs are used in cyclical redundancy character (CRC) checking, which checks the integrity of a data block on a transmission line, and in signature analysis, which is used to text VLSI ICs. These topics are fully discussed in more specialized books.

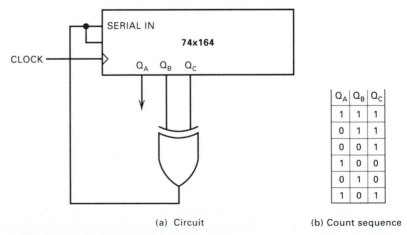

Q_A	Q_B	Q_C
1	1	1
0	1	1
0	0	1
1	0	0
0	1	0
1	0	1

(a) Circuit   (b) Count sequence

Figure 10-19   A 3-stage pseudo-random sequence generator: (a) circuit; (b) count sequence.

In this chapter, shift registers were introduced. The design of shift registers with special features such as LEFT-RIGHT and circular shifting was covered. Applications of shift registers to timing circuits, parallel-to-serial conversion, and PSRGs were discussed.

GLOSSARY

Circular shifting: Shifting in which the MSB is shifted into the LSB position or vice versa.

Johnson counter: A ring counter where the output is inverted and tied back to the input.

LEFT-RIGHT shift register: A shift register capable of shifting data in either direction.

Parallel loading: Simultaneously loading all bits of a shift register in parallel.

Pseudo-random sequence generator: A circuit that goes through 2^n-1 states in a random fashion.

Ring counter: A shift register circuit that circulates a single 1 (or 0) through the shift register.

Serial input: The input to the least significant stage of a shift register.

Shift register: A register of FFs that shift data in response to a clock.

PROBLEMS

Section 10-3.

10-1. If the contents of a 12-bit shift register is

100101101101

what does it contain:
(a) After three right shifts?
(b) After four left shifts?

Section 10-4.

10-2. Design a LEFT-RIGHT shift register using J-K FFs.

10-3. A 12-bit shift register contains the number 65.
(a) How would you multiply it by 16?
(b) How would you divide it by 16?

10-4. Find the average of the numbers 5, 7, 7, and 8 using the method of Example 10-5. Show what goes into the adder, what goes into the shift register, and what comes out of the shift register.

10-5. Make a simple improvement on the method of Example 10-5 to round the answer (i.e., if the true average is 6 or 6.25, the circuit output should be 6, but if the true average is 6.50 or 6.75, the circuit output should be 7).

Section 10-5.

10-6. Design a shift register that performs circular left or right shifts.

10-7. Design the Nth stage of a shift register that responds to SHIFT-LEFT, SHIFT-RIGHT, and asynchronous LOAD commands. Assume SHIFT LEFT/ RIGHT, SHIFT/LOAD, QN-1, QN + 1, and the input data bit are your inputs.

10-8. Design a synchronous SHIFT/LOAD shift register. If the register is in LOAD mode, data must be entered only on a positive clock transition.

10-9. Redesign the circuit of Fig. 10-4 to eliminate two inverters.

Section 10-7-1.

10-10. Design a shift register controlled timing circuit to produce the pulses of Fig. P7-16.

10-11. The following series of pulses are to be produced cyclically every 70 μs.
Time in μs
A. 0–15
B. 15–45
C. 30–65
D. 5–25
(a) Use one-shot timing. Identify the one-shots you use and give the values of the timing resistors and capacitors you use.
(b) Use shift register timing. It is not necessary to include a single cycle option.

10-12. For the circuit of Fig. P10-12, sketch the outputs as a function of time. They are repetitive. Start with the **74x164** in the clear state.

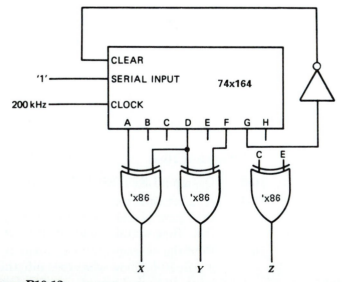

Figure P10-12

10-13. Design timing circuits to produce the pulses of Fig. P10-13.
    (a) Use one-shot timing. Show your resistors, capacitors, and their connections to the one-shots.
    (b) Use shift register timing.

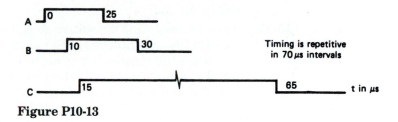

Figure P10-13

Section 10-7-2.

10-14. Design a 10-bit ring counter using a **74x164** and a **74x74**.

Section 10-7-4.

10-15. Design a 13-bit ring counter, circulating a 1 among twelve 0s, using two **74x164**s.

10-16.
    (a) Design an 8-stage Johnson counter using a **74x164**.
    (b) The count sequence, in hex, of this counter is 00, 01, 03 . . . Finish the count sequence in hex.

Section 10-7-5.

10-17. Given: Four lights in a row. Design a circuit to do the following:

T (sec)	L_1	L_2	L_3	L_4
0–1	OFF	OFF	OFF	OFF
1–2	ON	OFF	OFF	OFF
2–3	ON	ON	OFF	OFF
3–4	ON	ON	ON	OFF
4–5	ON	ON	ON	ON
5–6	OFF	OFF	OFF	OFF
0–1	OFF	OFF	OFF	OFF
REPEAT				

Assume your clock input is a 60 Hz square wave.

10-18. When power is applied to the circuit of Fig. 10-16, outputs A, C, and F come up LOW. Draw a timing chart showing how each of the eight outputs behaves as clocks are applied to the shift register.

10-19. Enlarge the timing chart of Fig. 10-18a by adding the outputs of the one-shots and the counter.

10-20. A device presents 16 bits on a serial input line along with a DATA READY pulse. Each bit lasts for 10 μs. The DATA READY pulse goes positive at the start of the first bit and remains HIGH until the end of the 16th bit. Design a circuit to accept the 16 serial bits and present them to a computer in parallel. The circuit must present a positive INPUT READY signal to the computer when it has the data, and the computer will reply with a negative-going ACKNOWLEDGE pulse when it has accepted the data. The ACKNOWLEDGE pulse must terminate INPUT READY.

10-21. Given 13 inputs and a negative-going synchronizing strobe that occurs occasionally and randomly. The data on the 13 input lines are valid only when the synchronizing strobe is low. Design a circuit to:

(a) Produce 0s as an output when the synchronizing strobe is low.

(b) Put out the 13 data bits in a 500 μs serial bit stream when the synchronizing pulse goes HIGH.

(c) Put out 1s after the data are finished.

10-22. This problem involves the simplified design of a Universal Asynchronous Receiver-Transmitter (UART). Design a UART circuit to the following specifications:

1. There is an 8-bit Transmit Data register. It will be loaded with data from the computer whenever the computer presents a $\overline{\text{STROBE}}$ pulse.

2. The status of the data register determines a Transmit Buffer Empty (TBMT) signal sent back to the computer. When it is HIGH, the computer may send data to the Transmit Data Register. The $\overline{\text{STROBE}}$ signal, which causes the register to accept the data, also causes TBMT to go LOW, so the computer will not send another byte of data.

3. The data register sends its data into the Transmit Shift Register, which will send the data out, one bit at a time, as its clock advances.

4. Transferring data into the Transmit Shift Register causes TBMT to go HIGH, because the data register is now empty. Data cannot be transferred into the shift register until it has finished shifting out the previous data.

Section 10-7-6.

10-23. Repeat Example 10-13 for the circuit of Fig. P10-23.

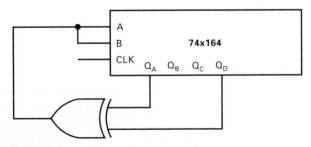

Figure P10-23

After attempting the problems, the student should return to section 10-2. If the student still cannot answer some of the self-evaluation questions, he or she should review the appropriate sections of the chapter to find the answers.

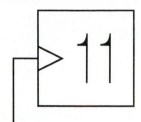

Multiplexers and Demultiplexers

11-1 Instructional Objectives

This chapter introduces multiplexers and demultiplexers, and presents some of their applications. After reading it, the student should be able to:

1. Digitally multiplex an n-line input to a 1-line output.
2. Understand and utilize **74x00** series multiplexers.
3. Design 1-line-to-n-line demultiplexers and utilize **74x00** series demultiplexers.
4. Design and utilize decoders.
5. Use multiplexers to generate combinatorial functions of several variables.
6. Use a priority encoder.
7. Design a keyboard decoder.

11-2 Self-Evaluation Questions

Watch for the answers to the following questions as you read the chapter. They should help you understand the material presented.

1. Define *multiplexing* and *demultiplexing*.
2. How many SELECT lines are required by an n-line-to-1-line multiplexer?

3. Is an unselected output of a TTL demultiplexer 0 or 1? Explain.

4. What is the level of the outputs of an unselected demultiplexer (a demultiplexer that is inactive because of a HIGH strobe input)?

5. What is the difference between a demultiplexer and a decoder? How can a demultiplexer be made to function as a decoder?

6. Explain why an n-1 input multiplexer is used to generate a function of n variables.

7. On a keyboard decoder, why is scanning essential?

11-3 Multiplexer Concepts

Multiplexing is the funneling of information from one of several input lines to a single output line. The inputs to a multiplexer are the several input lines, and one of the input lines is connected to the output. AND-OR-INVERT gates, discussed in section 5-9 are examples of inverting multiplexers. A **74x51** can connect one of two input lines to a single output line, but the output is inverted.

The basic multiplexer operation is shown in Fig. 11-1. Only one switch is closed at any one time, and connects the selected input to the output. There are many instances where one of several possible inputs to a circuit must be selected, and multiplexers are an ideal solution to the problem. The Electronic Control Module (ECM) in an automobile is an example. It is basically a microprocessor (μP) that controls the automobile's ignition. In order to optimize performance and gas mileage, it must monitor the *status* of the automobile. The μP uses a group of *sensors* to determine the temperature, pressure, and speed at various points in the engine to optimally control the car.

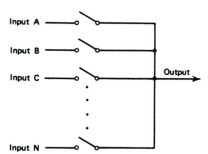

Figure 11-1 The basic multiplexer.

There are two possible ways to get the status of each of the several sensors to the μP. Each sensor could be directly connected, but this means running several wires (one for each sensor) to the μP. As an alternative, the sensor outputs could be multiplexed onto a single line. This is often done to save wiring and to transmit a large amount of information on a small number of lines.

Referring again to Fig. 11-1, the sensor outputs could be the inputs to the multiplexer. But there must be a way of choosing *which input is connected to the output.* Generally a multiplexer that has n inputs and one output requires k SELECT lines to select the particular input that is to be connected to the output. The relationship between the input lines and the select lines is typically $n = 2^k$, so an 8-input multiplexer will require 3 SELECT lines.

Multiplexers can be constructed out of AND and OR gates (AOI gates are an example), but there is no need to do so. Several multiplexers exist in the **74x00** series. They are listed in Table 11-1, and most of them will be discussed in this section. Further information is provided in the manufacturer's data books.

TABLE 11-1

Multiplexers in the 7400 Series

IC Number	Configurations	Number of Multiplexers in Each IC
74x150	16 line to 1 line	1
74x151	8 line to 1 line	1
74x153	4 line to 1 line	2
74x157	2 line to 1 line	4
74x158	2 line to 1 line	4
74x257	2 line to 1 line	4
74x258	2 line to 1 line	4

11-4-1   The 74x151 Multiplexer

Perhaps the simplest and most generic multiplexer is the **74x151**, whose pinout and function table are given in Fig. 11-2. It is an 8-line-to-1-line multiplexer in a 16-pin package. Figure 11-2a shows the pinout and Fig. 11-2b shows the logic. The 16 pins on the **74x151** are:

- Power
- Ground
- 8 data inputs
- 3 SELECT inputs
- A noninverting output (Y)
- An inverting output (W)
- A strobe

When the strobe is HIGH, the multiplexer is disabled; its Y output will always be 0 and its W output will always be 1.

EXAMPLE 11-1

The output of a **74x151** is to be the inverse of its sixth input (D6 on pin 13). How must it be set up to do this?

SOLUTION   The strobe must be LOW to enable the IC. The SELECT lines must be set to 6. This means the C and B inputs (pins 9 and 10) must be 1 and the A input (pin 11) must be 0. The output will be found on the W line (pin 6). If a noninverting output were required, it could be taken from the Y output.

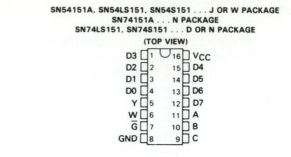

'151A, 'LS151, 'S151

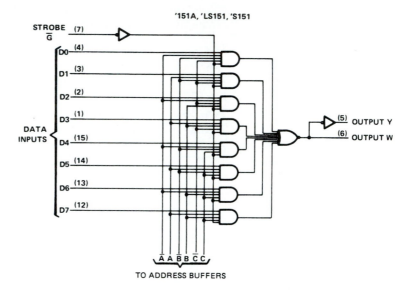

TO ADDRESS BUFFERS

ADDRESS BUFFERS FOR 'LS151, 'S151

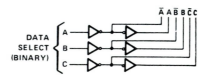

Figure 11-2 The **74x151** 8-line to 1-line multiplexer. (Reprinted by Permission of Texas Instruments.)

11-4-2 The 74150 Multiplexer

The **74150** is a 16-line-1-line multiplexer shown in Fig. 11-3. Because 16 inputs and 4 SELECT lines are required, it comes in a 24-pin package. The output is inverted and the IC has a STROBE input. If the STROBE is HIGH, the output is disabled and is always 1. The **74150** is strictly an inverting multiplexer; its output is the complement of the selected input.

11-4-3 The 74x153 4-line to 1-line Multiplexer

The functional block diagram and function table for the **74x153** are shown in Fig. 11-4. This is a dual 4-line-to-1-line multiplexer; it contains two 4-line-to-1-line multiplexers. The two multiplexers are not totally independent, however,

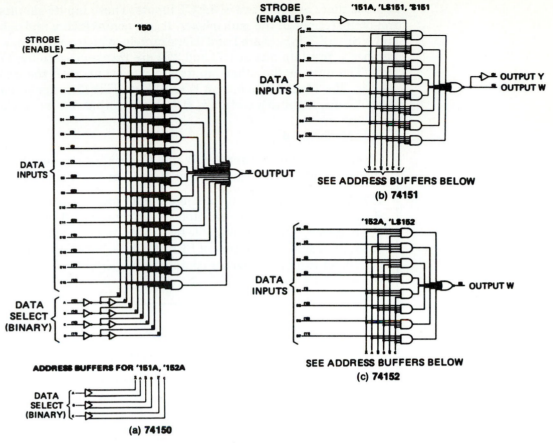

Figure 11-3 The **74150** 16-line-to-1-line multiplexer.

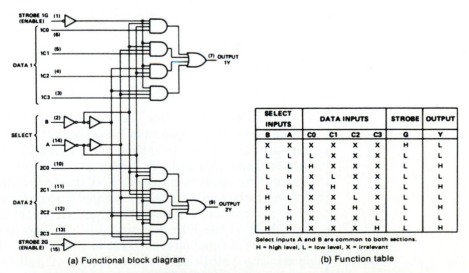

SELECT INPUTS		DATA INPUTS				STROBE	OUTPUT
B	A	C0	C1	C2	C3	G	Y
X	X	X	X	X	X	H	L
L	L	L	X	X	X	L	L
L	L	H	X	X	X	L	H
L	H	X	L	X	X	L	L
L	H	X	H	X	X	L	H
H	L	X	X	L	X	L	L
H	L	X	X	H	X	L	H
H	H	X	X	X	L	L	L
H	H	X	X	X	H	L	H

Select inputs A and B are common to both sections.
H = high level, L = low level, X = irrelevant

(a) Functional block diagram   (b) Function table

Figure 11-4 The **74x153** multiplexer. (From the *TTL Data Book,* 1988, Texas Instruments, Inc. Courtesy of Texas Instruments.)

because there are only two SELECT inputs. These inputs simultaneously control both sections of the multiplexer. It is a noninverting multiplexer. If the B and A SELECT inputs are 1 and 0 for example, which corresponds to the binary number 2, the data bits at 1C2 and 2C2 are transmitted to the 1Y and 2Y outputs. The data at the other inputs are irrelevant until the SELECT inputs change. The **74x153** also has an ENABLE or STROBE line for each section. If the STROBE input to a section is HIGH, the output of that section is LOW.

EXAMPLE 11-2

Given three 4-bit registers, A, B, and C, design a circuit whose output is either the contents of registers A, or B, or C, or the binary equivalent of 9.

SOLUTION   One of four possible inputs may appear on the output. This suggests a 4-line-to-1-line multiplexer. To choose 1 of 4 inputs, 2 SELECT bits are required. The selection table below was made by arbitrarily assigning an output to each combination of SELECT bits. If SELECT lines A and B are both 0, for example, then the contents of register A will be funneled through the multiplexer.

SELECT Bits		
B	A	Output
0	0	Register A
0	1	Register B
1	0	Register C
1	1	9

Because a 4-bit output is required, this circuit can be built using two **74x153**s, as shown in Fig. 11-5. The selection table requires that Register A is connected to the C0 inputs, Register B to the C1 inputs, Register C to the C2 inputs, and the number 9 is wired to the C3 inputs. The output is one of the four inputs depending on the SELECT line levels as indicated by the selection table.

EXAMPLE 11-3

Design a 64-line-to-1-line multiplexer. Show how it would select the 37th line.

SOLUTION   One solution is shown in Fig. 11-6. Since 1 of 64 lines is being chosen, 6 lines are required to make the selection. In Fig. 11-6, the 64 input lines are applied to four **74150**s, and 4 SELECT lines are applied to all the **74150**s. The outputs of the **74150**s are then connected to one section of a **74x153**. The two remaining SELECT lines are connected to the **74x153** and decide which of its 4 inputs is funneled through to the output. The final output is inverted to compensate for the inversion of the **74150**.

To select the 37th line, the SELECT lines must be binary 37, 100101. The 4 LSBs are applied to each **74150** and select the fifth input on each multiplexer. The 2 MSBs (10) are connected to the SELECT lines on the **74x153** and select the third **74150**. Thus, input 5 of the third **74150** is the line connected to the output. The selected data path is shown as a dark line on Fig. 11-6.

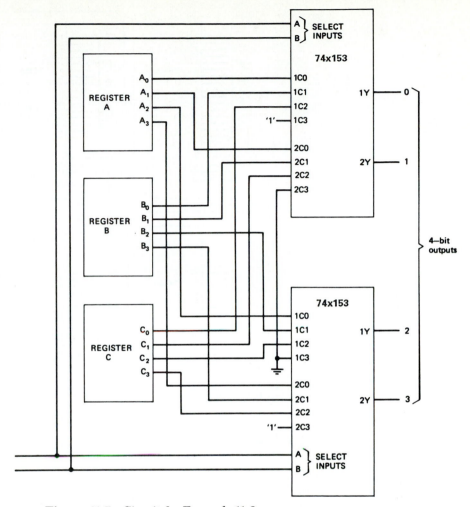

Figure 11-5 Circuit for Example 11-2.

11-4-4 The 74x157

The **74x157** is a quadruple 2-line-to-1-line multiplexer. Its functional block diagram is shown in Fig. 11-7. A HIGH input on the STROBE line disables the IC by forcing all outputs LOW. With the STROBE LOW, the level on the SELECT line determines whether the A or B inputs are connected to the outputs. The price paid for the larger number of output bits per IC (4) is the reduced number of input choices (2). The **74x157** is ideal for choosing one of two long data registers.

EXAMPLE 11-4

Two 12-bit registers are to be multiplexed together. Design the circuit.

SOLUTION Since each **74x157** accommodates 4 bits, three **74x157**s are needed. The first 12-bit register is tied to the A inputs of the three multiplexers, and the second 12-bit register is tied to the B inputs. A single SELECT line is tied to all the **74x157**s. If it is HIGH the second register is selected, and if it is LOW the first register is selected.

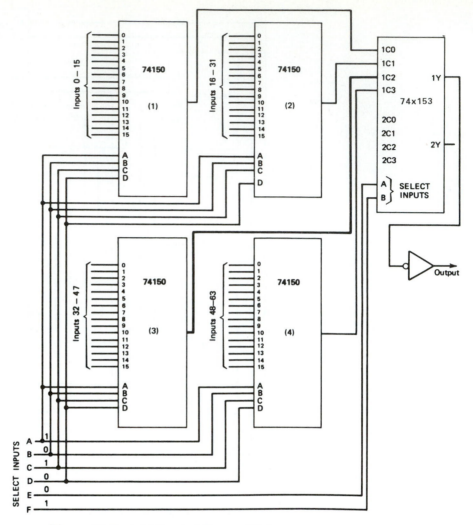

Figure 11-6 A 64-line to-1-line multiplexer.

The **74LS257** and **74LS258** are newer 2-line to 1-line multiplexers that function very much like the **74x157**. These multiplexers have 3-state outputs, however, and the strobe line is replaced by an OUTPUT CONTROL signal that must be LOW for the multiplexer to drive the bus. Otherwise, the multiplexer outputs are high impedance. These ICs are ideal for driving multiplexed inputs onto a 3-state bus. The difference between the **'257** and the **'258** is that the **'258** inverts its outputs.

11-5   Time Division Multiplexing

Multiplexing is often used when information from many sources must be transmitted over long distances, and it is less expensive to multiplex the data onto a single wire for transmission than to have a separate line for each source. There

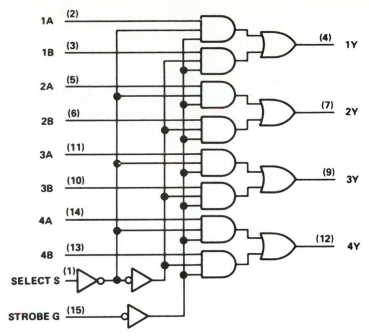

Figure 11-7 Functional block diagram of the **74x157**. (From the *TTL Data Book,* 1988, Texas Instruments, Inc. Courtesy of Texas Instruments.)

are two ways to multiplex information from many sources onto a single line: Frequency Division Multiplexing (FDM) and Time Division Multiplexing (TDM). Both are shown in Fig. 11-8. In both cases, the bandwidth of the transmitting line must be larger than the bandwidth (or frequency components) of the information to be transmitted.

In FDM, each signal is *modulated* up to a specific band and transmitted. Thus, if 3 kHz is the highest frequency of each signal (telephone conversations are an example), one signal may be sent from 3–6 kHz, another from 7–10 kHz, and so on until the bandwidth of the transmission media is exhausted. This is shown in Fig. 11-8a.

In TDM each signal gets a *specific time slot* on a transmission line. TDM is used more often with digital circuits and is shown in Fig. 11-8b. The bits on the line are divided into *frames*. 8-bit frames are shown in Fig. 11-8b. Message A is allocated bit 1 of each frame, message 2 gets bit 2, and so on. Example 11-5 shows the design of a time division multiplexer where each input is on the line for 1 μs during each 8-μs cycle.

EXAMPLE 11-5

Given eight input lines, A–H, design a circuit to connect each input line to the output line sequentially for 1 μs (i.e., input A is connected to the output for the first 1 μs, input B is connected to the output at 2 μs, etc.). The device is cycled continuously so that input A is connected to the output after input H.

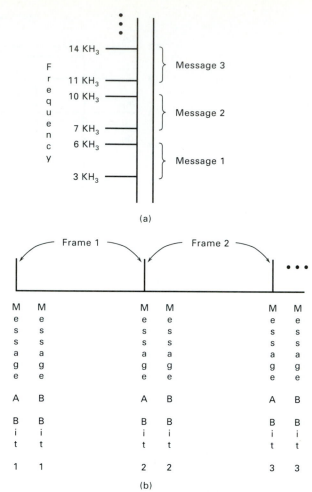

Figure 11-8 Frequency and time division multiplexing (a) Frequency division multiplexing (FDM); (b) time division multiplexing (TDM).

> **SOLUTION** One solution is shown in Fig. 11-9. The 1-μs oscillator clocks the **74x93** counter. The outputs of the counter are connected to the SELECT lines of the **74x151**. As the counter increments each μs, the SELECT lines change and each input is connected to the output in turn.

11-5-1 TDM in Microprocessors

Many μPs use time division multiplexing to reduce the number of lines that must be connected to the μP. Intel's **8085** and **8086** μPs use many of the same lines between the μP and its memory for both address and data. These μPs provide a signal labeled Address Latch Enable (ALE). When ALE is HIGH, the lines are being driven by the μP and provide addresses for the memory. When ALE is LOW, the line contains the data being transferred between the μP and its memory or peripheral ICs. At this time data can flow in either direction.

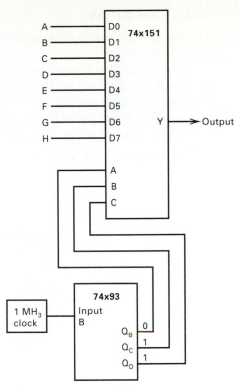

Figure 11-9 Time division multiplexing.

11-6 Demultiplexers

Demultiplexers reverse the multiplexing operation; they take a single input line and fan it out to one of many output lines. A demultiplexer, therefore, connects the input to the *output of the selected line*. All other output lines are *deselected*. For IC demultiplexers, *deselected outputs always assume a HIGH level*.

The demultiplexing principle is illustrated in Fig. 11-10, where it is assumed that only one switch is closed at any one time. Demultiplexers have 1 input and *n* output lines. They also require enough SELECT inputs to specify which one of the *n* output lines is connected to the input line. Like multiplexers, demultiplexers can be designed using SSI gates, but almost all engineers use IC demultiplexers.

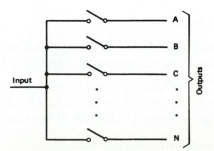

Figure 11-10 The basic demultiplexer.

11-6-1 The 74x138

The **74x138**[1] is the most commonly used 1-line-to-8-line demultiplexer. Its functional block diagram and function table are given in Fig. 11-11. They indicate that the **74x138** has three ENABLE inputs and three SELECT inputs. The ENABLE inputs must all be satisfied (G1 = 1, $\overline{G2A}$ = 0, $\overline{G2B}$ = 0) for the IC to function. Then the output corresponding to the number on the SELECT lines will be LOW. If any one of the ENABLE inputs is not satisfied, the IC is disabled and all the outputs of the **74x138** are HIGH.

logic diagram and function table

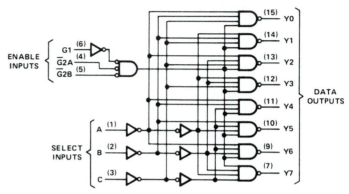

'LS138, SN54S138, SN74S138A

Pin numbers shown are for D, J, N, and W packages.

'LS138, SN54138, SN74S138A
FUNCTION TABLE

INPUTS					OUTPUTS							
ENABLE		SELECT										
G1	$\overline{G2}$*	C	B	A	Y0	Y1	Y2	Y3	Y4	Y5	Y6	Y7
X	H	X	X	X	H	H	H	H	H	H	H	H
L	X	X	X	X	H	H	H	H	H	H	H	H
H	L	L	L	L	L	H	H	H	H	H	H	H
H	L	L	L	H	H	L	H	H	H	H	H	H
H	L	L	H	L	H	H	L	H	H	H	H	H
H	L	L	H	H	H	H	H	L	H	H	H	H
H	L	H	L	L	H	H	H	H	L	H	H	H
H	L	H	L	H	H	H	H	H	H	L	H	H
H	L	H	H	L	H	H	H	H	H	H	L	H
H	L	H	H	H	H	H	H	H	H	H	H	L

Figure 11-11 The **74x138** 3-line to 8-line decoder. (Reprinted by Permission of Texas Instruments.)

EXAMPLE 11-6

Show the connections needed to use the **74x138** as a demultiplexer. Explain how it functions if the number on the SELECT line is 6.

SOLUTION The circuit is shown in Fig. 11-12. Here G1 is tied HIGH and $\overline{G2A}$ is connected to ground, so these two ENABLE inputs are always satisfied. The input is connected to $\overline{G2B}$. If the number on the SELECT lines is

[1]This IC is essentially the same as the Intel **8205**.

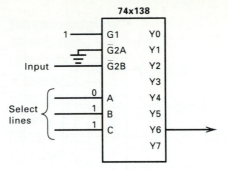

Figure 11-12 The circuit for Example 11-6.

6 and the input is HIGH, the $\overline{G}$2B ENABLE is not satisfied and all the outputs will be HIGH, the same as the input. If the input is LOW, however, all the ENABLE inputs are satisfied and Y6 will go LOW. Thus, the Y6 output will be the same as the input in either case and the **74x138** functions as a demultiplexer.

The **74x155** is an IC that can also function as a demultiplexer and a decoder. It is a little trickier to use because it often requires inverters to function properly. The **74x155** is discussed in the previous editions of this book.

11-6-2   The 74154

The **74154** is a 1-line-to-16-line demultiplexer in a 24-pin DIP whose functional block diagram and function table are shown in Fig. 11-13. The four SELECT inputs choose 1 of the 16 outputs. The selected output goes LOW only if inputs G1 and G2 are both LOW, so that one of the G inputs can act as a STROBE and the other can function as the input. If either one of the G inputs is HIGH, all 16 outputs of the **74154** will be HIGH.

EXAMPLE 11-7

Design a 1-line-to-32-line demultiplexer.

SOLUTION   A solution using two **74154**s is shown in Fig. 11-14. Five SELECT lines are required. Four of them are applied directly to the **74154**s. The fifth SELECT line is applied to the G1 input of one **74154** and its complement is applied to the G1 input of the other. Only the **74154** with the LOW G1 input is enabled. The two G2 inputs are tied together to become the demultiplexer input. If this input is LOW, the selected output is LOW.

11-6-3   TDM Multiplexing and Demultiplexing

When a circuit uses time division multiplexing (TDM) on the sending side of a transmission line, it must often time division demultiplex on the receiving end. The circuit of Fig. 11-15 illustrates the situation. Observe the line from the multiplexer to the demultiplexer. If this line is short, there may be no need for the circuit; the inputs could be directly connected to the outputs. If the line is long, several miles perhaps, the circuit is necessary. This might be the case where

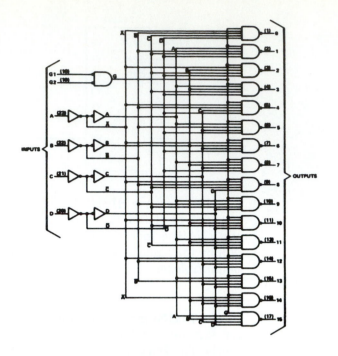

FUNCTION TABLE

INPUTS						OUTPUTS															
G1	G2	D	C	B	A	0	1	2	3	4	5	6	7	8	9	10	11	12	13	14	15
L	L	L	L	L	L	L	H	H	H	H	H	H	H	H	H	H	H	H	H	H	H
L	L	L	L	L	H	H	L	H	H	H	H	H	H	H	H	H	H	H	H	H	H
L	L	L	L	H	L	H	H	L	H	H	H	H	H	H	H	H	H	H	H	H	H
L	L	L	L	H	H	H	H	H	L	H	H	H	H	H	H	H	H	H	H	H	H
L	L	L	H	L	L	H	H	H	H	L	H	H	H	H	H	H	H	H	H	H	H
L	L	L	H	L	H	H	H	H	H	H	L	H	H	H	H	H	H	H	H	H	H
L	L	L	H	H	L	H	H	H	H	H	H	L	H	H	H	H	H	H	H	H	H
L	L	L	H	H	H	H	H	H	H	H	H	H	L	H	H	H	H	H	H	H	H
L	L	H	L	L	L	H	H	H	H	H	H	H	H	L	H	H	H	H	H	H	H
L	L	H	L	L	H	H	H	H	H	H	H	H	H	H	L	H	H	H	H	H	H
L	L	H	L	H	L	H	H	H	H	H	H	H	H	H	H	L	H	H	H	H	H
L	L	H	L	H	H	H	H	H	H	H	H	H	H	H	H	H	L	H	H	H	H
L	L	H	H	L	L	H	H	H	H	H	H	H	H	H	H	H	H	L	H	H	H
L	L	H	H	L	H	H	H	H	H	H	H	H	H	H	H	H	H	H	L	H	H
L	L	H	H	H	L	H	H	H	H	H	H	H	H	H	H	H	H	H	H	L	H
L	L	H	H	H	H	H	H	H	H	H	H	H	H	H	H	H	H	H	H	H	L
L	H	X	X	X	X	H	H	H	H	H	H	H	H	H	H	H	H	H	H	H	H
H	L	X	X	X	X	H	H	H	H	H	H	H	H	H	H	H	H	H	H	H	H
H	H	X	X	X	X	H	H	H	H	H	H	H	H	H	H	H	H	H	H	H	H

H = high level, L = low level, X = irrelevant

Figure 11-13 The **74154** 1-line-to-16-line demultiplexer. (From the *TTL Logic Data Book*, 1988, Texas Instruments, Inc. Courtesy of Texas Instruments.)

there are many computer terminals in a center-city location, and their data must be sent to a large, mainframe computer located in the suburbs or a different city. Then the data from the terminals are multiplexed in the center city, transmitted to the computer over a single line, and must be demultiplexed. This allows a large amount of data to be transferred over a single line and saves money because long transmission lines are expensive.

The circuit of Fig. 11-15 shows how eight terminals can be connected to a

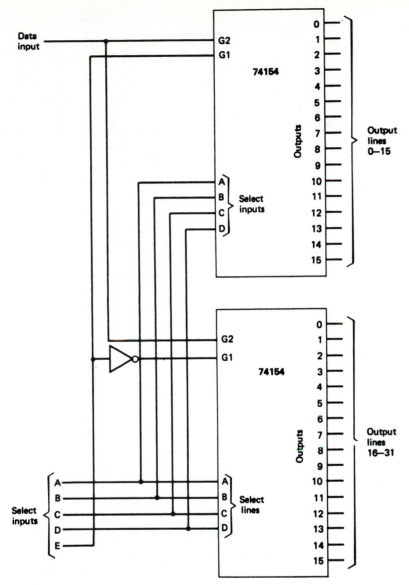

Figure 11-14 A 1-line-to-32-line demultiplexer.

74x151, multiplexed, and transmitted to the **74x138**, which acts as a time division demultiplexer at the receiving end. The **74x151** is used as the TDM, as discussed in sec. 11-5, and the **74x138** is the demultiplexer discussed in sec. 11-6-1.

For simplicity, the circuit is shown when the count is 6. The data entering the multiplexer on D6 is funnelled through the connecting line to the demultiplexer and appears at Y6. After the next clock pulse, the data on D7 will appear at Y7. The circuit is simplified because both ICs are getting the same count from the same counter. If the operations are separated by several miles, however, there must be a counter at each end and they must be synchronized. Several methods exist for synchronizing the transmitting and receiving ends of the circuit.

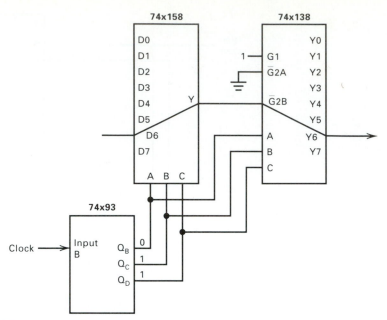

Figure 11-15 TDM multiplexing and demultiplexing.

11-7 Decoders

Demultiplexers can be used as digital decoders. A digital decoder has 2^n outputs and accepts n inputs. Only the output that corresponds to the binary number on the input lines is activated. Decoders are used in many digital circuits; they can be used to select memory addresses, and to decode instructions in a computer (see Section 16-6-2). They are used whenever one line out of several possible lines must be selected.

11-7-1 The 74x138 Used as a Decoder

The **74x138** can be used as a decoder simply by enabling it. Then the output that corresponds to the number on the SELECT lines will go low.

EXAMPLE 11-8

Figure 11-16 shows a **74x138** used in a memory circuit. What must the signals on the line be to make $\overline{\text{CSA}}$ LOW?

SOLUTION   First we will consider the ENABLE lines. For G1 to be HIGH, we must be attempting to read or write the memory, which means that either $\overline{\text{MEMR}}$ MEMory Read OR $\overline{\text{MEMW}}$ (MEMory Write) must be LOW. Also, INH MEM (Inhibit Memory) and A15 must both be LOW. When these all occur, the **74x138** will be enabled. $\overline{\text{CSA}}$ is connected to the fifth data output. It will go LOW if A12 = 1, A13 = 0, and A14 = 1 so the number on the SELECT lines is 5.

To sum up, $\overline{\text{CSA}}$ will go low, indicating the signal is selected, when we are trying to read or write the memory, the memory is not inhibited, and the top four bits (A15-A12) on the memory address bus are 0101.

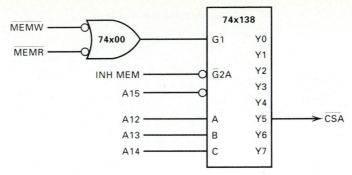

Figure 11-16 The circuit for Example 11-8.

11-7-2   The 74154 Used as a Decoder

The **74154** (Fig. 11-13) can be used as a 4-line-to-16-line decoder simply by tying both G inputs LOW. Then the output corresponding to the SELECT input is LOW and all other outputs are HIGH.

EXAMPLE 11-9

Design a 5-line-to-32-line decoder.

SOLUTION   The student must first realize that if we want to select one out of 32 possible output lines, there must be 5 inputs to determine the particular line. The decoder is built by tying the two G2 inputs of the 5-line-to-32-line demultiplexer (see Fig. 11-14) to ground. Four of the five required SELECT lines are tied to the **74154**s (they select one output on each chip) and the fifth SELECT input is tied to the G1 inputs (to disable one of the **74154**s). Consequently, for any combination of the 5 SELECT inputs, one and *only one* output line is decoded. Its output is LOW. The other 31 outputs are all HIGH.

11-7-3   The 74x155 as a Decoder

A 3-line-to-8-line decoder can be built from a **74155** by tying the data 1C and data 2C lines together. The three required SELECT lines consist of SELECT A, SELECT B, and the common data lines. Because section 1 inverts its data and section 2 does not, one of the outputs is always LOW, provided the STROBES are LOW.

11-7-4   The 74x42 BCD Decoder

The **74x42** is a Binary Coded Decimal (BCD) decoder. Its logic diagram and Function Table are given in Fig. 11-17. The **74x42** accepts four SELECT inputs in BCD form and places a 0 on the corresponding output line. If the number on the SELECT inputs is not a valid BCD number (it is greater than 9), all outputs remain HIGH.

logic diagram (positive logic)

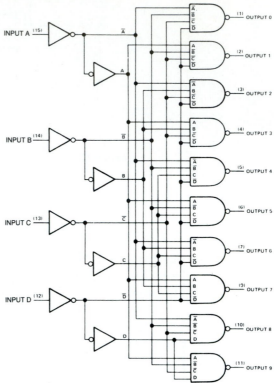

Pin numbers shown are for D, J, N, and W packages.

FUNCTION TABLE

NO.	BCD INPUT				DECIMAL OUTPUT									
	D	C	B	A	0	1	2	3	4	5	6	7	8	9
0	L	L	L	L	L	H	H	H	H	H	H	H	H	H
1	L	L	L	H	H	L	H	H	H	H	H	H	H	H
2	L	L	H	L	H	H	L	H	H	H	H	H	H	H
3	L	L	H	H	H	H	H	L	H	H	H	H	H	H
4	L	H	L	L	H	H	H	H	L	H	H	H	H	H
5	L	H	L	H	H	H	H	H	H	L	H	H	H	H
6	L	H	H	L	H	H	H	H	H	H	L	H	H	H
7	L	H	H	H	H	H	H	H	H	H	H	L	H	H
8	H	L	L	L	H	H	H	H	H	H	H	H	L	H
9	H	L	L	H	H	H	H	H	H	H	H	H	H	L
INVALID	H	L	H	L	H	H	H	H	H	H	H	H	H	H
	H	L	H	H	H	H	H	H	H	H	H	H	H	H
	H	H	L	L	H	H	H	H	H	H	H	H	H	H
	H	H	L	H	H	H	H	H	H	H	H	H	H	H
	H	H	H	L	H	H	H	H	H	H	H	H	H	H
	H	H	H	H	H	H	H	H	H	H	H	H	H	H

H = high level, L = low level

Figure 11-17 The **74x42** BCD decoder. (Reprinted by Permission of Texas Instruments.)

In Chapters 2 and 3, combinatorial logic functions of several variables were generated using Boolean algebra or Karnaugh mapping techniques. A combinatorial function can also be generated by a multiplexer:

1.   Write the required function in SOP (sum-of-products) form.
2.   If n input variables are specified, choose a multiplexer with $n - 1$ SELECT lines.
3.   The input lines to the multiplexer numerically correspond to the $n - 1$ MSBs of specified function.
4.   The LSB of the specified function determines the signal on each input line. This signal is obtained by examining the SOP form of the specification.

The procedure is illustrated in Example 11-10.

EXAMPLE 11-10

If $f(W,X,Y,Z) = \Sigma(0, 2, 3, 6, 9, 10, 14, 15)$, realize f using a multiplexer.

SOLUTION

1.   In SOP form:

$$f(W,X,Y,Z) = \overline{W}\,\overline{X}\,\overline{Y}\,\overline{Z} + \overline{W}\,\overline{X}Y\overline{Z} + \overline{W}\,\overline{X}YZ + \overline{W}XY\overline{Z} + W\overline{X}\,\overline{Y}Z \\ + W\overline{X}Y\overline{Z} + WXY\overline{Z} + WXYZ \tag{11-1}$$

2.   The given function depends on four variables. Therefore, the required multiplexer must have 3 SELECT lines (step 2); so a **74x151**, 8-line-to-1-line multiplexer is chosen.
3.   The W, X, and Y lines are connected to the SELECT inputs of the multiplexer.
4.   Each multiplexer input line is examined to determine what signal should be applied to it. The table of Fig. 11-18a was constructed to make this determination.
5.   The first column of this table lists all the W, X, and Y possibilities. Since these variables are connected to the SELECT lines, they select the input line corresponding to their binary value, as shown in column 2.
6.   Column 3 of the table lists the terms of the specified function covered by each input. The top line, for example, has $W = X = Y = 0$. This covers SOP terms 0 and 1.
7.   The top line represents $\overline{W}\,\overline{X}\,\overline{Y}$. The term $\overline{W}\,\overline{X}\,\overline{Y}\,\overline{Z}$ is part of the SOP form of the specifications, but $\overline{W}\,\overline{X}\,\overline{Y}Z$ is not. If $\overline{Z}$ *is connected to input line 0 of the multiplexer, it produces a 1 output when* W, X, and Y are 0 (selecting input 0) and Z is 0, so that $\overline{Z}$, the input to line 0, is 1.
8.   The second line of the table $(\overline{W}\,\overline{X}Y)$ selects line 1 of the multiplexer and covers terms 2 and 3 of the specification. These terms are both

W	X	Y	Input Number	Terms Covered	Input
0	0	0	0	0, 1	$\overline{Z}$
0	0	1	1	2, 3	$Z + \overline{Z} = 1$
0	1	0	2	4, 5	0
0	1	1	3	6, 7	$\overline{Z}$
1	0	0	4	8, 9	Z
1	0	1	5	10, 11	$\overline{Z}$
1	1	0	6	12, 13	0
1	1	1	7	14, 15	1

(a) Table

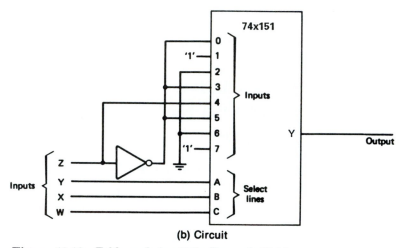

(b) Circuit

Figure 11-18   Table and circuit for Example 11-10.

1, so if input 1 is tied HIGH, a 1 output results whenever $W = 0$, $X = 0$, and $Y = 1$, as required.

9. The third line (input line 2) references SOP terms 4 and 5. Since they are both 0 in this example, input 2 is tied LOW. A 0 output occurs whenever $W = 0, X = 1$, and $Y = 0$.

10. The fourth line is $\overline{W}XY$. Since $\overline{W}XY\overline{Z}$ is part of the function, $\overline{Z}$ is tied to input 3.

11. The fifth line is $W\overline{X}\,\overline{Y}$. Here only $W\overline{X}\,\overline{Y}Z$ is part of the function so Z is tied to line 4.

12. The table is continued in this manner and specifies the input to each line of the multiplexer.

13. The final circuit is shown in Fig. 11-18b.

11-9 Encoding

Encoding could be considered the reverse of decoding. The process involves monitoring a group of lines or switches and providing a coded output when one of them is activated. The output code indicates which line has been activated.

There are two very important uses for encoding in computers: priority encoding and keyboard encoding.

11-9-1 The 74x148 Priority Encoder

A priority encoder monitors several lines, and assigns each of them a priority. If one of the lines is activated, it produces an output indicating which line it is. If several of the lines are activated, the output code indicates the number of the *highest priority* line that is activated. Priority encoders are used in computers where several devices can request interrupts. It enables the highest priority request to be serviced.

The **74x148** is the most commonly used priority encoder IC. Its pinout and function table are given in Fig. 11-19. It has eight prioritized input lines (0–7) and an ENABLE input, EI. It has an ENABLE output, EO, a SELECT output, GS, and three encoded outputs: A2, A1, and A0.

The operation of the **74x148** can be understood by studying the function table. On line 1 the IC is disabled because EI is HIGH. All its outputs are HIGH. On line 2 the IC is enabled, but there is no action; all the input lines are HIGH, indicating they are not trying to make a request. Output EO is LOW to allow the **74x148** to be cascaded if more than 8 inputs need to be prioritized (see Problem 11-14).

On lines 1 and 2, where the IC was disabled or no input was requesting service, the GS output was HIGH. A line makes a request for service by going LOW. The remaining lines cover the situation where the IC is enabled and at least one of the inputs is trying to make a request. This causes GS to go low, indicating that there is a valid request, and EO to go HIGH so that lower priority ICs can be disabled.

On line 3, input 7 is low. This is the highest priority request (the device awarded highest priority would have its request line tied to input 7). Therefore, the levels on the other input lines are irrelevant. The corresponding output lines encode the output using *negative logic* (LOW = logic 1). The encoded outputs on

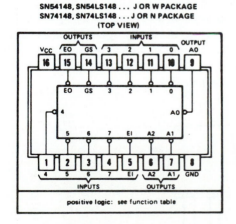

'148, 'LS148

FUNCTION TABLE

INPUTS										OUTPUTS				
EI	0	1	2	3	4	5	6	7		A2	A1	A0	GS	EO
H	X	X	X	X	X	X	X	X		H	H	H	H	H
L	H	H	H	H	H	H	H	H		H	H	H	H	L
L	X	X	X	X	X	X	X	L		L	L	L	L	H
L	X	X	X	X	X	X	L	H		L	L	H	L	H
L	X	X	X	X	X	L	H	H		L	H	L	L	H
L	X	X	X	X	L	H	H	H		L	H	H	L	H
L	X	X	X	L	H	H	H	H		H	L	L	L	H
L	X	X	L	H	H	H	H	H		H	L	H	L	H
L	X	L	H	H	H	H	H	H		H	H	L	L	H
L	L	H	H	H	H	H	H	H		H	H	H	L	H

(b) Function Table

(a) Pinout

Figure 11-19 The **74x148** priority encoder. (Reprinted by Permission of Texas Instruments.)

this line are LLL, which translates into 111 or 7, indicating that input 7 is making a request. The next line on the table, line 4 shows that input 7 is HIGH, but there is a request on input 6. This now becomes the highest priority. The output is LLH, which translates into 110 or 6.

EXAMPLE 11-11

A **74x148** is enabled and inputs 1, 3, and 5 are LOW. What are its outputs?

SOLUTION   Input 5 has the higher priority, so the requests on lines 1 and 3 will be ignored until the level on line 5 goes HIGH, after its request has been serviced. This situation is covered by line 5 of the function table. The outputs are LHL, which translates into 101 or 5, the number of the line making the request.

The **74x147** is a BCD encoder. It accepts 9 inputs on its input lines labeled 1–9, and produces 4 outputs that are the complement of the BCD equivalent of the highest priority input (9 is the highest priority input and the corresponding output is 0110). If no inputs are valid, the **74x147** places a 1 on each output.

11-9-2   Keyboard Encoding

Keyboards are often used to give commands and control the operation of electronic circuits. The keys on a hand calculator, for example, are divided into two groups; the numeric inputs (the digits 0–9) and the command inputs (add, subtract, etc.). Sophisticated calculators have many commands. The HP 21 calculator has 30 keys on its keyboard.

Even larger keyboards exist to provide alphabetic as well as numeric and command inputs. The terminal or keyboard connected to a computer or PC is the most common example of an alpha-numeric keyboard in current use.

The keys on a keyboard are a group of pushbutton switches, as shown in Fig. 11-20. They have two normally open contacts that close when the key is depressed. *Keyboard decoding* means determining which one of the keys on a keyboard is depressed, expressing this information digitally, and passing it on to the computer or other device.

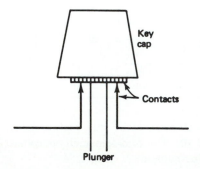

Figure 11-20   Diagram of a key switch.

Keyboard encoding can be accomplished by using multiplexers and demultiplexers working together. An encoding scheme for 64 switches is shown in Fig. 11-21. The **74x138** is operated as a 3-line-to-8-line decoder and only its selected output is LOW. Any 1 of the 8 decoder output lines can be connected to any 1 of the 8 multiplexer input lines by depressing the switch at the intersection of the lines.

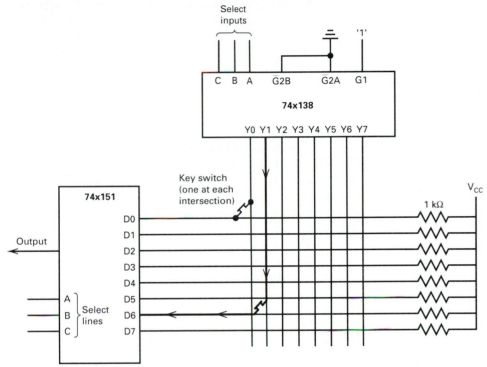

Figure 11-21   A method of keyboard decoding.

The **74x151** multiplexer produces a LOW output only if its selected input is LOW. A LOW output indicates a switch is closed, and the combination of the 3 SELECT inputs on the **74x151** and the 3 SELECT inputs on the **74x138** form a 6-bit code to determine which switch has been depressed.

EXAMPLE 11-12

The **74x151** output goes LOW when its SELECT inputs are 110 and the SELECT inputs on the **74x138** are 001. Which switch has been depressed?

SOLUTION   The current path is shown in dark in Fig. 11-21. Output Y1 of the **74x138** is LOW because of the state of its SELECT lines. This LOW level is fed through the depressed switch to input 6 of the **74x151** and causes its output to go LOW. The SELECT line bits indicate the switch between lines Y1 and D6 has been depressed.

If, for example, the given 6-bit code, 110 001, is to represent the letter X, a key cap with the letter X printed or engraved on it can be placed over the key switch at the intersection of lines Y5 and D6. When an operator presses the X key it is detected by a LOW output when the SELECT lines read 110001.

11-9-3 Practical Keyboard Encoding

Practical keyboard encoding systems are subject to the following constraints.

1. The keyboard must be *constantly scanned* to determine if any keys have been depressed.
2. Whenever any key is depressed, the scanning must stop and a code indicating which key was depressed must be available. A KEYSTROBE signal should also be generated to inform the computer or other receiving device that a key is depressed.
3. Key switches *bounce* and the circuit must effectively debounce them.
4. Fast typists tend to press a second key before they release the first key. This is called *2-key rollover* and should be allowed.

The circuit of Fig. 11-22 resolves all these difficulties for the 64-key circuit of Fig. 11-21. The time of the **74x123** one-shot is set to be somewhat longer than the bounce time of the key switches. As long as no keys are depressed, the CLOCKING FF remains SET and there is a steady stream of pulses out of the **74x08**. These pulses increment the counter and cause the keyboard to be scanned continually. They also clock the retriggerable one-shot so that $\overline{\text{KEYSTROBE}}$ is always HIGH, indicating no character is available.

When a key is depressed, the output from the **74x151** will go LOW when the scanning hits the key. This will clear the CLOCKING FF, which will cut off the clocks at the **74x08**. This will freeze the counter at the depressed key. The number in the counter indicates which key has been depressed.

When a key is depressed it first bounces, then it makes a solid contact. During the bounce time, the count will be interrupted when the scanner finds the depressed key, but scanning will resume when the switch bounces open. KEYSTROBE will remain HIGH because the one-shot's time is too long. When the key is finally closed firmly, the count will stop at the code for that key, the one-shot will stop getting trigger pulses, and $\overline{\text{KEYSTROBE}}$ will go LOW. If a second key is depressed during this time (2-key rollover), when the first key is released the circuit will scan until it stops at the code for the second key. $\overline{\text{KEYSTROBE}}$ will be HIGH for the one-shot time plus the scanning time and then will go LOW again, indicating a second key has been depressed.

EXAMPLE 11-13

For the circuit of Fig. 11-22, assume that the clock is 1 μs, the switches bounce for 1 ms, and the one-shot time is 2 ms. If a switch is depressed for 10 ms, how long is $\overline{\text{KEYSTROBE}}$ LOW?

SOLUTION During the 1 ms bounce time, $\overline{\text{KEYSTROBE}}$ is HIGH due to the intermittent nature of the interruptions. Note that every time the scan resumes it returns to the selected key every 64 μs, a relatively short time. After bouncing stops, it takes 2 ms for the one-shot triggers to expire, so $\overline{\text{KEYSTROBE}}$ will be LOW for 7 ms.

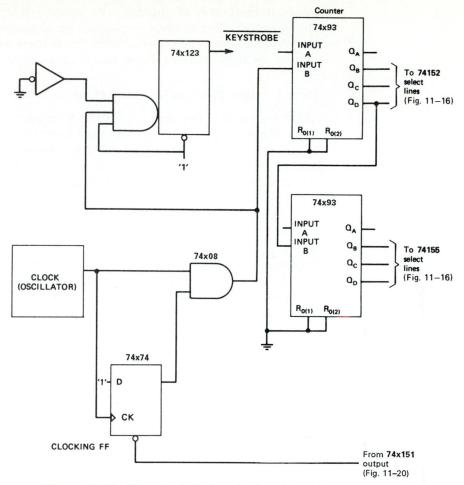

Figure 11-22 Scanning logic for a keyboard decoder.

11-9-4 IC Keyboard Decoders

The circuit of Fig. 11-22 forms the basis for *prepackaged IC keyboard encoders*. An example is the **MM5740** 90-key keyboard encoder manufactured by National Semiconductor, Inc. Most keyboard encoders are designed to work with standard typewriter keyboards. They have a 9×10 matrix for scanning the keys and provide inputs for SHIFT and CONTROL functions. Thus their outputs are the codes for both upper- and lower-case letters, numbers, punctuation, and a variety of special functions.

11-10 Display Drivers and Decoders

Throughout this book we have been describing digital circuits as circuits capable of producing logic levels (1s or 0s) at various points. In almost all circuits, the engineer must monitor these points to determine whether the circuit is functioning properly. Perhaps the simplest way to do this is to connect lights to these points. In most cases a light that is on indicates a logic 1, and off indicates a 0. The lights attached to these points can be incandescent, but these generally absorb much power and require buffer drivers (see section 5-6-2). Most modern circuits use light-emitting diodes as indicating lights.

11-10-1   Light-Emitting Diodes (LEDS)

LEDs are small diodes that emit a red glow when current is passed through them. When an LED is ON, it develops a forward voltage of 1.6 V, regardless of its current. Typical LEDs are specified to operate at 20 mA, but they will glow over a wide current range. For use with TTL circuits, a resistor (generally 150 Ω) is placed in series with the diode to limit the current and drop the excess voltage. Although the guaranteed output of a standard TTL inverter is only 16 mA, it is often used to drive an LED.

The circuit of an inverter driving an LED is shown in Fig. 11-23. When the inverter input is HIGH, current is driven through the resistor, the LED, and the lower transistor of the inverter's totem-pole output. Typically the resistor absorbs 3 V (20 mA × 150 Ω), the LED 1.6 V, and the inverter output voltage is 0.4 V. Together these add up to the 5 V of the supply. If the LED is ON, it indicates that the input to the inverter is a 1.

When the input to the inverter is a 0, the upper totem-pole transistor of the inverter turns ON. Now there is no current path through the LED and it remains dark. In this way the LED monitors the logic level at the inverter input, and the inverter is used as an LED driver.

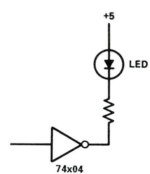

Figure 11-23   A TTL inverter driving an LED.

11-10-2   Seven-Segment Displays

A 7-segment display, which is designed to display a single *decimal* digit, is built of 7 LEDs. An example of converting a BCD decade to its corresponding 7-segment display digit has already been given in section 3-8-3. Fortunately, ICs designed to act as drivers for 7-segment displays are already in wide use. These decoder/drivers accept a BCD decade as input and convert it to outputs that illuminate the proper segments of a 7-segment display. Current limiting resistors are still needed with decoder/drivers.

11-10-3   The 74x47 Decoder/Driver

The most popular and versatile decoder/driver for 7-segment displays is the **74x47**, whose pin layout, numerical designation, and function table are given in Fig. 11-24. Inputs ABCD form a BCD decade and the seven outputs, a–g, are

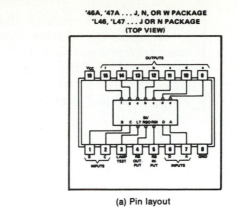

'46A, '47A . . . J, N, OR W PACKAGE
'L46, 'L47 . . . J OR N PACKAGE
(TOP VIEW)

(a) Pin layout

(b) Segment identification

(c) Numerical designations and resultant displays

DECIMAL OR FUNCTION	INPUTS						BI/RBO†	OUTPUTS							NOTE
	LT	RBI	D	C	B	A		a	b	c	d	e	f	g	
0	H	H	L	L	L	L	H	ON	ON	ON	ON	ON	ON	OFF	1
1	H	X	L	L	L	H	H	OFF	ON	ON	OFF	OFF	OFF	OFF	1
2	H	X	L	L	H	L	H	ON	ON	OFF	ON	ON	OFF	ON	
3	H	X	L	L	H	H	H	ON	ON	ON	ON	OFF	OFF	ON	
4	H	X	L	H	L	L	H	OFF	ON	ON	OFF	OFF	ON	ON	
5	H	X	L	H	L	H	H	ON	OFF	ON	ON	OFF	ON	ON	
6	H	X	L	H	H	L	H	OFF	OFF	ON	ON	ON	ON	ON	
7	H	X	L	H	H	H	H	ON	ON	ON	OFF	OFF	OFF	OFF	
8	H	X	H	L	L	L	H	ON	ON	ON	ON	ON	ON	ON	
9	H	X	H	L	L	H	H	ON	ON	ON	OFF	OFF	ON	ON	
10	H	X	H	L	H	L	H	OFF	OFF	OFF	ON	ON	OFF	ON	
11	H	X	H	L	H	H	H	OFF	OFF	ON	ON	OFF	OFF	ON	
12	H	X	H	H	L	L	H	OFF	ON	OFF	OFF	OFF	ON	ON	
13	H	X	H	H	L	H	H	ON	OFF	OFF	ON	OFF	ON	ON	
14	H	X	H	H	H	L	H	OFF	OFF	OFF	ON	ON	ON	ON	
15	H	X	H	H	H	H	H	OFF	OFF	OFF	OFF	OFF	OFF	OFF	
BI	X	X	X	X	X	X	L	OFF	OFF	OFF	OFF	OFF	OFF	OFF	2
RBI	H	L	L	L	L	L	L	OFF	OFF	OFF	OFF	OFF	OFF	OFF	3
LT	L	X	X	X	X	X	H	ON	ON	ON	ON	ON	ON	ON	4

(d) '46A, '47A, 'L46, 'L47 Function table

Figure 11-24 The **74x47** display driver. (From the *TTL Logic Data Book,* 1988, Texas Instruments, Inc. Courtesy of Texas Instruments.)

meant to be connected to a 7-segment display. Current limiting resistors must be connected between each display segment and the **74x47** output. Each output is an open-collector transistor. When it is ON, a current path exists and illuminates the segment connected to it. For example, if inputs DCBA are 0001, which is the BCD code for 1, only segments b and c glow. They form a single rightmost vertical line and give the appearance of the numeral 1. If a number greater than 9 is applied to the BCD inputs, an unrecognizable symbol appears on the output, as shown in outputs 10 through 15 of the resultant display (Fig. 11-24c).

EXAMPLE 11-14

A **74x47** has inputs of 1001. Show precisely how it drives a 7-segment display, and how the decimal number is displayed.

SOLUTION The result is shown in Fig. 11-25. With a BCD 9 (1001) applied to the inputs, the function table of the **74x47**, Fig. 11-24d, shows that all segments except d and e are ON. For each of the remaining 5 segments a current path exists through the segment, the current limiting resistor, and the open-collector transistor within the **74x47**. These segments glow as shown shaded on Fig. 11-25 and form the image of a 9.

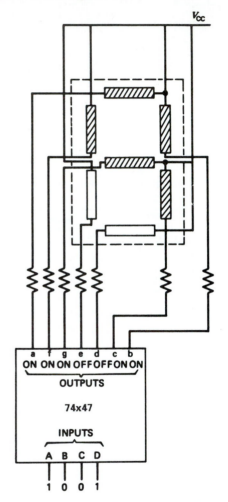

Figure 11-25 A **74x47** driving a 7-segment display.

11-10-4   Ripple Blanking

Ripple blanking is a feature of decoder/drivers that allows them to blank out leading 0s. If a 6-digit display is used, for example, most people prefer to see 2047, rather than 002047. The number 2047 can be displayed if the first two 7-segment displays are completely OFF. Note that while it is desirable to blank *leading* 0s, *imbedded* 0s (such as the 0 in 2047) cannot be blanked.

The **74x47** includes provision for *ripple blanking*. It has a RIPPLE BLANKING INPUT (RBI, pin 5) and a RIPPLE BLANKING OUTPUT (RBO, pin 4). If RBI and the BCD inputs ABCD are all LOW (see the RBI line in the function table), the RBO output is LOW and the display is *blanked* (all segments are OFF). If the RBO output of one stage is connected to the RBI input of the next stage, ripple blanking is accomplished. The next display is also blanked if its BCD input is 0. The display can also be blanked by placing a 0 on the BI/RBO output. This is the direct blanking input (BI) feature of the **74x47**. Whenever the BCD input to a **74x47** is not 0, RBO is HIGH, regardless of RBI and blanking terminates.

EXAMPLE 11-15

Show how the number 2047 can be displayed in a 6-digit display by making use of ripple blanking.

SOLUTION   The solution is shown in Fig. 11-26. The RBI of the first chip is grounded. For succeeding chips the RBO is connected to the RBI of the next IC. Consequently, the first two ICs, which have a BCD input of 0 and an RBI of 0, blank their displays by turning all the segments OFF. Since the third IC has an input of 2, it produces 2 on the display and causes RBO to be HIGH. The remaining ICs all receive a high RBI and display the digits on their ABCD inputs including the 0 in 2047.

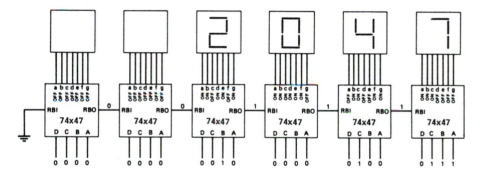

Figure 11-26   Display of the number 2047 using ripple blanking.

11-10-5   Common Cathode 7-segment Displays and the 74x48

It is now common practice to use LEDs with the cathode grounded. A single LED can be driven reasonably well from a TTL gate, as shown in Fig. 11-27. When the output of the TTL gate is HIGH, current through the upper transistor of the totem pole lights the LED. A LOW output shorts the LED and keeps it off. The resistor in the upper part of the totem pole limits the current so no pull-up is needed.

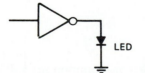

Figure 11-27   A TTL gate driving an LED.

Seven-segment common cathode displays, where each segment has its cathode tied to a common ground, are also becoming more popular. They can be driven by common cathode display drivers such as the **74x48**, which is similar to the **74x47** except that its outputs are reversed, so that it can drive common cathode instead of common anode displays.

A **74x48** is shown driving a 7-segment display in Fig. 11-28. If a 0 is to be displayed, for example, the **74x48** outputs for segments a through f are HIGH. This forces the current through the resistors to flow through the LEDs. The output of segment g is LOW, however, so current through the g resistor flows into the **74x48** instead of the LED and segment g remains dark.

Figure 11-28 also shows that the **74x48** can only absorb a small current. The specifications in the TTL data book limit this current to about 6 mA. This means that R_x must be creater than 650 Ω, so it is limited to driving 7-segment displays that require only a small current. For common LEDs that require 20 mA for reasonable brightness, an IC with a higher I_{OL}, such as the **DM8857** made by National Semiconductor, should be used.

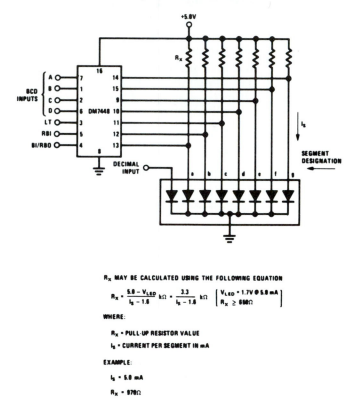

R_x MAY BE CALCULATED USING THE FOLLOWING EQUATION

$$R_x = \frac{5.0 - V_{LED}}{I_s - 1.6} \, k\Omega = \frac{3.3}{I_s - 1.6} \, k\Omega \quad \left[\begin{array}{l} V_{LED} = 1.7V @ 5.0 \, mA \\ R_x \geq 650\Omega \end{array} \right]$$

WHERE:

R_x = PULL-UP RESISTOR VALUE

I_s = CURRENT PER SEGMENT IN mA

EXAMPLE:

I_s = 5.0 mA

R_x = 970Ω

Figure 11-28 Nonmultiplex application of the **DM7448**. (From National Semiconductor Tech Note AN-99. Copyright 1974, National Semiconductor Corporation.)

11-11 Multiplexed Displays

It has recently become common to multiplex 7-segment displays so that many displays are driven by the same driver. In multiplexed displays each display is only driven for a portion of the time, but, because of the persistence of the human eye, it appears to be on constantly.

A typical multiplexing scheme is shown in Fig. 11-29. The digit drivers

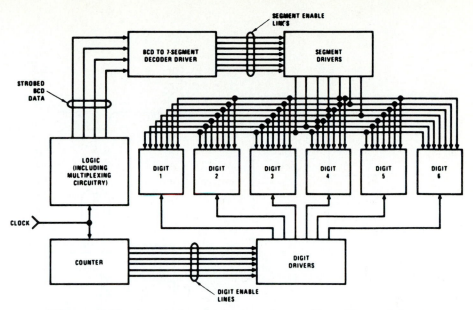

Figure 11-29 A typical multiplexing scheme. (From National Semiconductor Tech Note AN-99. Copyright 1974, National Semiconductor Corporation.)

turn on each driver in turn. While this is happening, the multiplexing circuitry presents the proper input to each of the segment drivers. The 7-segment displays on many microprocessor trainer kits, such as the SDK-85 and SDK-86 are controlled in this way.

EXAMPLE 11-16

Inputs from four registers are to be shown on four common cathode 7-segment displays using a **74x48**. Design the circuit.

SOLUTION   The circuit is shown in Fig. 11-30 and operates as follows.

1. The four input digits are multiplexed by the **74x153**s and passed on to the **74x48**.
2. The counter (the two **74x107**s) periodically changes the SELECT inputs to the multiplexer and simultaneously enables each digit driver in turn so the correct digit is shown on the corresponding display. For clarity, only the driving circuit for segment e of digit 1 is shown.
3. Because of the current limitations of the **74x48**, an *emitter follower* is used to provide the current necessary to drive the LED segment.
4. A transistor is also required in the digit driver circuit. The collector of that transistor is connected to the common cathode of the LED. Thus, if all segments are ON and drawing 20 mA, the current through the lower transistor can be 140 mA (160 mA if a decimal point is also used), which is far greater than a TTL output can sink.

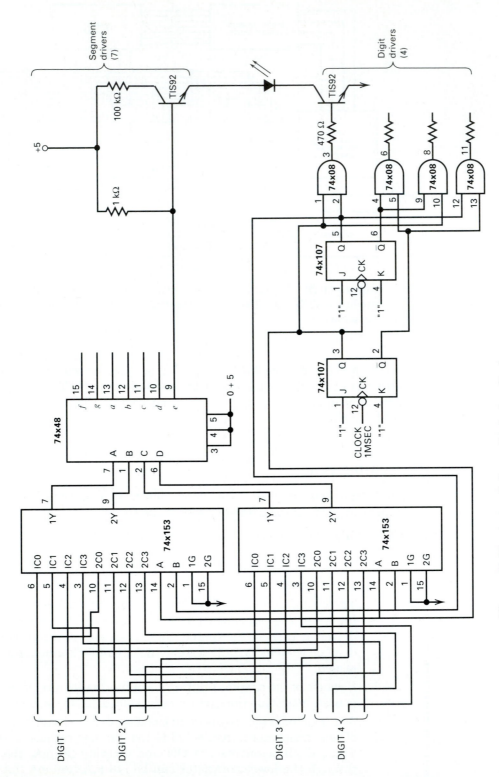

Figure 11-30 A circuit for multiplexing displays.

5. As shown, segment e of display 1 lights only if the *e* output of the **74x48** is HIGH, turning on the emitter follower, and pin 3 of the **74x08** is HIGH, providing base current to turn on the first digit driver.

6. This circuit requires 11 transistors, 7 individual segment drivers, and 4 digit drivers. A higher power decoder could eliminate the segment drivers. Current only flows through each LED one quarter of the time, but we have found experimentally that the current need only be slightly in excess of 20 mA, and the resistance values shown in Fig. 11-30 give a bright and pleasing display.

11-11-1 The Intel 8279

Most microprocessor kits that are used in colleges and universities use multiplexed displays. Intel manufactures the **8279** IC, which can be used for decoding small keyboards and for driving displays. This IC is used in the **SDK-85** and **SDK-86** kits. The **8279** is too complex to be described in detail here, but it accomplishes both keyboard encoding and multiplexed display driving essentially as described in this chapter. The Intel data books describe the **8279** thoroughly.

11-12 Liquid Crystal Displays

Liquid crystal displays (LCDs) operate by polarizing light so that it does or does not reflect ambient light. Like LEDs, they have been organized as 7-segment displays for numerical readout and can be multiplexed.

LCDs have one outstanding advantage over LEDs: they essentially act as a capacitor and consume almost no power. For this reason they are now being used in watches and as readouts for hand-held computers and electronic games. This is a significant advantage because in many circuits LEDs consume more power than all the ICs in the rest of the circuit. Unfortunately, LCDs reflect, rather than generate, light and must be in a well-lit environment to be seen clearly. They must also be driven by an a.c. voltage, but this can be generated by EXCLUSIVE-OR circuits (see section 14-8).

SUMMARY

In this chapter the concept of multiplexing and demultiplexing was explained and TTL multiplexers and demultiplexers were introduced. Methods of building larger multiplexers from the basic multiplexers were explored.

The use of a demultiplexer as a decoder and a multiplexer as a function generator were discussed. A keyboard decoder, which uses both multiplexers and demultiplexers working together, was described.

The most common methods of displaying digital data were discussed, and the use of multiplexed display drivers was explained. Liquid crystal displays were also introduced in this chapter.

GLOSSARY

Decoder: A circuit that decodes its inputs as a binary or BCD number and activates the corresponding output line.

Demultiplexer: A circuit that connects a single input line to one of several output lines.

Keyboard: A group of keys used for entering information, like a typewriter or teletype.

Keyswitch: A switch or button on a keyboard.

Multiplexer: A circuit that connects one of several input lines to a single output line.

STROBE: An input that disables multiplexers or demultiplexers when it is HIGH.

PROBLEMS

Section 11-4.

11-1.   Design a 32-line-to-line multiplexer using only 14- and 16-pin ICs.

11-2.   Design a 24-line-to-1-line multiplexer.

11-3.   For Problems 1 and 2, respectively, show the 1s and 0s if line 21 is selected.

11-4.   Place one of four 8-bit registers on a set of 8-output lines using:
    (a)   **74x153**s
    (b)   **74x157**s

11-5.   For Problem 11-4, which circuit requires the smallest number of ICs?

Section 11-5.

11-6.   Design a 16-line-to-1-line sequential multiplexer that places each input, in turn, on the output line for 10 μs. Use a **74150** in the circuit.

Section 11-6.

11-7.   Design a 1-line-to-8-line demultiplexer using SSI gates.

11-8.   Design a 1-line-to-64-line demultiplexer.

Section 11-7.

11-9.   Design a 64-line decoder using:
    (a)   **74154**s and SSI gates.
    (b)   **74x138**s only.

11-10.   Design a decoder with a 100-line output, assuming the inputs are given in two BCD decades.

Section 11-8.

11-11.   Given $F(v, w, x, y, z) = \Sigma(0, 3, 5, 6, 7, 10, 12, 13)$. Implement the function using a single IC.

11-12. Given $F(v, w, x, y, z) = \Sigma(0,2,3,6,7,8,11,14,17,18,20,25,26,31)$. Implement the function using multiplexers.

11-13. **(a)** Design a full adder using only a single **74x153**.

(b) Design a 3-bit adder using the circuit of part (a).

Section 11-9.

11-14. 16 lines are to be prioritized and encoded. Design the circuit.

11-15. In Fig. 11-22 assume no keys are depressed and the clock is 1 μs. If the clock starts at 0, which lines will be activated after 14 μs?

11-16. Design a keyboard decoder for a 128-key keyboard. Show your scanning logic.

11-17. In Problem 11-16, if the code for the letter B is 1000010, show the location of the B key on your board.

11-18. Assume a keyboard has the specifications given in Ex. 11-13. At $t = 0$, key 1 is depressed for 10 ms. At $t = 5$ ms a second key, 20 scan status away from key 1, is depressed for 10 ms. Draw a timing chart for $\overline{\text{KEYSTROBE}}$ during this period.

Section 11-10.

11-19. A 6-digit display uses remote blanking. In the event that all 6 digits are 0, only a single 0 the least significant digit should be displayed. Design the circuit.

11-20. A circuit must display any number from 0 to 108,000. The input number is available in 6 BCD decades, but only five 7-segment displays and **7447**s are available. Design a circuit to display each number unambiguously (i.e., the reader must be able to distinguish between 50 and 100,050).

After attempting the problems, return to Sec. 11-2 and be sure you can answer the self-evaluation questions. If any of them are still difficult, review the appropriate sections of the chapter to find the answers.

12

Programmable Logic

12-1 Instructional Objectives

This chapter introduces *fusible* logic. It concentrates on Programmable Array Logic (PALs) and also introduces gate arrays. These are ICs that come from the manufacturer with many small internal fuses, all intact. By selectively "blowing" or opening the fuses, the engineer can design an IC to meet the desired specifications. Because a single PAL or gate array IC can replace a large number of discrete ICs, they are being used in almost all digital circuits that are produced in large quantities. Modern personal computers use fusible logic extensively.

After reading the chapter, the student should be able to:

1. State the meaning in each number or letter in a PAL designation.
2. Select the proper PAL for a particular application.
3. Create a PALASM source file.
4. Create a function table and use it to predict the results of a design.
5. Control the 3-state gates on a PAL.
6. Implement circuits containing FFs on PALs.
7. Implement counters and shift registers using PALs.
8. Transform circuits designed using discrete logic into PALs.

12-2   Self-Evaluation Questions

Watch for the answers to the following questions. They should help you understand the material presented.

1. What is the advantage of a PAL compared to a discrete IC? What is the advantage of a discrete IC?
2. What is the difference between a combinatorial and a registered PAL?
3. Why is an input line with all fuses blown equivalent to a logic 1? Why is a line with all fuses intact equivalent to a logic 0?
4. How can an I/O line be used for input?
5. How can we determine whether a function table is correct? If it is incorrect?
6. Why may the first line of a function table for registered PALs be inaccurate?
7. What are the three types of logic structures within a gate array? What are their functions?

12-3   Introduction

Programmable Logic Devices (PLDs) are ICs that contain much logic within the package. They also contain internal fuses that are intact and make connections between the gates within the PLD. These fuses can be selectively blown open by a device called a *PLD programmer* to configure the IC to perform the logic operations required by the engineer.[1] To determine how a PLD functions, it is not sufficient to merely specify the part number of the PLD; the user must also know which fuses are blown and which have remained intact. Blowing fuses in a PLD is similar to putting a program into a computer; in both cases they determine how the device behaves.

The most commonly used PLD is the Programmable Array Logic (PAL) IC, that contains fused AND gates feeding OR gates.[2] They allow logic circuits to be implemented using PAL ICs, rather than discrete ICs such as **74x00**s, **74x08**s, and so on. PAL ICs, which come in 20- and 24-pin DIPs, contain far more logic than discrete ICs, and a single PAL can perform many and diverse logic functions; it is not limited to being only an AND gate or a NAND gate. Any circuit that is in a PAL can be implemented by discrete ICs, but one PAL can typically replace about five discrete ICs.

There are two types of PALs, *combinatorial* and *registered*. A combinatorial PAL contains only gates, while a registered PAL contains gates and flip-flops or registers. Registered PALs, also called *sequential* PALs, are more complex and are discussed in section 12-5.

12-3-1   The Gates in a Combinatorial PAL

Inside a combinatorial PAL there are a set of AND gates connected to an OR gate, as shown in Fig. 12-1. Actually there are several such configurations. The

[1]PLD programmers are manufactured by the Data I/O Corporation, Redmond, WA, Storey Systems, Mesquite TX, and others.

[2]Other PLDs that have different fuse configurations, such as Programmable Logic Arrays (PLAs), also exist. Because PALs are the most commonly used PLD, we will concentrate on them.

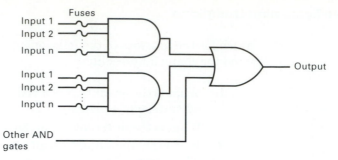

Figure 12-1 The typical configuration inside a PAL.

circuit is much like the AOI circuits of the **74x51** and **74x53**s discussed in section 5-9. There are two significant differences, however:

The PAL circuits are logically, but not physically, much larger than a standard IC. There are many more inputs to the AND gate, and more AND outputs that become inputs to the OR gate.

There is an *internal fuse* connected between each input and the AND gate. When a PAL is purchased, all the internal fuses are intact. By using a device called a *logic programmer*, the fuses can be selectively blown open, allowing the final PAL to perform the logic dictated by its fuse configuration. Blown fuses act as an open circuit or a logic 1.

12-3-2   The 16L8 PAL

The **16L8**, shown in Fig. 12-2, is the most commonly used combinatorial PAL.[3] The numbers describing a PAL are meaningful. The first number, 16 in this case, is the number of inputs. The letter is H or L and describes the active state of the outputs (High or Low). The second number, 8 in this case, is the number of outputs.

An examination of Fig. 12-2 shows that there are 8 AND-OR-INVERT (AOI) circuits. These circuits are similar, but not identical. Each AOI circuit has 8 horizontal lines connected to it. The horizontal lines are numbered 0–7 for the top circuit and on down to 56–63 for the bottom circuit. In addition there are 32 vertical lines, numbered 0–31 and shown in groups of 4. When the PAL is purchased, there is a fuse connecting each horizontal line to each vertical line. Thus, before the PAL is blown there are 32×64 or 2048 intact fuses. In this state the PAL is completely useless.

A single line on a PAL diagram is shown in Fig. 12-3a. This single line and symbol represents an AND gate that *ANDs together all the vertical lines connected to it*. There are three possibilities:

a.   All the fuses connected to the line are blown. This is shown in Fig. 12-3a. In this case the output of the AND gate will be a logic 1.

b.   All the fuses connected to the line are intact. This is shown by an X in the AND box. See Fig. 12-3b. The output of this gate will be a logic 0.

[3]Older combinatorial PALs, such as the **12H6** and **12L6**, may still be obtained, but these parts contain much less logic and are obsolete.

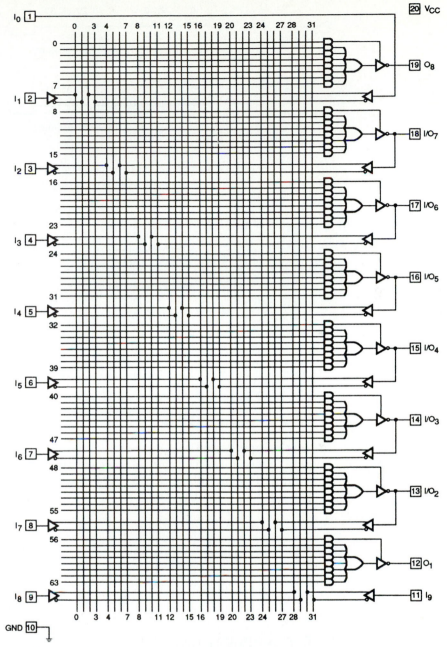

Figure 12-2 The **16L8** PAL. (Courtesy of Advanced Microdevices (AMD), Inc.)

c. Some of the fuses are blown open and some remain intact. This is shown in Fig. 12-3c. The Xs indicate intact fuses; where there are no Xs, the fuses are blown, which means their inputs are logic 1s. Fig. 12-3c is equivalent to the AND gate shown in Fig. 12-3d. Its output is the AND of the signals on vertical lines 0, 3, and 30. If the levels on these three lines are all 1s, the output will be a 1.

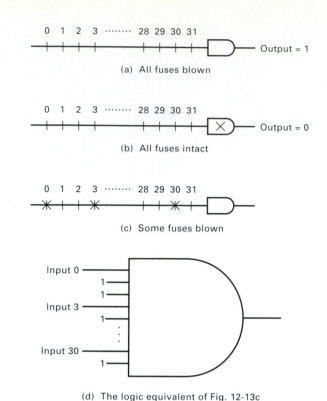

(a) All fuses blown — Output = 1

(b) All fuses intact — Output = 0

(c) Some fuses blown

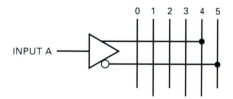

(d) The logic equivalent of Fig. 12-13c

Figure 12-3   A line on a PAL and its meaning.

12-3-3   The Input Gate

All the inputs to a PAL go through an internal gate, an inverting-noninverting gate. There are two outputs of the gate, the input variable and its complement, indicated by the bubble. In Fig. 12-4, the signal INPUTA is shown connected to vertical line 4 and the signal $\overline{\text{INPUTA}}$ is connected to vertical line 5.

Figure 12-4   An input gate on a PAL.

EXAMPLE 12-1

In Fig. 12-4, if a horizontal line is being used for logic, it will not have an intact fuse to vertical line 4 and vertical line 5. Why not?

SOLUTION   If the fuses to both lines 4 and 5 were left intact, the AND gate whose inputs are on that horizontal line would AND both INPUTA and its complement. The output would therefore be 0, regardless of any other signals on the horizontal line.

Many logic functions can be implemented using PAL gates. Figure 12-5 shows how to implement the simplest and most commonly used functions.

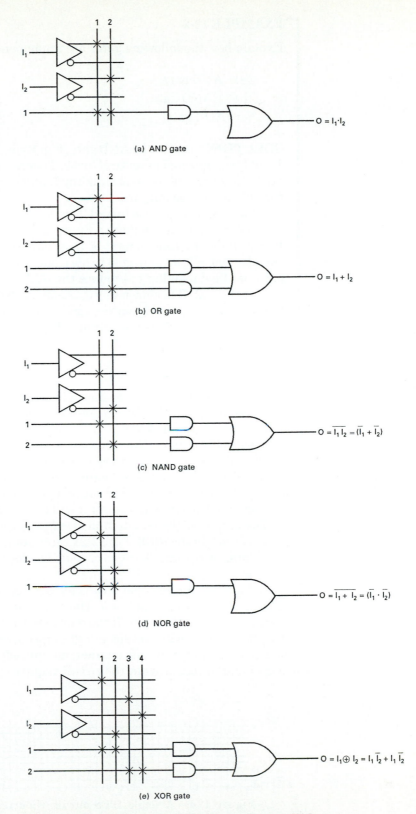

(a) AND gate

(b) OR gate

(c) NAND gate

(d) NOR gate

(e) XOR gate

Figure 12-5 Implementing the basic gates in a PAL.

EXAMPLE 12-2

Explain how the following gates are implemented in the PAL of Fig. 12-5

    (a)   AND gate

    (b)   NAND gate

    (c)   XOR gate

SOLUTION   (a) For the AND gate, Fig. 12-5a, I_1 is connected to vertical line 1 and I_2 is connected to vertical line 2. They are ANDed together because they are both connected to horizontal line 1, and come through the AND gate and the OR gate to the output.

    (b) NAND logic is achieved by using the negative logic equation $O = \overline{I_1} + \overline{I_2}$, which indicates that the output, O, will be HIGH if either input is LOW. In Fig. 12-5c the complement of I_1 is connected to horizontal line 1 and the complement of I_2 is connected to horizontal line 2. The output of the two AND gates is ORed together to produce the NAND of the inputs.

    (c) The XOR is shown in Fig. 12-5e. Horizontal line 1 ANDs together I_1 (on vertical line 1) and $\overline{I_2}$ (on vertical line 2). Similarly horizontal line 2 ANDs together $\overline{I_1}$ and I_2. Thus the output is $I_1\overline{I_2} + \overline{I_1}I_2$, which is the XOR of I_1 and I_2.

12-3-4   A Typical 16L8 Circuit

A typical, single **16L8** circuit is shown in Fig. 12-6. It consists of eight 32-input AND gates, a 7-input OR gate, a 3-state inverter, and a feedback gate.

    The top 32-input AND of each circuit gate controls the enable on the 3-state output inverter. In most cases this inverter is always ON, and this gate will have all fuses blown. Its output will then be a 1, the level that enables the inverter. The other seven AND gates are connected to the OR gate and can be used for AND-OR-INVERT (AOI) logic. If an input AND gate is not being used, it will have all fuses intact; its output will be 0 and it will not affect the OR gate.

    The output of the circuit is also *fed back and can be used as an input to another circuit*. In the **16L8** (Fig. 12-2), the six outputs labeled I/O can be used as either inputs or outputs. The two outputs labeled O (O_1 and O_8) can only be used as outputs.

    In Fig. 12-6, for example, I/O_7 is fed back on vertical line 6 and its complement is fed back on vertical line 7. The output line which is connected to pin 18, can also be used for an input. If the enable to the 3-state inverter in Fig. 12-6 is LOW, the inverter will present a high-impedance output. Then pin 18 can be connected to an input. In this manner, an unneeded output can be replaced by a needed input, if that is what the circuit requires (see section 12-4-3).

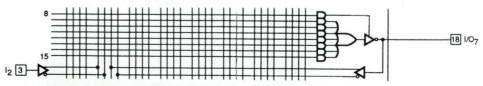

Figure 12-6 A single **16L8** circuit. (Courtesy of Advanced Microdevices (AMD), Inc.)

As previously mentioned, the **16L8** has 16 inputs. Ten of them, inputs I_0-I_9, are directly connected to input pins. The other 6 inputs are fed back from the output gates I/O_2–I/O_7. Gates 1 and 8 have no feedback connection and are therefore only labeled O_1 and O_8 respectively.

12-4   Implementing Combinatorial Functions

Consider the circuit of Fig. 12-7. To construct it using digital logic would require several gates of many different types. Some of these gates, like the 3-input OR gate, do not exist although they can be built using two 2-input OR gates. Other gates, like the 4-input AND gate, may not be available. This entire circuit can be implemented using a single PAL, however, and the problems of procuring the necessary ICs and wiring them together would disappear.

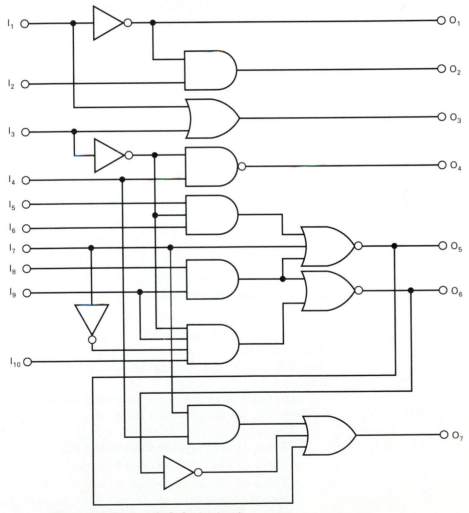

Figure 12-7   A sample logic circuit.

The first step in the PAL implementation is to write the equations. The following basic logic operators, chosen so they can be easily entered on a keyboard, are used in PAL logic equations.

The basic operators are defined to perform INVERT, AND, OR, and EXCLUSIVE-OR operations. These basic operators can be used to describe any logic function on the right side of the equation.

/ —INVERT operator
This operation is used whenever a signal has to be inverted. It precedes the signal to be inverted. For example: /A means "NOT A"

* —AND operator
This operation is used when ANDing two or more Boolean variables. The operation of ANDing of all the signals result in a product term. For example, A* /B* C means "A AND (NOT B) AND C"

+ —OR operator
This operator is used when ORing two or more product terms and/or signals. For example: A + /C means "A OR (NOT C)"

:+: —EXCLUSIVE-OR operator
This operator is used when EXCLUSIVE-ORing two or more product terms and/or signals. For example: A :+: E means "A EXCLUSIVE-OR E"

For the circuit of Fig. 12-7 there are seven equations, one for each output. They are:

$$O_1 = /I_1$$
$$O_2 = /I_1 * I_2$$
$$O_3 = I_1 + I_3$$
$$O_4 = /(/I_3 * I_4)$$
$$O_5 = /(/I_3 * I_5 * I_6 + I_7 + I_8 * I_9)$$
$$O_6 = /(I_8 * I_9 + /I_3 * /I_7 * I_9 * I_{10})$$
$$O_7 = O_5 + /O_6 + I_4 * I_7$$

12-4-1 Choosing the PAL

For any problem, the equations should first be developed. Then the proper PAL can be selected by consulting the manufacturer's data sheets. Two prominent manufacturers of PALs are Texas Instruments, Inc. (TI), Dallas, TX, and Advanced Micro Devices, Sunnyvale, CA. Both have data books on the PALs they manufacture.

The **16L8** is virtually the only combinatorial PAL that comes in a 20-pin package. The constraints on a **16L8** are:

- There can be no more than 16 inputs.
- There can be no more than 8 outputs.
- There can be no more than a total of 18 inputs and outputs. This is because there are only 18 logic pins on a 20-pin IC.
- There can be no more than 7 inputs to the OR gate.

EXAMPLE 12-3

Can the **16L8** be used to implement the equations we have been considering?

SOLUTION Fortunately, the problem only requires 10 inputs and 7 outputs. It satisfies all the constraints and can be used.

12-4-2 Constructing the PALASM File

After the equations have been written, a file for the problem must be constructed on a computer. This is similar to writing a source file for a computer program. The PAL file must be written for one of several *PAL languages*, which are similar to an assembler. The most popular languages are PALASM, ABEL, and CUPL. The PAL file and the selected language are both entered into a computer.[4] The language operates on the file and produces a JEDEC file and fuse plot.[5] The JEDEC file becomes the input to a device known as a PAL logic programmer. A blank PAL IC is inserted into the socket on the programmer, and the JEDEC file activates the programmer, which blows open the correct fuses. Some programmers will accept a file, convert it to fuse plot, and then blow the fuses.

Obviously, we cannot discuss all of the available languages. In this book we will focus on PALASM. This is perhaps the first language developed for PALs. It is readily available on diskette, and can be used with personal computers.

A PALASM file starts by identifying the PAL to be used in the upper left-hand corner of the first line of the file. The remainder of the first line, as well as lines 2, 3, and 4, are comments. These comments should include a brief description of the PAL, a designation for it, and the author's name and date. Any line that starts with a semicolon is also a comment line, and will not be included in the line count.

Lines 5 and 6 are the pin designations. They are then followed by the equations. In the problem we have been considering, we have listed positive outputs. Unfortunately, PALASM requires negative outputs for a PAL with low outputs like the **16L8**. The equations of section 12-4 can be transformed to low outputs using De Morgan's theorem, and become:[6]

$$/O1 = I1$$
$$/O2 = I1 + /I2$$
$$/O3 = /I1*/I3$$
$$/O4 = /I3*I4$$
$$/O5 = /I3*I5*I6 + I7 + I8*I9$$
$$/O6 = I8*I9 + /I3*/I7*I9*I10$$
$$/O7 = /O5*O6*(/I4 + I7) = /O5*O6*/I4 + /O5*O6*I7$$

[4]Digital Equipment's VAX computer at RIT had the programs PA20 and PA24 available. They used the PALASM language to operate on a source file for 20 and 24-pin PALs respectively.

[5]JEDEC (Joint Electron Device Engineering Council) has produced standard No 3-A. This is a standard for transferring a fuse plot from a PLD compiler, such as PALASM, to a PLD programmer, so the proper fuses can be blown.

[6]Although the previous input and output variables have been shown subscripted (e.g., I_2), the PALASM file, as entered from a keyboard, does not allow subscripted variables. Therefore, we are representing them as not subscripted (e.g., I2).

PALASM will not accept parentheses, so they must be broken open. The equation for O7 is an example; the rightmost form of the expression must be used.[7]

The input file for PALASM is shown in Fig. 12-8. It was created on a Digital Equipment VAX computer, but the file can be created on any computer or PC using EDLIN or any other editor program. The file starts by identifying the PAL and follows with three lines of data. All lines that start with semicolons are strictly comments. The fifth line, excluding the semicolon line, starts the pin list. Inputs in the equation must be on input pins and outputs on output pins. NC means no connection. The **16L8** has 6 pins that can be either input or output. We have chosen to connect input 10 (I10) to pin 13, which is one of those pins. The equations that have been developed for the PAL follow, and the file ends with the word DESCRIPTION followed by a short description.

```
PAL16L8
TEST PROBLEM FOR BOOK
FROM AMD PAL BOOK 1984
MAY  1991
;
NC   I1   I2   I3   I4   I5   I6   I7   I8   GND
I9   O7   I10  O6   O5   O4   O3   O2   O1   VCC
;
; EQUATION LIST
;
/O1  =  I1
/O2  =  I1  +  I2
/O3  =  /I1*/I3
/O4  =  /I3*I4
/O5  =  /I3*I5*I6  +  I7  +  I8*I9
/O6  =  I8*I9  +  /I3*/I7*I9*I10
/O7  =  /O5*O6*/I4  +  /O5*O6*I7
;
DESCRIPTION  THIS IS A TEST
```

Figure 12-8 Equations for the circuit of Figure 12-7.

12-4-3 The Assembled File

The PAL file can now be assembled using the PALASM program to produce a *pin diagram*, shown in Fig. 12-9, and a fuse plot, shown in Fig. 12-10. It also produces the JEDEC file that can be used by the programmer to blow the fuses. The pinout shows the input and output connections to the PAL as specified in the pin list.

The fuse plot shows all the fuses on the **16L8**. Blown fuses are identified by a dash (-) and intact fuses are identified by an X. Notice that most 3-state enables are connected to a line with all fuses blown. This provides a logic 1 and enables the outputs. The AND gate inputs that are unused are connected to a line with all fuses intact. This provides a logic 0 and does not affect the output. Pin 13, which can be either input or output, has been declared to be *input* in this example. *This means the output must be disabled.* PALASM disables the output by leaving all fuses to its 3-state gate, which is on horizontal line 48, intact. This provides a 0 on the enable, and disables the 3-state output gate, so the user can apply an input signal to this gate. This signal will control vertical lines 26 and 27.

The blown fuses are also shown in the **16L8** diagram of Fig. 12-11. They conform to the fuse plot of Fig. 12-10.

---

[7]More advanced versions of PALASM and other languages will accept parentheses and positive outputs.

```
              * * * * * * * * * * * * *   * * * * * * * * * * * * *
                    *                      *   *                 *
                  * * * *                                      * * * *
        NC        *  1 *            P A L              * 20 *        V_CC
                  * * * *                              * * * *
                    *                  1  6  LB          *
                  * * * *                              * * * *
        E1        *  2 *                               * 19 *        01
                  * * * *                              * * * *
                    *                                    *
                  * * * *                              * * * *
        I2        *  3 *                               * 18 *        02
                  * * * *                              * * * *
                    *                                    *
                  * * * *                              * * * *
        I3        *  4 *                               * 17 *        03
                  * * * *                              * * * *
                    *                                    *
                  * * * *                              * * * *
        I4        *  5 *                               * 16 *        04
                  * * * *                              * * * *
                    *                                    *
                  * * * *                              * * * *
        I5        *  6 *                               * 15 *        05
                  * * * *                              * * * *
                    *                                    *
                  * * * *                              * * * *
        I6        *  7 *                               * 14 *        06
                  * * * *                              * * * *
                    *                                    *
                  * * * *                              * * * *
        I7        *  8 *                               * 13 *        I10
                  * * * *                              * * * *
                    *                                    *
                  * * * *                              * * * *
        I8        *  9 *                               * 12 *        07
                  * * * *                              * * * *
                    *                                    *
                  * * * *                              * * * *
        GND       * 10 *                               * 11 *        I9
                  * * * *                              * * * *
                    *                                    *
              * * * * * * * * * * * * * * * * * * * * * * * * * * * *
```

Figure 12-9   The pin diagram for the circuit of Figure 12-7.

Figure 12-12 is a *brief fuse plot*; it does not show those lines where all fuses remain intact. It conforms to the logic equations. Observe horizontal lines 24 and 25, for example. Line 24 has all fuses blown, which enables the 3-state gate connected to O4. On line 25 the fuses that connect to vertical lines 9 and 12 are intact. These lines are connected to /I3 and I4 respectively, so the output is indeed /O4 = /I3*I4.

> **EXAMPLE 12-4**
>
> Describe the logic generated on lines 56, 57, and 58 of the brief fuse plot.
>
> **SOLUTION**   Line 56 has all fuses blown; this enables the 3-state gate that drives O7. Line 57 has intact fuses to vertical lines 13, 19, and 22. Line 13 is driven by /I4. Line 19 is controlled by the complement of O5, or /O5 and line 22 is directly connected to O6. Therefore, the logic on this line is /I4*/O5*O6. The logic on line 57 is found similarly; it is /O5*O6*I7. The final output, O7, is the OR of these two expressions.

```
PAL20 V1.7D  -  PAL16L8  -  FROM AMD PAL BOOK 1984

                        11  1111  1111  2222  2222  2233
               0123  4567  8901  2345  6789  0123  4567  8901

 0   ----  ----  ----  ----  ----  ----  ----  ----
 1   X---  ----  ----  ----  ----  ----  ----  ----        I1
 2   XXXX  XXXX  XXXX  XXXX  XXXX  XXXX  XXXX  XXXX
 3   XXXX  XXXX  XXXX  XXXX  XXXX  XXXX  XXXX  XXXX
 4   XXXX  XXXX  XXXX  XXXX  XXXX  XXXX  XXXX  XXXX
 5   XXXX  XXXX  XXXX  XXXX  XXXX  XXXX  XXXX  XXXX
 6   XXXX  XXXX  XXXX  XXXX  XXXX  XXXX  XXXX  XXXX
 7   XXXX  XXXX  XXXX  XXXX  XXXX  XXXX  XXXX  XXXX

 8   ----  ----  ----  ----  ----  ----  ----  ----
 9   X---  ----  ----  ----  ----  ----  ----  ----        I1
10   ----  -X--  ----  ----  ----  ----  ----  ----        /I2
11   XXXX  XXXX  XXXX  XXXX  XXXX  XXXX  XXXX  XXXX
12   XXXX  XXXX  XXXX  XXXX  XXXX  XXXX  XXXX  XXXX
13   XXXX  XXXX  XXXX  XXXX  XXXX  XXXX  XXXX  XXXX
14   XXXX  XXXX  XXXX  XXXX  XXXX  XXXX  XXXX  XXXX
15   XXXX  XXXX  XXXX  XXXX  XXXX  XXXX  XXXX  XXXX

16   ----  ----  ----  ----  ----  ----  ----  ----
17   -X--  ----  -X--  ----  ----  ----  ----  ----        /I1*/I3
18   XXXX  XXXX  XXXX  XXXX  XXXX  XXXX  XXXX  XXXX
19   XXXX  XXXX  XXXX  XXXX  XXXX  XXXX  XXXX  XXXX
20   XXXX  XXXX  XXXX  XXXX  XXXX  XXXX  XXXX  XXXX
21   XXXX  XXXX  XXXX  XXXX  XXXX  XXXX  XXXX  XXXX
22   XXXX  XXXX  XXXX  XXXX  XXXX  XXXX  XXXX  XXXX
23   XXXX  XXXX  XXXX  XXXX  XXXX  XXXX  XXXX  XXXX

24   ----  ----  ----  ----  ----  ----  ----  ----
25   ----  ----  -X--  X---  ----  ----  ----  ----        /I3*I4
26   XXXX  XXXX  XXXX  XXXX  XXXX  XXXX  XXXX  XXXX
27   XXXX  XXXX  XXXX  XXXX  XXXX  XXXX  XXXX  XXXX
28   XXXX  XXXX  XXXX  XXXX  XXXX  XXXX  XXXX  XXXX
29   XXXX  XXXX  XXXX  XXXX  XXXX  XXXX  XXXX  XXXX
30   XXXX  XXXX  XXXX  XXXX  XXXX  XXXX  XXXX  XXXX
31   XXXX  XXXX  XXXX  XXXX  XXXX  XXXX  XXXX  XXXX

32   ----  ----  ----  ----  ----  ----  ----  ----
33   ----  ----  -X--  ----  X---  X---  ----  ----        /I3*I5*I6
34   ----  ----  ----  ----  ----  ----  X---  ----        I7
35   ----  ----  ----  ----  ----  ----  ----  X-X-        I8*I9
36   XXXX  XXXX  XXXX  XXXX  XXXX  XXXX  XXXX  XXXX
37   XXXX  XXXX  XXXX  XXXX  XXXX  XXXX  XXXX  XXXX
38   XXXX  XXXX  XXXX  XXXX  XXXX  XXXX  XXXX  XXXX
39   XXXX  XXXX  XXXX  XXXX  XXXX  XXXX  XXXX  XXXX

40   ----  ----  ----  ----  ----  ----  ----  ----
41   ----  ----  ----  ----  ----  ----  ----  X-X-        I8*I9
42   ----  ----  -X--  ----  ----  ----  -XX-  --X-        /I3*/I7*I9*I10
43   XXXX  XXXX  XXXX  XXXX  XXXX  XXXX  XXXX  XXXX
44   XXXX  XXXX  XXXX  XXXX  XXXX  XXXX  XXXX  XXXX
45   XXXX  XXXX  XXXX  XXXX  XXXX  XXXX  XXXX  XXXX
46   XXXX  XXXX  XXXX  XXXX  XXXX  XXXX  XXXX  XXXX
47   XXXX  XXXX  XXXX  XXXX  XXXX  XXXX  XXXX  XXXX

48   XXXX  XXXX  XXXX  XXXX  XXXX  XXXX  XXXX  XXXX
49   XXXX  XXXX  XXXX  XXXX  XXXX  XXXX  XXXX  XXXX
50   XXXX  XXXX  XXXX  XXXX  XXXX  XXXX  XXXX  XXXX
51   XXXX  XXXX  XXXX  XXXX  XXXX  XXXX  XXXX  XXXX
52   XXXX  XXXX  XXXX  XXXX  XXXX  XXXX  XXXX  XXXX
53   XXXX  XXXX  XXXX  XXXX  XXXX  XXXX  XXXX  XXXX
54   XXXX  XXXX  XXXX  XXXX  XXXX  XXXX  XXXX  XXXX
55   XXXX  XXXX  XXXX  XXXX  XXXX  XXXX  XXXX  XXXX

56   ----  ----  ----  ----  ----  ----  ----  ----
57   ----  ----  ----  -X--  --X-  ----  ----  ----        /O5*O6*/I4
58   ----  ----  ----  ----  --X  --X-  X---  ----        /O5*O6*I7
59   XXXX  XXXX  XXXX  XXXX  XXXX  XXXX  XXXX  XXXX
60   XXXX  XXXX  XXXX  XXXX  XXXX  XXXX  XXXX  XXXX
61   XXXX  XXXX  XXXX  XXXX  XXXX  XXXX  XXXX  XXXX
62   XXXX  XXXX  XXXX  XXXX  XXXX  XXXX  XXXX  XXXX
63   XXXX  XXXX  XXXX  XXXX  XXXX  XXXX  XXXX  XXXX
```

OUTPUT POLARITY WORD XXXXXXXX

LEGEND: X : FUSE NOT BLOWN (L,N,O)   - : FUSE BLOWN (H,P,1)

NUMBER OF FUSES BLOWN = 583

SECURITY FUSE XX

Figure 12-10 The fuse plot for the circuit of Fig. 12-7.

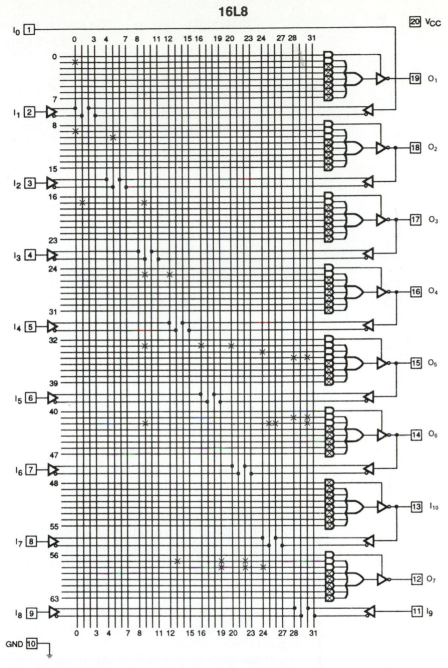

Figure 12-11 The blown fuses on the PAL for the circuit of Figure 12-7.

EXAMPLE 12-5

Logic function F1 is a function of four variables, W, X, Y, and Z. Given the function

$$F1 = \Sigma\ 2,4,5,8,A,B,D$$

show how to implement this function using a **16L8**.

ype pal11.b

PAL20 V1.7D – PAL16R4 – FROM AMD PAL BOOK 1984

```
                   11 1111 1111 2222 2222 2233
         0 123 4567 8901 2345 6789 0123 4567 8901

 0   ---- ---- ---- ---- ---- ---- ---- ----
 1   X--- ---- ---- ---- ---- ---- ---- ----   I1

 8   ---- ---- ---- ---- ---- ---- ---- ----
 9   X--- ---- ---- ---- ---- ---- ---- ----   I1
10   ---- -X-- ---- ---- ---- ---- ---- ----   /I2

16   ---- ---- ---- ---- ---- ---- ---- ----
17   -X-- ---- -X-- ---- ---- ---- ---- ----   /I1*/I3

24   ---- ---- ---- ---- ---- ---- ---- ----
25   ---- ---- -X-- X--- ---- ---- ---- ----   /I3*/I4

32   ---- ---- ---- ---- ---- ---- ---- ----
33   ---- ---- -X-- ---- X--- X--- ---- ----   /I3*I5*I6
34   ---- ---- ---- ---- ---- ---- X--- ----   I7
35   ---- ---- ---- ---- ---- ---- ---- X-X-   I8*I9

40   ---- ---- ---- ---- ---- ---- ---- ----
41   ---- ---- ---- ---- ---- ---- ---- X-X-   I8*I9
42   ---- ---- -X-- ---- ---- ---- -XX- --X-   /I3*/I7*I9*I10

56   ---- ---- ---- ---- ---- ---- ---- ----
57   ---- ---- ---- -X-- ---X --X- ---- ----   /O5*O6*/I4
58   ---- ---- ---- ---- ---X --X- X--- ----   /O5*O6*I7
```

OUTPUT POLARITY WORD XXXXXXXX

LEGEND: X : FUSE NOT BLOWN (L,N,0) – : FUSE BLOWN (H,P.1)

NUMBER OF FUSES BLOWN = 583

SECURITY FUSE XX

Figure 12-12   The brief fuse plot for Figure 12-11.

SOLUTION   Functions of this type have been discussed in section 3-3-3. The first problem is that the **16L8** demands low outputs. Perhaps the best approach is to find the complement of F1 and apply it to the **16L8**. For any equation, $\overline{F1}$ contains all the terms not in F1. Therefore:

$$\overline{F1} = \Sigma\ 0,1,3,6,7,9,C,E,F$$

Note, for example, that the number 2 is not in the equation. Therefore, if the input is the 2 term, $\overline{F1}$ will be 0, making the output 1. This conforms to the original specification.

Proceeding to the SOP form as we did in Chapter 3, we obtain the equation

$$\overline{F1} = \overline{W}*\overline{X}*\overline{Y}*\overline{Z} + \overline{W}*\overline{X}*\overline{Y}*Z + \overline{W}*\overline{X}*Y*Z + \overline{W}*X*Y*\overline{Z} + \overline{W}*X*Y*Z +$$
$$W*\overline{X}*\overline{Y}*Z + W*X*\overline{Y}*\overline{Z} + W*X*Y*\overline{Z} + W*X*Y*Z$$

Using the methods of Chapter 3, a minimal expression for this equation can be obtained by drawing the Karnaugh maps. With a little thought, however, we might realize that a minimal expression is not necessary in this problem. The **16L8** has 7 inputs to its OR gates, but the expression for $\overline{F1}$ has 9 terms. If this expression is reduced to 7 terms, it will be suitable for the PAL and there is no advantage to further simplification; it will not reduce the chip count of the circuit.

In the given expression, the first two terms and the last two terms can be combined by inspection, giving

$$\overline{F1} = \overline{W}*\overline{X}*\overline{Y} + \overline{W}*\overline{X}*Y*Z + \overline{W}*X*Y*\overline{Z} + \overline{W}*X*Y*Z +$$
$$W*\overline{X}*\overline{Y}*Z + W*X*\overline{Y}*\overline{Z} + W*X*Y$$

This expression can be implemented in a **16L8**. While we have *not* minimized the function, we have minimized the amount of effort required to find an expression suitable for the PAL.

12-4-4 Simulation

Simulation is another feature available on most PALASM programs. It allows the engineer to test a PAL *before* it is blown. The user creates a *function table* and assign values to the inputs. These values are sometimes called *test vectors*. The values are:

$$H = \text{High}$$
$$L = \text{Low}$$
$$X = \text{Don't care}$$

The user then lists the expected values of the output. PALASM checks the logic of the PAL, and determines whether the outputs will be generated by the inputs.[8]

Figure 12-13 is the original PAL file with a function table added to illustrate simulation. The function table consists of the following:

a. The words "FUNCTION TABLE."
b. A list of the input and output variables.
c. A dashed line to separate the list from the data.
d. Several lines of inputs and outputs to be tested. These are the test vectors. For this simple example, three lines were tested.
e. A dashed line to end the function table.

The first two lines of the function table were used to test the inverter connected to O1. On line 1 we specified that the input was HIGH and the output was LOW. This is correct operation. In line 2 we specified that both the input and output were HIGH. This is obviously an error, and is used to illustrate how the function table reacts to errors. Because we were only interested in the inverter part of the PAL, the rest of the inputs and outputs were given Xs (don't cares). In line 3, both I1 and I3 were LOW. The result, in accordance with the equation list, is that O1 should be HIGH and O3 should be LOW.

The results of the simulation are also shown in Fig. 12-13. The Xs, 1s, and 0s that apply to the variables are listed in *accordance with the pinout*, and not in accordance with the variable list in the function table. Because I1 is connected to pin 2 its inputs appear there, and the corresponding output, O1, appears at pin 19. Observe that pin 20 is VCC and listed in the function table as a logic 1.

[8]The testing of PALs is very important. The DATA I/O Corporation of Redmond, WA, a manufacturer of logic programmers, distributes an informative public domain diskette on PLD testing.

```
PAL16L8
TEST PROBLEM FOR BOOK
FROM AMD PAL BOOK 1984
MAY 1991
;
NC   I1   I2   I3   I4   I5   I6   I7   I8   GND
I9   O7   I10  O6   O5   O4   O3   O2   O1   VCC
;
; EQUATION LIST
;
/O1 = I1
/O2 = I1 + /I2
/O3 = /I1*/I3
/O4 = /I3*I4
/O5 = /I3*I5*I6 + I7 + I8*I9
/O6 = I8*I9 + /I3*/I7*I9*I10
/O7 = /O5*O6*/I4 + /O5*O6*I7
;
DESCRIPTION  THIS IS A TEST
;
; START OF SIMULATION
;
FUNCTION TABLE

I1  I2  I3  I4  I5  I6  I7  I8  I9  I10     O1  O2  O3  O4  O5  O6  O7
--------------------------------------------------------------------
H   X   X   X   X   X   X   X   X   X       L   X   X   X   X   X   X
H   X   X   X   X   X   X   X   X   X       H   X   X   X   X   X   X
L   X   L   X   X   X   X   X   X   X       H   X   L   X   X   X   X
--------------------------------------------------------------------
```

Simulation results

FROM AMD PAL BOOK 1984

```
1  X1XXXXXXXXXXXXXXXXXL1
2  X1XXXXXXXXXXXXXXXX?1
3  X0X0XXXXXXXXXXXXLXH1
```

NUMBER OF FUNCTION TABLE ERRORS = 1

Figure 12-13   The PALASM file for the circuit of Figure 12-7 including a function table and simulation results.

The simulation results of Fig. 12-13 indicate that lines 1 and 3 are correct, but line 2 is in error as shown by the question mark. The number of function table errors is then given. If there were no errors, the message "PASS SIMULATION" would have been printed.

Figure 12-14 shows a simulation for the same problem with a fourth line added to the function table. All the inputs were specified on this line and the corresponding outputs were calculated accordingly.

EXAMPLE 12-6

For the inputs given in Fig. 12-14, show that O5 and O7 should be LOW and O6 should be HIGH.

SOLUTION   The given inputs have I3 LOW and I5 and I6 HIGH. The term $\overline{I3}*I5*I6$ is in the equation for $\overline{O5}$. This term is a logic 1 for these values and therefore $\overline{O5}$ is a logic 1, which makes O5 LOW. Because I9 is 0, the equation for $\overline{O6}$ is a 0, which makes O6 a logic 1. Having determined values O5 and O6 have, we see that the term $\overline{O5}*O6*\overline{I4}$ is a 1. This makes $\overline{O7}$ a 1 and O7 a 0.

```
;
; START OF SIMULATION
;
FUNCTION TABLE

I1  I2  I3  I4  I5  I6  I7  I8  I9  I10    O1  O2  O3  O4  O5  O6  O7
-----------------------------------------------------------------
H   X   X   X   X   X   X   X   X   X     L   X   X   X   X   X   X
H   X   X   X   X   X   X   X   X   X     H   X   X   X   X   X   X
L   X   X   X   X   X   X   X   X   X     H   X   L   X   X   X   X
L   H   L   L   H   H   L   L   L   L     H   H   L   H   L   H   L
-----------------------------------------------------------------


FROM AMD PAL BOOK 1984

      1  X1XXXXXXXXXXXXXXXXL1
      2  X1XXXXXXXXXXXXXXXX?1
      3  X0X0XXXXXXXXXXXXLXH1
      4  X01001100X0L0HLHLHH1

NUMBER OF FUNCTION TABLE ERRORS = 1
```

Figure 12-14   The simulation table with a line added.

12-4-5   Control of the 3-State Output Gates

Most PALs have 3-state output gates. So far we have only allowed the 3-state output gates on the **16L8** to be always enabled or always high impedance (when the pin was used for an input). The general form for controlling a 3-state gate is:

$$\text{IF } (expression) \text{ equation}$$

If the expression within the parentheses evaluates to 1, the equation controls the gate outputs. If the expression evaluates to 0, the output is high impedance, and is listed as Z in the function table.

Figure 12-15 shows a PAL with a 3-state gate. The equation

$$\text{IF } (\overline{W})\,\overline{F3} = X + Y$$

means that $\overline{F3}$ will be HIGH, and the F3 output will be LOW, if X or Y is HIGH AND W is 0. If W is 1, the $\overline{F3}$ output will be high impedance (Z). Line 16 of the fuse plot shows that the enable on the 3-state gate is controlled by /W.

EXAMPLE 12-7

Figure 12-16 shows the result of the simulation for the circuit of Fig. 12-15. Why is there an error in line 9?

SOLUTION   Lines 7 and 8 of the function table (Fig. 12-15) show that W is 1. The expression /W is 0 and the 3-state output for F3 is disabled as indicated by the Z in the table. For line 9, however, W = 0, so the 3-state gate is enabled. The function table expects the output to be high-Z, but that is not possible because the gate is enabled; hence the error.

```
PAL16L8          PAL TEST PROBLEM
RIT
F1 = 2,4,5,8,A,B,D   F2 = 1,5,6,C,D,F
MAY 8, 1986
NC  W  Y  Z  NC  NC  NC  NC  GND
NC  NC  NC  NC  NC  F3  F1  F2  VCC
;
;
/F1 = /W*/X*/Y*/Z + /W*/X*/Y*Z + /W*/X*Y*Z
     + /W*X*Y + W*/X*/Y*Z + W*X*/Y*/Z + W*X*Y

/F2 = /W*/X*/Y*/Z + /W*/X*Y + /W*X*/Y*/Z + /W*X*Y*Z
     + W*/X + W*X*Y*/Z

IF (/W) /F3 = X + Y

FUNCTION TABLE

W   X   Y   Z      F1  F2  F3
--------------------------------
;
L   L   L   L      L   L   H
L   L   L   H      L   H   H
L   L   H   L      H   L   L
L   L   H   H      L   L   L
L   H   L   H      H   H   L
L   H   H   L      L   H   L
H   L   L   L      H   L   Z
H   H   H   H      L   H   Z
L   L   L   L      L   L   Z
--------------------------------

DESCRIPTION;  THIS IS STRICTLY A TEST
```

(a) Input file

```
PAL20  V1. – PAL16L8 – F1 = 2,4,5,8,A,B,D   F2 = 1,5,6,C,D,F
                      11 1111 1111 2222 2222 2233
      0123 4567 8901 2345 6789 0123 4567 8901
0     ---- ---- ---- ---- ---- ---- ---- ----
1     -X-- -X-- -X-- -X-- ---- ---- ---- ----   /W*/X*/Y*/Z
2     -X-- -X-- X--- ---- ---- ---- ---- ----   /W*/X*Y
3     -X-- X--- -X-- -X-- ---- ---- ---- ----   /W*X*/Y*/Z
4     -X-- X--- X--- X--- ---- ---- ---- ----   /W*X*Y*Z
5     X--- -X-- ---- ---- ---- ---- ---- ----   W*/X
6     X--- X--- X--- -X-- ---- ---- ---- ----   W*X*Y*/Z
7     XXXX XXXX XXXX XXXX XXXX XXXX XXXX XXXX

8     ---- ---- ---- ---- ---- ---- ---- ----
9     -X-- -X-- -X-- -X-- ---- ---- ---- ----   /W*/X*/Y*/Z
10    -X-- -X-- -X-- X--- ---- ---- ---- ----   /W*/X*Y*Z
11    -X-- -X-- X--- X--- ---- ---- ---- ----   /W*/X*Y*Z
12    -X-- X--- X--- ---- ---- ---- ---- ----   /W*X*Y
13    X--- -X-- -X-- X--- ---- ---- ---- ----   W*/X*/Y*Z
14    X--- X--- -X-- -X-- ---- ---- ---- ----   W*X*/Y*/Z
15    X--- X--- X--- ---- ---- ---- ---- ----   W*X*Y

16    -X-- ---- ---- ---- ---- ---- ---- ----   /W
17    ---- X--- ---- ---- ---- ---- ---- ----   X
18    ---- ---- X--- ---- ---- ---- ---- ----   Y
```

(B) Brief fuse plot

Figure 12-15   A PALASM file including a controlled 3-state gate.

F1 = 2, 4, 5, 8, A, B, D   F2 = 1, 5, 6, C, D, F

```
1  X0000XXXXXXXXXXXXHLL1
2  X0001XXXXXXXXXXXXHLH1
3  X0010XXXXXXXXXXXXLHL1
4  X0011XXXXXXXXXXXXLLL1
5  X0101XXXXXXXXXXXXLHH1
6  X0110XXXXXXXXXXXXLLH1
7  X1000XXXXXXXXXXXXZHL1
8  X1111XXXXXXXXXXXXZLH1
9  X0000XXXXXXXXXXX?LL1
```

NUMBER OF FUNCTION TABLE ERRORS = 1

Figure 12-16   Simulation results for Example 12-7.

12-5   Registered PALs

PALs that contain flip-flops (FFs) are called registered PALs, and generally have an **R** in their designation. The most commonly used registered PALs are the **16R4**, the **16R6**, and the **16R8**. The first number in the designation is the number of inputs (maximum) that the PAL can accommodate, the R indicates the PAL contains registers or FFs, and the last number indicates the number of FFs within each PAL.

The diagram of the **16R4** is shown in Fig. 12-17. It consists of four D-type FFs and four logic gates like those in a **16L8**. PALs with more FFs contain fewer logic gates. The **16R6**, for example, has 6 FFs but only 2 logic gates.[9]

The **16R4** is a typical registered PAL. All the FFs are clocked by a positive edge on the CLK input (pin 1); therefore, all the FFs change synchronously. The $\overline{OE}$ pin enables the outputs of the FFs when it is low. Usually it is connected to ground so that outputs are always enabled. There is an 8-input AND-OR gate connected to the D input of each FF. If the logic is such that the D input is a 1, the FF will set on the next clock edge; otherwise it will clear.

The FFs are connected to the outputs via an inverting 3-state gate. Consequently, if pin 17 is designated as O6, and its corresponding FF as Q6, we find O6 is the same as $\overline{Q6}$. The $\overline{Q}$ outputs are fed back through an inverting-noninverting gate, so the output of each FF can also become the inputs to other FFs or logic gates and the user can select either polarity.

EXAMPLE 12-8

Which lines are controlled by Q6 in a **16R4**?

SOLUTION   Figure 12-17 shows that the $\overline{Q}$ output of the top FF is connected to vertical line 10 via the noninverting side of the gate and to line 11 via the inverting side. Thus, if the FF is designated as Q6, vertical line 11 is the same as Q6 and vertical line 10 is the same as $\overline{Q6}$, which is the signal that appears on pin 17.

12-6   Examples Using Registered PALs

Perhaps the best way to study registered PALs is to give several examples and discuss their solutions.

[9] Diagrams of all the available PALs are available in the manufacturer's catalogs.

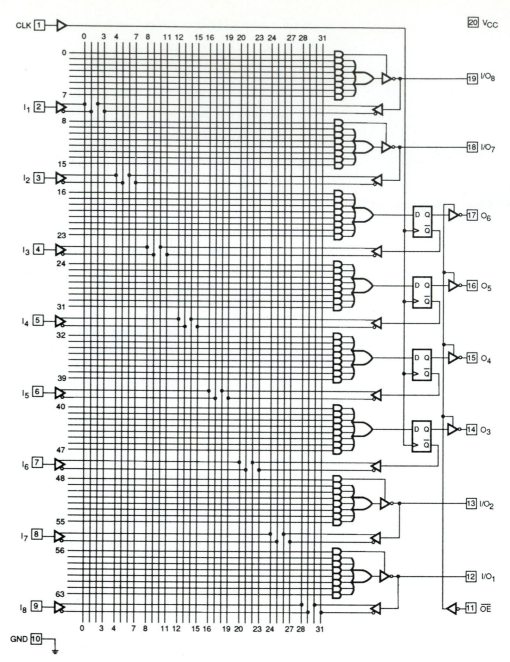

Figure 12-17 The **16R4** PAL. (Courtesy of Advanced Microdevices (AMD), Inc.)

12-6-1 An Introductory Example

This first example accomplishes nothing in particular. It is very simple, however, and is intended to provide an introduction to the programming of registered PALs.

The symbol := is used in registered PAL equations. It states that the value of Q *after* the next clock pulse is determined by the values of the variables *before* the clock pulse. For example, Q7:= Q0 + BD means that Q1 will be a 1 after the next clock pulse if Q0 is a 0 before the clock pulse or if B and D are both 1s.

EXAMPLE 12-9

Given the equations:

$$L1 = (A + BC)(D + \bar{E})$$
$$Q0 := \bar{B} + \overline{L1}$$
$$Q1 := \overline{Q0} + BD$$

Prepare the PALASM source file and discuss the results.

SOLUTION The source file is shown in Fig. 12-18a. The name of the source file, PALR4 in this example, is given on line 2 so the user can identify the disk file associated with the printout. Then the pinout is given. The original equations are listed on comment lines to clarify the problem. Because these PALs require low outputs, the original equations are then complemented using De Morgan's theorem. Finally, a function table is included in the source file.

The pinout is given in Fig. 12-18b, and the brief fuse plot is given in Fig. 12-18c. We would like to clarify a possible point of confusion. The output of FF0 has been declared to be Q0, on the schematic and the pinout and in the equations, but it is really the inverse of the Q output of FF0. Therefore, if Q0 is high the FF must be clear, and the $\overline{Q0}$ output will be the same as the output on the pin that is designated as Q0. This is verified by the fuse plot because the $\overline{Q0}$ output directly controls vertical line 10, which is connected to the inputs to Q1.

The simulation is shown in Fig. 12-18d. Note that a C is entered to indicate a clock pulse. The first line of the function table may also cause some confusion. Looking at the given inputs and the original equations we realize that L1 is surely LOW, and Q0 is HIGH. The third equation should cause Q1 to be LOW, but it is listed as HIGH. The reason for this is that, while L1 and Q0 depend only on the A, B, C, D, and E inputs, Q1 depends also on the previous state of Q0. But the first line doesn't know what the previous state of Q0 was, so it assumes 0, and the /Q0 term in the third equation causes Q1 to be HIGH. The second line of the vector table gives the proper results.

12-6-2 Design of a Simple Counter

In section 8-6-2 counters were designed using discrete logic. PALs can also be designed as counters using the principles developed there.

EXAMPLE 12-10

Design a 3-stage counter with a CLEAR line, so that when the CLEAR line is low, the counter will clear.

```
PAL16R4
PAL TEST PROBLEM PALR4
GIVEN ON FINAL EXAM MAY 1991
JD GREENFIELD
CLK A   B   C   D   E   NC  NC  NC  GND
GND NC  NC  NC  NC  Q1  Q0  NC  L1  VCC
;
;L1 = (A+BC) (D+/E)
;Q0 := /B + /L1
;Q1 := /Q0 + BD
;
/L1 = /A*/B + /A*/C + /D*E
/Q0 := B*L1
/Q1 := Q0*/B + Q0*/D

FUNCTION TABLE

A  B  C  D  E  CLK L1  Q0  Q1
---------------------------
L  L  L  L  L   C   L   H   H
L  L  L  L  L   C   L   H   L
H  L  L  H  L   C   H   H   H
H  H  L  H  L   C   H   L   H
---------------------------

DESCRIPTION
  THIS IS A TEST PROBLEM
```

(a) PALASM source file

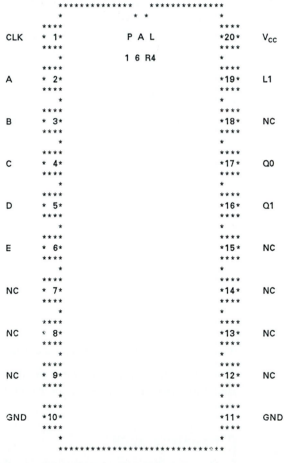

(b) Pinout

Figure 12-18

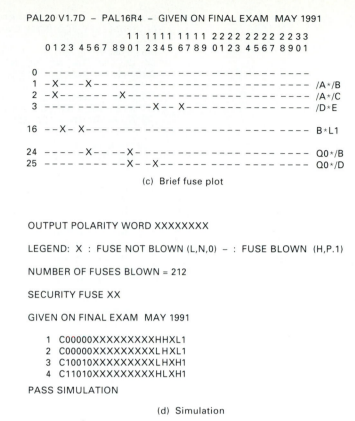

```
                         11 1111 1111 2222 2222 2233
              0 1 2 3  4 5 6 7  8 9 0 1  2 3 4 5  6 7 8 9  0 1 2 3  4 5 6 7  8 9 0 1

 0   – – – –  – – – –  – – – –  – – – –  – – – –  – – – –  – – – –  – – – –
 1   – X – –  – X – –  – – – –  – – – –  – – – –  – – – –  – – – –  – – – –   /A*/B
 2   – X – –  – – – –  – X – –  – – – –  – – – –  – – – –  – – – –  – – – –   /A*/C
 3   – – – –  – – – –  – – – –  – X – –  X – – –  – – – –  – – – –  – – – –   /D*E

16   – – X –  X – – –  – – – –  – – – –  – – – –  – – – –  – – – –  – – – –   B*L1

24   – – – –  – X – –  – – X –  – – – –  – – – –  – – – –  – – – –  – – – –   Q0*/B
25   – – – –  – – – –  – – X –  – X – –  – – – –  – – – –  – – – –  – – – –   Q0*/D
```

(c) Brief fuse plot

OUTPUT POLARITY WORD XXXXXXXX

LEGEND: X : FUSE NOT BLOWN (L,N,0) – : FUSE BLOWN (H,P.1)

NUMBER OF FUSES BLOWN = 212

SECURITY FUSE XX

GIVEN ON FINAL EXAM MAY 1991

```
    1   C00000XXXXXXXXXXHHXL1
    2   C00000XXXXXXXXXXLHXL1
    3   C10010XXXXXXXXXXLHXH1
    4   C11010XXXXXXXXXXHLXH1
```

PASS SIMULATION

(d) Simulation

Figure 12-18   The file for Example 12-9.

SOLUTION   The equations for a 3-stage counter built using synchronous D-type FFs were discussed in section 8-6-2. The first stage merely toggled; its equation can be written as

$$Q0 = \overline{Q0}*CLR$$

Q0 after the clock pulse will be 1 if Q0 before the clock pulse was 0 (the toggling function) AND if CLR was HIGH, indicating that *we are not attempting to clear*.

Similarly, the equations for Q1 and Q2 can be written as

$$Q1 = (Q0 \oplus Q1)*CLR$$
$$Q2 = (Q0*Q1 \oplus Q2)*CLR$$

To turn this circuit into a PAL, we realize that 3 stages of FFs are sufficient, so a **16R4** will suffice. The equations were then complemented (see Problem 12-11). The PAL file is shown in Fig. 12-19a. The simulation table shows that the counter counts properly from 0 to 7 and back to 0 as long as CLR is HIGH. The results of the simulation are given in Fig. 12-19b. There was a deliberate error introduced in vector 11; otherwise the results are correct.

```
PAL16K4              J. GREENFIELD
COUNTER TEST    COUNTER MADE 0F 3 FFS
RIT
MAY 1986
CK  CLR NC  NC  NC  NC  NC  NC  NC  GND
GND NC  NC  NC  Q2  Q1  Q0  NC  NC  VCC

/Q0 := Q0 + /CLR
/Q1 := Q1*Q0 + /Q1*/Q0 + /CLR
/Q2 := Q0*Q1*Q2 + /Q2*/Q0 + Q2*/Q1 + /CLR

FUNCTION TABLE
CK  CLR  Q0  Q1  Q2
- - - - - - - - - - - - - -
C   L    L   L   L
C   H    H   L   L
C   H    L   H   L
C   H    H   H   L
C   H    L   L   H
C   H    H   L   H
C   H    L   H   H
C   H    H   H   H
C   H    L   L   L
C   H    H   L   L
C   L    H   H   H
- - - - - - - - - - - - - -

DESCRIPTION
   THIS IS A 3 BIT COUNTER WITH A CLEAR
```

(a) The source file

```
RIT

 1  C0XXXXXXXXXXXXXLLLXX1
 2  C1XXXXXXXXXXXXXLLHXX1
 3  C1XXXXXXXXXXXXXLHLXX1
 4  C1XXXXXXXXXXXXXLHHXX1
 5  C1XXXXXXXXXXXXXHLLXX1
 6  C1XXXXXXXXXXXXXHLHXX1
 7  C1XXXXXXXXXXXXXHHLXX1
 8  C1XXXXXXXXXXXXXHHHXX1
 9  C1XXXXXXXXXXXXXLLLXX1
10  C1XXXXXXXXXXXXXLLHXX1
11  C0XXXXXXXXXXXX???XX1
```

NUMBER OF FUNCTION TABLE ERRORS = 3

(b) The simulation results

Figure 12-19 The PAL file for the 3-stage counter.

12-6-3 Design of a Truncated Counter

Truncated counters were designed in section 9-4-2. They can also be designed using PALs.

EXAMPLE 12-11

Design a 3-stage counter that rolls over at a count of 6, or counts 0,1, . . . 6,0.

SOLUTION   This circuit can be designed in one of two ways:

- State tables can be used to find the minimal circuit as discussed in Chapter 9. This circuit can then be blown into a PAL.

- The engineer can realize that the count of 6 means that Q2*Q1 = 1. This term can be added to each equation for the previous problem so that each stage clears after the count of 6 is reached.

The latter method requires less thought and more implicants. If it is to be blown into a PAL, however, it does not require more circuitry. The PAL file for this problem is shown in Fig. 12-20, along with a correct function table.

```
PAL16K4              GREG ZALEWSKI

RIT
MAY 13, 1991
CK  CLR  NC  NC  NC  NC  NC  NC  NC  GND
GND NC   NC  NC  Q2  Q1  Q0  NC  NC  VCC


/Q0 = Q0 + Q1*Q2 + /CLR
/Q1 = Q1*Q0 + /Q1*/Q0 + Q1*Q2 + /CLR
/Q2 = Q0*Q1*Q2 + /Q2*/Q0 + Q2*/Q1 + Q1*Q2 + /CLR


FUNCTION TABLE
CK  CLR  Q0  Q1  Q2
-  -  -  -  -  -  -  -
C    L    L    L    L
C    H    H    L    L
C    H    L    H    L
C    H    H    H    L
C    H    L    L    H
C    H    H    L    H
C    H    L    H    H
C    H    L    L    L
C    H    H    L    L
-  -  -  -  -  -  -  -
```

DESCRIPTION; THIS IS A HOMEWORK PROBLEM

Figure 12-20   A 3-stage counter that rolls over on a count of 6.

12-6-4   Shift Registers

Shift registers can also be designed using PALs. This allows the engineer to customize the design to fit any specifications.

The design of a shift register that functions like an 8-bit parallel-load, parallel-out shift register such as the **74x198** is shown in Fig. 12-21.[10] The shift register functions according to the function table of Fig. 12-21a. There are four data inputs, D3–D0, a clock input, two mode inputs, S1 and S0, and two serial input/output bits labeled SLISRO and SRISLO. There are four outputs from this shift register, Q3–Q0.

The two mode inputs allow four modes of operation. If S1 and S0 are both 0, the mode is *parallel load*. The data on the four D inputs, D3–D0, is loaded into the respective FFs. Of course this is a synchronous load because the FFs cannot change except at a clock edge.

If S1 is 1 and S0 is 0, the *shift-left* mode is invoked. On the next clock the data in Q0 will be shifted to Q1, and so on. The SLISRO (Serial Left Input–Serial Right Output) input will provide the level of the data being shifted into Q0. During left shifts SRISLO is enabled as an output. Its output is the same as Q3.

If S1 is 0 and S0 is 1, the *shift-right* mode is selected. Data from Q3 are shifted to Q2 and so on. The SRISLO (Serial Right Input–Serial Left Output)

[10]The design of this shift register was originally discussed in the Programmable Array Logic Handbook, published by Advanced Micro Devices, Sunnyvale, CA, 1984.

INPUTS				OUTPUTS						
S_1	S_0	SERIAL LEFT	SERIAL RIGHT	Q_3	Q_2	Q_1	Q_0	SERIAL RIGHT	SERIAL LEFT	
0	0	X	X	D_3	D_2	D_1	D_0	Z	Z	
0	1	X	0	0	$Q3_0$	$Q2_0$	$Q1_0$	$Q0_0$	Z	
0	1	X	1	1	$Q3_0$	$Q2_0$	$Q1_0$	$Q0_0$	Z	
1	0	0	X	$Q2_0$	$Q1_0$	$Q0_0$	0	Z	$Q3_0$	
1	0	1	X	$Q2_0$	$Q1_0$	$Q0_0$	1	Z	$Q3_0$	
1	1	X	X	$Q3_0$	$Q2_0$	$Q1_0$	$Q0_0$	Z	Z	

Figure 12-21(a) Function table.   03862A-64

```
PAL16R6                                    PAL DESIGN SPECIFICATION
PAT023                                     JENNY YEE      10/22/82
SHIFT REGISTER
ADVANCED MICRO DEVICES
CK   S1  S0  D3   D2   D1   D0  NC  NC  GND
OE   SRISLO  NC  Q3   Q2   Q1   Q0 NC  SLISRO VCC
;
;SHIFT REGISTER OUTPUT SIGNALS
;
/Q3   := /S1*/SO*/D3        +      /Q1   := /S1*/SO*/D1        +
         /S1* SO*/SRISLO    +               /S1* SO*/Q2        +
         S1*/SO*/Q2         +               S1*/SO*/Q0         +
         S1* SO*/Q3                         S1* SO*/Q1


/Q2   := /S1*/SO*/D2        +      /QO   := /S1*/SO*/DO        +
         /S1* SO*/Q3        +               /S1* SO*/Q1        +
         S1*/SO*/Q1         +               S1*/SLISRO*/SO     +
         S1* SO*/Q2                         S1* SO*/QO


IF(/S1*SO) /SLISRO = /QO


IF(S1*/SO) /SRISLO = /Q3


FUNCTION TABLE
CK  S1 SO D3 D2 D1 DO OE       SRISLO  SLISRO   Q3  Q2  Q1  QO
---------------------------------------------------------------------
;
; LOAD AND SHIFT RIGHT
C   L  L  L  L  L  L  L         Z       Z       L   L   L   L
C   H  H  X  X  X  X  L         Z       Z       L   L   L   L
C   L  H  X  X  X  X  L         H       L       H   L   L   L
C   L  H  X  X  X  X  L         L       L       L   H   L   L
C   L  H  X  X  X  X  L         L       L       L   L   H   L
C   L  H  X  X  X  X  L         L       H       L   L   L   H
C   L  H  X  X  X  X  L         L       L       L   L   L   L
;
; LOAD AND SHIFT LEFT
C   L  L  H  H  H  H  L         Z       Z       H   H   H   H
C   H  H  X  X  X  X  L         Z       Z       H   H   H   H
C   H  L  X  X  X  X  L         H       L       H   H   H   L
C   H  L  X  X  X  X  L         H       H       H   H   L   H
C   H  L  X  X  X  X  L         H       H       H   L   H   H
C   H  L  X  X  X  X  L         L       H       L   H   H   H
C   H  L  X  X  X  X  L         H       H       H   H   H   H
;
; HOLD
C   H  H  X  X  X  X  L         Z       Z       H   H   H   H
---------------------------------------------------------------------
DESCRIPTION
THIS DEVICE IMPLEMENTS A SHIFT REGISTER.  THE LAYOUT PROVIDED IS
A DEMONSTRATION OF HOW THE SHIFT REGISTER MAY BE DESIGNED USING
A PAL.                                              03862A-67
```

Figure 12-21(b) PAL source file and test vectors.

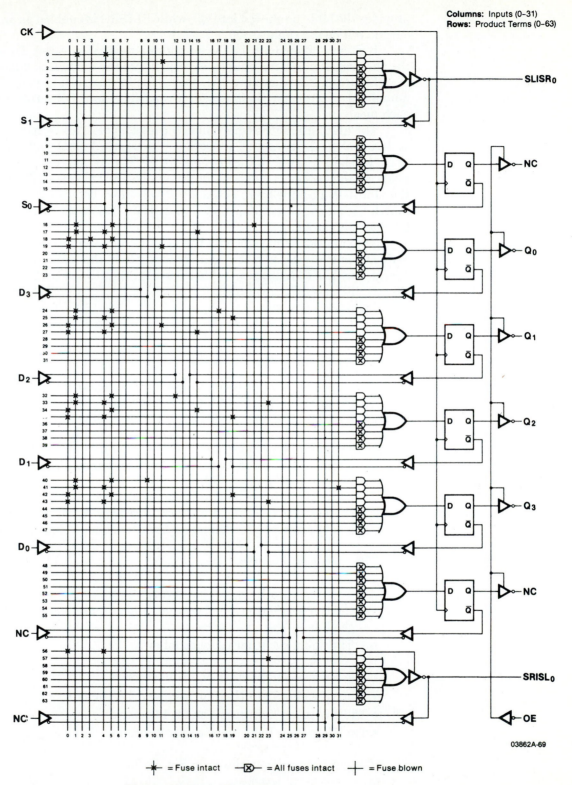

Figure 12-21(c) Fuse plot.

Columns: Inputs (0–31)
Rows: Product Terms (0–63)

$\ast$ = Fuse intact $\otimes$ = All fuses intact   + = Fuse blown

03862A-69

provides the bit to be shifted into Q3, while SLISRO is enabled as an output. Its output will be the same as Q0.

If S1 and S0 are both 1s, the *do-nothing* mode is selected. The contents of the shift register will not change in response to the next clock pulse.

Figure 12-21b shows the PAL source file for the shift register. We will examine the equation for $\overline{Q3}$ in detail. It consists of four OR terms; one term for each mode.

The first term in the equation is the load mode. This term contains $\overline{S1}*\overline{S2}$ and can only be 1 if S1 and S2 are both 0, which is the LOAD mode. $\overline{Q3}$ will be a 1, or Q3 will be a zero only if the level on D3 is 0. The second term is the shift-right mode. In this mode Q3 will be 0 if the signal on SRISLO is 0. The third term is the shift-left mode; Q3 will become 0 if Q2 was 0 before the clock edge. The fourth term is the do-nothing mode. Q3 will be a 0 after the clock edge if it was a 0 before the clock edge.

The equations conclude with the 3-state enable equations for SLISRO and SRISLO, and a series of test vectors to verify both the shift-right and shift-left operations. The PAL diagram is shown in Fig. 12-21c. Notice the intact fuses on the SLISRO and SRISLO lines.

12-6-5 A Simple Traffic Controller

Designing a circuit is an iterative procedure. The engineer examines the specifications and then designs a circuit to meet those specifications that can be implemented using the available technology. The circuit or device is then tested. Just because it works, however, does not mean that the design is optimum. It may not contain the minimum number of ICs or it may not minimize the cost, power dissipation, or other factors. A second or third look may often indicate ways in which the circuit can be improved.

This section discusses two iterations of a simple traffic controller problem that was presented to the students at my university. First, the circuit was to be designed using discrete ICs. Then the circuit was to be implemented using a PAL.

The specifications were:

a. There is a traffic signal consisting of red, green, and yellow lights in the North–South (NS) direction, and red, green, and yellow lights in the East–West (EW) direction.

b. The lights cycle normally as follows:

- The NS green light stays on for 10 seconds
- The NS yellow light then turns on for 2 seconds
- The EW green light turns on for 14 seconds
- The EW yellow light turns on for 2 seconds.

The light sequence is shown in Fig. 12-22.

c. As is normal for any traffic light, the NS red light is on whenever the EW green or EW yellow light is on, and vice versa.

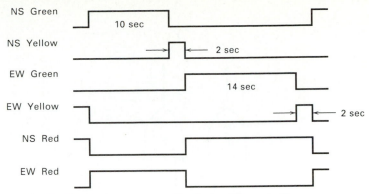

Figure 12-22 The light sequence for the traffic controller.

d. There is an NS traffic sensor and an EW traffic sensor embedded in the road to sense traffic in the NS or EW direction. If the NS green light is on and there is no EW traffic, as reported by the sensor, the NS green light must stay on until EW traffic appears. Then the NS yellow light must turn on within 2 seconds and normal cycling resumes. Corresponding specifications apply to the EW green light.

e. Add a PRESET input. If PRESET is low, the NS yellow light must be on and all others off.

The problem was originally assigned to be designed and built using ICs. Because the students lacked experience, the designs were overly complicated and used far too many ICs. Finally the professor, who was very aggravated, designed the IC circuit himself.

The circuit is shown in Fig. 12-23 and operates as follows:

1. The timing is controlled by an external 2-second clock and the **74x164** shift register. Either yellow light clears the shift register. When one of the green lights is on, 1s are shifted in. Q_E goes high after 10 seconds and Q_G goes high after 14 seconds. These signals can be used to turn the lights off at the proper times.

2. There are four FFs. They control the green and yellow lights. The red lights are controlled by the OR gates. When NS green OR NS yellow is on, EW red must be on and vice versa.

3. When the PRESET signal goes low, it directly sets the NS yellow light and clears the other FFs.

4. When the PRESET signal goes high, the NS yellow FF will clear on the next clock edge. The rising edge on $\bar{Q}$ will set the EW green FF.

5. If the EW green light is on AND 14 seconds have expired (Q_G is high) AND there is NS traffic, the next clock will set the EW yellow light. It will stay on for two seconds and also clear the shift register and the EW green FF.

6. With EW yellow set, the next clock edge will set NS green and turn EW yellow off. The NS sequence then proceeds similarly to the EW sequence.

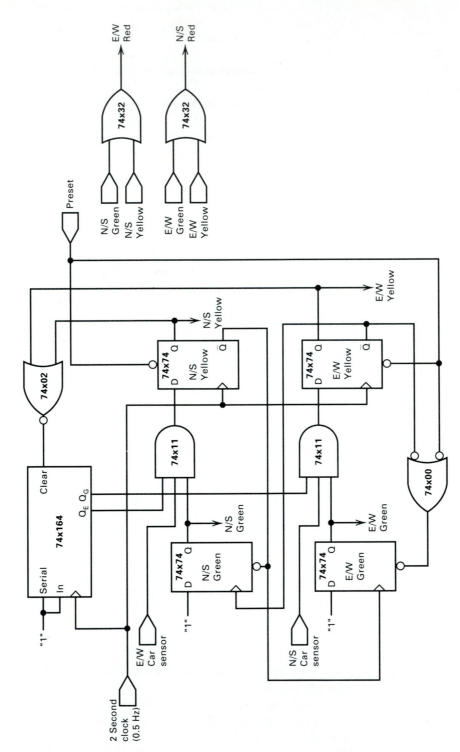

Figure 12-23 Traffic light controller.

The next step in the design was to blow a PAL for this circuit. Note that the FFs in Fig. 12-23 use the direct sets and clears. This had to be changed for PAL implementation. In the first PAL iteration we left the shift register as a discrete IC, but committed the rest of the logic in Fig. 12-23 to a single **16R4** PAL. Figure 12-24 shows the part of the schematic that was committed to the PAL.

The PAL source file is shown in Fig. 12-25a. The inputs are

- The clock
- Preset
- The two timing inputs taken from the shift register. E occurs at 10 seconds after the shift register is cleared and G occurs at 14 seconds.
- The two sensor inputs, NSEN and ESEN.

The outputs are the six outputs that control the lights and an output to clear the shift register. The equations should be examined in detail.

The first equation is for the NYEL FF. Notice that NYEL is a positive term, indicating that the NYEL FF will be set. The output, however, is $\overline{\text{NYEL}}$, which will be low if the FF is set. This is proper for driving an NYEL LED (see section 11-10-1). Now we must ask ourselves "When do want the NYEL FF to be set?" It must be set whenever PRESET is LOW, hence the term $\overline{\text{PRE}}$. It must also be set when the NGRN (North GReeN) FF is HIGH AND the time has expired (E = 1) AND there is EW traffic (ESEN = 1). Hence the equation:

$$\text{NYEL} := \overline{\text{PRE}} + \text{NGRN*ESEN*E}$$

EXAMPLE 12-12

When NYEL turns on, NGRN must turn off. Explain how the equations make this happen.

SOLUTION   We have already discussed the conditions that cause NYEL to turn on. They depend on NGRN being on, but NGRN must turn off at the same clock edge. The equation for NGRN consists of three terms. The first term depends on EYEL. Because both yellow lights should never be on at the same time, we can assume that EYEL is a logic 0. We have already seen that NYEL will not turn on unless E and ESEN are both 1s, but this makes the second and third terms in the equation for NGRN both 0. Therefore, all terms for NGRN are 0 at this time and the NGRN FF will turn off.

Now consider the EGRN equation. The FF must be off if PRESET is low, which explains the PRE literal in each term of the equation. The FF must turn on following the NYEL FF. Once it is on it must stay on until the time expires AND there is NS traffic. Unless both of these events have occurred, either the second or the third term in the equation will be 1 and the FF will stay on.

The EYEL and NGRN equations are similar to the NYEL and EGRN equations. The equations for the red lights use positive logic, but the outputs are the complements of these terms. The last equation indicates that CLR goes low whenever either yellow light is on; this clears the shift register.

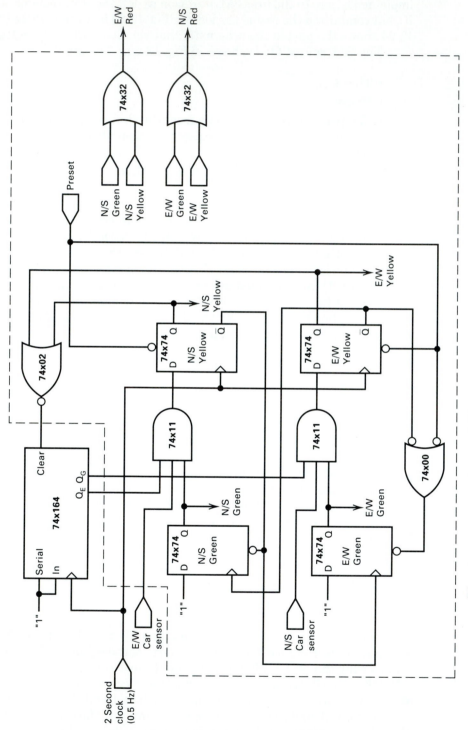

Figure 12-24 The part of Figure 12-23 within the **16R4** PAL.

```
TRAFFIC LIGHT CONTROLLER
RIT
J. GREENFIELD
CLK  PRE  NSEN  G  E  ESEN  NC  NC  NC  GND
NC  /NRED  /ERED  /NYEL  /EGRN  /EYEL  /NGRN  CLR  NC  VCC

NYEL := /PRE + NGRN*ESEN*E
EGRN := PRE*NYEL + PRE*EGRN*/NSEN + PRE*EGRN*/G
EYEL := PRE*EGRN*NSEN*G
NGRN := PRE*EYEL + PRE*NGRN*/ESEN + PRE*NGRN*/E
NRED = EGRN + EYEL
ERED = NYEL + NGRN
/CLR = NYEL + EYEL
```

FUNCTION TABLE

CLK	PRE	NSEN	G	E	ESEN	NYEL	EGRN	EYEL	NGRN	NRED	ERED	/CLR
L	L	L	L	L	L	H	L	L	L	L	H	H
C	L	L	L	L	L	H	L	L	L	L	H	H
C	H	L	L	L	L	L	H	L	L	H	L	L
C	H	L	L	L	L	L	H	L	L	H	L	L
C	H	H	L	L	L	L	H	L	L	H	L	L
C	H	L	H	L	L	L	H	L	L	H	L	L
C	H	H	H	L	L	L	L	H	L	H	L	H
C	H	H	H	L	L	L	L	L	H	L	H	L
C	H	L	L	L	L	L	L	L	H	L	H	L
C	H	L	L	H	L	L	L	L	H	L	H	L
C	H	L	L	L	H	L	L	L	H	L	H	L
C	H	L	L	H	H	H	L	L	L	L	H	H
C	H	X	X	X	X	L	H	L	L	H	L	L

DESCRIPTION
    THIS IS THE FIRST CUT AT THE TRAFFIC CONTROL PAL

(a) The source file

PAL20 V1.7D – PAL16R4 – RIT

```
                 11 1111 1111 2222 2222 2233
     0123 4567 8901 2345 6789 0123 4567 8901
```

```
 8  ---- ---- ---- ---- ---- ---- ---- ----
 9  ---- ---- ---- ---- ---- --X ---- ----      NYEL
10  ---- ---- ---- ---X ---- ---- ---- ----      EYEL

16  X--- ---- ---- ---X ---- ---- ---- ----      PRE*EYEL
17  X--- ---- ---- ---X ---- -X-- ---- ----      PRE*NGRN*/ESEN
18  X--- ---- ---- ---X -X-- ---- ---- ----      PRE*EGRN*/E

24  X--- X--- X--- ---- ---X ---- ---- ----      PRE*EGRN*NSEN*G

32  X--- ---- ---- ---- ---- --X ---- ----      PRE*NYEL
33  X--- -X-- ---- ---- ---X ---- ---- ----      PRE*EGRN*/NSEN
34  X--- ---- -X-- ---- ---X ---- ---- ----      PRE*EGRN*/G

40  -X-- ---- ---- ---- ---- ---- ---- ----      /PRE
41  ---- ---- ---X X--- X--- ---- ---- ----      NGRN*ESEN*E

48  ---- ---- ---- ---- ---- ---- ---- ----
49  ---- ---- ---- ---X ---- ---- ---- ----      NYEL
50  ---- ---- ---X ---- ---- ---- ---- ----      NGRN

56  ---- ---- ---- ---- ---- ---- ---- ----
57  ---- ---- ---- ---X ---- ---- ---- ----      EGRN
58  ---- ---- ---- ---X ---- ---- ---- ----      EYEL
```

OUTPUT POLARITY WORD XXXXXXXX

LEGEND: X : FUSE NOT BLOWN (L,N,0) – : FUSE BLOWN (H,P.1)

NUMBER OF FUSES BLOWN = 546
SECURITY FUSE XX
RIT

```
 1  000000XXXXXHLLHHHLX1
 2  C00000XXXXXL LHH H L X1
 3  C10000XXXXXLHHLHHHX1
 4  C10000XXXXXLHHLHHHX1
 5  C11000XXXXXLHHLHHHX1
 6  C10100XXXXXLHHLHHHX1
 7  C11100XXXXXLHHHLHLX1
 8  C11100XXXXXHLHHHLHX1
 9  C10000XXXXXHLHHHLHX1
10  C10010XXXXXHLHHHLHX1
11  C10001XXXXXHLHHHLHX1
12  C10011XXXXXHLLHHHLX1
13  C1XXXXXXXXXLHHLHHHX1
```

(b) the simulation and source file

Figure 12-25 The PALASM file for the traffic controller.

The equations are followed by a large function table designed to check out the traffic controller and ascertain that it is behaving correctly for a variety of inputs. In order to test the circuit, the PAL was blown using these equations. The circuit was built on a small circuit board containing only the PAL, the shift register, two red LEDs, two yellow LEDs, two green LEDs, and a resistor for each LED. The switches simulating the sensors and the 2-second clock were applied externally. The circuit functioned properly. The brief fuse plot and simulation file are shown in Fig. 12-25b.

In this design the shift register was left as a discrete component. The author was informed that the timing it controls can also be implemented in a PAL, perhaps using a **16R8**, thereby eliminating the shift register. This is still another iteration in the design.

12-7   Gate Arrays

Gate Arrays are a newer form of logic that can be used for more complex circuits, which require a large number of gates. They are larger and more flexible than PALs and can handle many more functions. They also require more Input/Output capability and therefore need more pins than a DIP can provide. Gate arrays are packaged in PLCC or Pin Grid Array cases.

This section is intended to be a brief introduction to gate arrays—to make the reader aware of their existence and some of their capabilities. Further information is available from the manufacturers. Two of the most prominent manufacturers of gate arrays are the Xilinx Corporation, San Jose, CA and the Altera Corp., San Jose, CA. No one should attempt to design with gate arrays without being familiar with the data books provided by these and other manufacturers.

12-7-1   The Use of Gate Arrays

Even the manufacturers of gate arrays recognize that they are not effective for all circuits. Xilinx recommends that PALs be used for circuits containing fewer than 100 gates.

There are two basic types of gate arrays: *programmable* and *custom*. Programmable gate arrays are like a larger version of a PAL. They can be obtained from the manufacturer immediately and configured by the user to his specifications. As with PALs, a program and a computer are required, as well as a logic programmer. Gate arrays are generally programmed by placing a charge on a floating gate (see section 13-11-3) instead of actually blowing fuses. This makes them erasable, and they can be reprogrammed.

Custom gate arrays are only for systems that must be manufactured in large quantities. They are built by the manufacturer to the users specifications. They incur NRE (Non-Recurrent Engineering) charges, other charges, and a waiting period. Xilinx estimates that for a 2000 gate circuit, the break-even point is about 18,000 units. For quantities less than this, programmable gate arrays should be used, but custom gate arrays will be less expensive for larger volumes.

12-7-2 The Structure of Gate Arrays

Internally, a gate array consists of three types of logic circuits:

- Configurable Logic Blocks (CLBs)[11]
- Input/Output blocks
- A Switching matrix

A CLB performs a logic operation on a set of inputs. The diagram of a small CLB is shown in Fig. 12-26. The box marked COMB (combinational) LOGIC can perform any logic function on the four input variables, A, B, C, and D, or any two logic functions on three variables. The output can be retained in the FF, which can be clocked by the K input, and SET, or RESET. Each of the two outputs can be F, G or the FF output. The choice can be made by the user.

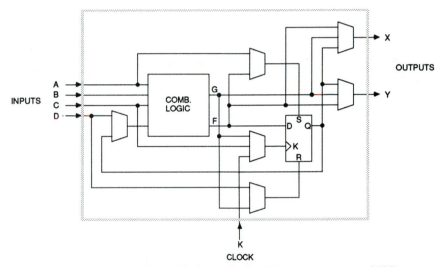

Figure 12-26 Configurable logic block. (Figure courtesy of Xilinx, Inc. © Xilinx, Inc. 1991. All rights reserved.)

A very simple I/O block is shown in Fig. 12-27. It is connected to a pin on the pin-grid array or PLCC. On output, the output signal can be 3-stated and controlled by the OUTPUT ENABLE signal. More sophisticated I/O blocks allow for inversion of the output and for controlling its slew rate.

A signal on the input pin can be directly connected to the IN line or it can be latched by the FF. Again the choice is specified by the user.

There are many CLBs and many I/O blocks on a gate array IC. The interconnections between these blocks is controlled by the switching matrix. Figure 12-28 shows the logic cell array structure. The CLBs are in the inner area. The I/O blocks are around the periphery where they have access to the external pins. The interconnect area is controlled by the switch matrix.

Gate arrays are very powerful logic structures. They should only be used, however, when the complexity of the problem warrants it.

---

[11]CLBs are Xilinx's term. Altera calls their main logic array a macrocell.

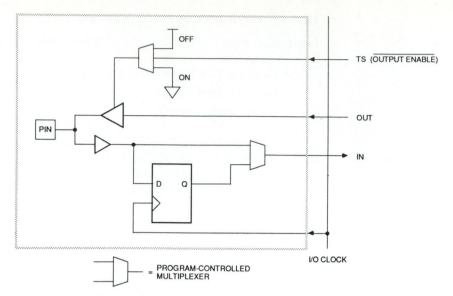

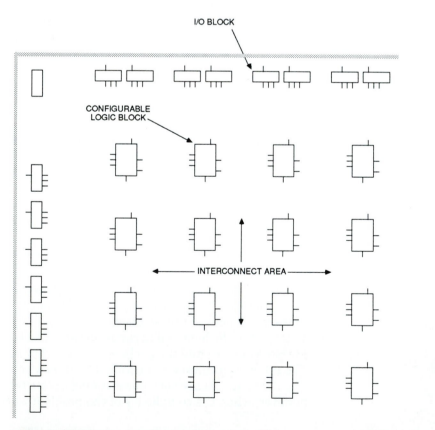

SUMMARY

In the near future, discrete ICs will be used mostly in experimental circuits, which will be built so they can readily be debugged. Once the bugs are out, these circuits will be burned into a fusible logic circuit such as PALs or gate arrays.

This chapter started by introducing the **16L8**, the most commonly used combinatorial PAL. The PALASM language for burning PALs was also introduced and examples of its use were presented. Next, registered PALs using FFs were discussed. Counters and shift registers were designed using them, and incorporating the principles developed in previous chapters. A detailed example showing the transformation of a discrete IC circuit to a PAL was presented. Finally gate arrays, which should be used for complex, high volume circuits, were discussed.

GLOSSARY

Blowing: Applying excessive current to a fuse in order to open it.

Combinatorial PAL: A PAL that contains only logic gates (no FFs).

Configurable logic block: A programmable block of logic within a gate array, that contains a FF for storage and also allows the user to specify logic functions on its inputs.

Gate array: A large IC containing many programmable configurable logic blocks, I/O blocks, and interconnections.

I/O block: A programmable block of logic within a gate array that connects the array to an input or output pin.

JEDEC: Joint Electron Device Engineering Council.

JEDEC file: A computer-generated file that can be put into a PLD programmer. It conforms to a standard developed by JEDEC, and tells the programmer which fuses to blow on a PLD.

PALASM: A computer language that enables PAL users to generate a file that can be used to blow a PAL.

Programmable Array Logic (PAL):   A PLD with fusible AND gates feeding an OR gate.

Programmable Logic Device (PLD):   An IC with many fusible links. Its logic function is obtained by selectively blowing some of its internal fuses.

Programmer: A hardware device used to blow open the fuses within a PLD.

Registered PAL: A PAL containing FFs within it.

Simulation table: A table used by a PLD language, such as PALASM, to calculate the expected outputs for a set of inputs.

PROBLEMS

Section 12-3-2.

12-1.   Draw the logic circuit for the line shown in Fig. P12-1.

Figure P12-1

Section 12-3-3.

12-2. Explain how the OR and NOR gates of Fig. 12-5 work.

12-3. Draw a circuit similar to Fig. 12-5 for
    (a) A 3-input AND gate
    (b) A 3-input XOR gate

Section 12-4.

12-4. Write the equations for a **16L8** for the circuit of Fig. P12-4.

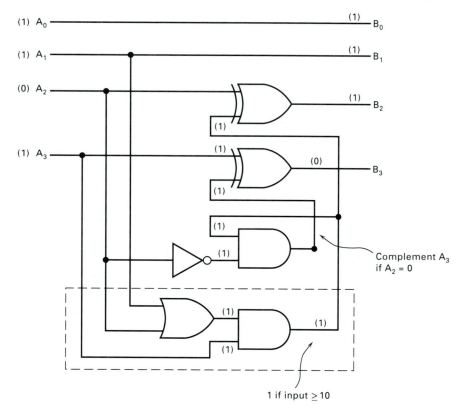

Figure P12-4

12-5. Write the equations for a **16L8** for the circuit of Fig. P12-5 (the SKIP-PAL circuit).

Section 12-4-3.

12-6. Implement the functions

$$F1 = \Sigma\, 2,4,5,8,A,B,D$$
$$F2 = \Sigma\, 1,5,6,C,D,F$$

on a **16L8**.

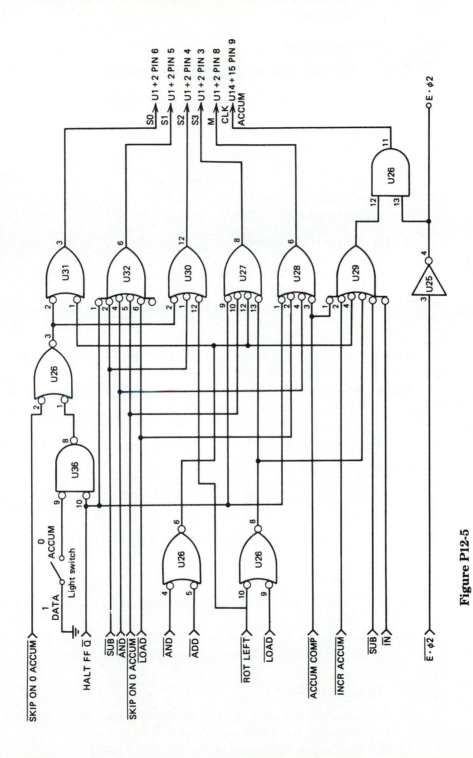

Figure P12-5

12-7. Using PALASM, generate the fuse plots for Problem 12-4. Using the fuse plot, show that the expression for A2 in Problem 12-4 has been implemented.

Section 12-4-5.

12-8. Write the PAL equation for the expression

$$IF\,(A\bar{B}C)\,M = WX\bar{Y}Z$$

Sketch the fuse plot for a **16L8**.

12-9. Repeat Problem 12-8 if the IF expression is

$$IF\,(A\overline{BC})$$

Section 12-6-1.

12-10. Given the equations:

$$L1 = (A + B)*C*D$$
$$Q1 := \bar{A}*L1 + B*C$$
$$Q2 := Q1*Q2*\bar{A} + \overline{Q1}*\bar{D}$$

Write the PALASM file for a **16R4**.

12-11. Construct a function table for Problem 12-10. If possible, use the simulation to verify the table.

12-12. Figure P12-12 is part of a fuse plot. Determine the equation for Q1 from it.

Section 12-6-2.

12-13. Show that

$$\overline{(Q1*Q0 \oplus Q2)*CLR} = Q0*Q1*Q2 + \overline{Q2}*\overline{Q0} + \overline{Q2}*\overline{Q1} + \overline{CLR}$$

12-14. Design a 4-stage counter with a CLEAR using a **16R4**.

Section 12-6-3.

12-15. Design a 4-stage counter that rolls over at a count of C(12) using a **16R4.**

12-16. Design a 4-stage down counter using a **16R4**. *Hint*: For a down counter, the Nth stage toggles whenever all N-1 stages are 0.

Section 12-6-4.

12-17. Compare the shift register designed in section 12-6-4 with a **74x198**. What are the advantages of each?

12-18. Implement the circuit of Fig. 10-14 in a **16R4**.

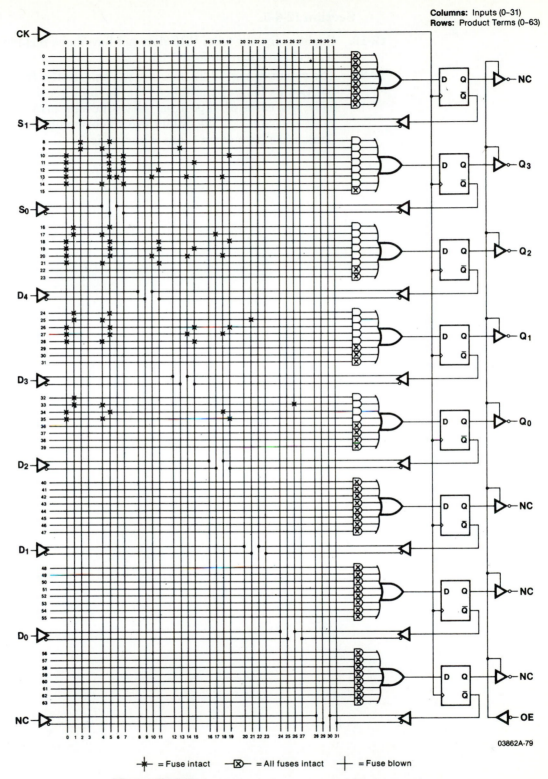

Columns: Inputs (0–31)
Rows: Product Terms (0–63)

—✳— = Fuse intact   —⊠— = All fuses intact   —┼— = Fuse blown

Figure P12-12

03862A-79

Section 12-6-5.

12-19. Explain why the term PRE does not have to be included in the second term of the equation for NYEL.

12-20. The circuit of Fig. 12-23 does not work perfectly. It will remain in one position as long as there is no traffic in either direction. Philosophically, who cares what position the traffic light is in if there are no cars? Nonetheless design the fix for the discrete circuit and for the PAL so the lights cycle even when there is no traffic.

Memories

13-1 Instructional Objectives

Electronic systems are often required to retain or remember many bits of information. When the amount of information exceeds several bits, this information is stored in a device called a *memory* rather than in individual FFs or a register. This chapter introduces the student to the semiconductor memories which are an essential part of all computers. After reading it he or she should be able to:

1. Explain the steps involved in reading or writing a memory word.
2. Use a semiconductor RAM to read or write information at a specific location.
3. Assemble several smaller RAMs together to make a larger RAM.
4. Utilize dynamic RAMs in a memory system.
5. Be able to design and build refresh circuits for DRAMs.
6. Design and use diode ROMs for small memories.
7. Understand and use ROMs and PROMs.

13-2 Self-Evaluation Questions

Watch for the answers to the following questions as you read the chapter. They should help you understand the material presented.

1. In memory terminology, the symbol K does not mean 1000. What number does it represent and why?
2. What are MAR and MDR? What factor determines their size?
3. What is a bus?
4. What is the difference between a memory chip configuration and a memory system configuration?
5. What is the difference between a ROM and a RAM? What input and output signals are connected to each?
6. Why are memory request signals unnecessary in a static RAM?
7. Define access time and cycle time. Why are they important?
8. In a dynamic RAM, why are refresh cycles necessary? How often must they be performed?
9. What is the difference between a PROM and a ROM?
10. Why is an erasable PROM the best choice when developing a microprocessor program?

13-3   Memory Concepts

Computers and many other sophisticated electronic circuits require the ability to retain or remember large amounts of information. A *memory* is an essential component of a computer and is used to remember that information in the form of binary bits. The information is used, as needed, to control the operation of the computer.

Flip-flops, introduced in Chapter 6, function as *1-bit memories;* they remember whether the last pulse that occurred was a SET pulse or a CLEAR pulse. When FFs are grouped into registers their capability increases; n FFs can remember 2^n different numbers. Even the smallest computers, however, require several thousand bits of memory. It is impractical to build large memories from discrete FFs. Modern memories are built using memory ICs. These chips are packaged in DIPs or even SIMMS (single in-line memory modules).

13-3-1   Memory Buses

In a computer or microprocessor (μP), data is transferred between the computer and the memory on a *bus.* A bus is a set of wires used for transmitting data. There is one wire on the bus for each bit being transferred. The data is transferred in *parallel;* all the bits are sent over the bus at the same time.

Figure 13-1 shows the three buses that connect a memory with a computer or μP.

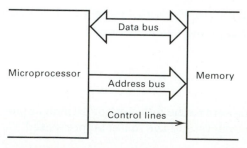

Figure 13-1   The buses connecting a μP to a memory.

- **The Data Bus**—The data bus is bidirectional; data can flow in either direction. It flows from the memory to the μP during a read cycle and from the μP to the memory during a write cycle (see section 13-4).
- **The Address Bus**—This bus carries the *memory address* (see section 13-4) from the μP to the memory.
- **The Control Bus**—This bus carries lines that control the operation of the memory from the μP to the memory. Usually there are several control lines, but we will only be concerned with one of them, the read/write line.

The μP and the memory have a *master-slave* relationship. The μP is the master; it determines what is on the address and control lines (and on the data lines in a write cycle). The memory is the slave. It responds to the address and control commands issued by the μP.

13-3-2   Bits and Words

Memories are subdivided into groups of bits called *words. A word consists of the number of bits involved in each data transfer.* The Intel **8085** and the Motorola **6800** series, for example, are 8-bit μPs. They transfer 8 bits of data over an 8-bit bus. The Intel **8086** or the Motorola **68000** are newer 16-bit computers. They are more powerful because they can transfer 16 bits at a time between the μP and the memory. Intel's **80386**, used in many personal computers (PCs), is still more powerful; it has a 32-bit word length.

13-3-3   K and M

The letter K is the most commonly used measure of the size of a memory. Unlike standard engineering terminology, where K is an abbreviation for kilo, or 1000, $K = 2^{10}$, or 1024, when applied to memories. This value of K is used because normal memory design leads to sizes that are powers of 2. For larger memories the letter M (mega) is used. Here $M = K^2$. This is precisely 2^{20} or approximately 1.04 million.

Although standard computer word sizes are 8, 16, or 32 bits, most memories are measured in terms of *bytes*. A byte is 8 bits long. Many personal computers (PCs), for example, are advertised as having 2 MB of memory. This means they have about 2.08 million bytes of memory. If, however, the PC contains a 16-bit μP, such as the **68000**, each word consists of two bytes and they have 1.04 million words.

13-4   Memory Structure

Each of the many words in memory is stored at a *particular address*. Data is *written* into the memory one word at a time and preserved for future use. The data is *read* at a later time when the information is needed. During the interval between writing and reading, other words may be stored or read at other locations. Thus, each memory word has *two attributes,* its *address,* which locates it within the memory, and the *data* that is stored at that location.

Memories also require *two registers,* one associated with *address* and one with *data. The Memory Address Register* (MAR) holds the address of the word currently being accessed, and the *Memory Data Register* (MDR) holds the data being written into or read out of the addressed memory location. The words stored within the memory are not immediately accessible to the system, but the MAR and MDR are available. In some memories the MAR and MDR are *not* part of the internal memory. In this case, they must be part of the user's circuitry. The MDR serves as a temporary storage area (buffer) for transferring words between the memory and the data bus.

EXAMPLE 13-1

The 1K by 8-bit memory of Fig. 13-2 is used as part of a μP system with a 16-bit address bus.

    a. How many bits are required to address all locations in this IC?
    b. How many address bits are used for the whole μP system?
    c. How many bits are required in the MDR?
    d. How many data bits are contained in this memory IC?

SOLUTION

    a. A memory specification of 1K × 8 means 1024 (2^{10}) words of 8 bits each. The MAR must hold an address value between 0 and 1023 (1024 total locations) to select any word in this IC. Thus, 10 bits are needed to address the memory.
    b. The μP uses 16 address lines (or bits in the MAR). Since 10 are used to select the locations in this memory IC, the other 6 are available to select other memory ICs.
    c. Each memory location contains 8 bits. Consequently, the MDR must be 8 bits long to accommodate one data word.
    d. A 1K × 8 memory contains 2^{10} words × 2^3 (8) bits per word = 2^{13} or 8192 bits. For a μC, this is a relatively small memory and several such ICs would typically be used.

In Fig. 13-2, the MAR is shown to contain address 2 and the MDR register contains the value that is in that location.

13-4-1 Reading Memory

Memories operate in two basic modes: READ or WRITE. A *memory is read when the information at a particular address is required by the system.* To read a memory:

    1. The location to be read is loaded into the MAR and placed on the address bus.
    2. A READ command is given.
    3. The data is transferred from the addressed word in memory to the MDR, and placed on the data bus where it is accessible to the μP.

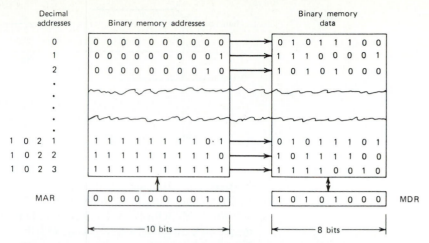

Figure 13-2 Typical 1024-word by 8-bit memory. (Joseph D. Greenfield/William C. Wray, USING MICROPROCESSORS AND MICROCONTROLLERS: The Motorola Family, 2e, © 1988. Reprinted by permission of REGENTS/PRENTICE HALL, Englewood Cliffs, New Jersey.)

Normally the word being read must not be altered by the READ operation, so that the word in a particular location can be read many times. The process of reading out of a location without changing it is called *nondestructive readout*.

13-4-2 Writing Memory

A memory location is written when the data must be preserved for future use. To write a memory:

1. The address is loaded into the MAR and placed on the address bus.
2. The data to be written is placed on the data bus.
3. The WRITE command is then given, which transfers the data from the data bus to the selected memory location.

One must be careful when writing memory. Unlike reading, writing is *destructive*. In the process of writing, the previous information in the specified location is destroyed (overwritten). Of course, it makes no sense to read a memory unless it has been previously written with data that should be read.

13-4-3 Memory Timing

Memory timing is very important. A READ cycle is limited by the access time of a memory, and a WRITE cycle is limited by the cycle time.

Access time is the time required for the memory to present valid data after the address and select signals are stable. **Cycle time** or WRITE time is the time the address and data must be held constant in order to write to the memory.

Access time and cycle time are illustrated in Fig. 13-3. Similar figures with specific times are supplied by the manufacturers of IC memories.

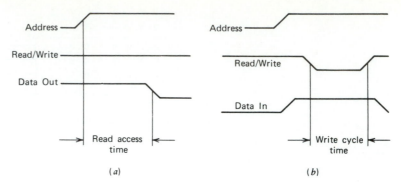

Figure 13-3 Memory timing. (a) Read cycle. (b) Write cycle. (Joseph D. Greenfield/William C. Wray, USING MICROPROCESSORS AND MICROCONTROLLERS: The Motorola Family, 2e, © 1988. Reprinted by permission of REGENTS/PRENTICE HALL, Englewood Cliffs, New Jersey.)

Figure 13-3a shows the normal situation where the memory is constantly in READ mode. The output data will change in response to an address change and the time for this response, *the access time,* is clearly visible.

Figure 13-3b shows the WRITE cycle. Note that the address and data are firm before the WRITE pulse occurs on the R/$\overline{W}$ line and do not change while the memory is in WRITE mode. This is necessary to prevent spurious writes that could enter unwanted information into the memory.

13-4-4   RAMs, ROMs, and Sequential Memories

The term RAM (Random Access Memory) is often applied to memories. Literally speaking, it means that each word in memory is *equally accessible* or that the time to access any word is the same regardless of its address. RAMs are generally fast; their access times are short enough so they can be used as the internal memory in a computer. The access time of IC memories is usually under 200 ns.

Perhaps the most common example of a non-RAM is the disk attached to a computer. A disk consists of a diskette spinning under a head that can both read and write. To read the disk, one must wait until the proper information is under the head. Thus, the time between issuing the read command and the proper positioning of the diskette is variable; it is called *latency time.* Once the head is positioned, however, a word can be read, followed by the next word and so forth. Therefore, non-RAM memories are usually sequential; once reading starts it can continue sequentially.

EXAMPLE 13-2

A diskette is spinning at 3600 rpm (revolutions per minute). What is its average latency time?

SOLUTION   3600 rpm translates to 60 rps, or 16.67 ms per revolution. Because, on average, the information required will be halfway across the diskette when the read command is given, the average latency time will be approximately 8 ms. This is far slower than the access time of a RAM.

Generally speaking, however, the term RAM is not used in its literal sense. Colloquially, RAM means a read/write memory; a memory that can be both read and changed by having data written into it. This distinguishes it from a ROM (Read Only Memory) that can only be read. ROMs are also fast and random access, but they are never called RAMs. They are discussed in section 13-9.

The characteristics of each type of memory are:

- **RAMs**—fast, can be read and written to. The disadvantage of a RAM is that it is *volatile*. This means it *loses its information when power goes down*. In PCs, for example, the power is first turned on. Then the programs and files to be used must be loaded into RAM from some nonvolatile memory (typically a disk) before they can be accessed. Some computers use battery backup to preserve the contents of their RAMs.
- **ROMs**—fast, random access, and nonvolatile. They retain their data when power is removed. ROMs cannot be written to so their contents never change. They are generally used to hold computer programs and data that may not vary.
- **Sequential Memories**—these memories can store large quantities of information and are non-volatile, but they are very slow compared to RAMs and ROMs. They are too slow to be used as the internal memory of a computer. Magnetic tapes and disks are sequential memories.

Personal computers make use of all three types of memories. When the computer is first turned on, the *monitor program,* which is in nonvolatile ROM, takes over and properly starts the computer. This gives the user control over the computer. Then the user selects the file of interest from the disk and loads it into RAM. The computer can now run and execute this file.

In the old days, circa 1980, Apple and Radio Shack came out with computers with 16 Kbytes of RAM and users thought they had more than enough. Now computers with 2 or 4 MB are common, but the programs have also expanded. Such programs as MICROSOFT WINDOWS, WORDPERFECT, or LOTUS need most of it. It seems that computer programs will always expand to the point where they exceed the capacity of the available memory.

13-5 Static RAMs

RAMs (read/write memories) are divided into two categories, *static* and *dynamic*. *Static* RAMs store their information in FFs and are very simple to use. *Dynamic* RAMs store their information as electrical charges on capacitors. But the charges leak off and the capacitors must be recharged periodically. The remainder of this section considers static RAMs. (Dynamic RAMs are discussed in section 13-6.) In the literature, static RAMs are often called SRAMs and dynamic RAMs are called DRAMs.

13-5-1 Interfacing with an SRAM Memory

SRAMs come in IC packages, mostly DIPS, and are arranged in matrices of n words by m bits. A semiconductor RAM normally has pins for the following inputs and outputs:

1. *m* output bits.
2. *m* input bits.
3. *n* address bits (for 2^n words).
4. A READ/WRITE input.
5. A CHIP SELECT or ENABLE input.

Because semiconductor ICs are often paralleled to build larger memories, most outputs are 3-state. When the IC is *not* selected, as determined by the CHIP SELECT signal, it cannot be written into, and all its outputs are turned OFF, or in the *high impedance* state.

There are no memory request signals in a static semiconductor memory. To read a location, the READ/WRITE line is simply set to the READ level. The memory then presents the data at the addressed word and continues to present this data until the address, READ/WRITE command, or CHIP SELECT signals change. *Reading is nondestructive;* the state of the internal FFs at the selected address is simply brought to the outputs.

A memory is written into by setting the READ/WRITE line to WRITE. For modern memories, the READ level is a logic 1 and WRITE is a logic 0. To write, the READ/WRITE line is pulsed low for one cycle time, and the data on the input lines is entered into the memory *at the addressed location*. One must be careful when writing, because writing changes or *overwrites* the previous contents of the memory, which are lost. The best way to write is to set up the address and data bus and then pulse READ/WRITE low. When the READ/WRITE line goes HIGH again, the information on the buses can be changed without changing the contents of the memory.

A typical advertisement for memory ICs, which appeared in a recent computer magazine, is shown in Fig. 13-4. It divides the memories into two groups, dynamic and static, and gives the salient characteristics of each chip. It starts with the IC number, then gives its access time, its configuration, and its price.

EXAMPLE 13-3

What lines would you expect to be connected to a **43256** static RAM?

SOLUTION   Figure 13-4 shows that the configuration of a **43256** IC is 256Kx1. The x1 means there is only one bit for each access. Thus, we would expect to find only one input line and only one output line in addition to the standard CHIP SELECT and READ/WRITE line. The 256K is the number of addresses on the IC. Because $256 = 2^8$ and $K = 2^{10}$, there are 2^{18} addresses, and 18 address input pins are required. There is only one bit at each address, so there are 2^{18} bits on this IC.

The **43256** also comes in two versions, the $-10L$ and the $-15L$. The $-10L$ is better; it is 50 ns faster and therefore more expensive.

13-5-2   Bipolar SRAMs

There are two types of memory technology for SRAMs: bipolar ICs and MOS ICs. The basic memory element within bipolar RAMs is the TTL type FF. Consequently, these memories are very fast and are used in high-speed systems.

Dynamic RAMs

TMS4416-12	120ns.	16K × 4	$ 2.25
TMS4416-15	150ns.	16K × 4	$ 2.00
4116-12	120ns.	16K × 1	$ 1.49
4116-15	150ns.	16K × 1	$ 1.09
4116-20	200ns.	16K × 1	$ .89
4164-100	100ns.	64K × 1	$ 2.75
4164-120	120ns.	64K × 1	$ 2.39
4164-150	150ns.	64K × 1	$ 2.15
4164-200	200ns.	64K × 1	$ 1.75
41256-60	60ns.	256K × 1	$ 5.25
41256-80	80ns.	256K × 1	$ 3.75
41256-100	100ns.	256K × 1	$ 3.15
41256-120	120ns.	256K × 1	$ 2.95
41256-150	150ns.	256K × 1	$ 2.59
41464-80	80ns.	64K × 4	$ 5.95
41464-10	100ns.	64K × 4	$ 4.95
41464-12	120ns.	64K × 4	$ 3.95
41464-15	150ns.	64K × 4	$ 3.59
511000P-70	70ns.	1M × 1	$13.95
511000P-80	80ns.	1M × 1	$12.95
511000P-10	100ns.	1M × 1	$12.35
514256P-80	80ns.	256K × 4	$13.45
514256P-10	100ns.	256K × 4	$12.95

Static RAMs

6116P-3	150ns.	16K × 1 (CMOS)	$ 2.79
6264LP-10	100ns.	64K × 1 (CMOS)	$ 6.95
6264LP-15	150ns.	64K × 1 (CMOS)	$ 4.95
43256-10L	100ns.	256K × 1	$10.95
43256-15L	150ns.	256K × 1	$ 9.95
62256LP-15	150ns.	256K × 1 (CMOS)	$10.95

Figure 13-4 Part of an advertisement showing memories for sale.

Bipolar SRAMs have two major disadvantages compared with MOS SRAMs:

- They are much smaller than MOS SRAMs in terms of the number of bits per IC.
- MOS memories (especially CMOS) consume less power. A bipolar memory requires about three times as much power per bit as an NMOS memory. CMOS memories require even less power.

Two older bipolar memory ICs that may still be found in some circuits are the **74x89** and the **74x170**.[1] Modern bipolar RAMs consist of the **74ALS870** and **'871** and the **74AS870** and **'871**. Because of their advanced technology, they are very fast. The **ALS** ICs have a 16 ns access time and the **AS** ICs have an 11 ns access time, but they only contain 128 bits. They find some use in specialized circuits that require their high speeds.

[1]These ICs are essentially obsolete. Their operation is described in the second edition of this book.

13-6   MOS SRAMs

SRAMs are static semiconductor memories that are built of thousands of MOS (metal-oxide-semiconductor) FFs. At present, CMOS is the dominant technology used in SRAM ICs. They are generally faster and easier to use than DRAMs and contain many more bits than a bipolar RAM. The most advanced SRAMs at this writing (circa 1992) are 1Mbit SRAMs available with access times about 20 ns.

13-6-1   The 2102 SRAM

The **2102** is a 1024-bit static RAM. This is a very small memory by modern standards, and the **2102** is no longer incorporated in new designs. It is, however, still available and is an excellent IC for introducing the principles and concepts of SRAM systems. It is also an excellent IC to use for laboratory experiments, because it comes in a 16-pin DIP and mating sockets are readily available.

The **2102** is arranged as a 1K-word-by-1-bit memory, as shown in Fig. 13-5. The pins on the **2102** are:

1. 10 address lines (A_0–A_9).
2. DATA IN
3. READ/WRITE (R/W)
4. DATA OUT
5. CHIP ENABLE ($\overline{CE}$)

The **2102** has 1 output, DATA OUT, which is 3-stated when the chip is not enabled ($\overline{CE} = 1$). To write, the IC must be enabled. The bit to be written is placed on the DATA IN pin and the addresses are placed on the address pins. The READ/WRITE line is then pulsed LOW. Larger memories can be built using several **2102**s, as Example 13-4 shows.

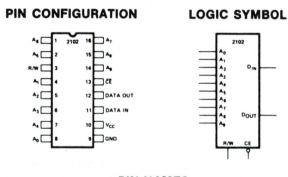

PIN NAMES

D_{IN}	DATA INPUT	$\overline{CE}$	CHIP ENABLE
A_0 – A_9	ADDRESS INPUTS	D_{OUT}	DATA OUTPUT
R/W	READ/WRITE INPUT	V_{CC}	POWER (+5V)

Figure 13-5   The **2102** 1024-bit static RAM memory. (From the *Intel Semiconductor Memory Book*. Copyright John Wiley & Sons, Inc. 1978. Reprinted by permission of John Wiley & Sons, Inc.)

EXAMPLE 13-4

Design a 4K-byte (4096-words-by-8-bit) memory using **2102**s.

SOLUTION Engineers often encounter a problem such as this one, which involves building a larger *memory system* from a group of ICs. To attack this type of problem, the engineer should first consider the *chip configuration* and the *system configuration*. In this problem, the chip configuration is 1Kx1 because we are using **2102**s. Each chip can accomodate one data bit and requires 10 address lines. The system configuration is specified as 4Kx8, so the memory system requires 8 data bits and 12 address lines. The solution proceeds as follows:

1. The number of **2102**s required must be determined. Since the total memory requires 4K words by 8 bits/word = 32K bits and each **2102** has 1K bit, thirty-two **2102**s are needed.

2. For a memory of this size, the computer or system driving the memory must provide 12 address bits, 8 data input bits, and a read/write line. It must be capable of accepting 8 data outputs from the **2102**s.

3. The thirty-two **2102**s are arranged in an 8-by-4 matrix, as shown in Fig. 13-6, to provide 4K words of 8 bits.

4. The 10 LSBs of the address are connected to the 10 address inputs of each **2102**. The two remaining address bits become the inputs to a decoder (**74x138** in Fig. 13-6) that selects which row of **2102**s will be enabled. Note that the decoder provides a single LOW output and $\overline{CE}$ must be LOW for the IC to be enabled, so the decoders can be directly connected to the $\overline{CE}$ inputs.

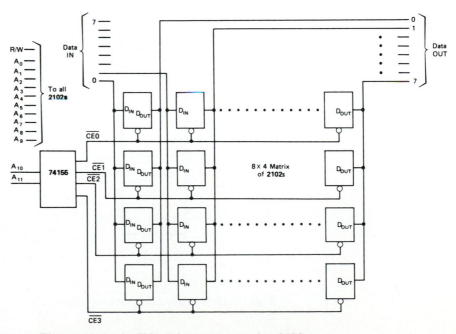

Figure 13-6 A 4K-by-8-bit memory using **2102**s.

5. The 8 data inputs are each connected to one column of **2102**s.

6. The 4 data outputs in each column are all connected together. Pull-up resistors are not needed since only one of the four **2102**s is enabled at any time and the rest are disabled, presenting a high impedance to the bus.

7. The read/write line is connected to all **2102**s.

8. To read the memory, the two high-order address bits select a row of **2102**s and the 10 low-order address bits select one of the 1024 words within the **2102**s. After the addresses are firm for the specified access time (>300ns), the 8 data output bits appear on the output lines.

9. Writing is similar. When a WRITE pulse is applied to the R/W line, the 8 data inputs are written into the **2102**s. Note that *writing destroys the previous contents of the memory* at the selected address, so the R/W line should always be kept in READ mode, except when actually trying to write.

10. The 10 address lines and the R/W line drive all 32 ICs in Fig. 13-6. Fortunately, these are MOS ICs with high-input impedances, so there is no loading or fanout problem.

13-6-2 Measuring Access Time[2]

It is not difficult to measure the access time of a memory IC in the laboratory. The circuit is shown in Fig. 13-7a. Using a **2102**, the following procedure can be followed:

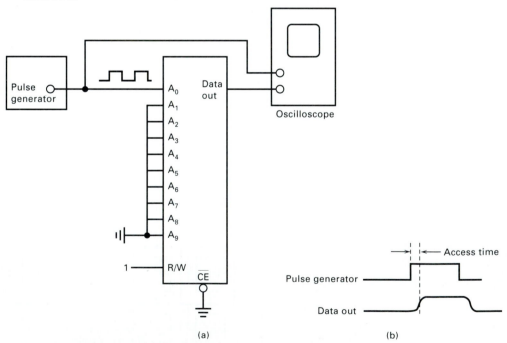

Figure 13-7 Measuring access time: (a) circuit; (b) oscilloscope traces.

[2]This experiment was done many times in the laboratory at the Rochester Institute of Technology.

1. Set addresses A1–A10 to all 0s, by connecting them to ground.
2. Set A0 to 0 and write a 0 into memory. This writes a 0 into address 0.
3. Change A0 to a 1 and write a 1 into memory. This writes a 1 into address 1.
4. Now apply a pulse generator to A0. As the generator pulses, it changes the address between 0 and 1. The DATA OUT line should also change. By observing the pulse generator input and the data output on a dual-trace oscilloscope, as shown in Fig. 13-7b, the time difference can be found. This is the access time; it should be about 100 ns.

The access time discussed in the previous paragraph was with respect to address changes; the IC was always selected and the address changed. There is another access time that is sometimes used—access time with respect to chip select ($\overline{CS}$). This is the time required to access data after the $\overline{CS}$ input has gone low.

13-6-3 The 2114 SRAM

The **2114** is a 1K-by-4-bit static RAM that is shown in Fig. 13-8. Thus the **2114** contains 4K bits, and has four times the capacity of a **2102**. Because each memory location holds four bits, it only requires two **2114**s to make a one byte memory.

The **2114** comes in a 18-pin package that has:

- 10 address inputs
- 4 data lines
- 1 chip select ($\overline{CS}$)
- 1 WRITE ENABLE input ($\overline{WE}$)

The four data lines are *bidirectional;* the data go either from or to the **2114** depending on whether the memory is being read ($\overline{WE} = 1$) or written ($\overline{WE} = 0$).

In a typical application, the **2114** is connected to a microprocessor that controls the addresses and read/write line as shown in Fig. 13-9. If $\overline{CS} = 1$, the **2114** is deselected; it presents a high impedance to the bus and cannot be written into. During read ($\overline{CS} = 0$, $\overline{WE} = 1$), the 3-state bus drivers within the **2114** turn on and the data at the selected address is sent from the **2114** to the µP. During a write operation the µP first selects the **2114** (causing $\overline{CS}$ to be LOW). It then sends a WRITE pulse, which causes $\overline{WE}$ on the **2114** to go LOW. This disables the **2114** bus drivers. At this time the bus drivers within the µP are enabled, and the data to be written flows from the µP to the **2114**.

EXAMPLE 13-5

Describe how a 4K-by-8-bit memory can be made of **2114**s.

SOLUTION   The solution is similar to the **2102** memory shown in Fig. 13-6, except that:

1. Only eight **2114**s arranged in a 4-row-by-2-column matrix are required.
2. The DATA-IN and DATA-OUT lines for each bit are merged to become a single bidirectional bus.

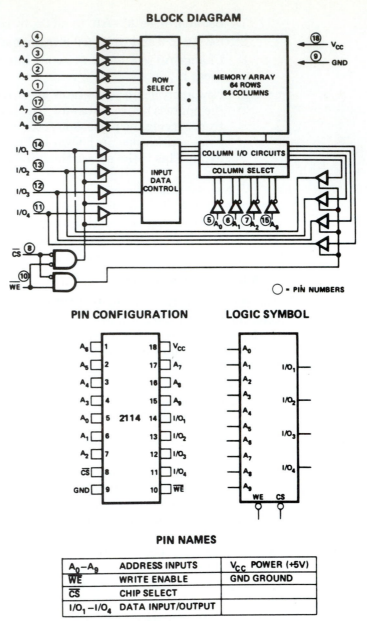

BLOCK DIAGRAM

○ = PIN NUMBERS

PIN CONFIGURATION

LOGIC SYMBOL

PIN NAMES

A_0-A_9	ADDRESS INPUTS	V_{CC} POWER (+5V)
$\overline{WE}$	WRITE ENABLE	GND GROUND
$\overline{CS}$	CHIP SELECT	
$I/O_1-I/O_4$	DATA INPUT/OUTPUT	

Figure 13-8 The **2114** static RAM. (From the *Intel Semiconductor Memory Book.* Copyright John Wiley & Sons, Inc. 1978. Reprinted by permission of John Wiley & Sons, Inc.)

13-6-4 The Intel 51256

Both the **2102** and **2114** are nearly obsolete because they contain relatively few bits. The Intel **51256**, shown in Fig. 13-10, is a more modern SRAM. It contains 256 Kbits in a 32Kx8 bit matrix. Many manufacturers are producing SRAMs in a byte-wide (8-bit) configuration because that is the most convenient size for microprocessors and other computers.[3]

[3]Some manufacturers are also producing 9-bit wide SRAMs, because the memory in the IBM-PC and compatibles is 9-bits wide, and they can be used directly in these computers.

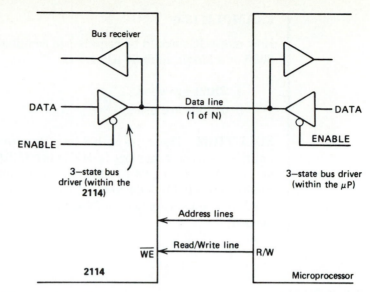

Figure 13-9 A microprocessor connected to a **2114**. (Courtesy of Motorola, Inc.)

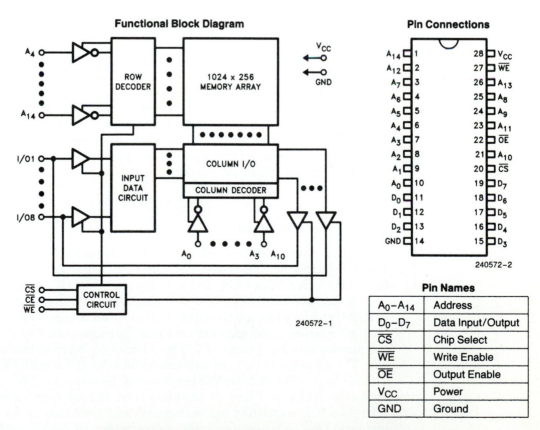

Pin Names

A_0–A_{14}	Address
D_0–D_7	Data Input/Output
$\overline{CS}$	Chip Select
$\overline{WE}$	Write Enable
$\overline{OE}$	Output Enable
V_{CC}	Power
GND	Ground

Figure 13-10 The Intel **51256** 32K × 8-bit SRAM. (Reprinted by permission of Intel Corporation, © Intel Corporation 1990.)

EXAMPLE 13-6

How many ICs would be required to completely fill the memory of an Intel **8085** or a Motorola **6800** μP if

    a. **2102**s are used.
    b. **51256**s are used.

SOLUTION   These are both 8-bit μPs with a 16-bit address bus. The maximum size of their memory is 64K bytes. If **2102**s were used, we would need 8 for each byte times 64 to fill the 64K possible addresses. Thus, 512 **2102**s would be needed in a 64 row by 8 column matrix. The **51256** is already byte-sized and can accommodate 32K addresses. Therefore, only two **51256**s are needed.

13-7   Dynamic RAMs

As previously mentioned, Dynamic RAMs (DRAMs) store their information as *charges on internal capacitors,* rather than in FFs. Unfortunately, data stored in the form of a charge on a capacitor leaks off as time progresses. Consequently, DRAMs must be *refreshed* periodically to restore the information. This requirement complicates the external circuitry of a DRAM. Refresh is discussed in section 13-7-3.

Although DRAMs are more complicated than SRAMs, because they require refresh circuitry, they dominate many applications requiring memory, and are used extensively in PCs. The reason is that capacitors require less "silicon real estate" than FFs. At any given time, DRAM ICs contain about four times as many bits as an SRAM. Of course, as the technology advances, the number of bits in both SRAMs and DRAMs increases. During the past decade, the number of bits in a DRAM has been quadrupling every three years, and it appears that this trend will continue.

13-7-1   The 4116 DRAM

The **4116** is a very small DRAM, but it is easy to understand and is presented as a prototype DRAM. The **4116** is a 16K-by-1-bit memory so it requires 14 address lines, DATA-IN, DATA-OUT, and a read/write ($\overline{\text{WRITE}}$) input.

A simplified drawing of the operation of the **4116** is shown in Fig. 13-11. The 16K (2^{14}) memory bits are stored on a *square matrix* of capacitors. There are 128 *rows of capacitors,* and each row contains 128 capacitors. The 14 address inputs are divided into two groups of 7. The 7 low-order address bits select an entire row. The contents of this row are driven into the 128 sense/refresh amplifiers. The 128 high-order address bits select one of the sense amplifiers and place that bit on the DATA-OUT line. At the same time, the information in all 128 bits is refreshed (the capacitors are recharged) and rewritten to the proper row in memory. *Thus, every read or write cycle refreshes all 128 bits in the selected row.*

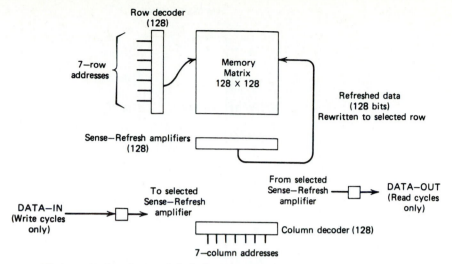

Figure 13-11 A simplified diagram of the **4116**. (Courtesy of Motorola, Inc.)

13-7-2 $\overline{\text{RAS}}$ and $\overline{\text{CAS}}$

The pin configuration for the **4116** is shown in Fig. 13-12. Notice that there are only 7 address pins on the chip. The 14 system addresses are *multiplexed* onto the chip by using the $\overline{\text{RAS}}$ (Row Address Select) and $\overline{\text{CAS}}$ (Column Address Select) signals. This saves pins and is the way most modern dynamic RAMs operate.

The timing for Read, Write, and Refresh cycles is shown in Fig. 13-13. The **4116** output is 3-stated (high impedance) when $\overline{\text{RAS}}$ and $\overline{\text{CAS}}$ are both HIGH. For a Read cycle, the following events occur:

1. The row addresses (the 7 LSBs of the 14 address bits) are put on the **4116**'s address inputs and $\overline{\text{RAS}}$ goes LOW. The **4116** latches in the row addresses.
2. After about 60 ns (see the manufacturer's specifications for the precise times), the row addresses are replaced by the column addresses.
3. $\overline{\text{CAS}}$ goes LOW and the column addresses are latched in.
4. Valid data appears at DATA-OUT after the access time measured from the start of $\overline{\text{CAS}}$ (T_{CAC} on Fig. 13-13) and remains valid until $\overline{\text{RAS}}$ and $\overline{\text{CAS}}$ go HIGH.

A write cycle is basically identical to the read cycle with the following differences:

1. The data to be written must be on the data bus before $\overline{\text{CAS}}$ goes LOW.
2. The WRITE pin must be held LOW (see Fig. 13-13 for the appropriate timing for this signal).
3. The output drivers in the **4116** are never enabled, and valid read data does not appear on the data bus.

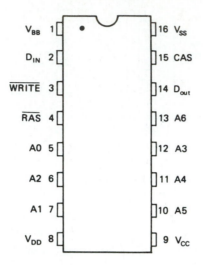

PIN ASSIGNMENT

V_{BB}	1	16	V_{SS}
D_{IN}	2	15	CAS
$\overline{WRITE}$	3	14	D_{out}
$\overline{RAS}$	4	13	A6
A0	5	12	A3
A2	6	11	A4
A1	7	10	A5
V_{DD}	8	9	V_{CC}

PIN NAMES

A0–A6   Address Inputs
$\overline{CAS}$ Column Address Strobe
D_{in} Data in
D_{out} Data out
$\overline{RAS}$ Row Address Strobe
WRITE   Read/Write Input
V_{SS} Power (−5 V)
V_{CC} Power (+ 5 V)
V_{DO} Power (+ 12 V)
V_{SS} Ground

Figure 13-12 Pin assignments for the **4116**. (Courtesy of Motorola Integrated Circuits Division.)

Looking at Fig. 13-13, the reader can see that many times are specified on the drawing. In the Memory Data Books, the manufacturers will provide drawings like this for each memory IC. They will also provide a table of all the times specified in the drawings. In the READ cycle in Fig. 13-13, for example, the most important times are:

- t_{RAS}—The minimum time $\overline{RAS}$ should be LOW.
- t_{RP}—This is the recovery time; the minimum time $\overline{RAS}$ must be high so the chip can recover before the next cycle.
- t_{RC}—The read cycle time. This is the minimum time of a read cycle. It should be about equal to $t_{RAS} + t_{RP}$.
- t_{CAS}—The minimum time $\overline{CAS}$ should be LOW.
- t_{RCD}—This is the minimum time for the RAS-to-CAS delay or the minimum time between the falling edge of $\overline{RAS}$ and the falling edge of $\overline{CAS}$.
- t_{CAC}—This is the minimum time from the start of $\overline{CAS}$ until the output data is valid. It is an access time for the DRAM measured from the start of $\overline{CAS}$.

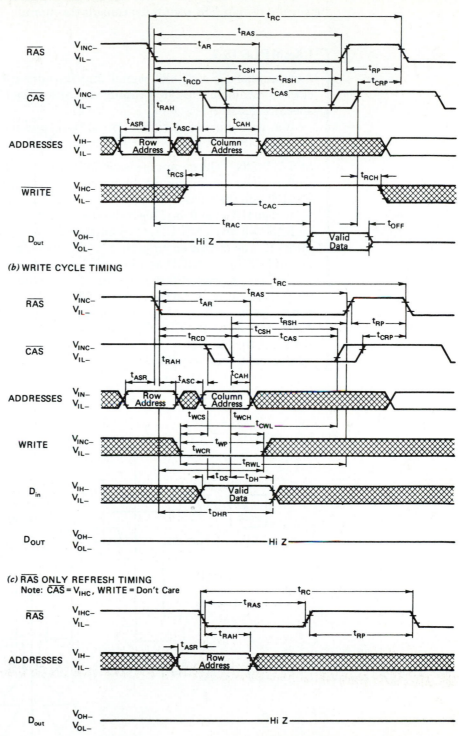

Figure 13-13 Read, write, and refresh cycle timing for a **4116**. (Courtesy of Motorola, Inc.)

There are many more times on the figure that should be taken into account for careful design. The user must consult the manufacturer's data to find them.

EXAMPLE 13-7

Assume there is a START signal that initiates a DRAM cycle and a 50 ns clock is available. Design a timing circuit for the DRAM with the following specifications:

    a. $\overline{RAS}$ must be down for 250 ns (t_{RAS}) and up for 100 ns (t_{RP}).
    b. $\overline{CAS}$ must be down for 150 ns (t_{CAS}).

SOLUTION   The circuit is shown in Fig. 13-14. The timing is similar to the shift register timing circuit discussed in section 10-7-1. The leading edge of START sets the START FF and starts shifting 1s through the shift register. Q_A will go HIGH at 0 ns. This is the start of $\overline{RAS}$. When Q_F goes HIGH, at 250 ns $\overline{RAS}$ will also go HIGH. $\overline{CAS}$ must be LOW during the last 150 ns of this interval. The circuit ensures that $\overline{CAS}$ goes LOW when Q_C goes HIGH (100 ns) and $\overline{CAS}$ goes HIGH at 250 ns.

    When Q_H goes HIGH at 350 ns, it clears the shift register and the start FF. The 100 ns between the end of $\overline{RAS}$ and the rise of Q_H is t_{RP}, the recovery time. The circuit can then respond to another START input (see Problem 13-7). Note also the SELECT output on Q_B. This occurs between the start of $\overline{RAS}$ and the start of $\overline{CAS}$. It can be used to change the addresses from row addresses to column addresses.

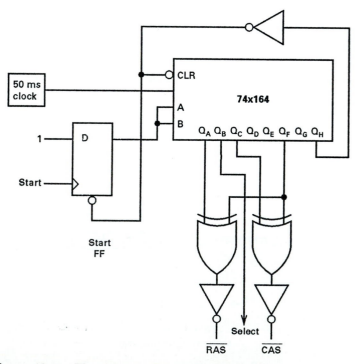

Figure 13-14   The timing circuit for Example 13-7.

13-7-3 Refresh Cycles

The charge on the capacitors in a DRAM must be refreshed periodically or it will leak off and data will be lost. For the **4116**, each location must be refreshed once every 2 ms (see the manufacturer's specifications). Fortunately, refreshing can be accomplished by reading a row of data into the Sense-Refresh Amplifiers and then rewriting the row to memory. All the capacitors in the row are refreshed during each refresh cycle.

The timing for a refresh cycle is shown in Fig. 13-13c. Note that only $\overline{\text{RAS}}$ is required for a refresh cycle. It causes all the data in the addressed row to be transferred to the Sense-Refresh Amplifiers, where it is refreshed, and then back to the row.

A **4116** must refresh all 128 rows during each 2-ms interval. This requires a 7-bit *refresh counter* to ascertain that each row is refreshed during the interval. The outputs of this counter become the memory address during each refresh cycle and the counter is incremented at the end of each cycle.

There are two modes of refresh: *burst* and *periodic*. In burst mode all 128 rows are refreshed at the beginning of each 2-ms interval; in periodic mode refresh cycles occur at approximately equal intervals.

EXAMPLE 13-8

If each refresh cycle takes 200 ns, how is the time divided between refresh and Read/Write memory cycles in:

(a) burst mode
(b) periodic mode

SOLUTION

(a) In burst mode the 128 refresh cycles require 25.6 µs. Thus, in each 2-ms interval, the first 25.6 µs are used for refresh, leaving the remaining 1974.4 µs available for Read/Write cycles.

(b) In periodic mode there is one refresh cycle every 15.6 µs. Thus, the memory periodically takes 200 ns out of each 15.6-µs interval for a refresh cycle.

13-7-4 The TMS 44C256 DRAM

The **TMS44C256** is a modern 1Mbit memory organized as 256 Kwords by 4 bits. The pinout and pin nomenclature are shown in Fig. 13-15.

The pins are, as we would expect:

- $\overline{\text{RAS}}$
- $\overline{\text{CAS}}$
- $\overline{\text{W}}$—Write Enable
- **AO–A8**—These 9 address inputs allow for 2^{18} (256K) addresses.
- **DQ1–DQ4**—These are the four data input/output lines. They output data from the DRAM during a READ cycle (W = 1), and hold the data to be written during a WRITE cycle (W = 0).

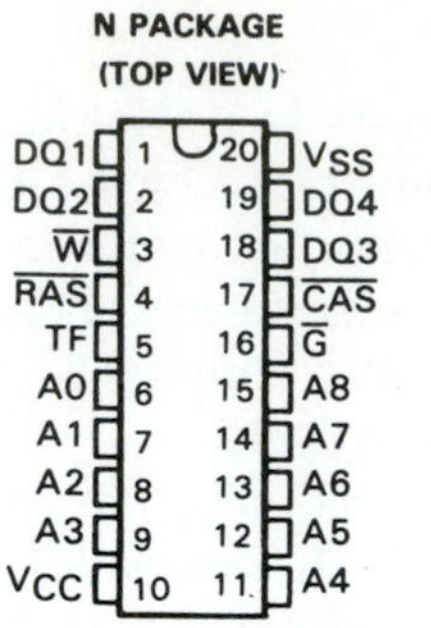

```
              N PACKAGE
             (TOP VIEW)

      DQ1 [ 1      20 ] VSS
      DQ2 [ 2      19 ] DQ4
       W̄ [ 3      18 ] DQ3
      R̄AS [ 4      17 ] C̄AS
       TF [ 5      16 ] Ḡ
       A0 [ 6      15 ] A8
       A1 [ 7      14 ] A7
       A2 [ 8      13 ] A6
       A3 [ 9      12 ] A5
      VCC [ 10     11 ] A4
```

PIN NOMENCLATURE	
A0–A8	Address Inputs
C̄AS	Column-Address Strobe
DQ1–DQ4	Data In/Data Out
Ḡ	Data-Output Enable
R̄AS	Row-Address Strobe
TF	Test Function
W̄	Write Enable
VCC	5-V Supply
VSS	Ground

Figure 13-15 The **TMS44C256** 256K × 4-bit DRAM. (Reprinted by permission of Texas Instruments.)

- **Ḡ**—This pin is used to enable the 3-state output buffers. During a normal read cycle it must be low so the output data can appear on the DQ1—DQ4 lines near the end of the cycle.
- **TF**—Test Function is used only to test the IC. It can be disconnected during normal operation.

The **TMS44C256** can retain its data for 8 ms. Thus, it requires 512 refresh cycles every 8 ms.

Figure 13-16 is the functional block diagram of the **TMS44C256**. It shows that there are two arrays of 256 rows whose address is determined at the start of R̄AS, and 512 columns whose address is determined at the start of C̄AS. The four data I/O lines go to a DATA IN register and a DATA OUT register. The operation of the DRAM is controlled by the signals at the top of the figure that go into the TIMING AND CONTROL section.

There are several ways the **TMS44C256** can be operated, but they operate predominantly as discussed in the previous paragraphs. The *MOS Memory Data Book* (published by Texas Instruments, 1988) gives three pages of timing specifications and seven pages of timing charts, similar to Fig. 13-13. They cannot be covered in detail here. The engineer should consult the manufacturer's data books before using these ICs.

13-7-5 SIMMs

SIMMs are Single In-line Memory Modules. They are a DRAM packaged with a single, long line of I/O pins instead of the more common DIP. They are being used extensively in modern PCs that use an **80386** processor. Figure 13-17 gives the pin diagram and pin nomenclature for a **TMS4256EL9**, which is a 256K by 9-bit DRAM designed for modern PCs. There are 8 bidirectional input/output

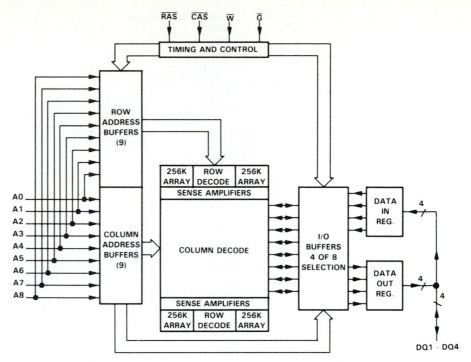

Figure 13-16   The functional block diagram of the **TMS44C256**. (Reprinted by permission of Texas Instruments.)

pins for the memory. The ninth bit, which isused for parity (see section 14-4), has separate input and output pins. At present the largest SIMMs commercially available are 4Mbx9, and cost $450, but one of these SIMMs can provide enough RAM for most PCs.

13-8   Refresh Controllers

A refresh controller is a circuit that simplifies the use of DRAMs by providing two essential functions: multiplexing and refreshing. A basic refresh controller circuit for a 256K DRAM is shown in Fig. 13-18. It consists of three major components: the address multiplexer, the refresh multiplexer, and the refresh counter.

A 256K DRAM has only 9 address pins but must respond to 2^{18} addresses, which means there must be at least 18 lines on the computer's address bus. The address multiplexer (mux) in Fig. 13-18 accepts 18 address inputs and multiplexes them down to 9 as required. It effectively divides the address bus into two groups of 9: the lower addresses and the upper addresses. The mux is controlled by the SELECT signal. At the start of $\overline{RAS}$ the 9 lower addresses are funnelled through to the output. The SELECT should then change so the 9 higher order addresses are output; they will then be on the bus at the leading edge of $\overline{CAS}$. During a nonrefresh cycle, the outputs of the address mux proceed through the refresh mux and become the address inputs to the DRAM.

The refresh controller must also make sure that *every location in the DRAM is refreshed periodically*. This is done by providing an address counter to

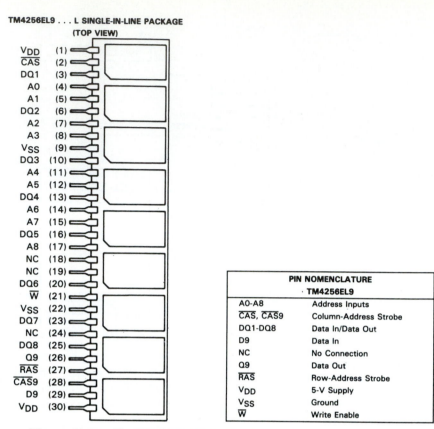

TM4256EL9 . . . L SINGLE-IN-LINE PACKAGE
(TOP VIEW)

V_{DD}	(1)
$\overline{CAS}$	(2)
DQ1	(3)
A0	(4)
A1	(5)
DQ2	(6)
A2	(7)
A3	(8)
V_{SS}	(9)
DQ3	(10)
A4	(11)
A5	(12)
DQ4	(13)
A6	(14)
A7	(15)
DQ5	(16)
A8	(17)
NC	(18)
NC	(19)
DQ6	(20)
$\overline{W}$	(21)
V_{SS}	(22)
DQ7	(23)
NC	(24)
DQ8	(25)
Q9	(26)
$\overline{RAS}$	(27)
$\overline{CAS9}$	(28)
D9	(29)
V_{DD}	(30)

PIN NOMENCLATURE	
TM4256EL9	
A0-A8	Address Inputs
$\overline{CAS}$, $\overline{CAS9}$	Column-Address Strobe
DQ1-DQ8	Data In/Data Out
D9	Data In
NC	No Connection
Q9	Data Out
$\overline{RAS}$	Row-Address Strobe
V_{DD}	5-V Supply
V_{SS}	Ground
$\overline{W}$	Write Enable

Figure 13-17 The **TMS4256EL9** SIMM. (Reprinted by permission of Texas Instruments.)

sequentially address each row. During a refresh cycle the REFRESH signal is asserted. The output of the refresh counter is then sent through the address multiplexer and onto the DRAM. The address counter is also incremented so that the next refresh cycle will refresh the next row on the DRAM.

EXAMPLE 13-9

The refresh controller of Fig. 13-18 is driving a **TMS44C256** DRAM. How many bits are in the refresh counter and how long can the interval between refresh pulses be?

SOLUTION   The refresh counter must be 9 bits long for a 256K DRAM. This means it can refresh 512 rows. According to the manufacturer's specifications, the **TMS44C256** has 512 rows and each row must be refreshed every 8 ms. If periodic refresh is used, the refresh time is 8 ms divided by 512 or 15.6 μs. There must be one refresh pulse every 15.6 μs. The rest of the time can be used for computer memory accesses.

13-8-1   A Burst Refresh Controller

Figure 13-19 is a burst refresh circuit for a 64K DRAM that requires 8 address lines and can wait 32 ms between refreshes. It uses many of the circuits and ICs discussed earlier in this book.

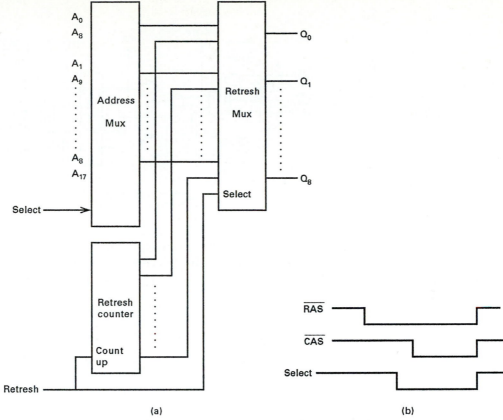

Figure 13-18 A basic refresh controller: (a) circuit (b) timing.

The **555** oscillator initiates refreshes every 32 ms by setting the **74x74** REFRESH ENABLE LATCH. The two **74HC123**s form a one-shot oscillator (see section 7-6-1). When the upper **'123** sets it provides $\overline{\text{RAS}}$, and the lower **'123** provides the recovery time (see Problem 13-10). The oscillator also increments the **74HC393** 8-bit counter that functions as the refresh counter.

The four **74HC153**s (see section 11-4-3) function as both the address mux and the refresh mux. During memory cycles, the B SELECT line is LOW and the A SELECT line chooses between the upper and lower halves of the 16-bit address bus. During refresh cycles, B is HIGH. The Q0 output of the counter is connected to both the 1C2 and 1C3 inputs to the first **74HC153**, so the A input is irrelevant. The 1Y output is the LSB of the RAS address. Similarly, output Q1 of the **74HC393** is connected to the 2C2 and 2C3 inputs to the **74HC153** and provides the next bit of the RAS address. The remaining three **74HC153**s provide the other six address bits during $\overline{\text{RAS}}$ (see Problem 13-11).

This circuit also provides a signal (EFI DISABLE LATCH) to the µP during refresh. This can be used as a READY signal, to force the µP to wait while the burst refresh is in progress.

13-8-2   The i3242 Dynamic RAM Controller

Manufacturers provide dynamic RAM controllers ICs. An example of an early, but simple, refresh controller is the **i3242**, manufactured by Intel, Inc., and shown in Fig. 13-20.

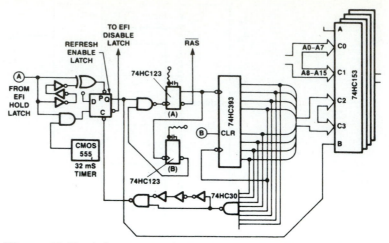

Figure 13-19 A burst refresh circuit. (Reprinted by permission of Intel Corporation, © Intel Corporation 1990.)

The **3242** has 14 address inputs and 7 address outputs. Figure 13-20 shows that its outputs are a series of AND-OR-INVERT (AOI) gates. ROW ENABLE determines whether the low-order or high-order 7 bits are fed through to the output, thus providing the correct addresses during RAS and CAS. REFRESH ENABLE disconnects the 14 system addresses and causes the internal counter to be placed on the outputs for a Refresh cycle. COUNT increments the counter.

The **3242** contains the equivalent of the address multiplexer, the refresh multiplexer, and address counter. The problems of keeping Data and Refresh requests separate and proper timing must still be handled by external circuits.

13-8-3 The 74ALS2967 Dynamic Memory Controller

The **74ALS2967** is a modern 256K dynamic memory controller available from Texas Instruments, Inc. The logic symbol for the **'2967** is shown in Fig. 13-21. The newer ICs, while becoming more powerful, also become more complex. We will attempt to explain the important features of the **'2967**.

The inputs to the **'2967**, which come from the μP and its associated circuitry, are shown in Fig. 13-21.

- **A0–A17**—These are the 18 address inputs to the memory.
- **MC1,MC2**—These are the mode inputs. Essentially they determine whether the controller is in refresh mode or normal mode.
- **RASI**—This is the RAS input to the IC. Its RAS output will be LOW when RASI is LOW. Its level is determined by timing external to the IC.
- **CASI**—Similar to RASI. It determines when the CAS output will be LOW.
- **MSEL**—This signal determines whether the internal address multiplexer selects the upper or lower half of the address bus.
- **CS**—Chip select disconnects the controller from the computer bus when HIGH.

LOGIC DIAGRAM

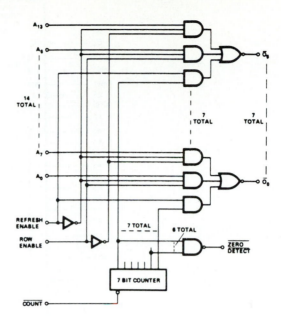

PIN CONFIGURATION

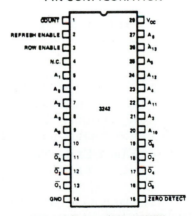

NOTE: A_0 THROUGH A_6 ARE ROW ADDRESSES.
A_7 THROUGH A_{13} ARE COLUMN ADDRESSES.

TRUTH TABLE AND DEFINITIONS

REFRESH ENABLE	ROW ENABLE	OUTPUT
H	X	REFRESH ADDRESS (FROM INTERNAL COUNTER)
L	H	ROW ADDRESS (A_0 THROUGH A_6)
L	L	COLUMN ADDRESS (A_7 THROUGH A_{13})

COUNT – ADVANCES INTERNAL REFRESH COUNTER.
ZERO DETECT – INDICATES ZERO IN THE FIRST 6
SIGNIFICANT REFRESH COUNTER
BITS (USED IN BURST REFRESH MODE)

Figure 13-20 The Intel **3242** dynamic RAM controller. (Reprinted by permission of Intel Corporation, © Intel Corporation 1990.)

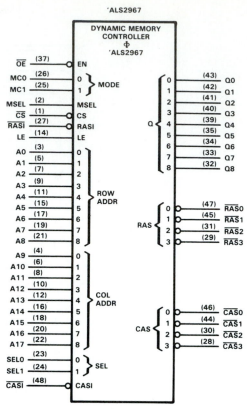

Figure 13-21 The **SN74ALS2967** DRAM memory controller. (Reprinted by permission of Texas Instruments.)

- **OE**—This signal enables the outputs when low.
- **LE**—Latch Enable. When HIGH the addresses are latched onto the controller. This signal could be connected to ALE (Address Latch Enable) on an Intel μP.
- **SEL1, SEL0**—These signals indicate which memory bank (see section 13-8) will be selected.

The outputs of the **'2967** are the 9 address inputs to the DRAM and the $\overline{RAS}$ and $\overline{CAS}$ outputs. Only one of the four $\overline{RAS}$ and $\overline{CAS}$ outputs will be active as determined by the SEL0 and SEL1 signals.

The logic diagram of a **'2967** is shown in Fig. 13-22. The refresh counter is in the upper left and the address latches are in the lower left. The multiplexer for both addresses and refresh is in the upper right. The lower right contains the bank decoding circuits for both $\overline{RAS}$ and $\overline{CAS}$.

13-9   Memory Bank Decoding

The reader may have noticed that there is no CHIP SELECT signal on a DRAM IC although there is a $\overline{CS}$ input on a DRAM controller. If the DRAM IC receives $\overline{RAS}$, it starts a memory or refresh cycle. Memories are usually arranged in

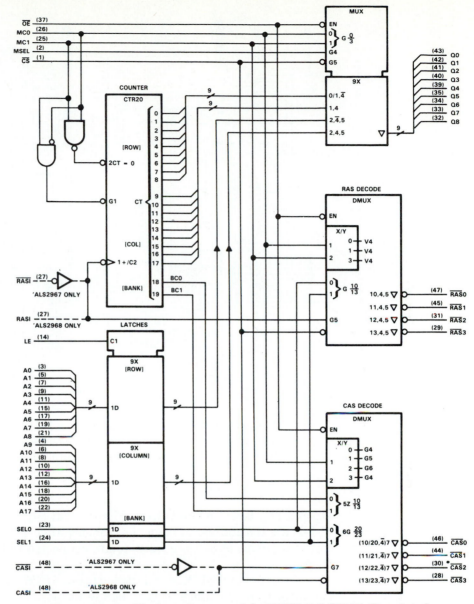

Figure 13-22 The logic diagram of the **TMS74ALS2967**. (Reprinted by permission of Texas Instruments.)

banks. Chip selection is accomplished by selecting the desired bank and sending $\overline{RAS}$ and $\overline{CAS}$ *to that bank only*. The **'2967** was set up for four banks of memory.

13-9-1 Memory on the IBM-PC

Memory bank selection and the operation of DRAMs can be best explained with reference to a specific system. We have chosen the IBM-PC as our first example.

Part of the memory of the IBM-PC is shown in Fig. 13-23. The **74LS245** in the upper left is a transceiver (see section 5-7-2) that connects the PC bus to the

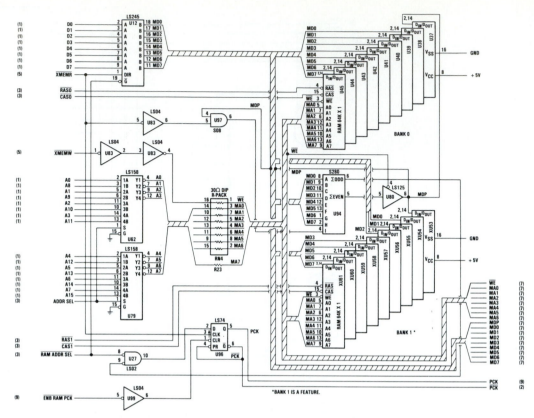

Figure 13-23 Part of the memory on an IBM PC. (Reprinted by permission from the "IBM PC Technical Reference Manual," © (1984) by International Business Machines Corporation.)

D_{in} and D_{out} lines on the DRAM. The two **74LS158** multiplexers (see section 11-3-2) are the address mux for the memory; they divide the 16-bit address bus into two groups of 8. Their outputs are connected to the address lines on the DRAMs.

The original IBM-PC was designed to use 64Kx1 DRAMs. Figure 13-23 shows two memory banks. Each bank is 64Kx9 for 64Kbytes of data (the ninth bit is strictly for parity). The upper bank is selected if $\overline{RAS0}$ and $\overline{CAS0}$ are active, the lower bank uses RAS1 and CAS1.

The IBM-PC uses an Intel **8088** µP. This limits the memory system to 1 megabyte. The system's address bus is therefore 20 bits wide to accommodate 1MB. Each 64K DRAM needs 16 address bits. Each DRAM, regardless of which bank it is in, uses the lower 16 bits of the address bus to select the address on the DRAM. The upper 4 bits on the address bus are used to select the bank. This allows the selection of up to 16 banks of memory. Of course, 16 memory banks times 64K addresses per bank equals 1MB, the capacity of the memory.

The situation described in the previous paragraph is typical. *The lower bits on the address bus select the byte on the memory IC, the upper bits are used to determine which ICs are used.*

EXAMPLE 13-10

Given the 20-bit memory address 2ABCD, explain which memory byte is selected in which bank.

SOLUTION The 16 LSBs, ABCD, go to all memory banks. The upper four address bits contain the number 2 in this example, and will select the bank. Thus, bank 2 will be selected, by providing $\overline{RAS2}$ and $\overline{CAS2}$ signals. Bank 2 is selected for all addresses between 128K and 192K.

EXAMPLE 13-11

Given a $\overline{RAS}$ input, design a circuit to provide $\overline{RAS}$ and $\overline{CAS}$, with proper timing, for four 64K banks of memory on an IBM-PC. Use the lower banks.

SOLUTION As explained in the previous paragraph, the 16 LSBs of the address bus (A0–A15) are used to select the byte on the DRAM IC. The four upper bits (A16–A19) can be used to select the bank. The memory bank is selected by choosing a particular $\overline{RAS}$ and $\overline{CAS}$. It is wise to construct a table to show the relationship between the bank select lines and the upper nibble of the address bus.

A19	A18	A17	A16	
0	0	0	0	$\overline{RAS0}$
0	0	0	1	$\overline{RAS1}$
0	0	1	0	$\overline{RAS2}$
0	0	1	1	$\overline{RAS3}$

The table shows which $\overline{RAS}$ signals are generated for each memory address.

The circuit can be implemented as shown in Fig. 13-24. It uses two **74x138** decoders (see section 11-6-1) and a **74x164** shift register for time delay (see section 10-7-3). The circuit is similar to the design used in the IBM-PC, but it is simplified. It operates as follows:

1. If A18 or A19 is HIGH, the address is not in the table, and the decoders are deselected; all the $\overline{RAS}$ outputs are HIGH and no DRAMs are selected.
2. If $\overline{RAS}$ is HIGH, the lower 4 outputs of the **74x138** are deselected and the shift register is clear.
3. When $\overline{RAS}$ goes LOW, one of the $\overline{RAS}$ signals, as determined by A16 and A17, goes LOW and sends $\overline{RAS}$ to the selected memory bank. It also starts shifting 1s through the **74x164**.
4. After the appropriate RAS-to-CAS delay, as determined by the clock and shift register, the lower decoder is enabled and provides the correct $\overline{CAS}$ pulse. Again, the SELECT pulse can be used for the address mux.

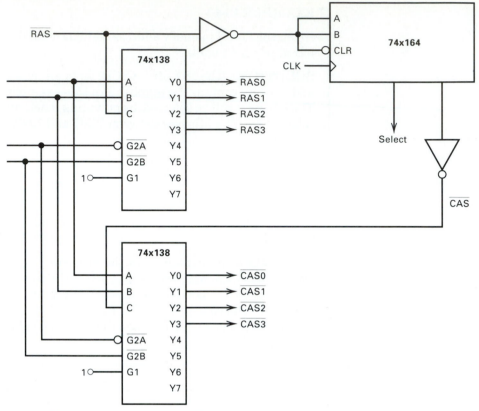

Figure 13-24 Circuit for Example 13-11.

13-9-2 DOS Decoding

The earliest versions of the DOS (Disk Operating System) program, the operating system that controls most PCs, had a 640K RAM limit. This is the limit of RAM on the IBM-PC and PC/XT. This limit seemed reasonable at the time because standard computers at that time were limited to 64K bytes of memory, so ten times that (640 K) seemed sufficient. The creators of DOS also reasoned that the **8088** μP had a 1MB limit on the address space and the upper addresses were reserved. They also felt then that 640K was infinite memory; no one could ever write a program that large. More modern PCs use **'386** and **'486** μPs that have longer address buses and can accommodate more memory.

EXAMPLE 13-12

Design, conceptually, a decoder to select the proper 64K bank in a 640K memory. It must also produce a $\overline{\text{RAMSELECT}}$ signal whenever a RAM is being addressed (RAMs are *not* being addressed if the address is higher than 640K).

SOLUTION The engineer must first realize that 640K equals ten banks of 64K. There are several approaches. One could use ordinary decoders such as two **74x138**s or a single **74154**. One might also think of PALs, but a **16L8** only has eight outputs and 11 are required.

In this example we are very lucky, however. The **74x42** (see section 11-7-4) will decode any number from 0 to 9 and produce all HIGH outputs if the input number is 10 or more. The circuit is shown in Fig. 13-25. The circuitry at the bottom is a BCD decoder. It will be LOW if the input number is 9 or less. This can be used to generate the RAMSELECT signal.

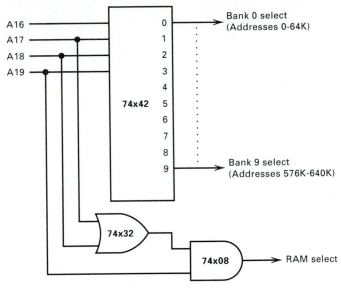

A16 — 0 → Bank 0 select (Addresses 0-64K)
A17 — 1
A18 — 2
A19 — 3
74x42   4
        5
        6
        7
        8
        9 → Bank 9 select (Addresses 576K-640K)

74x32
74x08 → RAM select

Figure 13-25   The memory bank decoder for Example 13-12.

13-9-3   Memory on the Apple II Computer

Figure 13-26 shows the memory of an Apple II Computer. It uses three banks of 16K memory with **4116** DRAMs. The figure shows that all memory banks receive the same $\overline{RAS}$ signal, from the clock generator. There is one $\overline{CAS}$ signal for each bank, however. Only the selected bank receives $\overline{CAS}$; the other banks perform a RAS-only refresh cycle. Note also that the **74LS174** FFs are clocked on the trailing edge of $\overline{RAS}$, which is when the output data can be read.

Figure 13-27 shows the timing for the Apple. Each phase of the clock consists of 7 intervals of 70 ns. $\overline{RAS}$ is down for 5 cycles ($t_{RAS} = 350$ ns) and up for two ($t_{RP} = 140$ ns). $\overline{CAS}$ is down for the last 3 cycles, giving a RAS-to-CAS delay of 140 ns. The AX signal changes between $\overline{RAS}$ and $\overline{CAS}$ and is used to control the memory mux (not shown).

13-10   Read Only Memories

Read only memories (ROMs) are prewritten in some permanent or semi-permanent form. They are not written during the course of normal device operation. ROMs are nonvolatile; the data will remain after power has been turned off. They are used to perform code conversions, table look-up, and to hold the start up programs (BIOS) in microprocessors.

The concept of a ROM is extremely simple; the user supplies an address and the ROM provides the data output of the word prewritten at that address.

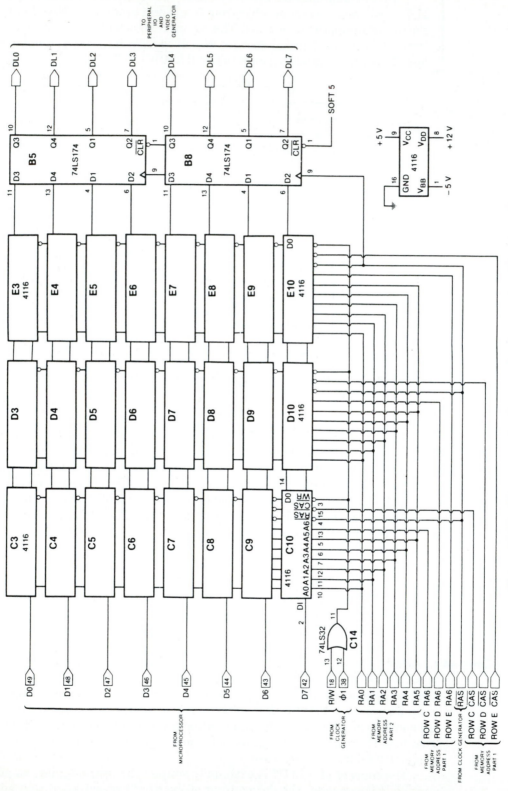

Figure 13-26 The memory banks on an Apple computer. (Courtesy of Apple Computer, Inc.)

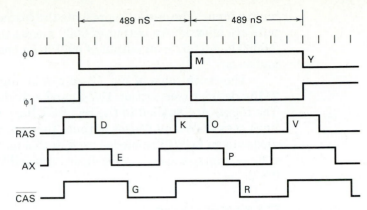

Figure 13-27 The timing of an Apple computer. (Courtesy of Apple Computer, Inc.)

As with RAMs, ROMs are organized on an *n*-word-by-*m*-bit basis. Supplying the proper address results in an m-bit output. Access time for a ROM is the time between the address input and the appearance of the resulting data word.

13-10-1 Diode Matrices

A major consideration when using a ROM is how to write it; how to get the data into the ROM. A small, simple ROM can be made from a decoder and a diode matrix. The placement of the diodes determines the output of the ROM, which can easily be changed by moving a diode.

The first step in building any ROM is to construct a truth table relating the inputs (addresses) to the required outputs. For example, consider the table of Fig. 13-28, which has a 4-bit input, a 4-bit output, and can be used to convert an angle between 0 and 90 degrees to its cosine. The angle is represented by a 4-bit

Angle in Degrees	Input in Binaty				Output (cos θ)			
0	0	0	0	0	1	1	1	1
6	0	0	0	1	1	1	1	1
12	0	0	1	0	1	1	1	1
18	0	0	1	1	1	1	1	1
24	0	1	0	0	1	1	1	0
30	0	1	0	1	1	1	0	1
36	0	1	1	0	1	1	0	0
42	0	1	1	1	1	0	1	1
48	1	0	0	0	1	0	1	0
54	1	0	0	1	1	0	0	1
60	1	0	1	0	1	0	0	0
66	1	0	1	1	0	1	1	0
72	1	1	0	0	0	1	0	0
78	1	1	0	1	0	0	1	1
84	1	1	1	0	0	0	0	1
90	1	1	1	1	0	0	0	0

Figure 13-28 A cosine conversion table.

binary input. The output represents the cosine of the angle. With 4 bits, 16 outputs are allowed. An output of 0000 means the cosine of the angle is between 0.0000 and 0.0625; an output of 0001 is for cosines between 0.0625 and 0.1250, and so on.

The truth table of Fig. 13-28 can be implemented as a ROM by using a **74154** decoder (see section 11-7-2) and a diode matrix, as shown in Fig. 13-29. The inputs are applied to the SELECT lines of the **74154**. The corresponding lines goes LOW and causes outputs that are connected to the selected line by diodes to go LOW. The diodes used in this matrix should have a small forward voltage drop so the sum of the diode drop and V_{OL} of the decoder does not exceed 0.8 V.

EXAMPLE 13-13

What output does the ROM of Fig. 13-29 produce if the input is 48 degrees?

SOLUTION   An input of 48 degrees results in a 1000 input to the SELECT lines of the **74154** and line 8 of the decoder goes LOW. Since output lines 0 and 2 are connected to line 8 via diodes, the output is 1010. This corresponds to the function table of Fig. 13-28 for the ROM, and indicates that the cosine of 48 degrees is between 0.625 and 0.6875. Actually, the cosine of 48° is 0.6691.

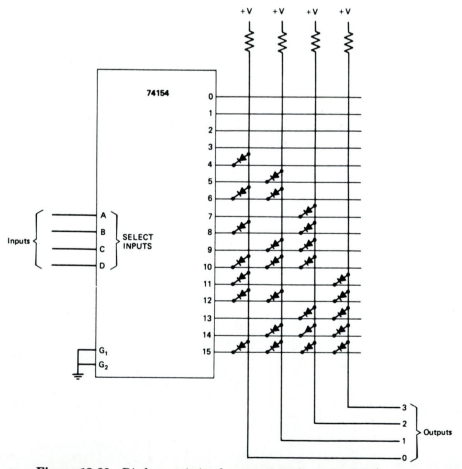

Figure 13-29   Diode matrix implementation of a cosine ROM.

The diode matrix ROM discussed in section 13-9-1 is easily changed, but it is only used for very small memories. Most ROMs used today are DIP ICs that are readily available from many manufacturers. Figure 13-30 shows the various types of ROMs available. They are divided into Mask ROMs and various types of PROMs.

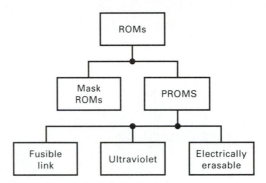

Figure 13-30   Types of ROMs.

13-11-1   Mask ROMs

Mask programmed ROMs are built by the manufacturer. The ROM is made by including or excluding a small semiconductor jumper to produce a 1 or 0 for each bit. The user must supply the manufacturer of the ROM with the code each word is to contain. Because of the custom programming involved, a *mask* is made that can be expensive. The user should be sure of the inputs before incurring the mask charge because the ROM is usually useless if a single bit is wrong. Once the mask is made, identical ROMs can be produced inexpensively.

Some ROM patterns are very commonly used and are available at low cost. The BIOS (Basic I/O System) that controls the IBM PC, and the monitor systems for µP kits such as the **SDK-85** and **SDK-86**, use mask programmed ROMs. Character generators for CRT displays, such as the **2513**, are another example of commonly used and readily available mask programmed ROMs.

Figure 13-31 shows the ROMs used to hold the BIOS on the IBM-PC. There are six 8K byte ROMs for a total of 48K. This means 13 bits of the address bus must be used to select the byte in the ROM, and 7 bits are available to select the ROM. To select any ROM IC the upper nibble of the address bus must be F (A19, A18, A17, A16 must all equal 1). This leaves A15, A14, and A13 available to determine which of the 6 ROM ICs are selected. These bits are connected to a **74LS138** decoder that provides outputs $\overline{CS0}$ through $\overline{CS7}$. These outputs are used to select the ROM.

13-11-2   Fusible Link PROMs

A PROM (Programmable Read Only Memory) is a ROM whose program is written by the user. This saves the user the mask cost and the turnaround time—the time between sending a mask program to the manufacturer and the delivery of the ROM. The user must enter the program into the PROM at the facility before it can be used.

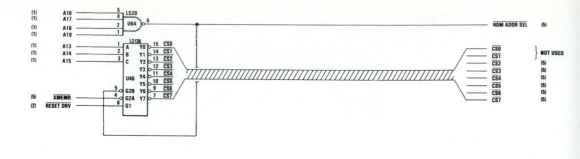

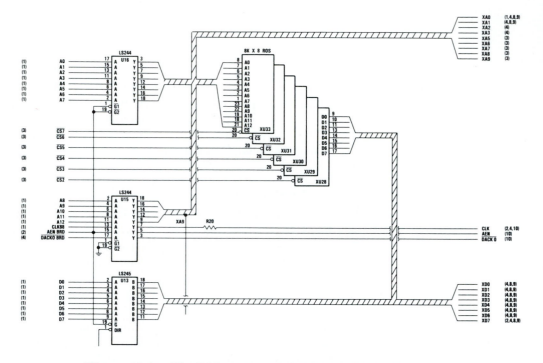

Figure 13-31 The ROM on the IBM PC (simplified). (Reprinted by permission from the "IBM PC Technical Reference Manual," © (1984) by International Business Machines Corporation.)

The fusible link PROM comes with a small nichrome or silicon fuse in each bit position. All outputs are initially 0. The user can then drive an excessive current through the fuses that blows the fuses open and changes the output to a 1. Specifications on the programming pulse (the required voltages, currents, and pulse durations) are supplied by the manufacturer. Many users will purchase a device called a PROM *programmer,* or PROM blaster. This facilitates blowing the fuses, and means the user does not have to design the necessary circuitry for blowing the fuses. The operation is similar to blowing the fuses on a PAL.

13-11-3   EPROMs

One problem with both mask and fusible line ROMs is that a single incorrect bit may make the memory useless. EPROMs (Erasable Programmable Read Only

Memories) that can be both programmed and erased at the user's facility are a solution to this problem.

The internal structure of an EPROM is shown in Fig. 13-32. EPROMs contain a *floating gate,* electrically insulated from the rest of the circuit, as part of each bit. Initially there is no charge on the floating gate, which makes it a logic 1. To write a 0, a relatively high voltage (the programming voltage) is applied to the select gate. This causes electrons to tunnel from the substrate to the floating gate, where they are trapped by the insulation. The presence of electrons on the floating gate can be sensed as a logic 0.

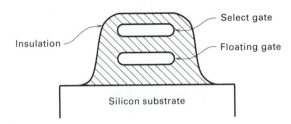

Figure 13-32   The internal structure of an EPROM (simplified).

EPROMs can be erased by exposing them to ultraviolet light. The ICs are equipped with a transparent lid to allow the ultraviolet light to enter. In normal operation this lid is often closed off by a piece of tape to prevent ambient ultraviolet light from entering and causing erasures.

Modern EPROMs are all byte-wide and are available from 2K to 128K bytes. The **TMS27C256**, whose pinout and pin nomenclature are shown in Fig. 13-33, is a typical EPROM. The C in its label indicates CMOS technology and the 256 indicates 256K bits, arranged as 32Kx8. The pin nomenclature shows that there are 8 data pins, 15 address pins (for 32K bytes), and 3 control pins. The control pins are $\bar{E}$, the chip enable, $\bar{G}$, the output enable, and V_{PP}, the programming voltage. Normally V_{PP} is at the V_{CC} level, but it must be raised to about 12.5 volts when programming the IC.

The Texas Instrument Data Book shows that the **TMS27C256** operates in several modes. We will only consider the normal, or read, mode and the write, or program, mode. The IC can be read simply by taking $\bar{E}$ LOW, enabling the chip, and taking $\bar{G}$ LOW, enabling the output. Then the byte at the specified address will appear on the output lines.

To program or write the **TMS27C256**:

1. $\bar{G}$ must be HIGH to disable the outputs. $\bar{E}$ must also be HIGH.
2. The programming voltages must be applied. V_{CC} must be raised to 6.5 V and V_{PP} must be raised to 13 V.
3. The address to be written must be placed on the address lines and the data to be written must be placed on the data lines.
4. Programming is now accomplished by pulsing $\bar{E}$ low for 100 μs. The location is then read. If it reads out the same data that was sent in, the write was successful, and the next location is programmed. If not, the $\bar{E}$ line is pulsed again for 100 μs. If the data is not valid after 10 such pulses, the IC is defective.

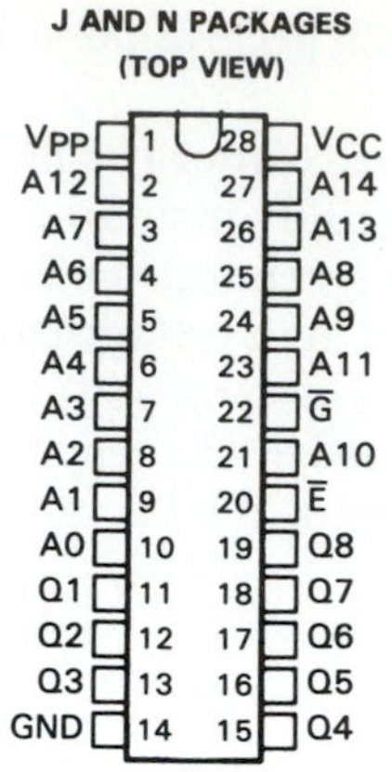

J AND N PACKAGES
(TOP VIEW)

VPP	1	28	VCC
A12	2	27	A14
A7	3	26	A13
A6	4	25	A8
A5	5	24	A9
A4	6	23	A11
A3	7	22	$\overline{G}$
A2	8	21	A10
A1	9	20	$\overline{E}$
A0	10	19	Q8
Q1	11	18	Q7
Q2	12	17	Q6
Q3	13	16	Q5
GND	14	15	Q4

PIN NOMENCLATURE	
A0-A14	Address Inputs
$\overline{E}$	Chip Enable/Power Down
$\overline{G}$	Output Enable
GND	Ground
NC	No Connection
NU	Make No External Connecction
Q1-Q8	Outputs
VCC	5-V Power Supply
VPP	12-13 V Programming Power Supply

Figure 13-33 The **TMS27C256** 32K × 8-bit EPROM. (Reprinted by permission of Texas Instruments.)

All EPROMs are programmed in essentially the same manner: by raising the programming voltage and pulsing the enable. There are minor differences in programming other EPROMs and the user should consult the manufacturer's manuals for details.

13-11-4 EEPROMS

EEPROMs (Electrically Erasable Programmable Read Only Memories) are memories that have an internal charge pump to produce the programming voltages. They can be programmed while remaining in the circuit, instead of having to be taken out and placed in a programmer.

The **68HC11** is a Motorola μP with a small internal EEPROM, which can be used to retain data or short programs. The EEPROM can be read with normal access times, but it requires 10 ms to write data into the EEPROM. Thus, the EEPROM must be programmed *off-line;* it cannot be programmed while the computer is running.

SUMMARY

This chapter started with an introduction to memory concepts, such as reading, writing, the MAR, and the MDR. The rest of the chapter concentrated on semiconductor memories. Memory configurations and memory systems were discussed.

Static RAMs were introduced first. This was followed by a section on dynamic RAMs, including the use of the dynamic RAM controller. Finally, the various types of ROMs and PROMs that are currently used were presented.

GLOSSARY

Access time: The time required to read valid data after the addresses and chip enable are firm.

Bipolar: Circuits that are essentially TTL.

Cycle time: The time required to write data into a static RAM (or to read and rewrite a dynamic RAM).

Dynamic memory: A memory where the data is stored on capacitors, and must be refreshed periodically.

EEPROM: Electrically Erasable PROM.

EPROM: Erasable Programmable ROM.

Masking: The process of building a custom ROM.

PROM: Programmable read only memory, a ROM that can be programmed by the user.

RAM: Random access memory.

Refresh: The act of restoring the data on the capacitors within the memory.

ROM: Read only memory.

Scratch-pad memory: A small memory generally less than 256 bits.

Semiconductor memory: A memory whose basic element is a semiconductor FF or gate.

Static memory: A memory where the data is stored in FFs.

PROBLEMS

Section 13-5.

13-1. A memory module is 16K by 32 bits. Determine:
   (a) How many bits are in the MAR.
   (b) How many bits are in the MDR.

13-2. A PC is advertised as having a 4MB RAM. Precisely how many bytes and how many bits does this memory contain?

Section 13-6.

13-3. An 8K-by-12-bit memory is made up of **2102**s.
   (a) How many data inputs are required?
   (b) How many data outputs are required?
   (c) How many address bits are required?
   (d) How many read/write lines are required?
   (e) How many different chip select lines are required?
   (f) Draw that portion of the circuit that selects particular chip select lines.

13-4. Repeat Problem 13-3 using **2114**s.

13-5. Devise a way to measure access time with respect to $\overline{\text{CS}}$.

13-6. Design a 256Kx16 memory using **51256**s.

Section 13-7.

13-7. In the circuit of Fig. 13-14, two START pulses arrive within 150 ns. Explain how the circuit locks out the second pulse, so that it is ineffective.

13-8. Design a 1 Mbyte DRAM using **TMS44256** DRAM ICs. Show the interface between the μP and the memory.

13-9. Figure P13-9 shows the pinouts for three Motorola RAMs. For each memory IC determine:
  (a) whether it is a dynamic or static RAM.
  (b) what is the number of words and the number of bits/word.

Section 13-8.

13-10. Assume a dynamic RAM controller exists for a 64Kx1 dynamic RAM with multiplexed addresses.
  (a) How many input addresses are required? Where do they come from?
  (b) How many output addresses are on the chip? Where do they go?
  (c) What are the address outputs during $\overline{RAS}$?
  (d) What are the address outputs during $\overline{CAS}$?
  (e) What are the address outputs during a refresh cycle?

13-11. If the **74HC123**s in Fig. 13-19 use a 10 K resistor, find the capacitors for each if t_{RAS} is 100 ns and t_{RP} is 50 ns.

13-12. In the circuit of Fig. 13-19, show exactly where output Q5 of the **74HC393** counter goes.

13-13. Figure P13-13 shows an array of **2104** memories. Each **2104** is a 4K-by-1-bit dynamic RAM.
  (a) How many **2104**s are in the 8K-by-8-bit array?
  (b) **8205** is a decoder. What memory addresses cause it to select $\overline{CSA}$ and $\overline{CSB}$?
  (c) What is the function of the **8216**s?
  (d) What is the function of FFs 6 and 7? (*Hint:* see section 6-13-6).
  (e) The **3242** has no connections to inputs A6 or A13. Why?
  (f) Draw a timing chart showing the response of the circuit to a $\overline{MEMW}$ request.

  Show $\overline{RAS}$, $\overline{CAS}$, $\overline{WE}$ and the outputs of each of the 7 FFs.

13-14. A **74ALS2967** is used as a controller for a 64K memory bus that uses 64K DRAMs. Show how the system address lines are connected to the refresh controller.

13-15. Design a 1Mbyte DRAM memory system using 256 Kbit ICs. Include refresh. Use ICs discussed in this chapter. Draw the interface between the μP and the memory.

Section 13-9.

13-16. Describe, as far as possible, the function of each IC in Fig. 13-23.

13-17. To make a maximum DOS memory, 640 Kbytes for an IBM PC, you need
  18 41256
  4 41464
  2 4164-120

  Show how to set up this memory, and how to decode the address bus to select each memory IC in your system. See table P13-17.

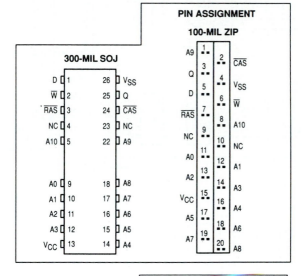

PIN NAMES

A0–A10 . . . . Address Input
D . . . . Data Input
Q . . . . Data Output
$\overline{W}$ . . . . Read/Write Enable
$\overline{RAS}$ . . . . Row Address Strobe
$\overline{CAS}$ . . . . Column Address Strobe
V_{CC} . . . . Power Supply (+ 5 V)
V_{SS} . . . . Ground
NC . . . . No Connection

PIN ASSIGNMENT

100-MIL ZIP

300-MIL SOJ

D	1	26	V_{SS}
$\overline{W}$	2	25	Q
$\overline{RAS}$	3	24	$\overline{CAS}$
NC	4	23	NC
A10	5	22	A9
A0	9	18	A8
A1	10	17	A7
A2	11	16	A6
A3	12	15	A5
V_{CC}	13	14	A4

PIN NAMES

A0–A9 . . . . Address Input
DQ0–DQ3 . . . . Data Input/Output
$\overline{G}$ . . . . Output Enable
$\overline{W}$ . . . . Read/Write Enable
$\overline{RAS}$ . . . . Row Address Strobe
$\overline{CAS}$ . . . . Column Address Strobe
V_{CC} . . . . Power Supply (+ 5 V)
V_{SS} . . . . Ground

PIN ASSIGNMENT

NC	1	32	V_{CC}
NC	2	31	A14
A8	3	30	E2
A7	4	29	$\overline{W}$
A6	5	28	A13
A5	6	27	A9
A4	7	26	A10
A3	8	25	A11
A2	9	24	$\overline{G}$
A1	10	23	A12
A0	11	22	$\overline{E1}$
DQ0	12	21	DQ8
DQ1	13	20	DQ7
DQ2	14	19	DQ6
DQ3	15	18	DQ5
V_{SS}	16	17	DQ4

PIN NAMES

A0—A14 . . . . Address Input
DQ0—DQ8 . . . . . . Data Input / Output
$\overline{W}$ . . . . Write Enable
G . . . . Output Enable
$\overline{E1}$, E2 . . . . Chip Enable
NC . . . . No Connection
V_{CC} . . . . Power Supply (+ 5 V)
V_{SS} . . . . Ground

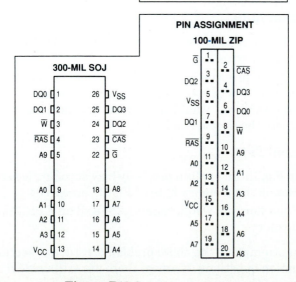

PIN ASSIGNMENT

100-MIL ZIP

300-MIL SOJ

DQ0	1	26	V_{SS}
DQ1	2	25	DQ3
$\overline{W}$	3	24	DQ2
$\overline{RAS}$	4	23	$\overline{CAS}$
A9	5	22	$\overline{G}$
A0	9	18	A8
A1	10	17	A7
A2	11	16	A6
A3	12	15	A5
V_{CC}	13	14	A4

Figure P13-9

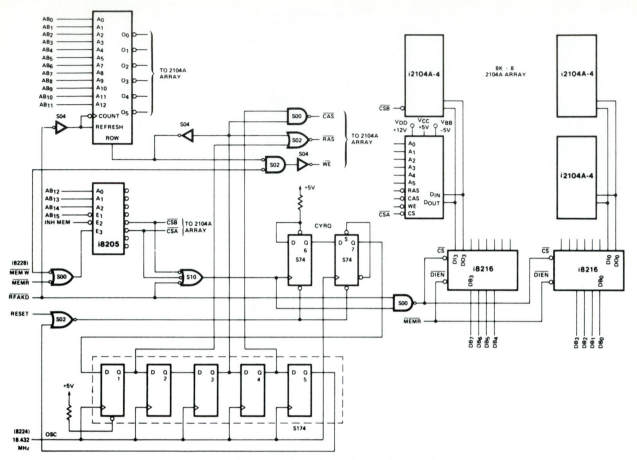

Figure P13-13 An 8K-word × 8-bit synchronous refresh memory system using the Intel® **2104A** dynamic RAM. (From the *Intel Semiconductor Memory Book*. Copyright John Wiley & Sons, Inc. 1978. Reprinted by permission of John Wiley & Sons, Inc.)

Section 13-10.

13-18. Why are diodes necessary in Fig. 13-29? (Why can't the connections be made by wire?)

13-19. Explain how you would design a diode matrix ROM to drive a common cathode 7-segment display for a HEX output. Specifically state what hardware you would use. Draw the diode configuration for the line where the inputs are 1 1 0 0.

Section 13-11.

13-20. In Fig. 13-31, explain precisely how decoding is achieved in the IBM-PC. How is the memory byte at FC123 accessed?

13-21. Show how you would connect 256Kx8 bit EPROMs to form a 1Mbx16 bit memory.

After attempting the above problems, return to section 13-2 and review the self-evaluation questions. If any of them are still unclear, reread the appropriate sections of the text to find the answers.

Add-On RAM and IC Insertion/Extraction Tools for PC/XT/AT Compatible Kits

✱ RAMs subject to frequent price changes

TABLE P13-17

Part No.	Description	Price	Part No.	Description	Price
4164-120✱	65,536 × 1 (120ns)	$3.75	41256-150✱	262,144 × 1 (150ns)	$12.95
4164-150✱	65,536 × 1 (150ns)	$3.25	41464-12✱	65,536 × 4 (120ns)	$10.95
41256-120✱	262,144 × 1 (120ns)	$13.95	K1416	14-20 Pin IC Insertion/	$6.95
				Extraction Tools	

14

Exclusive OR Circuits

14-1   Instructional Objectives

This chapter explores the use of the EXCLUSIVE OR (XOR) gate, and ICs based on XOR gates, in building comparators, parity checkers, and code converters. After reading it, the student should be able to:

1. Build comparison circuits using XOR gates or **74x85**s.
2. Build parity checkers and generators using XOR gates.
3. Build parity checkers and generators using **74x280**s.
4. Parity check long words by cascading **74x280**s.
5. Design and build a Gray code-to-binary converter.
6. Design an LRC generator and checker.

14-2   Self-Evaluation Questions

Watch for the answers to the following questions as you read the chapter. They should help you to understand the material presented.

1. Explain the differences between XOR and equality gates. What is their relationship?
2. What is the advantage of a wire-AND output for a comparison circuit?

3. Why is parity useful?

4. Explain the difference between parity generation and parity checking.

5. Why does parity generation always add one bit to the length of a word?

6. How can a single **74x280** generate a tenth odd parity bit if a 9-bit word is supplied?

7. Give one advantage and one disadvantage of the Gray code compared to the binary code.

8. How are XOR gates used to generate a.c. for LCDs?

14-3 Comparison Circuits

The EXCLUSIVE OR (XOR) gate was introduced in section 5-10. Some of the circuits that use XOR gates are described in this chapter. This section considers the use of XOR gates and other circuits to compare two binary numbers and take some action if they are equal. A comparison circuit is used in a logic analyzer (see section 19-8-5), for example, to compare the incoming data with a particular data pattern and generate a trigger when they match. A comparison circuit is digital equivalent of a COMPARE instruction in a microprocessor.

14-3-1 Comparison Circuits Using XOR Gates

The use of discrete XOR gates to build a 4-bit comparator was discussed in section 5-10 (see Example 5-12). To briefly review, the single XOR gate can only have two inputs. Its output is HIGH if the two inputs are unequal and LOW if the two inputs are equal. (Both are 0 or both are 1.)

The **74x86** quad XOR is currently the most popular XOR IC. The comparator of Fig. 5-34 was designed by making a bit-by-bit comparison of the registers, using **74x86** gates. A LOW output indicated that both inputs to the gate were equal. *If the outputs of all the **74x86** gates were LOW, it indicated that each pair of bits in the registers were equal and, therefore, the numbers in both registers were equal.*

14-3-2 The EXCLUSIVE NOR Gate

The EXCLUSIVE NOR or EQUALITY gate has two inputs and produces a HIGH output whenever the two inputs are equal. The symbol $\odot$ is used to indicate equality. The basic equation is:

$$Y = A \odot B = AB + \bar{A}\bar{B} \qquad (14\text{-}1)$$

which indicates Y is 1 if the two inputs, A and B, are equal (both 0 or both 1).

The **74LS266** is a QUAD X-NOR gate with open-collector outputs so that many bits can be wire-ANDed together for a multiple bit comparison. It is a low power Schottky device and is not as commonly used or as readily available as the more popular **74x86** XOR.

EXAMPLE 14-1

Show that an equality gate can be constructed by inverting the output of a **74x86** as in Fig. 14-1.

SOLUTION *Proof 1. DeMorgan's Theorem*
The inverter complements the function by DeMorgan's theorem:

$$Y = \overline{(A \oplus B)}$$
$$= \overline{(A\bar{B} + B\bar{A})}$$
$$= (\bar{A} + B)(\bar{B} + A)$$
$$= \bar{A}\bar{B} + AB \text{ or } A \odot B$$

Proof 2. Pure Reasoning
With two inputs, only two possibilities exist: either the inputs are equal or they are not. Since the XOR produces a HIGH output when the two inputs are unequal, its complement must produce a HIGH output when the inputs *are* equal. Note that if the inverter had been placed on one of the inputs to the **74x86**, instead of the output, the result would have been the same (see Problem 5-1).

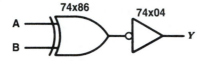

Figure 14-1 An equality gate made by inverting the output of an XOR gate. Y is HIGH only if A = B.

EXAMPLE 14-2

Given two *n*-bit words, A and B, design a comparison circuit using **74LS266** EQUALITY gates to produce a HIGH output when the two words are equal.

SOLUTION The solution is shown in Fig. 14-2. Note the symbol for the EQUALITY gate. Each pair of bits is compared by a **74LS266** gate and the outputs are wire-ANDed together. If the inputs to any gate are unequal, the output of that gate is 0, and it causes the wire-ANDed output to go LOW. When each pair of inputs is equal, all the **74LS266** gates produce a HIGH output, allowing the final output to remain HIGH. The wire-ANDing feature of the **74LS266** simplifies the comparison circuit (see Problem 14-2).

14-3-3 The 74x85 4-Bit Comparator

The **74x85** is an MSI IC specifically designed to compare two binary numbers. The two 4-bit numbers to be compared are designated as $A_3A_2A_1A_0$ and $B_3B_2B_1B_0$. The function of the **74x85** is to determine if the two numbers are equal and, if not, which is larger. The pin layout and function table are shown in Fig. 14-3. The inputs are:

1. Four A inputs, $A_3A_2A_1A_0$.
2. Four B inputs, $B_3B_2B_1B_0$.
3. Three cascading inputs: A > B, A < B, and A = B.

The function of the IC is to compare two 4-bit numbers, A and B. Three outputs are provided, A > B, A < B, and A = B, to give the result of the comparison.

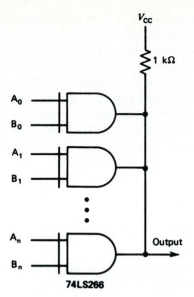

Figure 14-2 An *n*-bit comparator using **74LS266** equality (EXCLUSIVE-NOR) gates.

74LS266

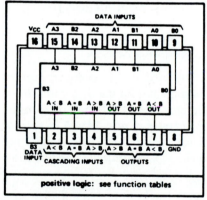

(a) Pin Layout

	COMPARING INPUTS				CASCADING INPUTS			OUTPUTS		
A3, B3	A2, B2	A1, B1	A0, B0	A > B	A < B	A = B	A > B	A < B	A = B
A3 > B3	X	X	X	X	X	X	H	L	L
A3 < B3	X	X	X	X	X	X	L	H	L
A3 = B3	A2 > B2	X	X	X	X	X	H	L	L
A3 = B3	A2 < B2	X	X	X	X	X	L	H	L
A3 = B2	A2 = B2	A1 > B1	X	X	X	X	H	L	L
A3 = B3	A2 = B2	A1 < B1	X	X	X	X	L	H	L
A3 = B3	A2 = B2	A1 = B1	A0 > B0	X	X	X	H	L	L
A3 = B3	A2 = B2	A1 = B1	A0 < B0	X	X	X	L	H	L
A3 = B3	A2 = B2	A1 = B1	A0 = B0	H	L	L	H	L	L
A3 = B3	A2 = B2	A1 = B1	A0 = B0	L	H	L	L	H	L
A3 = B3	A2 = B2	A1 = B1	A0 = B0	L	L	H	L	L	H
A3 = B3	A2 = B2	A1 = B1	A0 = B0	X	X	H	L	L	H
A3 = B3	A2 = B2	A1 = B1	A0 = B0	H	H	L	L	L	L
A3 = B3	A2 = B2	A1 = B1	A0 = B0	L	L	L	H	H	L

(b) Function table

Figure 14-3 The **74x85** 4-bit comparator. (From the *TTL Logic Data Book,* 1988, Texas Instruments, Inc. Courtesy of Texas Instruments.)

The function table shows how the **74x85** operates. The comparison starts with the most significant bits, A_3 and B_3. If they are unequal, the comparison is already determined (i.e., if $A_3 = 1$ and $B_3 = 0$, then A is larger than B *regardless* of the values of the less significant bits). If $A_3 = B_3$ (A_3 and B_3 are either both 0 or both 1), the results of the comparison depend on A_2 and B_2. If these are also equal, the comparison depends on A_1 and B_1 and finally on A_0 and B_0. If all A and B bits, respectively, are equal, the two numbers are equal and the output $A = B$ should be HIGH. This occurs if the cascading inputs, $A > B$ and $A < B$, are grounded and the $A = B$ input is tied HIGH, as shown by the eleventh line of the function table.

EXAMPLE 14-3

Design a 4-bit comparator using a single **74x85**. Show exactly how it works if $A = 0100$ and $B = 0110$.

SOLUTION The solution is shown in Fig. 14-4. Note that the cascading inputs are connected to produce a HIGH on the $A = B$ output if the two inputs are equal.

For the given numbers, $A_3 = B_3$, $A_2 = B_2$, and $A_1 < B_1$. Therefore, the sixth line of the function table is used. A_0, B_0, and the cascading inputs do not matter. The $A < B$ output goes HIGH, which is correct since $A = 4$ and $B = 6$.

14-3-4   Cascading 74x85s

74x85s can be cascaded to compare numbers of any length. Figure 14-5 shows how two **74x85**s can be used to compare two 8-bit numbers (X_7–X_0 to Y_7–Y_0). Numbers longer than 4 bits can be compared by connecting the $A > B$, $A = B$, and $A < B$ outputs of one **74x85** to the corresponding inputs of the *next more significant* set of four bits. The output must be taken from the **74x85** whose inputs are the *most significant bits* of the comparison.

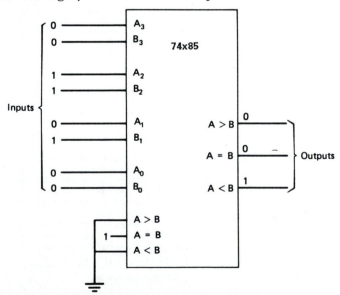

Figure 14-4 A **74x85** comparing 4 to 6 as specified in Example 14-3.

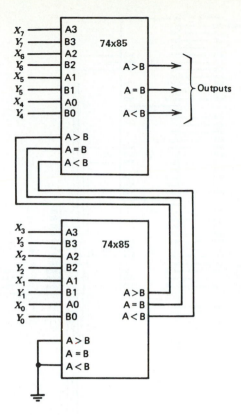

Figure 14-5 Cascading two **74x85**s to compare 8-bit numbers.

EXAMPLE 14-4

Design a circuit to compare an 8-bit number to $(195)_{10}$.

SOLUTION $(195)_{10} = 11000011$. This is an 8-bit number so the circuit of Fig. 14-5 can be used. If the Y inputs (Y_7-Y_0) are connected to 11000011, respectively, and the unknown number is brought into the X inputs, the **74x85**s will compare the numbers and produce the proper output.

A clever circuit for comparing two 24-bit words is shown in Fig. 14-6. Although this circuit does not reduce the IC count, it is faster. When comparing two 24-bit words, an extension of the circuit of Fig. 14-5 would require six **74x85**s and six **74x85** delays; whereas the circuit of Fig. 14-6 would also require six **74x85**s, but only take two gate delays to make the comparison.

14-4 Parity Checking and Generation

Parity is the addition of a bit to a binary word, to help ensure the integrity of the data. The concept is best made clear by an example, so consider the writing and reading of a magnetic tape. Typically data are written on the tape by a set of *write heads*, which magnetize the tape in accordance with the bits being written. Data is read by the *read heads* that detect the magnetization of the tape and convert it to 1s and 0s. Data is written on tape in character format. If there

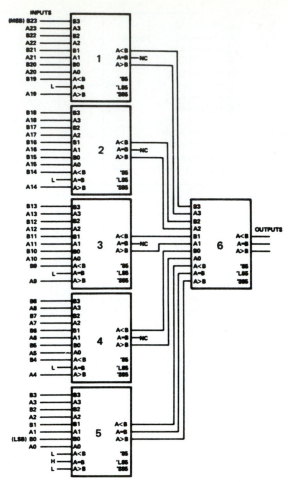

Figure 14-6 Cascading **74x85**s to produce a 24-bit comparator. (From the *TTL Logic Data Book,* 1988, Texas Instruments, Inc. Courtesy of Texas Instruments.)

are eight bits to a character, eight write heads are required and all eight bits are written simultaneously across the tape. The tape continues on and the next character is written. Character format is illustrated in Fig. 14-7.

Often data is written on a tape and then the tape is stored until needed, perhaps as long as several years. Parity is a means of checking the data so that we may be confident *the data read is the same as the data originally written* to the tape.

To ensure the integrity of the data, a *parity bit* is added to each character and is written on the tape along with the bits that determine the character. There are two types of parity: odd and even. To *generate odd parity, a bit is added to each character so that the number of 1s in the character is odd.* If, for example, a character, as supplied from a computer or other device, consists of 7 bits, an eighth bit is added to make the number of 1s in the 8-bit character odd. This 8-bit, odd parity character is then written to the tape.

Even parity is the complement of odd parity. If even parity is used, the number of 1s in each character is even. This process of adding a bit to make the parity either even or odd is called *parity generation.*

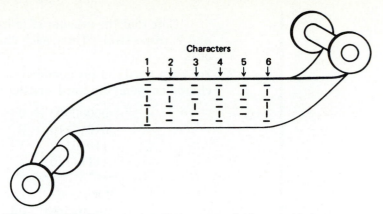

Characters

1 2 3 4 5 6

Figure 14-7 Characters written on a reel of magnetic tape.

The IBM-PC uses parity in its memory. It operates on bytes (8-bits), but it has a 9-bit memory. During write cycles, 8 bits are placed on the data bus by the μP, and circuits within the PC generate a ninth bit for odd parity. Whenever the memory is read, each word is checked for odd parity. This verifies the integrity of the data. If a word is read with even parity, it generally indicates the memory is faulty.

EXAMPLE 14-5

Characters A, B, C, and D are received from the **8088** μP in an IBM-PC. Generate a ninth bit to maintain:

1. Odd parity.
2. Even parity.

A	00000000
B	01010110
C	11000100
D	11111110

SOLUTION

1. Characters A and B, as received, contain an even number of 1s. Character A contains no 1s and character B contains four 1s. Characters C and D contain an odd number of 1s (3 and 7, respectively). To generate odd parity, 1s are added to characters A and B and 0s are added to characters C and D. The 9-bit words written on the tape look like this:

00000000	1
01010110	1
11000100	0
11111110	0

Original characters   Parity bit

Note that the number of 1s in each 8-bit character are 1, 5, 3, and 7, respectively. Thus, each character contains an odd number of 1s.

2. For even parity, a bit is added to make the number of 1s in each character even. The word written to tape looks like this:

$$
\begin{array}{ll}
00000000 & 0 \\
01010110 & 0 \\
11000100 & 1 \\
11111110 & 1
\end{array}
$$

$$
\underbrace{} \quad \underbrace{}
$$

Original   Parity
characters   bit

Note that the number of 1s in each character are now 0, 4, 4, and 8, respectively. Each character contains an even number of 1s.

14-4-1   The Parity Karnaugh Map

Karnaugh maps (see Chapter 3) can be used to help design parity generator and parity checker circuits. Consider the function $f(A,B,C,D) = \Sigma(1,2,4,7,8,11,13,14)$. Its Karnaugh map is shown in Fig. 14-8 and resembles a checkerboard. No 1s can be combined into subcubes. The function is:

$$
\begin{aligned}
f(A,B,C,D) = \bar{A}\bar{B}\bar{C}D + \bar{A}\bar{B}C\bar{D} + \bar{A}B\bar{C}\bar{D} + \bar{A}BCD + A\bar{B}\bar{C}\bar{D} \\
+ A\bar{B}CD + AB\bar{C}D + ABC\bar{D}
\end{aligned}
\tag{14-2}
$$

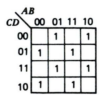

Figure 14-8   Karnaugh map for $f(A,B,C,D)$ $= \Sigma(1,2,4,7,8,11,13,14)$.

It requires eight 4-input NAND gates and an 8-input NAND gate if NAND implementation is used.

Fortunately, there is a better way to implement Eq. (14-2). Functions that produce *checkerboard* Karnaugh maps are easily implemented using XOR gates. Equation (14-2) reduces to:

$$
f(A,B,C,D) = A \oplus B \oplus C \oplus D
$$

as shown in Appendix C, and the circuit to implement it is shown in Fig. 14-9.

Considering $f(A,B,C,D)$ further, we discover that the number of 1s in each term is odd. This leads to the conclusion that there is a direct relationship between parity circuits and XOR gates. It can be shown (Appendix D) that if *all* the outputs of a register are XORed together, the output of the XOR circuit is HIGH *only* if the number of 1s in the input word is odd.

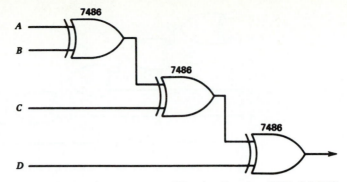

Figure 14-9   Implementation of Eq. (14-2) using XOR GATES.

14-4-2   Parity Checking Using XOR Gates

When a word is read out of memory, or a character is read from a disk, it should have the proper parity. *Parity checking* is the process of examining all n bits of a word to determine if the number of 1s in the n bits is odd or even, and reporting an error if the parity is wrong.

EXAMPLE 14-6

Design an odd parity checker for 8-bit words.

SOLUTION   An 8-bit parity checker can be built by XORing the 8 inputs together. Two circuits to accomplish this are shown in Fig. 14-10. In each case the output is HIGH *only* if there is an *odd* number of 1s among the 8 inputs. Therefore, any LOW output indicates an error because we are checking for odd parity.

Although *both* circuits of Fig. 14-10 contain the *same* number of gates, Fig. 14-10b is more commonly used. It is faster because it contains only 3-gate delays, whereas Fig. 14-10a has 7-gate delays.

14-4-3   Parity Generation Using XOR Gates

Parity generation involves adding an extra bit to an n-bit word in order to produce the proper parity in the $(n + 1)$-bit word. The concept was explained in section 14-4 and Example 14-5. A circuit to generate the proper parity bit can be built using XOR gates. If, for example, *odd* parity is *required,* the original n-bit word is checked for odd parity. If the parity of the n bits is odd, a 0 is written in the $n + 1$ bit (the parity bit) and the *odd* number of 1s in the word is still preserved. If the parity of the n bits is even, however, a 1 is written as the parity bit so that the parity of the $(n + 1)$-bit word is odd.

EXAMPLE 14-7

(a) Design a parity generating circuit to add an eighth bit to a 7-bit word so that the word has odd parity.

(b) Show how it operates if the 7-bit input word is 1101111.

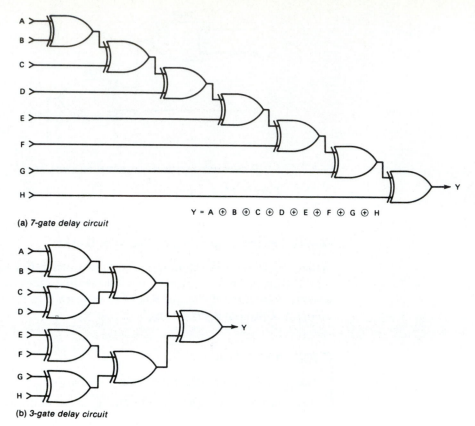

$$Y = A \oplus B \oplus C \oplus D \oplus E \oplus F \oplus G \oplus H$$

(a) *7-gate delay circuit*

(b) *3-gate delay circuit*

Figure 14-10   Two ways of using XOR gates as parity checker/generators. (From Morris and Miller, *Designing with TTL Circuits,* copyright 1971. Courtesy of Texas Instruments, Inc.)

SOLUTION

(a) The original 7 bits are XORed together to form a 7-bit odd parity checker. If the parity checker output is HIGH, the eighth bit should be a 0 because there are an odd number of 1s in the original 7-bit word, but if the parity checker output is LOW, indicating even parity, the eighth bit must be a 1 to cause odd parity. Therefore, *the eighth bit is generated by inverting the output of the 7-bit parity checker.*

(b) The parity generating circuit is similar to the **parity checker** of Fig. 14-10, and is shown in Fig. 14-11. The 1s and 0s specified in the example are also written on the figure. Note that the 8-bit output consists of the 7 input bits plus bit H, which causes the parity of the 8-bit output to be odd.

14-4-4   Parity Checking in Microprocessors

Some microprocessors (the **8080**, **8085**, and **Z80**) check the parity of an 8-bit word after each arithmetic operation. They set a *parity bit* in their *condition code register* to indicate whether the parity of the word is even or odd. The par-

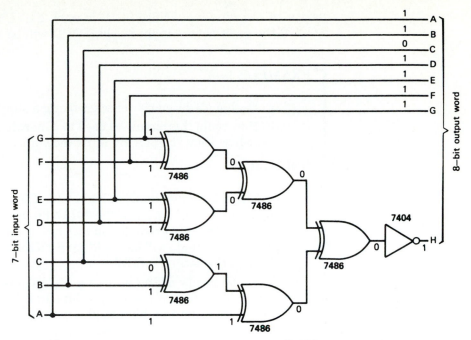

Figure 14-11 An 8-bit parity generator. All XORs are **74x86**s.

ity bit, or flag, can then be used to control conditional jump instructions so that the proper action is taken depending on the parity of the word.

14-5   Parity Checking and Generation Using the 74x280

The **74x280** is a modern parity checker/generator IC.[1] Its pinout and logic diagram is shown in Fig. 14-12. It has nine inputs (A–I) and two outputs, an ΣODD output and an ΣEVEN output. The outputs are always complementary. If the

FUNCTION TABLE

NUMBER OF INPUTS A THRU I THAT ARE HIGH	OUTPUTS	
	Σ EVEN	Σ ODD
0, 2, 4, 6, 8	H	L
1, 3, 5, 7, 9	L	H

H = high level, L = low level

logic symbol[†]

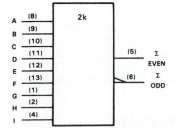

Figure 14-12 The function table and logic symbol for the **74x280**. (Reprinted by permission of Texas Instruments.)

[1]The **'280** is a modern IC which is not available in the standard series. It is available in the **LS**, **S**, **HC**, and **AS** series.

parity on the nine input lines (A–I) is even, ΣEVEN is HIGH and ΣODD is LOW. If the parity of the input is odd, ΣODD will be HIGH.

EXAMPLE 14-8

(a) Design a 7-bit parity checker using a single **74LS280**.

(b) How could this circuit be used to generate an eighth bit for odd parity? Show what the bits are if the incoming word is 1101101.

SOLUTION

(a) The solution is shown in Fig. 14-13. The seven inputs are applied to the A through G inputs of the **74LS280**. Since inputs H and I are not required, they are grounded so they are read as 0s and do not affect the parity. The inputs will have odd parity if the ΣODD output is HIGH and the ΣEVEN output is LOW.

(b) If the parity of the incoming word is even, the ΣEVEN output is HIGH. Therefore, if the 7 input bits and the ΣEVEN output are combined, they form an 8-bit, odd parity word. Consequently, the 7-bit parity checker also acts as a parity generator. The 8-bit output word is shown in dashed lines on Fig. 14-13. If the incoming word is 1101101, its parity is odd. Therefore, ΣEVEN will be LOW, and the 8-bit output word will be 11011010, which has odd parity. The 1s and 0s for this case are also shown in Fig. 14-13.

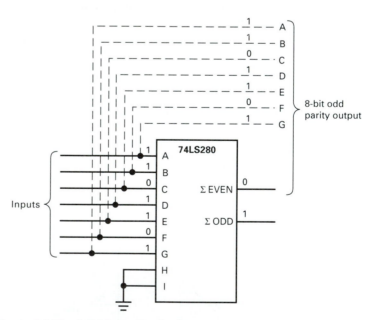

Figure 14-13 A 7-bit parity checker.

14-5-1 Cascading 74x280s

To check the parity of words longer than 9 bits, **74x280s** can be cascaded. We suggest connecting the ΣODD output of the first **74x280** to the I input of the second **74x280**. The first **74x280** can accept 9 inputs and the second **74x280** can accept 8 additional input bits. Therefore, two **74x280**s can check a 17-bit word. Each additional **74x280** added in this manner increases the checking capability by 8 bits.

EXAMPLE 14-9

(a) Design a 16-bit odd parity checker using two **74x280**s.

(b) Show how your parity checker works if the input bits are 0110111011010111.

SOLUTION

(a) Perhaps the most straightforward way to build this parity checker is shown in Fig. 14-14. The first 8 inputs are applied to the first **74x280**, which has its unused I input tied to ground. The ΣODD output is tied to the I input of the second **74LS280**, which also receives the remaining 8 bits.

To determine the state of the outputs, assume all 16 input bits are 0. Then the ΣODD output of the first IC will be LOW, so that all 9 inputs to the second **74x280** will be LOW. This is even parity and the ΣEVEN output of the second IC will be HIGH. This is the error indicator. If the ΣEVEN output is LOW, the parity of the 16-bit word is odd.

(b) The 1s and 0s for the given inputs are also shown in Fig. 14-14. For the first **74x280**, the input parity is odd (five 1s), making the ΣODD output HIGH. For the second **74x280** the input parity is also odd because its input has seven 1s. Therefore, its ΣODD output is HIGH, and its ΣEVEN output must be LOW, indicating odd parity (there are eleven 1s in the 16-bit word).

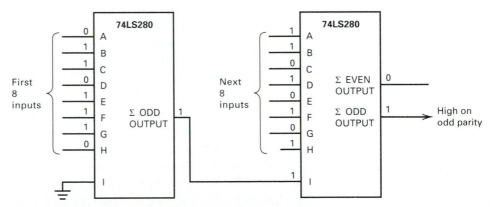

Figure 14-14 A 14-bit parity checker.

14-5-2 Parity Checking on the IBM-PC

A single **74x280** can perform the dual functions of both generating and checking parity and does so in many computers. Figure 14-15 is a simplified version of the way a **74LS280** generates and checks parity on the IBM-PC. The actual circuit is shown as part of Fig. 13-23.

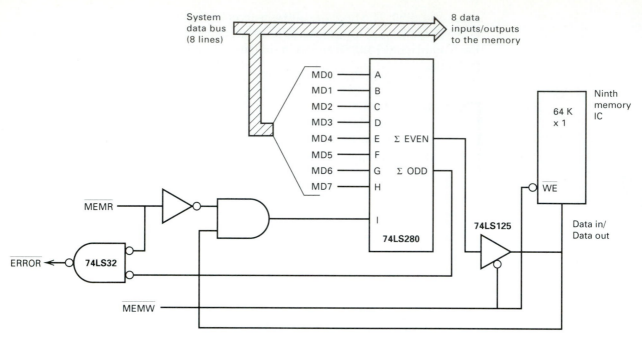

Figure 14-15 A simplified drawing of the parity generator/checker used on the IBM-PC.

In Fig. 14-15 all memory ICs are 64Kx1, and only the ninth memory IC is shown. All nine chips have their DATA-IN and DATA-OUT lines tied together. The IBM-PC provides two control signals to the memory: a $\overline{\text{MEMR}}$ line that goes LOW when the memory is being read, and a $\overline{\text{MEMW}}$ line that goes LOW during a WRITE cycle. During a READ cycle the 8 memory data ICs place their data on the 8 DATA-IN/DATA-OUT lines and the data goes to the system data bus. During a WRITE cycle, data on the bus becomes the inputs to the 8 memory ICs and is written. Note that the 3-state drivers connected to the DATA-OUT pins on the memory ICs are enabled during a READ cycle, but never during a WRITE cycle.

The ninth, or parity, memory IC is shown in Fig. 14-15. It does not receive data directly from the system data bus, and it never puts data out on the system data bus. During a WRITE cycle the **74LS280** must generate parity. It operates as follows:

1. The 8 inputs from the system data bus are applied to inputs A–H of the **74LS280**.
2. Because $\overline{\text{MEMR}}$ is HIGH during a WRITE cycle, the I input to the **74LS280** is 0. Therefore, it generates a parity bit based on the 8 inputs it receives from the system data bus.
3. The **74LS125** 3-state gate (see section 5-7-1) is enabled during a write cycle. The ΣEVEN output of the **74LS280** goes through the **74LS125** and becomes the data input to the memory. It is written into the parity memory IC.

During a READ cycle the **74LS125** is disabled and the DATA-OUT pin on the memory drives the line. It becomes the I input to the **74LS280**. The $\overline{\text{ERROR}}$ signal goes LOW only if there is a READ cycle in progress ($\overline{\text{MEMR}}$ is LOW) and an error has been detected (ΣODD is LOW).

EXAMPLE 14-10

(a) Show that the circuit of Fig. 14-15 generates odd parity.

(b) Show that $\overline{\text{ERROR}}$ only occurs when even parity is detected during a READ cycle.

SOLUTION

(a) If the data inputs from the system bus are all 0s, the ΣEVEN output of the **74LS280** will be HIGH. The 8 memory data ICs will have 0s written into them, but the parity IC will have a 1 on its DATA-IN line. Because the memory receives eight 0s and a 1, it is receiving odd parity (one 1).

(b) During a READ cycle, if the memory reads out nine 0s, for example, the **74LS280** will receive nine inputs of 0. This is even parity and ΣODD will go LOW. It is gated with $\overline{\text{MEMR}}$ to produce the $\overline{\text{ERROR}}$ signal.

14-5-3 The 74180

The **74180** is a parity generator/checker IC that preceded the **74LS280**. The **74180** is somewhat more complex and is only available as a standard series IC. This chip is discussed in the earlier editions of this book.

14-6 More Sophisticated Error-Correcting Routines

Odd or even parity checking schemes are highly effective, but they have two drawbacks: they will not detect *two errors* in the same word or character, and they provide no capability for *correcting an error* if one is found. When high reliability is required, additional checks such as *longitudinal redundancy checking (LRC) or cyclical redundancy checking* (CRC) are used.

Often a group or block of words is being read (the words stored on one sector of a disk is an example) and the user must be sure that every one of the words is correct. It is now common practice to *add* an additional character or two to ensure the integrity of the data. These are the longitudinal or cyclical redundancy characters. They are redundant because they contain no new information and cause more bits to be transmitted for the same message, but their value in checking the data compensates for the additional block length.

14-6-1 Longitudinal Redundancy Checking

Longitudinal redundancy checking is accomplished by adding an LRC (Longitudinal Redundancy Character) to each block of data. Bit 1 of the LRC is a 1 if all the bit 1s in each byte of the transmitted block taken together contain an *odd* number of 1s. The other bits are similarly determined.

The situation is shown in Fig. 14-16 where, for simplicity, a 6-word-by-4-bit data block is assumed. Vertical or odd parity is added to each of the 6 words, and the 5-bit LRC character is calculated by looking horizontally across the 6 bits and determining if the number of 1s is odd or even.

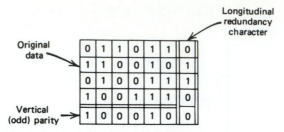

Figure 14-16 Addition of vertical parity and an LRC to a 6-word-by-4-bit data block.

As with vertical parity, LRCs *are generated when writing* and appended as the last character of a block. The LRC character must be *checked* whenever the block is read.

EXAMPLE 14-11

Design a circuit to generate an LRC character.

SOLUTION A circuit using J-K FFs is shown in Fig. 14-17. For an N-bit word, it requires N FFs. At the start of transmission the CLEAR line is pulsed LOW, clearing all the FFs. The data bit to be written is also connected to the J and K inputs of the corresponding FF. The word clock occurs as each word is transmitted. If the data bit is a 1, the FF toggles; otherwise it does not change. When the entire block is transmitted, the LRC character is on the outputs of the FFs. This character must then be transmitted.

When reading the data, the LRC can be checked by a circuit similar to that of Fig. 14-17. The clock that clocks in the received data bits is also used to

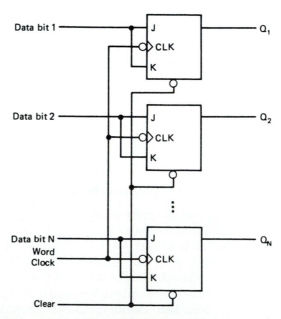

Figure 14-17 A circuit for generating an LRC.

toggle the N FFs if the received data bit is a 1. When the LRC is received, it can be compared to the FFs and a match indicates that a valid data block has been received. It is simpler, however, to allow the LRC character to be clocked in and toggle the FFs also. If the block has been received correctly, the FF outputs after the LRC has been received will all be 0s.

14-6-2 Error Correction

If a block is received with only a single bit in error, the LRC scheme of Fig. 14-16 allows us to find and correct the error. If only one bit is wrong, it will cause one vertical parity error and one bit in the LRC word to be wrong. The bit in error is at the intersection of the vertical and horizontal parity errors and should be complemented (see Problem 14-22).

14-6-3 Cyclical Redundancy Checking

This checking is achieved by appending a cyclical redundancy check character (CRC) to each data block. Most disk systems use a CRC to check the validity of the data. PC systems divide the files stored on the disk into 256-byte blocks and append a 16-bit CRC after each block to check its validity.

   The CRC is calculated by taking the data and dividing it by a given polynomial. This division can be accomplished in hardware by a circuit using a shift register and XOR gates for feedback. CRC circuits are also available as ICs from several manufacturers. Unfortunately, the theory of CRC is too long and complex to be presented here. The reader should consult other books for more information.

14-7 The Gray Code

The Gray code is a code for representing numbers that is often used when analog-to-digital conversions (see section 18-5) are required. The Gray code has the advantage that *only one bit in the numerical representation changes between successive numbers*. Unfortunately, it is difficult to perform arithmetic operations on numbers in Gray code, so Gray-to-binary and binary-to-Gray conversions are often performed. XOR gates can be used to perform these conversions.

14-7-1 Construction of the Gray Code

Figure 14-18 shows the Gray code equivalent of the first 16 numbers. The Gray code table was obtained by using the following algorithm:

1. For G_0 write one 0 followed by two 1s, two 0s, two 1s, and so on.
2. For G_1 write two 0s followed by alternating groups of four 1s and four 0s.
3. In general, for the G_n column, start by writing 2^n 0s. Then write alternating groups of 2^{n+1} 1s and 0s.

Decimal Number	Binary Equivalent				Gray Code Equivalent			
	B_3	B_2	B_1	B_0	G_3	G_2	G_1	G_0
0	0	0	0	0	0	0	0	0
1	0	0	0	1	0	0	0	1
2	0	0	1	0	0	0	1	1
3	0	0	1	1	0	0	1	0
4	0	1	0	0	0	1	1	0
5	0	1	0	1	0	1	1	1
6	0	1	1	0	0	1	0	1
7	0	1	1	1	0	1	0	0
8	1	0	0	0	1	1	0	0
9	1	0	0	1	1	1	0	1
10	1	0	1	0	1	1	1	1
11	1	0	1	1	1	1	1	0
12	1	1	0	0	1	0	1	0
13	1	1	0	1	1	0	1	1
14	1	1	1	0	1	0	0	1
15	1	1	1	1	1	0	0	0

Figure 14-18 Gray code and binary equivalents of decimal numbers.

Note that as the count progresses, only a single bit in the Gray code representation changes. When a count goes from 7 to 8, for example, all 4 bits in the binary representation change, but only G3 in the Gray code representation changes.

14-7-2 Uses of the Gray Code

One of the common uses of the Gray code is to determine the position of a rotating shaft. Typically, this is done by attaching a code wheel to the shaft as shown in Fig. 14-19. The code wheel contains transparent and opaque segments. Light transmitted through the code wheel is detected by phototransistors or other

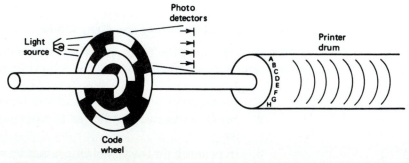

Figure 14-19 A code wheel used on a drum printer.

photosensitive sensors. The position of the shaft is determined by the 1s and 0s (transparent and opaque segments) detected by the photoelectric sensors. This scheme is often used on rotating drum line printers, where the code wheel output determines when the hammers strike the drum and print the character.

A 4-bit binary code wheel is shown in Fig. 14-20a and a 4-bit Gray code wheel is shown in Fig. 14-20b. Each code wheel produces 16 codes, so the shaft position resolution is $^{360}/_{16} = 22.5$ degrees. For greater resolution, additional concentric segments must be added.

If a radius is drawn straight up on the binary code wheel of Fig. 14-20a, the segments to the left of the radius are shaded and all the segments to the right are clear. *When the detector or sensor is placed in exactly this middle position, it could produce either a 0 or a 1 output for each sector. Consequently, any number between 0 and 15 might appear on the output.* This ambiguity is avoided by the Gray code wheel, where only one segment changes at a time.

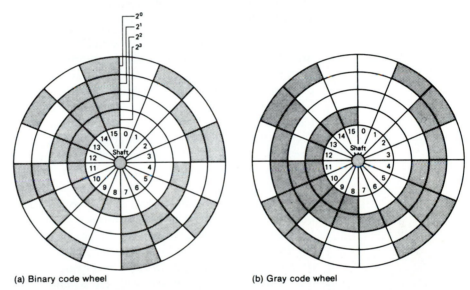

(a) Binary code wheel   (b) Gray code wheel

Figure 14-20 Binary and Gray code wheels. (From Barna and Porta, *Integrated Circuits in Digital Electronics*. Copyright John Wiley & Sons, Inc. 1973. Reprinted by permission of John Wiley & Sons, Inc.)

EXAMPLE 14-12

(a) Assume a sensor is right on the division between 11 and 12 and produces a random output for each sector that changes. What numbers might it select if the code wheel is binary?

(b) Repeat for a Gray code disk.

SOLUTION

(a) On the binary code wheel the three least significant (outermost) sectors change between 11 and 12. Only the innermost sector (the MSB) is a 1 on both sides of the line. Consequently, any number between 8 and 15 may appear as an output.

(b) On the Gray code disk only the second sector changes between 11 and 12. Consequently, the output is either 1110 (the Gray equivalent of 11) or 1010 (the Gray equivalent of 12). This improves the shaft resolution considerably.

14-7-3 Gray-to-Binary Conversion

In determining shaft positions, it is advantageous to pick the position off a disk in Gray code and convert the Gray code output to binary for numerical processing within a computer. Fortunately, Gray-to-binary converters are very simple to design and build and consist entirely of XOR gates.

EXAMPLE 14-13

Design a converter to change a 4-bit Gray code to its equivalent 4-bit binary representation.

SOLUTION Code conversion problems were introduced in section 3-8-4, and the methods developed there are applicable to this problem. The inputs consist of the 4 Gray code bits ($G_3G_2G_1G_0$) and there are 4 separate binary outputs ($B_3B_2B_1B_0$). The solution is shown in Fig. 14-21. The decimal numbers are listed in column 1 of the truth table (Fig. 14-21a) and the corresponding Gray code entries are listed in column 2, using the procedure developed in section 14-7-1. The Gray code entries, which are the inputs, do not progress in the normal manner of truth table entries; they do not increment. Therefore, column 3 was written. *The numbers in column 3 are the decimal equivalent of the Gray code entries of column 2, read as binary numbers.* This is done to allow us to develop the equations and Karnaugh maps. The corresponding binary outputs are listed in column 4.

In Fig. 14-21b, the equations for the binary outputs are developed. *These equations are obtained for each binary bit by listing the numbers in column 3 that correspond to the 1s in each binary output column.*

Once the equations are written, Karnaugh maps can be drawn for each bit, as shown in Fig. 14-21c, and the equations relating the binary bits to the Gray code bits developed. The map for B_0 is a checkerboard. Fortunately, checkerboard Karnaugh maps were encountered in section 14-4-1 and since the maps are the same, the solution is the same:

$$B_0 = G_3 \oplus G_2 \oplus G_1 \oplus G_0$$

For B_1 we have, by combining subcubes

$$B_1 = \bar{G}_3\bar{G}_2G_1 + \bar{G}_3G_2\bar{G}_1 + G_3G_2G_1 + G_3\bar{G}_2\bar{G}_1$$

which reduces (with a little inspiration) to

$$B_1 = \bar{G}_3(\bar{G}_2G_1 + G_2\bar{G}_1) + G_3(G_2G_1 + \bar{G}_2\bar{G}_1)$$
$$= \bar{G}_3(G_2 \oplus G_1) + G_3(\overline{G_2 \oplus G_1})$$
$$= G_3 \oplus G_2 \oplus G_1$$

The maps for B2 and B3 are even simpler.

The final step, the design of the converter, becomes very simple, as shown in Fig. 14-21d. The 4 Gray code input bits are converted to the corresponding binary bits simply by using 3 XOR gates.

Decimal Number	Gray Code				Binary Output			
	G_3	G_2	G_1	G_0	B_3	B_2	B_1	B_0
0	0	0	0	0	0	0	0	0	0
1	0	0	0	1	1	0	0	0	1
2	0	0	1	1	3	0	0	1	0
3	0	0	1	0	2	0	0	1	1
4	0	1	1	0	6	0	1	0	0
5	0	1	1	1	7	0	1	0	1
6	0	1	0	1	5	0	1	1	0
7	0	1	0	0	4	0	1	1	1
8	1	1	0	0	12	1	0	0	0
9	1	1	0	1	13	1	0	0	1
10	1	1	1	1	15	1	0	1	0
11	1	1	1	0	14	1	0	1	1
12	1	0	1	0	10	1	1	0	0
13	1	0	1	1	11	1	1	0	1
14	1	0	0	1	9	1	1	1	0
15	1	0	0	0	8	1	1	1	1

(a) Truth table

$B_3 = \Sigma\ (12,\ 13,\ 15,\ 14,\ 10,\ 11,\ 9,\ 8)$

$B_2 = \Sigma\ (6,\ 7,\ 5,\ 4,\ 10,\ 11,\ 9,\ 8)$

$B_1 = \Sigma\ (3,\ 2,\ 5,\ 4,\ 15,\ 14,\ 9,\ 8)$

$B_0 = \Sigma\ (1,\ 2,\ 7,\ 4,\ 13,\ 14,\ 11,\ 8)$

(b) Equations

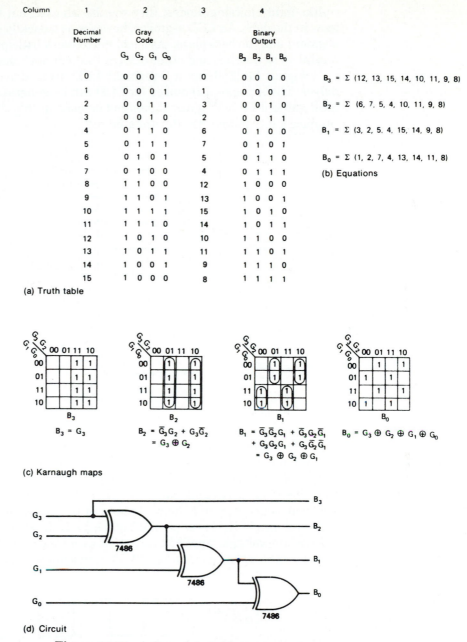

(c) Karnaugh maps

$B_3 = G_3$

$B_2 = \overline{G}_3 G_2 + G_3 \overline{G}_2 = G_3 \oplus G_2$

$B_1 = \overline{G}_3 \overline{G}_2 G_1 + \overline{G}_3 G_2 \overline{G}_1 + G_3 G_2 G_1 + G_3 \overline{G}_2 \overline{G}_1 = G_3 \oplus G_2 \oplus G_1$

$B_0 = G_3 \oplus G_2 \oplus G_1 \oplus G_0$

(d) Circuit

Figure 14-21 A Gray code to binary converter.

14-8 Liquid Crystal Displays

Liquid Crystal Displays (LCDS) were introduced in section 11-12. The circuits that drive them use XOR gates. To turn on they require a low frequency a.c. drive voltage between the crystal segment and the backplane because a d.c. component of more than 50 mV will tend to shorten their lives. Fortunately, the a.c. voltage is between the backplane and the LCD segment. An ON LCD segment merely reflects ambient light, whereas an OFF segment does not. Thus, unlike light-emitting diodes, they require an external light source; they don't glow in the dark. An LCD segment, however, is basically a capacitor between the segment and the backplane. As such, it uses very little power. LCDs are used in digital wristwatches and other devices that cannot consume significant power.

Figure 14-22 shows a 7-segment LCD being driven by a **4511** latch and driver. The a.c. signal required for LCD can be generated using **4070B** CMOS XOR gates. TTL XOR gates should not be used in this circuit because they may produce a d.c. voltage greater than 50 mV.

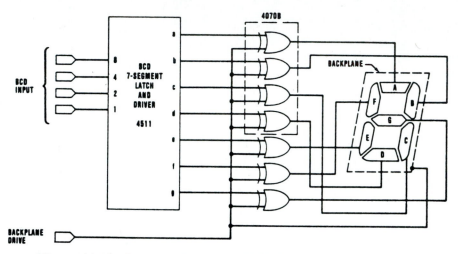

Figure 14-22   Driving a 7-segment LCD. (Courtesy of Beckman Instruments, Inc.)

The backplane drive signal is connected to a 50 percent duty cycle clock (40 Hz is a suggested frequency), as shown in Fig. 14-23. Assume the segment a output of the **4511** is a 1. Then, when the backplane voltage is HIGH, the voltage to segment a is LOW, and vice versa. Thus, an a.c. voltage of 3 V to 5 V is created between segment a and the backplane of the LCD, turning the segment ON. This is shown in the lower part of Fig. 14-23. If the segment a output of the **4511** is LOW, the voltages applied to segment a and the backplane will be in phase as shown in the upper part of Fig. 14-23. Consequently, there will be no effective

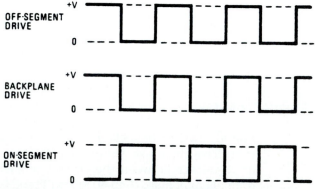

Figure 14-23   Phase shifting an LCD. (Courtesy of Beckman Instruments, Inc.)

voltage between them and the segment will remain OFF. Seven-segment LCD displays may also be multiplexed. The reader should consult more advanced literature for details.

SUMMARY

In this chapter, the uses of XOR gates and ICs that depend on them were introduced. The **74x85** was presented as a TTL comparator and the **74x280** was presented as the TTL parity checker/generator. Circuits illustrating the use of these ICs were also presented.

The Gray code was also introduced and its use in shaft position sensing explained. A binary-to-Gray code converter was designed. Finally, the operation of LCDs was explained.

GLOSSARY

Code wheel: A wheel consisting of transparent and opaque segments, used to give an encoded representation of a shaft position.

Comparator: A circuit that compares two numbers and produces an output indicating whether they are equal. It may also indicate which number is greater if they are unequal.

EQUALITY gate: A 2-input gate that produces a high output when the two inputs are equal.

Even parity: A word has even parity if the number of 1s in the word is even.

EXCLUSIVE NOR: Synonym for EQUALITY gate.

Gray code: A code where only one bit changes between successive numbers.

Odd parity: A word has odd parity if the number of 1s in the word is odd.

Parity checker: A circuit that checks the number of 1s in a word to determine if the parity is proper.

Parity generator: A circuit that causes the number of 1s in a word to be even or odd (depending on which parity is chosen) by adding a bit to a word.

PROBLEMS

Section 14-3-1.

14-1.   Given two n-bit registers A and B, composed of FFs, show that for the ith stage, $Q_A \oplus \bar{Q}_B$ produces a high output when the ith bits are equal. Use this principle to design an n-bit comparator without using EQUALITY gates.

Section 14-3-2.

14-2.   Design a 5-bit comparator circuit using hypothetical EQUALITY gates having totem-pole outputs.

Section 14-3-3.

14-3. Design a 5-bit comparator using only a single **74x85** and some SSI gates.

Section 14-3-4.

14-4. Suppose someone, using the circuit of Example 14-4, erroneously connected the outputs of the more significant **74x85** to the inputs of the less significant **74x85**, and took the outputs from the less significant **74x85**. Find a number that would give an incorrect result when compared to $(195)_{10}$.

14-5. Design a circuit to determine if a 9-bit binary number is greater than, equal to, or less than 400.

14-6. Two 24-bit numbers differ only in that $A_6 = 1$ and $B_6 = 0$. Show how the comparator of Fig. 14-6 would compare them.

14-7. Repeat Problem 14-6 if $A_9 = 0$ and $B_9 = 1$.

Section 14-4.

14-8. What is the parity of each of the following bytes:
(a) FF
(b) OO
(c) AB

14-9. Add a bit to each of the bytes in Problem 14-8, to create a 9-bit, odd parity word.

Section 14-4-2.

14-10. Show how the parity checkers of Fig. 14-10 operate, by showing all the 1s and 0s, if the input word is:
(a) 11011001
(b) 11001111

Section 14-4-3.

14-11. Show how the parity generator of Fig. 14-11 operates if the input word is:
(a) 0001000
(b) 1101100
(c) 1111111

Section 14-5.

14-12. Show how the 7-bit parity checker and 8-bit parity generator of Fig. 14-13 work for the inputs of Problem 14-11.

14-13. How can the circuit of Fig. 14-13 be used to generate a tenth bit for odd parity?

Section 14-5-1.

14-14. Design a 10-bit odd parity checker, using:

(a) Two **74x280**s.

(b) One **74x280** and additional gates (no more than 4).

For each case, show how your checker works if the inputs are:

$$1001100110$$

14-15. Given 15 input bits, design a circuit to produce a 16-bit, odd parity word.

14-16. Design a parity checker to check a 15-bit word and produce a high output on odd parity. Show how your circuit works if the 15 input bits are:

Bit 1 2 3 4 5 6 7 8 9 10 11 12 13 14 15

0 1 1 0 1 1 1 0 0 1 1 0 1 0 1

14-17. You are given 8-input bits and a clock. When the clock is HIGH, the data are valid and must be checked for odd parity. A LOW output should indicate an error (even parity). When the clock is LOW, the data are changing and must not be checked. Consequently, the error output must always be HIGH. Design the circuit.

14-18. Design a 17-bit parity checker using only two **74x280**s.

Section 14-5-2.

14-19. Modify Fig. 14-15 so it would work if there were only one READ/WRITE line coming from the memory.

14-20. In Fig. 14-15, why can't the **74LS125** be enabled continually?

14-21. Design a circuit to accept a 16-bit word and generate a 17th bit for odd parity on WRITE. On READ the circuit must check the 17-bit output word and produce an $\overline{ERROR}$ signal if it detects even parity. Use only two **74x280**s, and a few small-scale gates.

Section 14-6-1.

14-22. The data of Fig. P14-22 were received and consist of a 7-word by 4-bit data block with odd parity and an LRC appended. One of the received bits is wrong. Find the vertical parity error, the LRC error, and the erroneous bit.

							LRC
0	0	1	0	0	1	0	0
1	0	1	0	0	0	1	0
0	0	1	1	1	1	1	1
1	0	0	1	1	0	0	1
1	1	1	1	1	1	1	1

Vertical parity   **Figure P14-22**

14-23. A disk is being read. The data block consists of 128 words of data (each word consists of 7 bits plus an eighth for odd parity) and an LRC word. The disk presents a START pulse and 129 data clocks. Data is valid during the positive portion of the data clock. Design a circuit to check the vertical and horizontal parity of the received data, to set a FF after the entire block has been received, and to set a second FF if there is an error in the block.

Section 14-7-1.

14-24. Construct a table similar to Fig. 14-18 for 5-bit numbers.

14-25. Design a binary-to-Gray converter for 4-bit numbers.

When you have finished the problems, return to section 14-2 and reread the self-evaluation questions. If you have difficulty answering them, return to the appropriate sections of the chapter to find the answers.

15

Arithmetic Circuits

15-1 Instructional Objectives

This chapter discusses the most common arithmetic circuits that can be designed and built using digital ICs. Circuits to add, subtract, and multiply are presented. In addition, the arithmetic-logic unit, a single IC that can perform many arithmetic and logic operations, is introduced. After reading the chapter, the student should be able to:

1. Design binary adders and subtracters.
2. Build an adder/subtracter using **74x283**s.
3. Design overflow and underflow detectors for 2's complement adder/subtracters.
4. Design BCD adders and subtracters.
5. Utilize arithmetic-logic units such as the **74x181**.
6. Build a multiplier circuit.

15-2 Self-Evaluation Questions

Look for the answers to the following questions as you read the chapter. They should help you understand the material presented.

1. If the inputs to an adder are n-bit numbers, how many bits must the *sum register* contain?
2. What is the difference between a *half-adder* and a *full-adder*?
3. How does a subtracter circuit react if the difference is negative (the subtrahend is greater than the minuend)?
4. In a binary adder/subtracter, how is the operand 2's complemented for subtraction and presented unchanged for addition?
5. In 2's complement arithmetic, what are the criteria for overflow or underflow?
6. What is the significance of the carry in a BCD adder?
7. How does an arithmetic-logic unit select the function to be performed?
8. What is the advantage of look-ahead carry? When should it be used?
9. If an m-bit number is multiplied by an n-bit number, how long is the product register?
10. What is a coprocessor? What is its function?

15-3 The Basic Adder

The simplest arithmetic circuit is the *basic adder*. It was introduced in section 3-8-2. For binary numbers it accepts two n-bit binary numbers as inputs, and produces an $(n+1)$-bit binary number as the sum. Essentially it consists of a half-adder, $n-1$ full-adder circuits, and the $n+1$ stage (the MSB) that only receives the carry from the nth stage.

15-3-1 The Half-Adder

The basic adder stage accepts inputs from the *addend*, *augend*, and a *carry-in* from the less significant stage. It generates a *sum* output and a *carry* output, which it sends to the next more significant (succeeding) stage. A *half-adder* is simpler. It only accepts addend and augend inputs. Because it need *not* accept a carry-in, it can be used for the least significant stage of an adder, where there is never a carry input.

The truth table and circuit for the basic half-adder are shown in Fig. 15-1. The sum is 1 if *either* the addend or augend (A or B) are 1. If both A and B are 1, it is adding 1 plus 1 and the result is 10(2). The 0 is the LSB of the sum output and

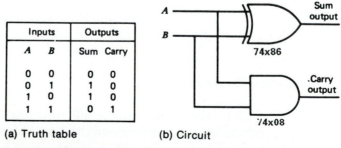

Inputs		Outputs	
A	B	Sum	Carry
0	0	0	0
0	1	1	0
1	0	1	0
1	1	0	1

(a) Truth table (b) Circuit

Figure 15-1 The half-adder.

the 1 is the carry output. From the truth table it is apparent that the sum is merely an XOR gate while the carry output is generated by an AND gate. The circuit is shown in Fig. 15-1b.

15-3-2　The Full-Adder

The *full-adder* accepts the ith bit of the augend and addend (A_i and B_i) and produces the ith bit of the sum. It must also accept a carry-in from the $i - 1$ stage, and generate a carry-out to the $i + 1$ stage.

The truth table, Karnaugh maps, and circuit of the full adder are shown in Fig. 15-2. The truth table (Fig. 15-2a) is constructed by considering the addition of three binary bits, A_i, B_i, and the carry-in.[1] The sum is 1 only if one or all three of the inputs are 1. This is the same as a 3-bit odd parity checker (section 14-4-1). The carry output must be 1 if two or three of the inputs are 1.

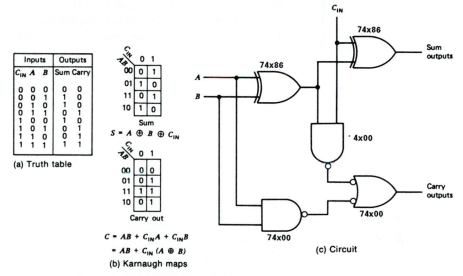

(a) Truth table

$$S = A \oplus B \oplus C_{IN}$$

$$C = AB + C_{IN}A + C_{IN}B$$
$$= AB + C_{IN}(A \oplus B)$$

(b) Karnaugh maps

(c) Circuit

Figure 15-2　The full-adder.

The Karnaugh maps for the sum and carry are shown in Fig. 15-2b. The map for the sum is a checkerboard (as it would be for an odd-parity checker) and

$$S = A \oplus B \oplus C_{IN}$$

The map for the carry-out yields:

$$
\begin{aligned}
C_{OUT} &= AB + C_{IN}A + C_{IN}B \\
&= AB + C_{IN}A(B + \bar{B}) + C_{IN}B(A + \bar{A}) \\
&= AB + C_{IN}AB + C_{IN}A\bar{B} + C_{IN}AB + C_{IN}B\bar{A} \\
&= AB(1 + C_{IN} + C_{IN}) + C_{IN}(A\bar{B} + B\bar{A}) \\
&= AB + C_{IN}(A \oplus B)
\end{aligned}
$$

The latter form is generally preferred to the form

$$C_{OUT} = AB + C_{IN}(A + B)$$

[1]The reader may wish to refer to the binary addition example presented in section 1-8-1.

because the term $A \oplus B$ can be used in both the sum and carry outputs. A full-adder circuit designed on the basis of these equations is shown in Fig. 15-2c.

EXAMPLE 15-1

Design an adder for 3-bit numbers. Show how your adder works if $A = 7$ and $B = 6$.

SOLUTION The adder circuit is shown in Fig. 15-3. It can be understood best by referring to the block diagram, Fig. 15-3a, which shows that it is composed of a half-adder and two full-adders. *The MSB of the output is simply the carry of the second full-adder.*

The circuit is shown in Fig. 15-3b, where the half- and full-adders are taken from Figs. 15-1 and 15-2. The 1s and 0s for inputs of 7 and 6, respectively ($A = 111, B = 110$) are also shown in the figure. By continuing to follow the logic, we see that the output is 1101 or 13.

Adders for larger numbers can be built in this manner simply by adding a full-adder stage for each additional bit (see Problem 15-2).

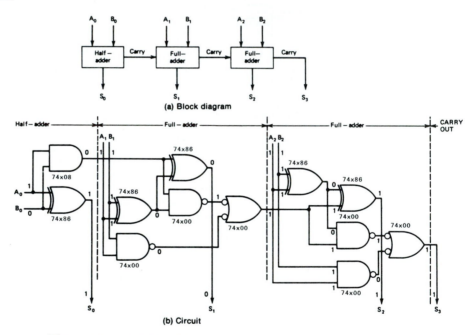

(a) Block diagram

(b) Circuit

Figure 15-3 A 3-bit adder.

15-4 Subtraction

Binary subtraction is the process of finding the *difference* (A minus B) between two binary numbers, A and B. Mathematically it is performed using 2's complement arithmetic, as explained in Chapter 1, by complementing the subtrahend and adding it to the minuend. A circuit to perform binary subtraction can be built in the same manner as binary addition, by using a *half-subtracter* for the least-significant stage, followed by a series of full-subtracters. Each subtracter stage generates a *borrow-out* if it must borrow from a more-significant stage.

Each stage except the least significant stage (the half-subtracter) may receive a borrow-in signal from the preceding stage.

The truth table for a full-subtracter can be constructed by considering the rules for binary subtraction (section 1-8-2). The inputs to the full-subtracter are A, B, and borrow-in, and the full-subtracter generates a difference output and a borrow-out. The truth table is shown in Fig. 15-4. The entries are obtained by considering the conditions on each line of the truth table.

Line number	Inputs			Outputs	
	B_{IN}	A	B	D	B_{OUT}
0	0	0	0	0	0
1	0	0	1	1	1
2	0	1	0	1	0
3	0	1	1	0	0
4	1	0	0	1	1
5	1	0	1	0	1
6	1	1	0	0	0
7	1	1	1	1	1

Figure 15-4   Truth table for a full-subtracter.

15-5   The 74x283 4-Bit Adder

Arithmetic operations are performed so often that special ICs have been designed for them. The **74x283**[2] is essentially a hexadecimal adder. It accepts two 4-bit numbers (A and B) and a carry-in (C_0) as inputs. The 4 bits of the A input are connected to the A_1, A_2, A_3, and A_4 inputs on the **74x283**. The 4 bits of the B input are similarly connected. The **74x283** produces a 4-bit sum output (Σ) and a carry output (C_4). A functional drawing of the **74x283** is shown in Fig. 15-5.

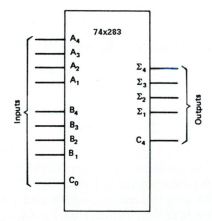

Figure 15-5   The **74x283** 4-bit adder.

[2]The **74x283** replaces the older **7483**. The two ICs function identically, but the **'x283** has power and ground on pins 8 and 16 (the corner pins). The **'283** is also available in the **HC** series.

The IC operates as follows:

1. If the sum of the two inputs plus the carry-in is between 0 and 15, the sum appears in the Σ outputs and the carry-out (labeled C_4) is 0.

2. If the sum is between 16 and 31, carry-out C_4 is 1 and the Σ outputs are 16 less than the sum. Note that in a 4-bit adder, the carry-out has a value or weight of 16.

EXAMPLE 15-2

A **74x283** has the following inputs:

$$A_4A_3A_2A_1 = 0111 \quad (7)$$
$$B_4B_3B_2B_1 = 1010 \quad (A)$$
$$C_0 = 1$$

Find the outputs.

SOLUTION   The inputs are seen to be 7 and 10, with a carry input. The sum of the inputs is therefore 18. According to the above rules, a carry-out is produced and the 1 outputs are 2 (18 minus 16). Therefore, $C_4 = 1$ and $\Sigma_4\Sigma_3\Sigma_2\Sigma_1 = 0010(2)$.

15-5-1   Cascading 74x283s

To build an adder for words longer than 4 bits, **74x283**s can be cascaded simply by tying C_4 of one **74x283** to the C_0 input of the next most significant **74x283**, as shown in Fig. 15-6. Note that the least-significant stage is placed on the right and the addition proceeds from *right* to *left*. Although this is opposite to the usual directions of data flow, it allows us to read the input numbers and the output in the normal manner, from *left* to *right*.

EXAMPLE 15-3

Design a circuit to add 202 and 231 using two **74x283**s.

SOLUTION   The 1s and 0s for this example are also shown in Fig. 15-6. First 202 and 231 are converted to binary or hex:

$$202 = 11001010 = CA$$
$$231 = 11100111 = E7$$

These become the A and B inputs to the **74x283**s. The four LSBs of each number are connected to the least significant **74x283**, which sees inputs of A and 7 and no carry. Its sum output is 1 and it produces a carry-out. The more significant **74x283** sees inputs of C and E and a carry input. Therefore, it produces an output of B, or 1011 plus a carry output. If the carry output plus the 8-bit output of the two **74x283**s are read as a single binary number, we have

$$\text{Sum} = 1B1 = 110110001 = (433)_{10}$$

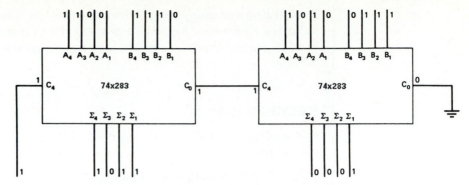

Figure 15-6 An 8-bit adder using two **74x283**s.

15-5-2   The 2's Complement Adder/Subtracter

An *adder/subtracter* for 2's complement numbers can be built from **74x283**s, as shown in Fig. 15-7. The mode of the adder/subtracter is controlled by a toggle switch or an add-subtract line. When the line is LOW, the circuit is an adder. The carry-in is 0 and the XORs act as straight-through gates (see Example 5-13). The circuit simply sums the A and B inputs.

In subtract mode, $C_0 = 1$ and the output of the XOR gates is the inversion of the B inputs. Hence, B is complemented and 1 is added to it, effectively 2's complementing the operand. Consequently, if B is the subtrahend, subtraction is accomplished by 2's complementing and adding B to A.

EXAMPLE 15-4

Show how the circuit of Fig. 15-7 subtracts 6 from 11.

SOLUTION   The 1s and 0s for this circuit are shown on Fig. 15-7. The switch is shown in the subtract position and the A and B inputs are 11 and 6. Because of the inversion caused by the XORs, the **74x283** receives inputs of 11, 9, and a carry. Since these sum to 21, the **74x283** produces a carry-out and a sum output of 5. The result of 5 is correct and the carry-out is ignored.

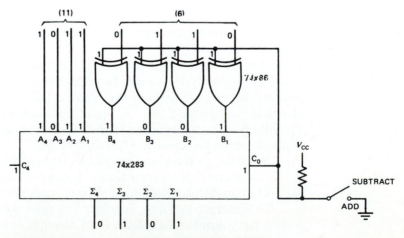

Figure 15-7 A 4-bit ADDER/SUBTRACTER.

In this and many other subtractor circuits, a carry-in or carry-out of 1 indicates no *borrow*. A carry-in or carry-out of 0 indicates that the subtrahend is greater than the minuend, giving a negative result and requiring a borrow.

EXAMPLE 15-5

Add and subtract 89 from -35 (decimal) using a 2's complement adder/subtracter.

SOLUTION Since the operands take more than 4 bits, a 2-stage adder/subtracter is required. The solution is shown in Fig. 15-8, where the A input is -35 (11011101) and the B input is $+89$ (01011001). The numbers in bold type indicate the addition operation. Here the least significant IC, which is on the right, sees inputs of 13 and 9. They sum to 22, or 6 plus a carry-out. The more significant chip sees inputs of 13 and 5, plus a carry-in, which sums to 19, or 3 plus a carry-out. The result of the addition is 00110110 or $(54)_{10}$, which is correct.

The numbers in italics apply to the subtraction of 89 from -35. The least significant chip sees 13 plus 6 plus a carry-in and produces a sum of 4 and a carry-out. The more significant chip sees inputs of 13, 10, and a carry-in. It therefore produces an output of 8 (1000) and a carry-out. The results of the subtraction operation are 10000100, or -124, which is correct.

This problem can be shortened considerably by using hex notation. Inspecting the binary numbers above, we see that $(-35)_{10} = DD$ and $(+89)_{10} = 59$. Their sum is $DD + 59 = (36)_{16}$ or $(54)_{10}$ and their difference is $DD - 59 = (84)_{16}$ or $(-124)_{10}$. The hex results, 36 and 84, are the outputs of the two **74x283**s in addition and subtraction.

Clearly this circuit can be extended to handle longer numbers. It is ideal for use in computers where the numbers are stored in memory in 2's complement form and only need to be brought out and passed through the adder/subtracter to achieve the desired results.

15-6 Overflow and Underflow in 2's Complement Arithmetic

Digital registers or registers within a microprocessor have a fixed number of bits. Consequently, *the range of numbers they can accommodate is limited*. The limitations on the numbers that can be handled by an n-bit register are, as shown in Section 1-9-2, $2^{n-1} - 1$ positive numbers, and 2^{n-1} negative numbers. The numbers in the registers of an 8-bit microprocessor, like the Intel **8085**, for example, must be between $+127$ and -128.

However, 2's complement arithmetic operations can produce a result that is "out-of-bounds"; a number that is higher or lower than the register can contain. This condition is called *overflow* or *underflow*.

To illustrate overflow, consider the decimal number 100 expressed as an 8-bit binary number, 01100100. If an attempt is made to add 100 plus 100, the

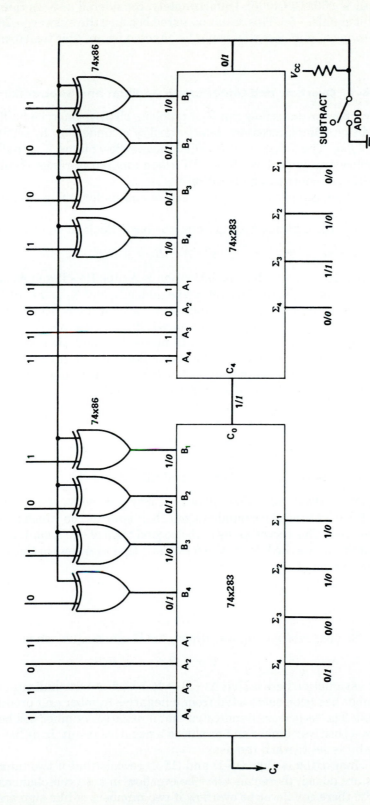

Figure 15-8 An 8-bit ADDER/SUBTRACTER.

result is 200 (11001000). Unfortunately, considered as a 2's complement number, it equals -56. This result occurred because the answer, $+200$, was *beyond the range* of numbers that could be handled by an 8-bit register.

15-6-1 Overflow and Underflow in Addition and Subtraction

If overflow or underflow can be a problem, circuits must be built to detect this condition. Microprocessors detect overflow or underflow by setting a FF called the V *flag*. The V flag or bit is called the overflow bit but it actually detects both overflow and underflow. It is SET when any out-of-range result occurs in an addition or subtraction operation.

There are two criteria for overflow and underflow in microprocessors:

1. For *addition* instructions, the basic Boolean equation for overflow is

$$V = \bar{A}_7\bar{B}_7R_7 + A_7B_7\bar{R}_7 \qquad (15\text{-}1)$$

where it is assumed that the operation is A plus B $\rightarrow$ R and A_7 is the MSB of A (the augend), B_7 is the MSB of B (the addend), and R_7 is the MSB of the result. The plus sign in the equation indicates the logical OR.

If the first term of the equation is 1, it indicates that two positive numbers have been added (because A_7 and B_7 are both 0) and the result is negative (because $R_7 = 1$). This possibility has been illustrated in the preceding paragraph.

The second term indicates that two negative numbers have been added and have produced a positive result.

EXAMPLE 15-6

Show how the hex numbers 80 and C0 are added.

SOLUTION   80 + C0 = 40 plus a carry (see section 4-3-3). Note that 80 and C0 are both negative numbers, but their sum (as contained in a single byte) is positive. This corresponds to the second term of equation 1. Fortunately, this addition sets the V bit to warn the user that overflow (in this case underflow) has occurred.

---

2. For *subtraction* operations, the Boolean equation is

$$V = A_7\bar{B}_7\bar{R}_7 + \bar{A}_7B_7R_7 \qquad (15\text{-}2)$$

The assumption here is that A $-$ B $\rightarrow$ R. The first term indicates that a positive number has been subtracted from a negative number and produced a positive result. The second term indicates that a negative number has been subtracted from a positive number and produced a negative result. In either case, the overflow bit is set to warn the user.

Inspection of Eqs. (15-1) and (15-2) reveals that if two numbers of *unlike sign* are added, there can *never* be overflow in a 2's complement system. Similarly, there can never be overflow if two numbers of like sign are subtracted.

For certain applications, such as calculators, it is advantageous to perform all arithmetic operations in BCD format. This eliminates the need for conversions. Unfortunately, BCD arithmetic circuits are more complex and costly than binary circuits. This fact should be carefully considered before using BCD circuits.

15-7-1   The BCD 4-Bit Adder

A BCD 4-bit adder accepts a single BCD digit from the A operand, a single BCD digit from the B operand, and a carry-in. *It produces a single* BCD *digit as the sum output. It must also produce a carry-out if the sum is greater than 9.*

Perhaps the simplest way to design a BCD adder is to base its operation on the **74x283**. If the sum of the three inputs (the A digit, the B digit, and the carry-in) is 9 or less, the **74x283** produces the correct output. If the sum is greater than 9, a carry-out must be generated and the sum corrected.

A single-decade BCD adder is shown in Fig. 15-9. It consists of three parts:

1.   The basic adder.
2.   The carry detection circuit.
3.   The correction circuit.

The basic adder is simply a 4-bit *binary* adder. Note, however, that the A and B inputs can never be greater than 9 because they are in BCD form, and the output of the **74x283** can never be greater than 19.

The carry detection circuit produces a carry-out if the sum is greater than 9. Sums between 10 and 15 are detected by the logic gates and sums between 16 and 19 produce a carry-out of the **74x283**. The equation is

$$\text{Carry-out} = C_4 + \Sigma_4(\Sigma_3 + \Sigma_2)$$

which is implemented by the logic gates.

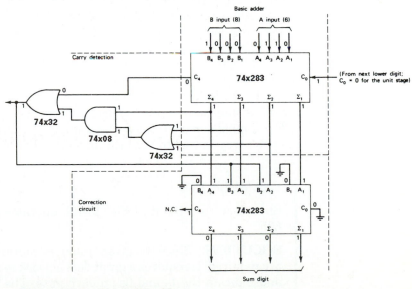

Figure 15-9   The BCD adder.

A second **74x283** is used in the correction circuit, and the final sum is the output of this IC. If the sum of the inputs is 9 or less, there is no carry-out and the lower **74x283** adds 0 to the output of the upper **74x283**, giving the correct sum. If the sum is greater than 9, the carry-out has a weight of 10, and 10 must be subtracted from the sum. This is equivalent to adding the 2's complement of 10 (i.e., 6) to the output of the first adder. Therefore, whenever a carry-out is produced, 6 is added to the first sum to correct and give the proper sum output in BCD form.

Most microprocessors are essentially hex computers, but they can do BCD addition by using the DAA (Decimal Adjust Accumulator) instruction. This instruction functions much like the circuit of Fig. 15-9 and adds 6 whenever an adjustment must be made to convert a hex sum into a BCD sum.

BCD adders can be cascaded to add numbers several digits long simply by tying the carry-out of a stage to the carry-in of the next stage. Of course, the least-significant stage never has a carry-in, so its carry input is grounded.

EXAMPLE 15-7

The inputs to a BCD adder are 6, 8, and a carry-in. Show how it performs the addition.

SOLUTION   The 1s and 0s for this example are shown in Fig. 15-9. The upper **74x283** accepts the inputs and produces an output of 15. The carry circuit produces a carry-out and causes the lower **74x283** to add 6 to the 15 it receives from the upper **74x283**. The Σ output of the lower **74x283** is 5 (0101). Thus, the circuit has successfully added the inputs by producing an output of 15 (a sum output of 5 plus a carry output).

15-7-2   The BCD Subtracter

BCD subtraction is more difficult (and painful) than BCD addition, especially if the possibility of a negative result is allowed. The simplest BCD subtracter that assumes a 0 or positive result is shown in Fig. 15-10. Figure 15-10a shows a single stage of the subtracter and Fig. 15-10b shows how the stages can be cascaded for numbers of several digits.

The circuit functions by adding the 15's complement of the B input (produced by the 4 inverters) to the A input. A correction of 10, when necessary, is made by the lower **74x283**. Note that a carry of 1 into a stage indicates that the previous stage is *not* attempting to borrow, and a *stage generates a carry-out of 0 when it must borrow from the next stage*. Therefore, the least significant stage has its carry-in tied to 1, which indicates that *there is no borrow into it*.

EXAMPLE 15-8

Show how the circuit of Fig. 15-10b operates when subtracting 3627 from 8353.

SOLUTION   Since the given numbers contain 4 digits, 4 subtracter stages are required. It requires a great deal of work to draw the 4 stages individually and indicate the 1s and 0s on each gate. The work can be reduced using a

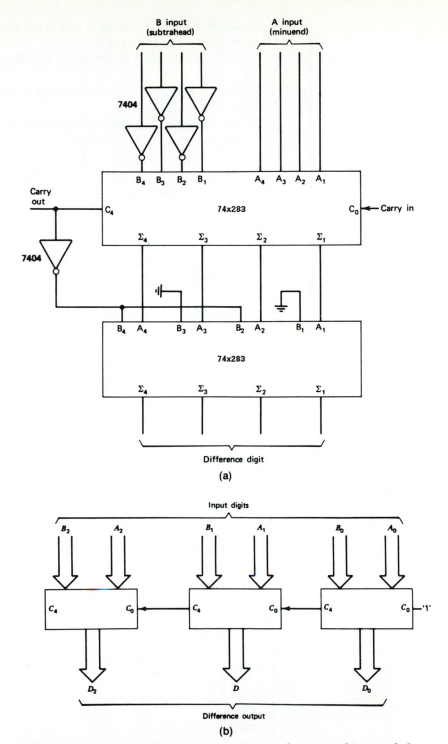

Figure 15-10 The BCD subtracter: (a) single stage; (b) cascaded stages.

table, shown in Fig. 15-11, where A and B are the minuend and subtrahend, and $\overline{B}$ is the 15's complement of B.

The table of Fig. 15-11 is developed as follows:

1. Start at the least-significant digit (stage 1), where the carry-in is known to be 1.
2. The inputs to the upper adder are 3, 8, and 1 (the carry-in). It produces a sum of 12 and no carry-out.
3. The inputs to the lower adder are 12 and 10. The sum output of the upper adder is connected to the A inputs of the lower adder, and the B inputs to the lower adder are either 10 (when there is no carry) or 0 (when carry-out = 1).
4. The lower adder produces a result of $12 + 10 = 22 = 6 +$ carry-out. The difference digit is 6.
5. The inputs to the second stage are now seen to be 5, 13, and 0 (no carry-in from the first stage). It produces a sum of 2 and a carry-out.
6. The rest of the table is constructed by continuing with the procedure developed here.

Note: The most significant stage must have a carry-out of 1 for a positive result.

The table gives the numbers at all the *intermediate* stages of the subtracter and simplifies debugging.

It is possible to build BCD adder/subtracters by modifying and expanding on the designs presented here. These circuits become complex, however, and the

	Stage			
	4	3	2	1
A	8	3	5	3
B	3	6	2	7
$\overline{B}$	12	9	13	8
C_{IN}	0	1	0	1
Σ	4	13	2	12
C_{OUT}	1	0	1	0
B''	0	10	0	10
Result	4	7	2	6

Figure 15-11 Subtracter table for Example 15-8.

reader is advised to investigate commercially available BCD arithmetic units in IC form before attempting to build these circuits using SSI and MSI chips.

15-8   Arithmetic-Logic Units

The discussions of the previous sections have shown a need for circuits that perform more complex arithmetic operations. Such circuits are called *arithmetic-logic* units (ALUs).

15-8-1   The 74x181

The **74x181**[3] is the basic ALU in the **7400** TTL series. A simplified functional layout of the **74x181** is shown in Fig. 15-12a and the function table is given in Fig. 15-12b.

The **74x181** accepts two 4-bit words, A and B, as data inputs and a CARRY-IN, which acts as an *inverted carry* during addition operations because it is LOW when a carry-in occurs. There are also *five control inputs* that determine the *operation* performed on the inputs. The MODE input determines whether the output is a *logical* or *arithmetic* function of the inputs. *The carry-in does not affect the logic functions.* The four SELECT lines select 1 of 16 possible logic operations or arithmetic operations.

The outputs of the **74x181** include the 4-bit result (the F outputs), a CARRY-OUT labeled C_{n+4}, an A = B output, and the GENERATE and PROPAGATE outputs. The outputs are determined by carefully following the function table. Note that the $+$ sign means logical OR and the word *plus* means the sum of the inputs.

The advantage of the **74x181** is that it can perform the operations of addition, subtraction, shifting (one place), AND, OR, XOR, and many others on the input variables simply by changing the SELECT and MODE inputs. Logic operations are performed when the M (Mode) input is HIGH. They are performed on a bit-by-bit basis, and the carry is irrelevant. The most important logic operations and their corresponding S values, as shown in Fig. 15-12b, are:

S	Operation
1	NOR
4	NAND
6	XOR
9	EQUIVALENCE
11	AND
14	OR

Also, the F outputs can be set to be equal to the A or B inputs or their complements by selecting other values of S.

The arithmetic operations are performed when the M input is LOW. The carry-in *does* affect these operations and there are two columns in Fig. 12−5b, to

---

[3]The **74x181** is available in the LS series and in the advanced CMOS (ACT) and advanced Schottky (AS) series.

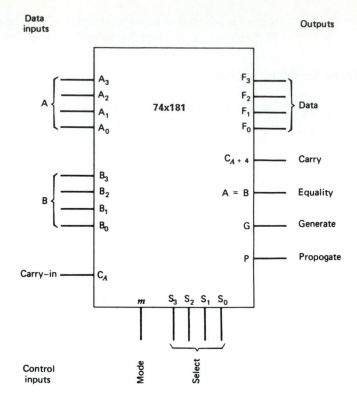

Data inputs

Outputs

74x181

A_3 A_2 A_1 A_0 — A

B_3 B_2 B_1 B_0 — B

Carry-in — C_A

m S_3 S_2 S_1 S_0

F_3 F_2 F_1 F_0 — Data

C_{A+4} — Carry

$A = B$ — Equality

G — Generate

P — Propogate

Control inputs

Mode

Select

(a) Functional layout

TABLE 1

SELECTION	M = H LOGIC FUNCTIONS	ACTIVE-HIGH DATA M = L; ARITHMETIC OPERATIONS	
S_3 S_2 S_1 S_0		C_n = H (no carry)	C_n = L (with carry)
L L L L	F = $\bar{A}$	F = A	F = A PLUS 1
L L L H	F = $\overline{A+B}$	F = A + B	F = (A + B) PLUS 1
L L H L	F = $\bar{A}B$	F = A + $\bar{B}$	F = (A + $\bar{B}$) PLUS 1
L L H H	F = 0	F = MINUS 1 (2's COMPL)	F = ZERO
L M L L	F = $\overline{AB}$	F = A PLUS A$\bar{B}$	F = A PLUS A$\bar{B}$ PLUS 1
L M L H	F = $\bar{B}$	F = (A + B) PLUS A$\bar{B}$	F = (A + B) PLUS A$\bar{B}$ PLUS 1
L M H L	F = A $\odot$ B	F = A MINUS B MINUS 1	F = A MINUS B
L M H H	F = A$\bar{B}$	F = A$\bar{B}$ MINUS 1	F = A$\bar{B}$
M L L L	F = $\bar{A}$ + B	F = A PLUS AB	F = A PLUS AB PLUS 1
M L L H	F = $\overline{A \odot B}$	F = A PLUS B	F = A PLUS B PLUS 1
M L H L	F = B	F = (A + $\bar{B}$) PLUS AB	F = (A + $\bar{B}$) PLUS AB PLUS 1
M L H H	F = AB	F = AB MINUS 1	F = AB
M H L L	F = 1	F = A PLUS A*	F = A PLUS A PLUS 1
M H L H	F = A + $\bar{B}$	F = (A + B) PLUS A	F = (A + B) PLUS A PLUS 1
M H H L	F = A + B	F = (A + $\bar{B}$) PLUS A	F = (A + $\bar{B}$) PLUS A PLUS 1
M H H H	F = A	F = A MINUS 1	F = A

*Each bit is shifted to the next more significant position.

(b) Function table

Figure 15-12   The **74x181** arithmetic/logic unit. (From The *TTL Data Book for Design Engineers,* 2nd ed., Texas Instruments, Inc. Courtesy of Texas Instruments, copyright 1976.)

give the results when the carry-in is HIGH and when it is LOW. Many of the arithmetic operations are rarely used or not useful. The useful operations are:

S	Operation
0	Increment
3	Set to all 1s or all 0s (depending on the carry-in).
6	Subtraction
9	Addition
12	Shifting (1 place)
15	Decrement

In arithmetic operations, the **74x181** operates in one of two submodes—addition or subtraction. All S (Select) numbers excepts 6 and 15 operate in addition mode. For addition operations, a 0 on the carry line (C_n) indicates a carry input and increments the result. Any addition operation performed by the **74x181** produces a LOW carry-out if the sum is greater than 15.

In subtraction operations (S = 6 or S = 15), a positive or 0 result will produce a 0 on the carry-out line. If the carry-in line is HIGH, it will decrement the result. *If the result of a minus operation is negative, it is presented as a 4-bit, 2's complement number, and the carry-out is HIGH.* For example, if the result is −4, the F outputs read 1100 and the carry-out is 1.

EXAMPLE 15-9

Find the F output of the **74x181** for the shift operation if the A input is 6 and the carry-in is:

(a)   HIGH

(b)   LOW

SOLUTION   The shift operation is selected if M = 0 and S = 12. The function table shows the output to be A plus A if the carry-in is HIGH, indicating no carry. Here the result would be 6 + 6 = 12 (or C). Notice that C is 6 shifted by one bit.

If the carry-in is LOW, the function table gives F as A plus A plus 1. If A is 6, the result is 13 or D. Again this is 6 shifted, but a 1 has been brought into the LSB. This allows **74x181**s to be cascaded for larger numbers (see Problem 15-12). Note that the B inputs to the **74x181** are irrelevant for this function; only the A inputs can be shifted. In both cases the carry-out of the **74x181** is HIGH because the sum is less than 16.

EXAMPLE 15-10

Find the F outputs of the **74x181** if M = 0, S = 15, C_{in} = 1, and the A inputs are:

(a)   5

(b)   0

SOLUTION This is the decrement mode. The table indicates that F = A − 1. If A = 5, F = 4 and the carry-out is 0. If A = 0, however, the result of the decrement process is −1 or F and the carry-out is 1, which indicates a negative result.

A table of the outputs of a **74x181**, when A = 9 (1001) and B = 10 (1010), is shown in Fig. 15-13 to help explain its operation. Using the given inputs and the function table (Fig. 15-12b), the outputs presented in the table can be calculated.

EXAMPLE 15-11

Explain how the outputs of line 14, Fig. 15-13, were determined.

SOLUTION

(a) The logic output is given in Fig. 15-12b as F = A + B (A or B) for line 14. Since A = 1001 (9) and B = 1010 (10), A + B = 1011 as shown.

(b) With M = 0 and no carry, (C_{in} = 1), F = (A + $\bar{B}$) plus A. Here A + $\bar{B}$ is A, 1001, ORed with $\bar{B}$, 0101, which equals 1101, or numerical 13. This is added to the number on the, A inputs (9), as the *plus* operation specifies, and the result is 22, or 6 and a carry-out. Because the sum is greater than 15, the carry-out is LOW and is shown as a 0 on the C_{n+4} line.

(c) With a LOW carry-in (Cn = 0), F = (A + $\bar{B}$) plus A plus 1 = 23. The results as shown in Fig. 15-13 are 7 and a LOW carry-out.

Line	Selection S_3 S_2 S_1 S_0	M = 1 Logic Functions F_3 F_2 F_1 F_0	M = 0 Arithmetic Operations C_{IN} = 1 (no carry) F_3 F_2 F_1 F_0	C_{n+4}	C_{IN} = 0 (with carry) F_3 F_2 F_1 F_0	C_{n+4}
0	0 0 0 0	0 1 1 0	1 0 0 1	1	1 0 1 0	1
1	0 0 0 1	0 1 0 0	1 0 1 1	1	1 1 0 0	1
2	0 0 1 0	0 0 1 0	1 1 0 1	1	1 1 1 0	1
3	0 0 1 1	0 0 0 0	1 1 1 1	1	0 0 0 0	0
4	0 1 0 0	0 1 1 1	1 0 1 0	1	1 0 1 1	1
5	0 1 0 1	0 1 0 1	1 1 0 0	1	1 1 0 1	1
6	0 1 1 0	0 0 1 1	1 1 1 0	1	1 1 1 1	1
7	0 1 1 1	0 0 0 1	0 0 0 0	0	0 0 0 1	0
8	1 0 0 0	1 1 1 0	0 0 0 1	0	0 0 1 0	0
9	1 0 0 1	1 1 0 0	0 0 1 1	0	0 1 0 0	0
10	1 0 1 0	1 0 1 0	0 1 0 1	0	0 1 1 0	0
11	1 0 1 1	1 0 0 0	0 1 1 1	0	1 0 0 0	0
12	1 1 0 0	1 1 1 1	0 0 1 0	0	0 0 1 1	0
13	1 1 0 1	1 1 0 1	0 1 0 0	0	0 1 0 1	0
14	1 1 1 0	1 0 1 1	0 1 1 0	0	0 1 1 1	0
15	1 1 1 1	1 0 0 1	1 0 0 0	0	1 0 0 1	0

Figure 15-13   Outputs at a **74181** if A = 9 and B = 10.

EXAMPLE 15-12

Design an adder/subtracter using cascaded **74x181**s. Show how it works if A = 57 and B = 28 under:

(a) Addition.

(b) Subtraction.

(c) Repeat using the hex equivalent of the given numbers.

SOLUTION The **74x181** can be cascaded in the normal manner, by tying the carry-out of a stage to the carry-in of the succeeding stage.

(a) Addition can be performed by setting S to 9, M to 0, and the carry-in of the least significant stage to 1.

$$\text{If A} = 57 = (00111001)_2$$
$$\text{and B} = 28 = (00011100)_2$$

The least significant **74x181** sees inputs of 1001, 1100, and a HIGH carry-in. Its output, as specified by the function table, is 9 plus 12 = 5. Since the sum is greater than 15, the carry-out is LOW.

The second stage sees inputs of A = 3, B = 1, and a LOW carry-in. Its output is A plus B plus 1 = 5. The final output is, therefore, $(01010101)_2 = (85)_{10}$, which is the correct answer.

(b) Subtraction can be performed by setting S to 6, M to 0, and the carry-in of the least significant stage to 0. Then the least significant **74x181** sees an A input of 9, a B input of 12, and a LOW carry-in. Its output is 3 or 1101 (the 4-bit 2's complement of 3). It also produces a HIGH carry-out because the result is negative.

The next stage sees inputs of 3, 1, and a HIGH carry-in. The results are A minus B minus 1 = 3 − 1 − 1 = 1. The final answer is, therefore, 00011101 or 29, which is also correct.

(c) Using the hex equivalent of the given numbers, A = 39 and B = 1C. Their sum is $(55)_{16}$ or $(85)_{10}$. For subtractions we first have 9 − C = −3 or D (the 4-bit 2's complement of D). This also produces a borrow out so the most significant nibble becomes 3 − 1 − 1 = 1, and the final result is $(1D)_{16} = (29)_{10}$. It is wise to use the hex arithmetic because the **74x181**s essentially operate in hex mode and it greatly simplifies debugging if a problem should arise.

15-8-2 The 74LS381

The reader may have observed that many of the functions provided by the **74x181** have little practical value. The function of Example 15-11 is an example. The manufacturers have also seen this and have produced the **74LS381**, which is a somewhat smaller and simpler ALU. It is only available in the LS series.

The pinout, pin designations, and function table of the **74LS381** are shown in Fig. 15-14. There are only three SELECT pins and no MODE input so the **74LS381** can only perform eight operations. These are the most commonly used arithmetic and logic operations and are listed in the function table.

DESIGNATION	PIN NOS.	FUNCTION
A3, A2, A1, A0	17, 19, 1, 3	WORD A INPUTS
B3, B2, B1, B0	16, 18, 2, 4	WORD B INPUTS
S2, S1, S0	7, 6, 5	FUNCTION-SELECT INPUTS
C_n	15	CARRY INPUT FOR ADDITION, INVERTED CARRY INPUT FOR SUBTRACTION
F3, F2, F1, F0	12, 11, 9, 8	FUNCTION OUTPUTS
$\overline{P}$ ('LS381A 'S381 ONLY)	14	ACTIVE-LOW CARRY PROPAGATE OUTPUT
$\overline{G}$ ('LS381A 'S381 ONLY)	13	ACTIVE-LOW CARRY GENERATE OUTPUT
C_{n+4} ('LS382A ONLY)	14	RIPPLE-CARRY OUTPUT
OVR ('LS382A ONLY)	13	OVERFLOW OUTPUT
V_{CC}	20	SUPPLY VOLTAGE
GND	10	GROUND

SN54LS381A, SN54S381
. . . J OR W PACKAGE
SN74LS381A, SN74S381
. . . DW OR N PACKAGE
(TOP VIEW)

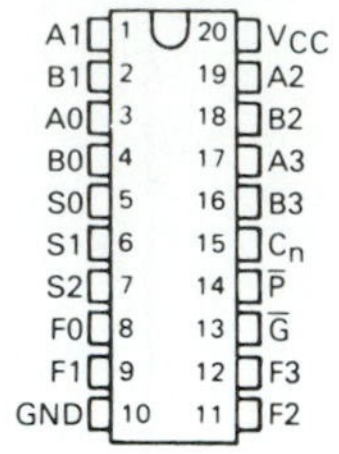

```
      A1 [ 1   U  20 ] Vcc
      B1 [ 2      19 ] A2
      A0 [ 3      18 ] B2
      B0 [ 4      17 ] A3
      S0 [ 5      16 ] B3
      S1 [ 6      15 ] Cn
      S2 [ 7      14 ] P̄
      F0 [ 8      13 ] Ḡ
      F1 [ 9      12 ] F3
     GND [ 10     11 ] F2
```

FUNCTION TABLE

SELECTION			ARITHMETIC/LOGIC
S2	S1	S0	OPERATION
L	L	L	CLEAR
L	L	H	B MINUS A
L	H	L	A MINUS B
L	H	H	A PLUS B
H	L	L	A $\oplus$ B
H	L	H	A + B
H	H	L	AB
H	H	H	PRESET

H = high level,   L = low level

Figure 15-14   The **74LS381**: a) pinout; b) PIN designations; and c) function table. (Reprinted by Permission of Texas Instruments.)

15-8-3   Look-Ahead Carry

In the adders discussed thus far, the output of each stage depended on its inputs and the carry-in from the previous stage. These adders are relatively slow because the carry must ripple through all the stages before the result is correct. *Look-ahead carry* is a method of speeding up the carry generation on an adder. It requires additional hardware and should only be used when the *speed* of the adder circuit is extremely critical.

The **74x182** is an IC that contains all the logic required for look-ahead carry, and is used in conjunction with the **74x181**. The ICs are connected using the GENERATE and PROPAGATE outputs of the **74x181**s.

Because look-ahead carry is rarely used, and its explanation is fairly complex, we feel it does not warrant further discussion here. This subsection is intended to make the user aware of its existence, should he encounter it. The topic is explored further in other books, including the earlier editions of this book.

15-9   Binary Multiplication

The ability to perform binary multiplication and division is not as important as the ability to add, subtract, and logically manipulate data. Some microprocessors, such as the Intel **8085**, do not multiply and divide (these operations are performed by a program, or "in software").

Binary multiplication is achieved by a succession of shifts and additions. In this section, a multiplier for 7-bit words is designed using the shift-and-add algorithm developed in the next paragraph.

15-9-1 The Multiplication Algorithm

Multiplication starts with a multiplier, multiplicand, and PRODUCT REGISTER. The PRODUCT REGISTER, which is initially 0 and eventually contains the product, must be as long as the number of bits in the multiplier and multiplicand added together. In computers, the multiplier and multiplicand are typically one n-bit word each, and two n-bit words must be reserved for the product.

Multiplication can be performed in accordance with the following algorithm:

1. Examine the least significant bit of the multiplier. If it is 0, shift the PRODUCT REGISTER one bit to the right. If it is 1, add the multiplicand to the MSB of the PRODUCT REGISTER and then shift.
2. Repeat step 1 for each bit of the multiplier.

At the conclusion, the product should be in the PRODUCT REGISTER.

It is common to shift the multiplicand one bit *left* for each multiplier bit and then add it to the PRODUCT REGISTER. This is analogous to multiplication as taught in grade school. The algorithm presented here, however, holds the *position* of the multiplicand *constant* but shifts the PRODUCT REGISTER *right*. This algorithm is more easily implemented in hardware (see section 15-9-2).

EXAMPLE 15-13

Multiply 22 × 26 using the above algorithm.

SOLUTION   The solution is shown in Fig. 15-15. The multiplier bits are listed in a column with the LSB on top. The multiplication proceeds in accordance with these bits, as the leftmost column shows.

Consider line 5 as an example. The multiplier bit is 1, so the 5-bit multiplicand (26) is added to the 5 MSBs of the PRODUCT REGISTER that appear on line 4 as 13. The result, 39, appears on line 5. The product moves steadily to the right. The final product appears on line 9 and is a 10-bit number in this case. Note that the MSB of the PRODUCT REGISTER is reserved for carries that may result from the additions.

```
Multiplier = 22 =    1  0  1  1  0

Multiplicand = 26 = | 1  1  0  1  0 |
                                                          Multiplier
                                                              Bit
Line Number

      1             | 0  0  0  0  0 |                                   Initial product
      2             | 0  0  0  0  0 | 0                         0       Shift product register
      3             | 1  1  0  1  0 | 0                         1       Add multiplicand
      4             | 0  1  1  0  1 | 0  0                               Shift product register
      5           1 | 0  0  1  1  1 | 0  0                      1       Add multiplicand
      6             | 1  0  0  1  1 | 1  0  0                            Shift product register
      7             | 0  1  0  0  1 | 1  1  0  0                0       Shift product register
      8           1 | 0  0  0  1  1 | 1  1  0  0                1       Add multiplicand
      9             | 1  0  0  0  1 | 1  1  1  0  0                      Shift — final product
```

Figure 15-15   Multiplying 22 × 26.

15-9-2 Hardware Implementation

A hardware implementation of a multiplier for two 7-bit words is shown in Fig. 15-16. It uses many circuits that were described in earlier sections of this book. The basic parts of the circuit are:

1. The MULTIPLICAND REGISTER. This contains the multiplicand.

2. The ADDER. It consists of **74x283**s and adds the multiplicand register and the product register.

3. The PRODUCT REGISTER. It is built of a **74198** and a **74164**. The more significant stages of the register must have both parallel load and parallel output capability; hence a **74198** is chosen. Two 4-bit shift registers such as **74x195**s could also be used. The most significant bit of the PRODUCT REGISTER must be available to hold any carry-out of the adder (see lines 5 and 8 of Fig. 15-15). Since the less significant stages of the PRODUCT REGISTER only need serial inputs, a **74x164** parallel-output shift register is selected.

4. The **555** timer. This produces a continuous clock. Other oscillators discussed in Chapter 7 would serve as well.

5. The CONTROL COUNTER. The more significant stage of this counter controls the SHIFT/LOAD input to the **74198**, and changes the mode on alternate pulses. The clock pulses to the **74198** are supplied by the least significant stage.

6. The gating of the **74x32** and **74x08**. This gating supplies a clock to the **74198** on each pulse of the CONTROL COUNTER *if it is in shift mode OR Q_H of the MULTIPLIER SHIFT REGISTER is 1*. It eliminates LOAD clocks when Q_H of the MULTIPLIER REGISTER is 0.

7. The SHIFT COUNTER. This is a **74x193** that is loaded with the number of shifts required. For a 7-bit multiplier, the **74x193** is loaded with the number 6, as shown.

8. The MULTIPLIER SHIFT REGISTER. This register is loaded with the multiplier (Q_H is the LSB) that is shifted out during its operation.

9. The INACTIVE FF. After the proper number of shifts the BORROW-OUT of the **74x193** sets the INACTIVE FF, which then holds the CONTROL COUNTER clear and stops the circuit from changing further. This retains the answer in the PRODUCT REGISTER until the next start pulse occurs.

10. The START switch. Depressing the debounced START switch initializes the circuit. When it is released, the multiplication begins.

The operation of the circuit is briefly described:

1. Depressing the START switch places a 0 at point A. This clears the PRODUCT REGISTER and the INACTIVE FF, loads the MULTIPLIER SHIFT REGISTER with the multiplier, and the SHIFT COUNTER with the count. It prevents the CONTROL COUNTER from counting by holding the J input to the **74x107** LOW.

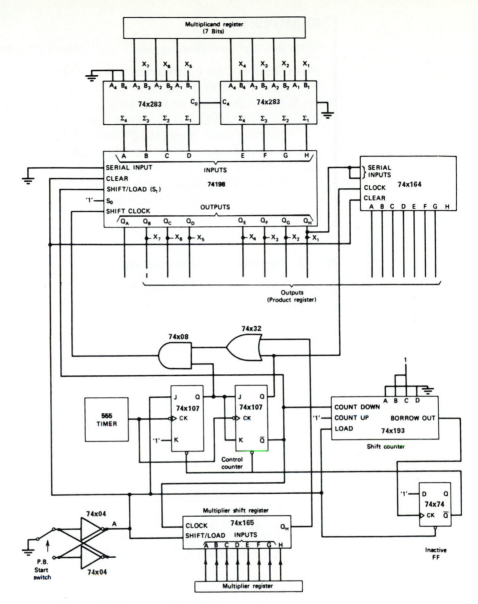

Figure 15-16 A 7-bit multiplier.

2. When the START switch is released, the CONTROL COUNTER starts. It clocks the MULTIPLIER SHIFT REGISTER and SHIFT COUNTER and provides the proper SHIFT and LOAD pulses to the PRODUCT REGISTER. The LOAD pulses load the sum of the multiplicand and the PRODUCT REGISTER into the PRODUCT REGISTER, effectively implementing the add-and-shift algorithm.

3. When the SHIFT COUNTER counts down the proper number of pulses, it produces a BORROW OUT. This sets the INACTIVE FF and stops the circuit from pulsing further.

EXAMPLE 15-14

Show how the circuit of Fig. 15-16 multiplies 75 $(01001011)_2$ by 127 $(01111111)_2$.

SOLUTION The solution is shown by the timing chart and data table of Fig. 15-17. The shift clocks occur at $t = 3, 7, 11$, and so on. Load clocks occur at $t = 1, 5, \ldots$ There are no load clocks at $t = 9, 17$, and 21 because the MULTIPLIER SHIFT REGISTER output is 0.

The data table shows the PRODUCT REGISTER contents as the multiplication progresses. At $t = 1$, for example, the PRODUCT REGISTER is loaded with the multiplicand, and it is shifted at $t = 3$. At $t = 5$, the PRODUCT REGISTER is loaded with the sum of its own contents plus the multiplicand, or $63 + 127 = 190$. The shifts and loads continue until the final product occurs at $t = 27$.

15-9-3 Practical Considerations

The multiplier of Fig. 15-16 was designed so that the two halves of the PRODUCT REGISTER do not shift at the same time. The **74x164** shifts at $t = 2, 6,$

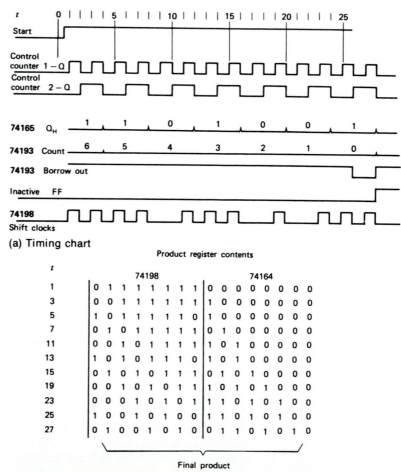

(a) Timing chart

Product register contents

t	74198								74164							
1	0	1	1	1	1	1	1	1	0	0	0	0	0	0	0	0
3	0	0	1	1	1	1	1	1	1	0	0	0	0	0	0	0
5	1	0	1	1	1	1	1	0	1	0	0	0	0	0	0	0
7	0	1	0	1	1	1	1	1	0	1	0	0	0	0	0	0
11	0	0	1	0	1	1	1	1	1	0	1	0	0	0	0	0
13	1	0	1	0	1	1	1	0	1	0	1	0	0	0	0	0
15	0	1	0	1	0	1	1	1	0	1	0	1	0	0	0	0
19	0	0	1	0	1	0	1	1	1	0	1	0	1	0	0	0
23	0	0	0	1	0	1	0	1	1	1	0	1	0	1	0	0
25	1	0	0	1	0	1	0	0	1	1	0	1	0	1	0	0
27	0	1	0	0	1	0	1	0	0	1	1	0	1	0	1	0

Final product

Figure 15-17 Multiplying 75×127 using the circuit of Fig. 15-16.

10 . . . while the **74198** shifts at $t = 3, 7, 11, \ldots$ This prevents race and sneak problems that may arise if the same clock drives both shift registers and a clock delay occurs somewhere in the circuit. It also simplifies the circuitry.

The circuit of Fig. 15-16 is very easy to test in the laboratory. It is best tested by replacing the start switch with a PULSE GENERATOR that provides repetitive short negative pulses. Multiplication occurs during the long positive portion of the wave. It would also be wise to synchronize the oscillator with the PULSE GENERATOR (possibly using the reset on the **555**). Operation of the multiplier can be observed on a CRO and any faults corrected.

15-9-4 Signed Multiplication

The *multiplication of signed* numbers is accomplished by converting both operands to positive numbers, multiplying them and then converting, if necessary, so the product has the proper sign.

If the numbers are in 2's complement form and a 2's complement answer is required, the *Booth algorithm* can be used to produce the result directly. The circuit of Fig. 15-16 can be converted to a *Booth algorithm multiplier* with minor modifications.

15-10 Arithmetic Processing Units and Coprocessors

The Arithmetic Processing Unit (APU) is an IC that can perform many complex arithmetic functions such as:

> ADD
> SUBTRACT
> MULTIPLY
> DIVIDE
> SQUARE ROOT
> SINE, COSINE, TANGENT
> INVERSE SINE, COSINE, TANGENT
> LOGARITHMS

The APU can perform these operations on fixed or floating point data and very long words. The APU is driven by a clock (just as the multiplier circuit of Fig. 15-16). Each operation takes a number of clock cycles and more complex operations take more clock cycles.

The APU is designed to get data and commands from a microprocessor. When the APU receives a command, it becomes busy for the number of clock cycles required to complete the arithmetic operation. It then outputs an $\overline{\text{END}}$ signal to signify the completion of the operation and its ability to accept another command.

15-10-1 The 8087 Coprocessor

The Intel Corporation has designed an **8087** coprocessor to work with its **8086/88** microprocessors to calculate complex mathematical expressions such as exponentiation, square roots, and trigonometric functions. The **8086** can perform all of these functions without the **8087** by using software subroutines. There are three major reasons for using the coprocessor, however:

1. The **8087** can perform a complex mathematical operation in about 1 percent of the time it would take an **8086** using a software subroutine.
2. The **8087** contains eight 80-bit registers. Thus, arithmetic operations can be performed on much longer words, and with much greater accuracy. These long words give the system a distinct advantage even for such simple operations as multiplication and division.
3. The **8087** can perform operations on 16-, 32-, or 64-bit integers, but it can also perform *floating-point* arithmetic. Floating point is used in many scientific calculations.

The **8087** operates in conjunction with the **8086**. If it is installed, the **8086** will turn over complex calculations to the **8087** whenever it encounters them. Of course, Intel has also produced coprocessors such as the **80287** and **80387** to work with their more advanced '**286** and '**386** μPs.

Personal computers generally contain a socket for a coprocessor. If the PC user is mainly doing word processing or other simple operations, the coprocessor is probably unnecessary. If the PC is being used for scientific work, with many calculations involved (commonly called "number-crunching"), then a coprocessor may be indispensable.

SUMMARY

This chapter discussed various circuits for performing binary arithmetic. It began with half-adders, full-adders, and subtracter circuits built from SSI gates.

Larger scale integrated (LSI) circuits for performing binary arithmetic, specifically the **74x283**, **74x181**, and **74381**, were introduced. Use of the **74x283** in 2's complement and BCD ADDER/SUBTRACTERS was considered. Use of the **74x181** ALU to perform arithmetic and logic functions was described in some detail.

We then proceeded to more complex circuits. A multiplier circuit was designed, which should give the reader a feel for what goes into an APU or coprocessor. Finally, coprocessors were discussed.

At this point, the reader should be able to design circuits to perform any simple arithmetic or logical operations he or she requires. More sophisticated arithmetic operations are generally done by computers, using programmed algorithms, or by arithmetic processing units when high speed is required.

GLOSSARY

2's complement: A representation of numbers where negative numbers are obtained by complementing their positive equivalent and adding 1.

ADDER/SUBTRACTER: A circuit that can either add or subtract 2 operands, depending on the mode.

Arithmetic-logic unit (ALU): A circuit capable of performing a variety of arithmetic or logical functions.

BCD arithmetic: Arithmetic involving decimal digits, expressed in BCD form.

Complementation: The process of taking the difference between a given number and an input number.

Full-adder: A circuit that accepts 2 binary bits and a carry-in as inputs and produces their sum and carry as outputs.

Full-subtracter: A circuit that accepts 2 inputs and a borrow-in and produces their difference and a borrow as outputs.

Half-adder: A circuit that accepts 2 binary bits as inputs and produces their sum and carry as outputs.

Half-subtracter: A circuit that accepts 2 inputs, computes their difference, and determines whether there is a borrow-out.

Look-ahead-carry: Using special circuitry to rapidly generate the carry-in of an adder circuit.

Overflow: Overflow occurs when the result of an arithmetic operation is a larger number than the output register can accommodate.

PRODUCT REGISTER: The register used to hold the product in a multiplier.

Underflow: Underflow occurs when the result of an arithmetic operation is a more negative number than the output register can accommodate.

PROBLEMS

Section 15-3-2.

15-1. Add 5 and 3 using the circuit of Fig. 15-3. Show all your 1s and 0s.

15-2. Add 25 and 14 using a half-adder and full-adders. You may use symbolic blocks as in Fig. 15-3a, but show all the 1s and 0s.

Section 15-4.

15-3. Design a subtractor circuit to subtract 3-bit binary numbers. Show how it works if the numbers are 6 minus 5.

Section 15-5-1.

15-4. Add the numbers 59 and 66 using the circuit of Fig. 15-6.

Section 15-5-2.

15-5. Use the circuit of Fig. 15-8 to perform the following operations:
    (a)   65 + 57
    (b)   65 − 57
    (c) −35 + 42
    (d) −35 − 42
    (e) −35 − (−42)

15-6. For the circuit of Fig. P15-4:

    (a) What are we trying to do? (Consider your inputs as 2's complement numbers.)

    (b) Find the two defective circuits.

    (c) Redraw the circuit to show it working without defective gates.

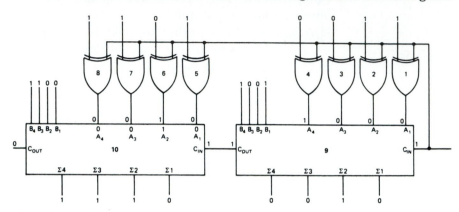

Figure P15-4

Section 15-6-1.

15-7. Add the 8-bit hex numbers 90 and 9C. Do they produce overflow?

15-8. The following numbers are decimal. Which operations produce overflow?

    (a) $100 + 43$

    (b) $100 + (-43)$

    (c) $100 - (-43)$

    (d) $-90 + -90$

    (e) $-100 + (-43)$

    (f) $-100 - (-43)$

15-9. Design a circuit to detect overflow in addition.

Section 15-7-1.

15-10. Add 32 and 69 using a BCD adder. Show all the 1s and 0s.

Section 15-7-2.

15-11. Subtract 676 from 2432 by using the BCD subtracter of Fig. 15-10. Make a table similar to Fig. 15-11.

Section 15-8-1.

15-12. Two **74x181**s are cascaded. Show the S and M inputs for the increment, decrement, complement, and shift operations. Show the carry-in, carry-out, and F outputs of each **74x181**, if the A inputs are

    (a) CD

    (b) DO

    (c) 4F

15-13. Explain line 5 of Fig. 15-13.

15-14. Perform the following operations using **74x181**s (all given numbers are decimal numbers):

(a) -87
      $+(+43)$

(b) -87
      $-(+43)$

(c) -43
      $+(-17)$

(d) -43
      $-(-17)$

(e) $+65$
      $+(-37)$

(f) $+65$
      $-(-37)$

(g) $+65$
      $+(-100)$

(h) $+65$
      $-(-100)$

15-15. Which results in Problem 15-14 produce overflow?

15-16. How can **74x181**s be used to decrement a number? Show the 1s and 0s if the number is 80.

15-17. Add 1 to 271 by using **74x181**s. Show the 1s and 0s.

15-18. Design a single circuit to perform the following functions of two 8-bit operands using two **74x181**s.
  (a) Addition
  (b) Subtraction
  (c) AND
  (d) OR
  (e) NAND
  (f) XOR
Use a rotary switch to select the function.

15-19. Two **74x181**s are connected together. Finish table P 15-19. For any irrelevant answers, use X.

15-20. The circuit of Fig. P15-20 can be used to determine if all 8 bits of A are 0 (if A = 0). Explain how it works. Where is the output taken from?

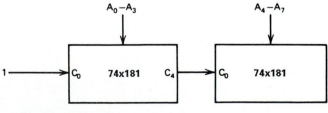

For the **74181**s: S = 15 (F)
                M = 0

Figure P15-20

Section 15-8-2.

15-21. Redo Problem 15-14 using **74LS381**s.

15-22. A shift is not an explicit **'381** function. Show how to attach a multi-

TABLE P15-19

	A_0	B_0	F_0	C_0		A_1	B_1	F_1	C_1
S = 12 M = 0 C_N = 0	B	C				5	3		
S = 6 M = 0 C_N = 0	A	D				6	A		
S = 9 M = 1 C_N = 1	A	D				3	9		
S = 15 M = 0 C_N = 1	0	7				5	7		

plexer to a **'381** so that when an external SHIFT line is HIGH, it will shift the A inputs. It must function normally when SHIFT is LOW.

Section 15-9-1.

15-23. Multiply 37 $\times$ 45 using the algorithm.

Section 15-9-2.

15-24. Multiply 63 $\times$ 107 using the circuit of Fig. 15-16. Produce a timing chart and data table similar to that of Fig. 15-17.

15-25. Design a multiplier for 8-bit words. (*Hint*: Modify Fig. 15-16 by adding a FF and some SSI gates.)

After working the problems, return to the self-evaluation questions of section 15-2. If you cannot answer some of them, review the pertinent sections of the chapter to find the answers.

16

The Basic Computer

16-1   Instructional Objectives

The objective of this chapter is to describe the operation of a computer and to show the reader how to design a rudimentary minicomputer using ICs. This knowledge should clarify the operation of modern microprocessors, which are also introduced.

After reading this chapter the student should be able to:

1. List each basic part of the computer and describe its function.
2. Write a simple computer program.
3. Draw a flowchart.
4. Explain how an arithmetic or logic instruction is executed.
5. Explain how a BRANCH or JUMP instruction is executed.
6. Design a rudimentary computer.
7. Modify the computers presented here to add or delete specific instructions.

16-2   Self-Evaluation Questions

Watch for the answers to the following questions as you read the chapter. They should help you understand the material presented.

1. What is an Op code?
2. Is there any distinction between data and instructions when they are in memory? How does the computer tell the difference?
3. What is a decision box? How many entries and exits does it have? What instructions does it correspond to?
4. What is a self-modifying program? What are its drawbacks?
5. What is the function of the FETCH/EXECUTE FF?
6. Why is the MAR loaded during both FETCH and EXECUTE cycles?
7. What is the difference between indexed and indirect instructions?

16-3 Introduction to the Computer

A computer is a complex digital electronic circuit, but it is composed of the simpler circuits discussed in the earlier chapters of the book, and, after a reasonable explanation, the student should be able to understand how it is designed and how it functions. The computer consists of four basic parts, as shown in Fig. 16-1:

1. The memory
2. The arithmetic-logic unit (ALU)
3. The input-output system
4. The control unit

Fortunately, the memory and ALU have been covered in Chapters 13 and 15, respectively, and the input-output system is only required so the computer can communicate with external devices such as TTYs, disks, and printers. These will be discussed in Chapter 17. This chapter concentrates on the Control Unit that regulates the operation of the computer.

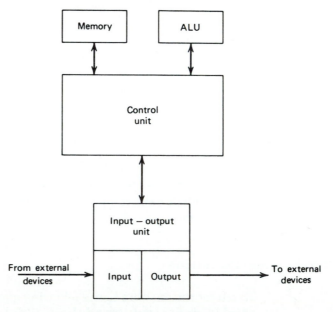

Figure 16-1 Block diagram of a basic computer.

Unfortunately, the programs that run the computer (the software) are not shown in the diagram of Fig. 16-1. It is necessary, therefore, to present an introduction to programming before we can consider the Control Unit further.

16-3-1   Introduction to Programming

Computers are controlled by the *programs* written for them. This makes computers very versatile because *their operation can be changed simply by changing the program.*

A program is a *group of instructions* that cause the computer to perform a given task. Each instruction causes the computer to do something specific. Most instructions affect the *memory* and the *accumulator*, a register within the control unit that contains the basic operand used in each instruction. When an arithmetic or logical instruction that requires two operands is executed, one of the operands is in the accumulator and the other operand is in memory. The results are generally placed in the accumulator.

As a simplified introduction to a computer, assume that each instruction consists of two parts: an Op code and an address. The *Op code* tells the computer *what to do* when it encounters that instruction. The *address is the memory location involved.* The most common instructions are listed in Table 16-1. The symbol (M) means the *contents of memory* at the address contained in the instruction. If memory location 500 contains the number 25, for example, the instruction LOAD 500 causes the memory to be read at address 500 and puts the 25 it finds there in the accumulator.

TABLE 16-1

Some Common Instructions

Instruction	Symbol	Explanation
LOAD	$(M) \rightarrow (A)$	The contents of memory (at the address specified in the instruction) are written to the accumulator.
STORE	$(A) \rightarrow (M)$	The contents of the accumulator are written to memory.
ADD	$(A) + (M) \rightarrow (A)$	The contents of memory are added to the accumulator. The results go to the accumulator.
SUB	$(A) - (M) \rightarrow (A)$	The contents of memory are subtracted from the accumulator. The results go to the accumulator.

EXAMPLE 16-1

If the accumulator contains 35 and 500 contains 25, what will the accumulator contain after the instruction ADD 500 is executed?

SOLUTION   An ADD instruction adds two numbers. One is the number already in the accumulator (35) and the other is the number in memory at location 500 (25). As required by the ADD instruction, the computer adds them and places the result (60) in the accumulator.

16-3-2 A Simple Introductory Program

To make a computer do something useful, a program must first be written and entered into memory. The group of instructions that comprise the program *reside in memory*. Thus, a computer's memory is divided into two areas: a *program area* that contains the instructions, and a *data area* that contains the data used by the program.[1] To run the program, the instructions are generally executed sequentially. Figure 16-2 is a sample program to add the numbers 5, 6, and 7, that are in locations 20, 21, and 22, and store the result in location 23.

Location	Instruction	
	Op code	Address
4	LOAD	20
5	ADD	21
6	ADD	22
7	STORE	23

Figure 16-2   A program to add the contents of locations 20, 21, and 22.

The program proceeds as follows.

1. First the program area had to be selected. For this simple program it was arbitrarily chosen as locations 4 to 7.

2. Next the data area was chosen at memory locations 20 to 23. Note that program and data occupy different areas of memory. Figure 16-3 shows the contents of memory for this problem.

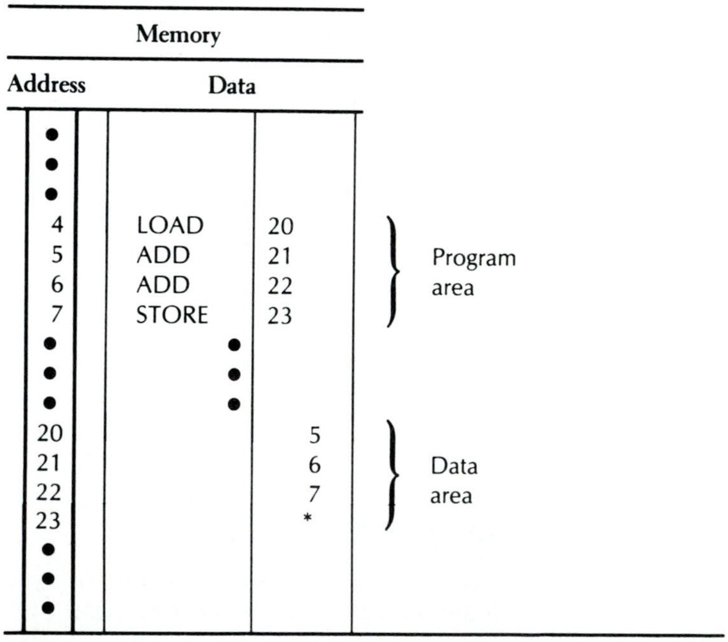

Memory		
Address	Data	
• • •		
4	LOAD	20
5	ADD	21
6	ADD	22
7	STORE	23
• • •	• • •	
20		5
21		6
22		7
23		*
• • •		

} Program area

} Data area

*Reserved for result.

Figure 16-3   Memory contents for the program of Fig. 16-2.

[1]Computers also contain a stack area, but that is beyond the scope of this book.

3. The data (the numbers 5, 6, and 7) are written into locations 20, 21, and 22, respectively. Address 23 is reserved for the result.

4. The program is placed in locations 4 through 7 of memory, and execution is started in location 4.

5. The first instruction LOAD 20 causes the contents of location 20 (5) to be placed in the accumulator.

6. The next instruction adds the contents of 21 (6) to the accumulator. It now contains 11.

7. Next the contents of location 22 are added to the accumulator.

8. The last instruction stores the contents of the accumulator into location 23. At the end of the program it contains 18, the sum of the numbers in 20, 21, and 22.

16-4 Flowcharts

Flowcharts are used by programmers to show the progress of their programs graphically. They are a clear and concise method of presenting the programmer's approach to a problem. They are often used as a part of programming documentation, where the program must be explained to those unfamiliar with it. Since good documentation is essential for the proper use of any computer, the rudiments of flowcharts are presented in this section.

16-4-1 Flowchart Symbols

The flowchart symbols used in this book are shown in Fig. 16-4.

1. The *oval* symbol is either a *beginning or termination box*. It is used simply to denote the start or end of a program.

2. The *rectangular block* is the *processing or command block*. It states what must be done at that point in the program.

(a) Beginning or termination block

(b) Processing or command block

(c) Input/Output block.

(d) Decision block

Figure 16-4 The most common standard flowchart symbols. (From Greenfield and Wray. *Using Microprocessors and Microcomputers: The Motorola Family, 2e,* © 1988. Reprinted by permission of REGENTS/PRENTICE HALL, Englewood Cliffs, New Jersey.

3. The *parallelogram* is an *input-output block*. Such commands as READ or WRITE, especially from an external device such as disk or card reader, are flow-charted using these boxes.

4. The *diamond box* is a *decision box*. It usually contains a question within it. There are typically two output paths; one if the answer to the question is yes, and the other if the answer is no. Sometimes when a comparison between two numbers is made, there might be three exit paths corresponding to the greater than, less than, and equal possibilities.

EXAMPLE 16-2

Draw a flowchart to add the numbers 1, 4, 7, and 10 together.

SOLUTION The solution is shown in Fig. 16-5. It consists simply of a start box, four command boxes, and a stop box. This is an example of *straight-line programming* since *no decisions were made*. It was also assumed that the numbers 1, 4, 7, and 10 were available in the computer's memory and were not read from an external device.

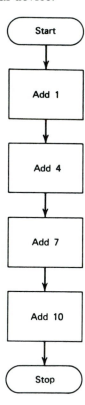

Figure 16-5 Flowchart for Example 16-2. (From Greenfield and Wray. *Using Microprocessors and Microcomputers: The Motorola Family, 2e,* © 1988. Reprinted by permission of REGENTS/PRENTICE HALL, Englewood Cliffs, New Jersey.

16-5 JUMP Instructions and Loops

The program for the flowchart of Example 16-2 can be written very simply as:

        LOAD 20
        ADD 21
        ADD 22

ADD 23

HALT

where locations 20, 21, 22, and 23 contain the numbers 1, 4, 7, and 10, respectively. This program is extremely simple; indeed, the user can compute the answer more quickly than he or she can write the program. Suppose, however, the program is expanded so that we are required to add the numbers 1, 4, 7, 10, . . . 10,000. Conceptually, this could be done by expanding the program and flowchart of section 16-4. But the program and data areas would then require 3,333 locations each and just writing them would become very tedious. Obviously, something else must be done.

16-5-1   JUMP Instructions

JUMP instructions provide the solution to the above problem. Normally instructions are executed one after the other as the program proceeds through the list of instructions. *A JUMP instruction alters the normal sequence of program execution* and is used to create *loops* that allow the same sequence of instructions to be executed many times.

An example of a JUMP instruction is:

JMP 500

where JMP is the abbreviated Op code for the JUMP instruction and 500 is the *jump address*. It causes the jump address to be written into the PC (Program Counter, see section 16-6-2). Thus, the location of the next instruction to be executed is the jump address (500 in this case), rather than the sequential address.

There are two types of JUMP instructions—unconditional and conditional.

The unconditional JUMP always causes the program to go to the JUMP address. It is written (using Intel's terminology) as JMP.

The conditional JUMP causes the program to jump only if a specified condition is met. For this introductory chapter, two conditional JUMP instructions are used:

JP (JUMP Positive)—JUMP on positive accumulator (0 is considered as a positive number).

JN (JUMP Negative)—JUMP on negative accumulator.

Therefore, should the computer encounter one of these instructions, it can simply test the MSB of the accumulator. This determines whether the number in the accumulator is positive (since we are using a 2s complement μP) and whether the branch should be taken.

EXAMPLE 16-3

A computer is to add the numbers 1, 4, 7, 10, . . . 10,000.[2] Draw a flowchart for the program.

[2]Since the object of this section is to teach programming and not mathematics, ignore the fact that this is an arithmetic progression whose sum is given by a simple formula. We also ignore the fact that 10,000 cannot be contained within a single byte.

SOLUTION The flowchart is shown in Fig. 16-6. We recognize that we must keep track of two quantities. One is the number to be added. This has been labeled N in the flowchart and progresses 1, 4, 7, 10 . . . The second quantity is the sum S, which progresses $1, 1 + 4, 1 + 4 + 7$ . . . or $1, 5, 12, . . .$ The first box in the flowchart is an *initialization* box. It sets N to 1 and S to 0 at the beginning of the program. The next box (S = N + S) sets the new sum equal to the old sum plus the number to be added. The number to be added is then increased by 3. At this point the flowchart loops around to repeat the sequence. This is accomplished by placing an unconditional jump instruction in the program. The quantities S and N will then progress as specified.

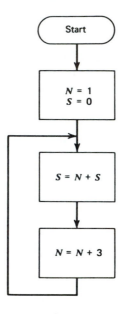

Figure 16-6 Flowchart for Example 16-3. (From Greenfield and Wray. *Using Microprocessors and Microcomputers: The Motorola Family, 2e,* © 1988. Reprinted by permission of REGENTS/PRENTICE HALL, Englewood Cliffs, New Jersey.

16-5-2 Decision Boxes and Conditional Branches

The reader has probably already realized that there is a serious problem with Example 16-3; the loop never terminates. Actually, putting a program into an endless loop is one of the most common programming mistakes.

There are two common ways to determine when to end a loop: *loop counting* and *event detection*. Either method requires the use of decision boxes in the flowchart and corresponding conditional branch instructions in the program. Loop counting is considered first.

Loop counting means determining the number of times the loop should be traversed, counting the actual number of times through and comparing the two.

EXAMPLE 16-4

Improve the flowchart of Fig. 16-6 so that it terminates properly.

SOLUTION The program should terminate not when N = 10,000 but when 10,000 is added to the sum. For the flowchart of Fig. 16-6, N is increased to 10,003 immediately after this occurs. At the end of the first loop N = 4, the second loop N = 7, and so on. It can be seen here that N = 3L + 1, where L is the number of times through the loop. If N is set to 10,003, L = 3334. The loop must be traversed 3334 times.

The correct flowchart is shown in Fig. 16-7. The loop counter L has been added and set initially to −3334. It is incremented each time through the loop and tested to see if it is positive. After 3,334 loops, it becomes 0. Then the YES path from the decision box is taken and the program halts.

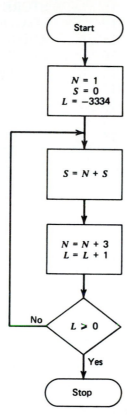

Figure 16-7 Flowchart for Example 16-4. (From Greenfield and Wray. *Using Microprocessors and Microcomputers: The Motorola Family, 2e*, © 1988. Reprinted by permission of REGENTS/PRENTICE HALL, Englewood Cliffs, New Jersey.

EXAMPLE 16-5

Write the code for the flowchart of Fig. 16-7.

SOLUTION The program is arbitrarily started at location 10 and the data area is set at location 80. In the data area three variables N, S, and L are needed. These should be initialized to 1, 0, and −3334, respectively, before the program starts. In addition, two constants, 1 and 3, are needed during the execution of the program. Before the program starts, the data should look like this:

Location	Term	Initial value
80	N	1
81	S	0
82	L	−3334
83	Constant	3
84	Constant	1

The program can now be written directly from the flowchart.

Location	Instruction	Comments
10	LOAD 81	
11	ADD 80	} S = N + S
12	STORE 81	
13	LOAD 80	
14	ADD 83	} N = N + 3
15	STORE 80	
16	LOAD 82	
17	ADD 84	} L = L + 1
18	STORE 82	
19	JN 10	
1A	HALT	

Note that the instructions follow the flowchart. The decision box has been implemented by the JN instruction. The program loops as long as L remains negative.

Check: As a check on the program, we can write the contents of N, S, and L at the end of each loop in the following table.

Times through the Loop	N	S	L
1	4	1	− 3333
2	7	5	− 3332
3	10	12	− 3331

The chart shows that each time around the loop:

$$(N - 1)/3 + [L] = 3334$$

Therefore, when N = 10,003, L indeed equals 0 and the loop terminates.

Event detection terminates a loop when an event occurs that should make the loop terminate. In Example 16-5, that event could be the fact that N is greater than 10,000. Using event detection, locations 16 through 19 could be replaced by:

16) LOAD N
17) SUBTRACT 105
18) Not needed (NOP)
19) JN 10

where location 105 contains 10,001. This program branches back until N = 10,003. In this problem, the use of event detection is conceptually simpler and saves one instruction.

16-5-3 A More Complex Problem

As a final introductory programming example, consider the problem of adding together all the numbers in memory between locations 100 and 200. This problem has many practical applications. If, for example, a firm has 101 debtors, it can store the amount owed by each debtor in one of the locations. The sum of these debts is the firm's accounts receivable.

The flowchart for this example is shown in Fig. 16-8. At the start, the sum is set to 0 and the loop counter is 101 because there are 101 addresses between 100 and 200.[3] The loop counter and the addresses are then incremented. Note the box $S = S + (A)$. The parentheses around A mean that the *contents of location A are being added to the sum.*

The difficult part of this problem is to find a way to increment the addresses. One way to do this is to increment the ADD instruction. This is done in Example 16-6.

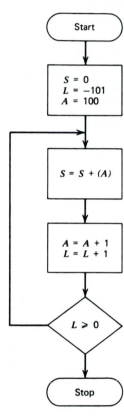

Figure 16-8 Flowchart for adding together the contents of locations 100 to 200. (From Greenfield and Wray. *Using Microprocessors and Microcomputers: The Motorola Family, 2e,* © 1988. Reprinted by permission of REGENTS/PRENTICE HALL, Englewood Cliffs, New Jersey.

EXAMPLE 16-6

Write the code for the flowchart of Fig. 16-8.

SOLUTION Again, the program is started at location 10. Because locations 100 to 200 contain data, the constant area is set up at 300. It looks like this:

Location	Term	Initial Value
300	S	0
301	L	−101
302	Constant	1

[3]In this introductory problem, decimal addresses are assumed for simplicity.

The coding can then proceed as follows.

Location	Instruction	Comments
10	LOAD 300	
11	ADD 100	S = S + (A)
12	STORE 300	
13	LOAD 11	Increment address portion of instruction 11 and resume.
14	ADD 302	
15	STORE 11	
16	LOAD 301	
17	ADD 302	L = L + 1
18	STORE 301	
19	JN 10	If minus, go back to location 10.
20	HALT	

Note particularly instructions 13, 14, and 15. These instructions treat location 11, which contains an instruction, as data. The assumptions are that the instructions consist of an Op code and address, that the address occupies the LSBs of the instruction, and that adding 1 to the instruction increments the address without affecting the Op code or other parts of the instruction. These assumptions are valid for most computers.

This method of solving this problem is not generally used, and *we do not recommend it*, because more powerful methods, such as indexing, are preferable (see section 16-10). It was presented, however, to emphasize that *instructions occupy memory space and can be treated as data*. In this example, the instruction in location 11 is treated as data when the instructions in locations 13, 14, and 15 are executed. This is a *self-modifying program*, which means it *changes its own contents as it executes*. Most programmers *avoid* self-modifying programs. They are complex and difficult to document or understand, and the program cannot be run successively without reentering the initial values between each run. Also, they cannot be used in ROM.

16-6 The Control Unit

The control unit is the hardware core of a computer. It consists of the timing circuits and registers necessary to control the operation of the computer.

16-6-1 FETCH, EXECUTE, and Timing

The execution of a computer instruction generally requires two cycles: a FETCH cycle and an EXECUTE cycle. During the FETCH cycle, the computer reads memory to get the instruction; during the EXECUTE cycle, it actually executes the instruction. The computer then starts the next instruction by performing another FETCH cycle. Thus, a computer operates basically by alternately doing FETCH, EXECUTE, FETCH, EXECUTE . . . until it steps through the entire program and encounters a HALT instruction.

The FETCH and EXECUTE cycles are subdivided into time slots. For the first computer to be described we will assume four time slots or phases, labeled T_0, T_1, T_2, and T_3, occur during each cycle.

16-6-2   A Rudimentary Computer

The diagram of a rudimentary computer is shown in Fig. 16-9. The dimensions of the memory are assumed to be 4K words by 16 bits. An instruction word is assumed to consist of a 4-bit Op code and a 12-bit address as shown in Fig. 16-10. The memory in Fig. 16-9 is shown containing a simple program to add the numbers in locations 100 and 101 and store the results in 101. It also shows that data are at 100 and 101.

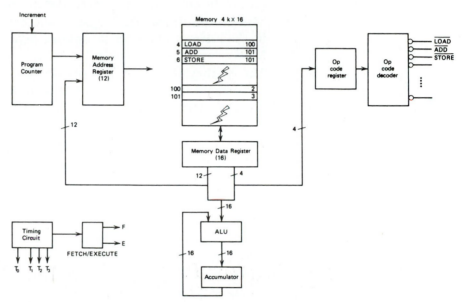

Figure 16-9   A rudimentary computer.

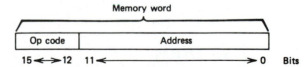

Figure 16-10   Division of an instruction word into Op code and address portions.

Beside the memory, the computer consists of the following parts.

1.   **The FETCH/EXECUTE FF.** This FF determines whether the computer is in FETCH or EXECUTE mode.
2.   **The Timing Generator.** This develops the four-phase clock used to synchronize the operations during the FETCH and EXECUTE cycles.
3.   **The MAR (Memory Address Register).** This contains the memory address (see section 15-3).

4. **The MDR (Memory Data Register).** This contains the data going to or from the memory (see section 15-3).

5. **The Program Counter (PC).** This register is as long as the MAR. It contains the *address* of the next instruction to be executed. Before each FETCH cycle begins, the contents of the PC are transferred to the MAR so that the next instruction can be fetched from memory. The PC is incremented during each instruction so that the next instruction will be fetched from the next sequential location.

6. **The Op Code Register.** This register retains the Op code portion of the instruction word.

7. **The Op Code Decoder.** This IC decodes the information in the Op code register and presents a single active line (LOAD, ADD, STORE, etc.) corresponding to the particular instruction being executed.

8. **The Accumulator.** This register holds data, one of the operands used by the instruction. Since it contains data, it must be as long as the MDR.

9. **The Arithmetic-Logic Unit (ALU).** The ALU generally gets an operand from memory, performs an operation (ADD, SUBTRACT, AND, OR, etc.) on the memory data, and returns the results to the accumulator.

The interconnecting wires or buses between the various components are also shown in Fig. 16-9.

EXAMPLE 16-7

For the above computer, how many bits are in:

(a) The MAR
(b) The MDR
(c) The PC
(d) The Op code register
(e) The accumulator
(f) The ALU

SOLUTION

1. Because the memory is 4K words by 16 bits per word, the MAR must contain 12 bits (2^{12} = 4K) and the MDR must contain 16 bits.

2. The PC essentially contains an address (the address of the instruction to be executed). It contains as many bits as the MAR (12).

3. Because the Op code portion of the instruction word is 4 bits, the Op code register only needs 4 bits.

4. Both the accumulator and the ALU contain data. Because a data word is as long as a memory word, the accumulator and ALU both contain 16 bits.

16-6-3   The FETCH Cycle

Because the instructions that comprise a program are retained in memory, but must be executed in the control unit, instruction execution starts with a FETCH cycle that reads (or FETCHes) the instruction from memory to the control unit. In many computers, including our sample computer, all FETCH cycles are identical. A FETCH cycle occurs while the FETCH/EXECUTE FF is HIGH. Remember there are four time slots or phases during each cycle. Table 16-2 shows the events that occur during the FETCH cycle. It proceeds as follows:

1. The MAR is loaded with the contents of the PC. This happens before or just at the beginning of the FETCH cycle.
2. During $F \cdot T_0$ (F means FETCH mode or the FETCH FF is HIGH and T_0 means the first time slot in the cycle), the memory is read. This brings the instruction into the MDR.
3. The Op code portion of the instruction (bits 12–15 in our computer) is loaded into the Op code register at $F \cdot T_1$. The output of the Op code register goes to the Op code decoder, which activates the one output line on the decoder that corresponds to the instruction being executed.
4. At $F \cdot T_2$, the Program Counter is incremented.
5. At $F \cdot T_3$, the address portion of the instruction (bits 0–11) is placed in the MAR.

TABLE 16-2

The FETCH Portion of an Instruction

$F \cdot T_0$	Memory is read to FETCH the instruction.
$F \cdot T_1$	OP code portion of the instruction goes to the Op code register.
$F \cdot T_2$	Program Counter is incremented to point to the address of the next instruction.
$F \cdot T_3$	The address portion of the instruction word is sent to the MAR.

At the end of the FETCH cycle, the proper output of the Op code decoder is activated and the operand address is in the MAR. If the instruction was LOAD 500, for example, the LOAD line out of the decoder would be active and the address (500) would be in the MAR.

16-6-4   The EXECUTE Cycle

When the FETCH cycle ends, the FETCH/EXECUTE FF toggles and the EXECUTE cycle begins. What occurs during the EXECUTE cycles depends primarily on the instruction decoded by the Op code decoder. The events that occur during the LOAD, ADD, and STORE instructions are shown in Table 16-3. The LOAD is actually accomplished by a CLEAR and ADD procedure; the accumulator is first *cleared*, then it is *added* to the contents of memory. The results are the memory contents and they are placed in the accumulator. Note that the ADD instruction is identical to the LOAD instruction, except that the accumulator is not cleared at $E \cdot T_0$.

TABLE 16-3

Execution of the LOAD, ADD, and STORE Instructions

LOAD

$E \cdot T_0$	Accumulator is cleared. ALU commanded to ADD.
$E \cdot T_1$	ALU adds memory contents to accumulator contents (0 because accumulator is CLEAR).
$E \cdot T_2$	Sum goes into the accumulator.
$E \cdot T_3$	(PC) goes into MAR.

ADD

$E \cdot T_0$	ALU commanded to ADD.
$E \cdot T_1$	ALU adds memory contents to accumulator contents.
$E \cdot T_2$	Sum goes into the accumulator.
$E \cdot T_3$	(PC) goes into the MAR.

STORE

$E \cdot T_0$	Accumulator output set to MDR.
$E \cdot T_2$	Memory placed in WRITE mode.
$E \cdot T_3$	(PC) goes into the MAR.

16-6-5   Execution of a Simple Program

In this section we will consider how our rudimentary computer can perform the simple program shown in Fig. 16-9.

Location	Instruction
4	LOAD 100
5	ADD 101
6	STO 101

Table 16-4 has been prepared to show the contents of the registers in the control unit during each step in the program. An X in a register indicates the data are irrelevant or unknown at that time.

The program starts with the PC and the MAR locating the first instruction. At $F \cdot T_1$, the 4 bits indicating LOAD are transferred to the Op code register. The PC is incremented to 5 at $F \cdot T_2$ and the instruction address, 100, is transferred to the MAR at $F \cdot T_3$.

The accumulator is cleared at $E \cdot T_0$ of the LOAD instruction. The ALU then adds the 2 (read from location 100) to the 0 in the accumulator and puts the sum in the accumulator at $E \cdot T_2$. At $E \cdot T_3$ the PC contents, now 5, are transferred to the MAR, so the computer is ready to fetch the next instruction.

The execution of the ADD and STORE instructions are similar. The FETCH cycles are identical. A WRITE pulse, not shown on the table, occurs at $E \cdot T_2$ of the STORE instruction.

16-7   The Hardware Design of a Computer

Section 16-6 described the operation of a computer *conceptually*. In this section we will go through the design of a computer in greater detail. We will identify the ICs and other components that comprise each part of the computer, and finish with a rudimentary but viable computer.

TABLE 16-4

Step-by-Step Execution of the Program of Section 16-6-5

Time	PC	MAR	MDR	Op Code	Accum.	ALU
LOAD 100						
$F \cdot T_0$	4	4	LOAD 100	X	X	X
$F \cdot T_1$	4	4	LOAD 100	LOAD	X	X
$F \cdot T_2$	5	4	LOAD 100	LOAD	X	X
$F \cdot T_3$	5	100	LOAD 100	LOAD	X	X
$E \cdot T_0$	5	100	2	LOAD	0	X
$E \cdot T_1$	5	100	2	LOAD	0	2
$E \cdot T_2$	5	100	2	LOAD	2	2
$E \cdot T_3$	5	5	2	LOAD	2	X
ADD 101						
$F \cdot T_0$	5	5	ADD 101	LOAD	2	X
$F \cdot T_1$	5	5	ADD 101	ADD	2	X
$F \cdot T_2$	6	5	ADD 101	ADD	2	X
$F \cdot T_3$	6	101	ADD 101	ADD	2	X
$E \cdot T_0$	6	101	3	ADD	2	5
$E \cdot T_1$	6	101	3	ADD	2	5
$E \cdot T_2$	6	101	3	ADD	5	5
$E \cdot T_3$	6	6	3	ADD	5	X
STORE 101						
$F \cdot T_0$	6	6	STO 101	ADD	5	X
$F \cdot T_1$	6	6	STO 101	STO	5	X
$F \cdot T_2$	7	6	STO 101	STO	5	X
$F \cdot T_3$	7	101	X	STO	5	X
$E \cdot T_0$	7	101	X	STO	5	X
$E \cdot T_1$	7	101	5	STO	5	X
$E \cdot T_2$	7	101	5	STO	5	X
$E \cdot T_3$	7	7	5	STO	5	X

16-7-1 The Timing Circuit

The basic timing of most computers and microprocessors is derived from a high-frequency clock. Often a crystal-controlled clock is used to keep the timing precise. The timing for our computer is shown in Fig. 16-11. The clock feeds a 2-bit counter, which goes to a **74x138** decoder to produce the required 4-phase clock. The leading edge of T_0 also toggles the FETCH-EXECUTE FF so the computer alternates between FETCH AND EXECUTE cycles. The times are expressed as $\overline{T}_0, \overline{T}_1, \overline{T}_2$, and $\overline{T}_3$ because they are LOW active. In some figures we occasionally use uncomplemented times such as T_1. T_1 is simply the inverse of $\overline{T}_1$.

EXAMPLE 16-8

Draw a timing chart for the basic computer timing.

SOLUTION The solution is shown in Fig. 16-12. Note that each time is LOW for one quarter of each cycle because the active output of a decoder is LOW.

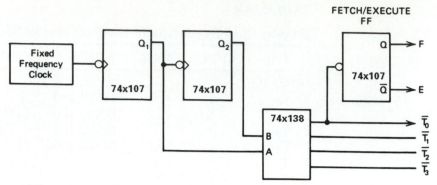

Figure 16-11 The basic timing for the computer.

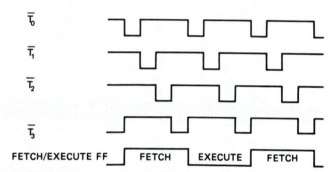

Figure 16-12 Timing waveforms for the computer.

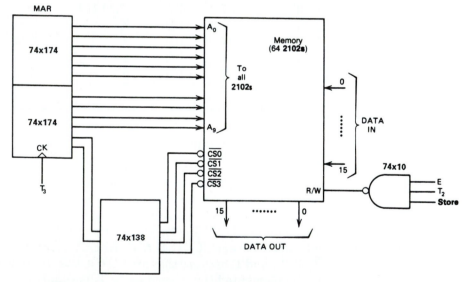

Figure 16-13 The memory and MAR.

16-7-2 The Memory, MAR, and MDR

The memory is shown in Fig. 16-13. For simplicity, **2102**s (see section 13-6-1) were chosen as the basic memory IC. A 4K-word-by-16-bit memory requires 64 **2102**s in a 4 × 16 matrix. A 12-bit MAR is also required. The MAR is made up of two **74x174**s. Ten of its outputs go directly to the **2102**s. The other two addresses go into a decoder, which selects one of the 4 rows of memory by driving the proper CS line LOW.

Because **2102**s are used, the MDR really consists of the DATA OUT and DATA IN lines. The memory is almost always in READ mode, so that its contents are not changing. The DATA OUT lines contain the memory contents of its selected address.

The only time the memory is written is at $E \cdot T_2$[4] of a STORE instruction. The DATA IN lines are driven by the accumulator to provide the data to be stored in memory.

16-7-3   The Op Code Register and Decoder

The Op code register, shown in Fig. 16-14, only needs 4 bits and a single **74x174** or **74x175** is sufficient. Its inputs are the Op code, the 4 MSBs of the MDR, and are clocked in at $F \cdot T_1$. The four outputs go to the SELECT lines on a **74154** (see section 11-6-2), which functions as the Op code decoder. The single LOW output from the **74154** is the instruction being executed.

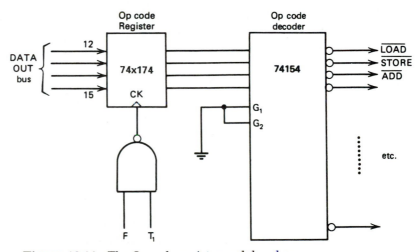

Figure 16-14   The Op code register and decoder.

16-7-4   The Accumulator and the ALU

The accumulator, the ALU, and its interconnections are shown in Fig. 16-15. The accumulator uses three **74x174**s (see section 6-7-2). The instruction inputs ($\overline{\text{LOAD}}$, $\overline{\text{ADD}}$, etc.) come directly from the Op code decoder. The accumulator is cleared at $E \cdot T_0$ by CLEAR and LOAD instructions. It is loaded at $E \cdot T_2$ by LOAD instructions and all the arithmetic or logic instructions. Those instructions that do not change the accumulator, such as STORE or JUMP, are not inputs to the NAND gate controlling the accumulator clock.

The Arithmetic-Logic Unit (ALU) is built using four **74x181**s (see section 15-8-1) and works closely with the accumulator. The A inputs to the ALU come directly from the accumulator, and the B inputs come from the memory. The

---

[4]$E \cdot T_2$ means phase 2 of the EXECUTE cycle.

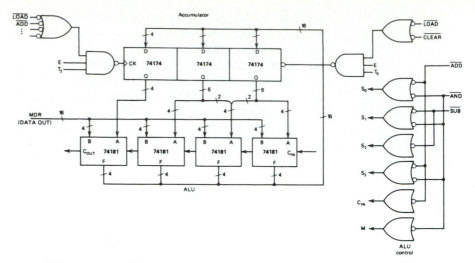

Figure 16-15 The accumulator and ALU.

ALU outputs go to the D inputs of the accumulator where they are clocked in at $E \cdot T_2$ of those instructions that affect the accumulator.

The ALU is controlled by the six NAND gates shown on the right side of Fig. 16-15. They send the proper S, M, and carry inputs to the **74x181**s.

EXAMPLE 16-9

In Fig. 16-15 explain how the following instructions control the ALU:

(a) ADD

(b) AND

SOLUTION

(a) To ADD, the ALU must have a 9 on its S inputs, C_{in} must be HIGH and M must be LOW. An ADD instruction causes the Op code decoder to set $\overline{ADD}$ LOW, leaving all other outputs HIGH. Because $\overline{ADD}$ is connected to the S_0, S_3, and C_{in} inputs, these will be HIGH and the other S and M inputs will be LOW. This combination of inputs causes the ALU to add.

(b) Because AND is a logic instruction, Fig. 15-12 shows that M, S_0, S_1 and S_3 must be HIGH. This occurs if $\overline{AND}$ goes LOW. The CARRY IN is irrelevant for logic instruction and is not connected for simplicity.

In most computers other instructions that affect the accumulator will also be used, and they will provide additional inputs to the NAND gates (see Problem 16-9).

16-8   The Complete Computer

The complete computer is shown in Fig. 16-16. The parts of this computer that were not discussed in section 16-7 are:

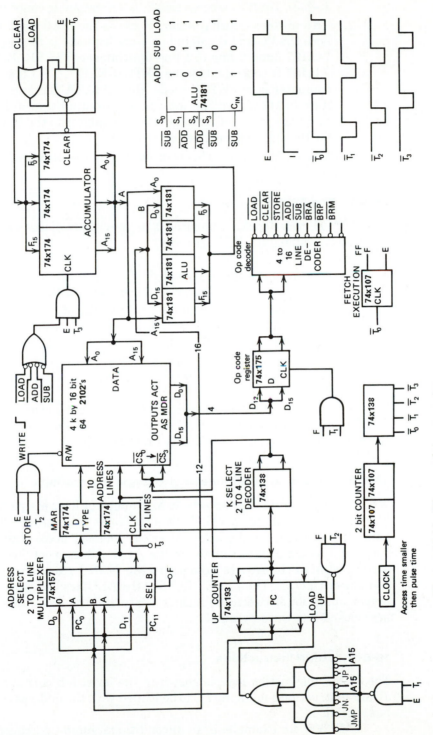

Figure 16-16 The complete computer.

1. **The program counter.** The PC is a 12-bit counter. Because it must be capable of being incremented and parallel loaded (see section 16-8-3), **74x193**s (see section 8-9-1) presettable counters were chosen.

2. **The address multiplexer.** Because the MAR must be loaded from both the PC and the data bus, a multiplexer is required. We used the **74x157** quad 2 line to 1 line multiplexer. The SELECT input to the **74x157** is controlled by the FETCH-EXECUTE FF so that when the computer is in FETCH mode, the DATA OUT inputs from the **2102**s (bits 0–11) are on the multiplexer's output and during EXECUTE mode the PC is fed through the multiplexer to the MAR. The reasons for this are explained in section 16-8-1.

3. **The JUMP logic.** This is discussed in section 16-8-3.

16-8-1 The FETCH Cycle

The FETCH cycle for this computer is as previously discussed. The following events occur during the FETCH cycle.

1. The Op code register is loaded at $F \cdot T_1$. It retains the instruction for the rest of the FETCH cycle and throughout the EXECUTE cycle.

2. The PC is incremented at the trailing edge of $F \cdot T_2$ by the NAND gate connected to the COUNT-UP input of the **74x193**s.

3. The MAR is loaded with the address on the DATA-OUT lines at the beginning of T_3. This happens because in FETCH mode the Data bus comes through the Address Multiplexer.

Note that the MAR is loaded at the start of every T_3 time slot. During EXECUTE mode it is loaded with the PC contents; this causes the memory to read the next instruction for the following FETCH cycle.

16-8-2 Execution of Arithmetic and Logic Instructions

When the Op code decoder detects an arithmetic or logic instruction, it sets the S, M, and C_{in} inputs to the ALU. This occurs at $F \cdot T_1$. At $F \cdot T_3$ the operand address is clocked into the MAR. After one memory access time, the memory operand appears at the B inputs of the ALU. All instructions that cause the accumulator to change are brought to the NAND gate connected to the accumulator CLOCK. Thus, at $E \cdot T_2$, after the ALU has had time to perform the proper operation between the memory data and the accumulator, the results are clocked into the accumulator.

16-8-3 JUMP Instructions

The JUMP instruction logic is shown in the lower left corner of Fig. 16-16. A JUMP instruction causes the PC to be loaded with the address contained in the instruction.

JMP 500, for example, is an unconditional jump that causes the next instruction to be executed at location 500. It is executed as follows:

1. At $F \cdot T_1$, $\overline{JMP}$ is decoded by the Op code decoder.
2. At $F \cdot T_3$, the MAR is loaded with the address in the instruction (500). The contents of memory at location 500 will be placed on the DATA OUT lines, but nothing will be done with it during this instruction. The MAR outputs, however, are also connected to the parallel load inputs to the **74x193**s. At $E \cdot T_1$ a LOAD pulse is generated that loads the MAR contents (500) into the PC. At $E \cdot T_3$ the contents of the PC (now 500) are loaded into the MAR and the next FETCH occurs with the MAR containing 500. Thus, the instruction in location 500 is executed next.

Conditional jumps are executed similarly. The condition is tested. If it is satisfied, the PC is loaded with the memory address. Otherwise the PC is not loaded and the computer executes the next sequential instruction. The logic for two conditional jumps, JP and JN, is shown in Fig. 16-16. $\overline{JN}$ from the Op code decoder is gated with $\overline{A15}$. If A15 is a 1, the number in the accumulator is negative, $\overline{A15} = 0$, and a LOAD pulse is produced. The JP instruction is similar except that $\overline{JP}$ is gated with A15.

EXAMPLE 16-10

A JZ (jump on zero accumulator) and its complement, JNZ (jump on not zero accumulator) are instructions that are often used to compare two numbers, and to jump, depending upon whether the two numbers are equal.

How can a JZ be added to this computer?

SOLUTION The simple solution is to use a 16-input AND gate to determine if all bits of the accumulator are 0. However, a more elegant solution that requires fewer gates exists:

1. Set up the ALU so that $S = 15$ (the function is $F = A$ minus 1), the CARRY-IN is 1, and $M = 0$.
2. The CARRY OUT of the ALU will be 1 only if all bits of the accumulator are 0 (see Problem 16-10).
3. The CARRY OUT can be inverted and gated with the $\overline{JZ}$ output of the Op code decoder to provide a load pulse to the PC. This occurs only if the accumulator contains a 0.

Other instructions, in addition to the JZ, can be added to this computer as required. The length of the Op code register and decoder, however, limit it to 16 different instructions.

16-9   Building the Computer

A rudimentary computer can be designed and built using the ideas developed in the previous sections. Such a computer was designed, built, and tested by the students at RIT. We felt this was the best way to learn how computers actually work, and it also presented the students with nontrivial debugging problems.

Unfortunately, this computer makes demands on two of the student's most limited resources: time and money. Both constraints dictate that the computer be designed with as few ICs as possible. The first decision, consequently, was to go to an 8-bit computer with a 3-bit Op code and a 5-bit address. Two **2114**s were used to give a 1K-by-8-bit memory.

The results of these constraints led to a computer that could execute 24 instructions, and had only 256 bytes of memory. In addition, practical problems involved in building a computer, such as reading and writing memory, designing the I/O, and determining how to halt the computer or have it run in single-step mode (executing one instruction at a time), had to be solved. This computer is discussed further in the second edition of this book.

16-10   The Operation of Modern Microprocessors

The computer described in section 16-9 operates as a minicomputer or mainframe (large) computer does. They generally consist of one instruction word that provides both the Op code and the address used in the instruction. The instruction word for a mainframe contains more sections than just the Op code and address and is more powerful. But each instruction requires only one memory fetch cycle and each location referenced by the PC contains exactly one instruction.

Microprocessors (μPs) do not operate in this manner. In a μP the first byte of each instruction is an Op code fetch. This fetch gives the μP an Op code only. The μP then examines this Op code to determine if it requires additional information, such as an address or immediate information. This additional information is found in the next byte or bytes of the program.

Some instructions require only the Op code fetch to completely specify them. These are called *inherent* instructions. Examples are increment or shift the accumulator. Because the number to be operated on is already in the accumulator, no further memory references are necessary; the μP has all the information it needs and can proceed to execute the instruction. Inherent instructions are 1-byte instructions.

Other instructions, such as a LOAD, require more information. A LOAD copies the contents of memory into the accumulator, but the memory address must be specified. The Op code fetch only tells the μP that the instruction is a LOAD. In general, the address is contained in the next two bytes. The μP must then fetch the next two bytes in the program, to obtain the address; then it can place the address on the address bus, read the data at that address, and put it in the accumulator.

EXAMPLE 16-11

The instruction LOAD 12E4 is written into a program at location C000. How does it appear in memory?

SOLUTION   It appears as:

C0000	LOAD	Op code for LOAD
C001	E4	Lower byte of the address

| C002 | 12 | Upper byte of the address |
| C003 | | Op code of the next instruction |

This is for an Intel μP, such as the **8085**. For a Motorola μP it would appear as LOAD 12E4, because the Motorola writes the MS bype of the address in the next location. To start the instruction, the PC would have to contain the address C000. In either case the instruction requires three bytes, and increments the PC three times before it finishes.

16-10-1 Overview of a μP's Registers and Accumulators

Most μPs have the control unit and ALU within them. They do not contain the memory or input/output drivers (I/O is considered in chapter 17). These must be provided by external ICs, as they are on the IBM-PC (see section 17-11-1).

The control unit contains the program counter (PC), the Op code register and decoder, the stack pointer, and other registers and accumulators. The interface between the μP and the memory consists essentially of the MDR, which drives the bidirectional data bus, the MAR, which controls the memory addresses, and the READ/WRITE line. The control unit controls the interaction between all these registers.

Figure 16-17 is a block diagram of the Intel **8085** μP. It is, of course, more complex than the simple computer discussed in section 16-8, and space precludes a thorough discussion of the IC.[5] Nevertheless, many of the circuits discussed in this chapter are incorporated in the **8085**.

The upper part of Fig. 16-17 contains the interrupt and serial I/O control. These are beyond the scope of this book. The middle left section has the accumulator and ALU, which are similar to those presented in section 16-7. The timing and control section, at the lower left, controls the timing and execution of the various instructions. We will only point out here that the unit accepts a crystal input, and generates the clocks required by the μP, and also generates the $\overline{RD}$ (read) and $\overline{WR}$ (write) signals required by the memory.

The center section of the figure shows the instruction register and decoder. The instruction decoder is more complicated in a μP because it must decide if the instruction needs more data from memory and whether to make additional memory references.

The register set is shown on the right side of Fig. 16-17. These consist of the PC, the stack pointer, and the B, C, D, E, H, and L registers. Thus, the **8085** has six 8-bit registers and one 8-bit accumulator. Both the registers and the accumulator can be loaded from memory, be stored in memory, and be incremented or decremented. The accumulator, however, must hold one of the operands in an arithmetic or logic instruction (the other operand can be in one of the other registers or memory) and the results are always stored in the accumulator. The accumulator is also the only register that can be shifted. The accumulator *must* be used in many of the **8085** instructions.

The MAR and MDR (see section 13-4) are at the bottom right of the figure. Thus, most of the logic blocks that comprise the **8085** have been discussed in this book and the student should understand how they function.

[5]For a thorough discussion of this μP, the reader should consult one of the many books on it.

BLOCK DIAGRAM

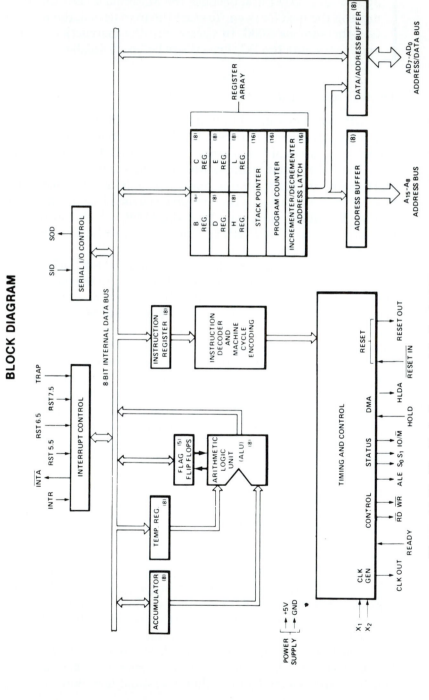

Figure 16-17 The block diagram of the Intel **8085** microprocessor. Reprinted by permission of Intel Corporation, © Intel Corporation.

16-10-2 Indirect Addressing

In section 16-5-3 we showed how to add all the numbers in a memory area (sometimes called a buffer) from locations 100 to 200. We also advised against using this type of program because it is extremely awkward and it is self-modifying; it changes itself as it progresses. Unfortunately, with the extremely simple instruction set of section 16-5, and the extremely simple computer of section 16-8, this is the only way to handle this problem. Modern μPs incorporate two intelligent ways to attack this type of problem: indirect addressing and indexing.

Indirect addressing means retaining the memory address of the operand either in memory or in a register. Many mainframe computers use indirect addressing on a memory location. The @ symbol is used to indicate indirect addressing in the program. Computers that use indirect addressing usually have an *indirect bit in their instruction word*. If this bit is set, it indicates indirect addressing.

EXAMPLE 16-12

Memory location 500 contains 35 and memory location 35 contains 25. Show how the following instruction works:

LOAD @ 500

SOLUTION Because this is an indirect instruction, it must be executed in two steps as shown in Fig. 16-18. In the first step, 500 is placed in the MAR. The memory reads the contents of 500 (35) and sends it to the computer via the data bus and MDR. When the computer receives the number, it must place it in the MAR. In the second step, the 35 on the address bus causes the memory to place 25 on the data bus. This number is placed in the accumulator completing the instruction.

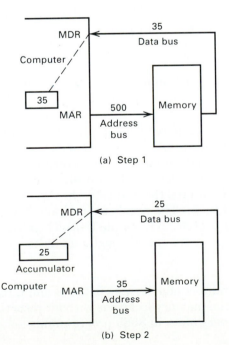

(a) Step 1

(b) Step 2

Figure 16-18 Solution to Example 16-12.

At this time the reader, with a little thought, should be able to design a computer that can execute indirect instructions. With so many μPs available, however, it is probably not worth the effort.

16-10-3   Indirect Addressing in the 8085

The Intel **8085** μP uses direct addressing via its H and L registers. When an instruction references memory without specifying an address, the 16-bit memory address is taken from the H (HIGH) and L (LOW) registers. For example, there are two instructions that can load the contents of a memory location, say location 100, into memory. One is the *direct instruction*, LDA 100. This instruction specifies the address directly. It takes three bytes: an Op code followed by two additional bytes to specify the 16-bit memory address. The second method is to use the instruction MOV A,M. This is a one byte instruction that says "Move the contents of memory into the accumulator." Because the address is not specified in the instruction, it is taken from the H and L registers.

EXAMPLE 16-13

Write a program for the **8085** to add all the numbers in memory locations BUFFER to BUFFEREND. This is the same as Example 16-6 if BUFFER is 100 and BUFFEREND is 200.

SOLUTION   The program might be:

```
        LDX     H,L     BUFFER
        LDA     A       0
LOOP    ADD     A,M
        INX     H,L
        CMP     H,L     BUFFEREND +1
        JNZ     LOOP
        HALT
```

The first instruction puts the starting address of BUFFER into H and L. Next the accumulator is set to 0. The ADD instruction adds memory and the accumulator. The memory address is taken from H and L. They are then incremented and compared with the end of the buffer. JNZ stands for JUMP in Not Zero. If the result of the comparison is not 0, the end of the buffer (actually one location beyond the end of the buffer) has not been reached and the program jumps back to LOOP and adds the next location. (For example, H and L might contain 100 the first time through the loop, but they will contain 101 the second time.) The numbers in H and L increment until they reach BUFFEREND + 1 (201 in Example 16-6). Then the results of the comparison are 0, and the program falls through and halts.

The use of indirect addressing via the H and L registers permits the **8085** to access sequential memory locations without resorting to self-modifying programs. It occurs very often in **8085** programming.

16-10-4 Indexing

At about the time Intel brought out the **8085**, its chief rival, Motorola, produced the **6800** μP. The two corporations evidently decided to do everything as differently as possible, so while Intel used indirect addressing, Motorola used indexing.

In indexing, a special register called the Index Register (or X for short) is set aside to hold memory addresses. Instructions called indexed instructions take their address from this register. A number, called an *offset*, can be added to the contents of the X register to give the desired memory address, but the offset is most often 0, and will be ignored in this simplified introduction.

A program to add all the numbers between BUFFER and BUFFEREND using indexing looks very much like the program of Example 16-13. It is:

```
         LDX       BUFFER
         CLR       A
LOOP     ADD    A  0,X
         INX
         CPX       BUFFEREND + 1
         JNZ       LOOP
         HALT
```

The program starts by loading the starting address into A and clearing the sum. The instruction ADD A 0,X says "Add to accumulator A the number in memory at the Index register." Thus, the first time through the loop X will contain 100 and the contents of 100 will be added to A.

The X register will then be incremented by the INX instruction and compared to the address of the end of the buffer. As long as there is no match, the program will loop back and add the next number.

16-11 An Overview of Modern Microprocessors

This section presents a "thumbnail sketch" of some of the most popular modern μPs. Newer and more powerful μPs are constantly being developed, and the reader should read the manufacturer's literature for the latest innovations. Books on many of these μPs are mentioned in the reference section at the end of this book.

16-11-1 8-bit Microprocessors

8-bit μPs are not as powerful as their larger successors. They have limited address space and are not as adept as mathematical manipulations (number-crunching). Nevertheless, millions of them are still used because they are small, simple, and inexpensive. They are used primarily in control applications, such as in washing machines and automobiles.

The characteristics of some of the 8-bit µPs are:

- **Intel 8085:** 8-bit data bus
  16-bit address bus (limit—64K bytes)
  Six 8-bit registers
  One 8-bit accumulator
- **Motorola 6800:** 8-bit data bus
  16-bit address bus (limit—64K bytes)
  Two 8-bit accumulators
  One 16-bit index register
- **Motorola 68HC11:** This is a more advanced version of the **6800**. It has two 16-bit index registers and a more powerful Central Processing Unit. Its two 8-bit accumulators can be concatenated for 16-bit data. Perhaps the most important innovation is the many I/O capabilities of the **68HC11**.

16-11-2   16-bit Microprocessors

The characteristics of the most popular 16-bit microprocessors are:

- **Intel 8086:** 20-bit address bus (1-Mbyte limit)
  16-bit data bus
  Fourteen 16-bit registers
- **Intel 8088:** Like the **8086** except that it only has an 8-bit data bus. It is used in the IBM-PC, PC/XT, and clone computers.
- **Motorola 68000:** 23-bit address bus (16-Mbyte limit)
  16-bit data bus
  Eight 32-bit accumulators. They can operate on 8, 16, or 32-bit words.
  Eight 24-bit address registers
  The **68000** is used in the Macintosh and Amiga computers.
- **Intel 80286:** 24-bit address bus (16-Mbyte limit)
  16-bit data bus
  16-bit registers
  The **80286** is used in the AT series of personal computers.

16-11-3   32-bit Microprocessors

The following are some of the more popular 32-bit µPs.

- **Intel 386:** 32-bit address bus
  32-bit data bus
  32-bit accumulators
  The **386** comes in a 132 pin pin-grid array package. It is used in many IBM-PC and clone personal computers. The **386SX** is a **386** with only a 16-bit data bus. Thus, it is somewhat slower because it requires two memory cycles to access a 32-bit word.

- **Motorola 68020:** 32-bit address bus
  32-bit data bus
  32-bit accumulators

The **68020** comes in 169 pin pin-grid array package (114 pins used) and is used in some of the more advanced Macintosh computers.

Newer 32-bit μPs include the Intel **486** and the Motorola **68030** and **68040**. All of these μPs are *upward-compatible*. That means they use the same basic instruction set as the basic 16-bit μPs, and code written for the 16-bit μPs will run on the larger computers. Thus, the Intel **286, 386**, and **486** have the same instruction set as the **8086**, with some additions, and all the Motorola μPs have basically the same instruction set as the **68000**.

At present there does not seem to be any need to go beyond a 32-bit data bus or a 32-bit address bus. For the data bus, 32 bits is adequate for most operations. It is not adequate for serious number-crunching, but those types of problems are handled by a μP with a coprocessor that has 80-bit registers. Address limitations can be overcome by using *virtual memory*. More addresses create other problems, as Example 16-14 shows.

EXAMPLE 16-14

How many memory ICs would it take to fully populate a memory of 32 address bits by 32 data bits if 4-Mbit ICs are used?

SOLUTION   The address space is 4 gigawords (2^{32}) words. (One giga is 2^{30} or a little more than 1 billion.) The memory size, in bits, is 2^{32} words by 2^5 (32) bits per word or 2^{37} bits. A 4-Mbit IC contains 2^{22} bits. Therefore, it would take 2^{15} ICs to populate this memory. The cost of 32,000 ICs is daunting, but the problem of packaging this number of ICs is even more daunting. It is impractical to fully populate the memory of an advanced μP, especially in a personal computer.

SUMMARY

The object of this chapter was to show the reader how to design a rudimentary computer. The reader was first introduced to the basic instructions used in a computer and simple programs built from them. Flowcharts to help document and clarify program flow were discussed.

Then each basic part of the computer was discussed. The operations during the Fetch and Execute cycles were explained. A computer was designed using the ICs and circuits discussed earlier in the book. Theoretically the student should be able to build this computer if he or she had infinite time and money.

We built on this design to outline the operation of modern μPs. Finally, an overview of the most popular μPs was presented.

GLOSSARY

Accumulator: A register or registers in a computer that hold the operands used by the instructions.

Bus: A group of interrelated wires; generally a bus carries a set of signals from one digital device to another.

Conditional jump: An instruction that causes the program to branch only if a certain condition is met.

Control unit: The internal parts of a computer that control and organize its operations.

Event detection: Using the occurrence of an event to terminate a loop or program.

Execute: The portion of an instruction cycle where the instruction is executed.

Fetch: The portion of an instruction cycle where the instruction is sent from memory to the instruction register.

Flowchart: A graphic method used to outline or show the progress of a program.

Giga: A prefix meaning 2^{30} or approximately one billion.

Increment: To add 1 to a number.

Input/Output (I/O) system: The part of a computer that communicates with external devices.

Instruction: A command that directs the computer to perform a specific operation.

JUMP instruction: An instruction that alters the normal course of a program by causing it to jump to another instruction.

Load: An instruction that causes data to be brought from memory into an accumulator register.

Loop: Returning to the start of a sequence of instructions so that the same instructions may be repeated many times.

Op code: The portion of an instruction that tells the computer what to do.

Program: A group of instructions that control the operation of a computer.

Program counter (PC): A register in a computer that contains the address of the next instruction to be executed.

Rotate: A circular shift where the bits from one end of the word move to the opposite end of the word.

Software: Refers to the programs used in a computer.

Store: An instruction that causes data in the accumulator to be moved to memory or a peripheral register.

PROBLEMS

Section 16-3-2.

16-1. Write a simple program to add the numbers in locations 30 and 31 and then subtract the numbers in locations 32 and 33. Assume a SUBTRACT instruction is available in your computer.

Section 16-4.

16-2. Draw a flowchart for Problem 16-1.

Section 16-5-2.

16-3. Write a program to add the numbers 1, 6, 11, 16, . . . 20,001. Use the standard instructions.

16-4. The positive integer N is in location 500. Write a program to store (N + 2)! in location 501. Be sure to identify any constants you use.

Section 16-5-3.

16-5. There is a set of numbers in locations 100 to 300. Some of them are 22. Write a program to count the number of 22s in the area, and store the result in location 99. Use a self-modifying program and the instructions we discussed.

16-6. Write a program to transfer the contents of locations 2000 to 2047 to 2410 to 2457.

Section 16-6-5.

16-7. The program in memory is

Location	Instruction
4	LOAD 100
5	ADD 101
6	SUB 102
7	STO 103

Locations 100, 101, and 102 contain 7, 8, and 9. Show the contents of each part of the computer by finishing the chart of Fig. P16-7.

Section 16-8.

16-8. Assume the memory data register and accumulator are made up of **74x74**s. Describe how the instruction "OR 500" is fetched and executed and show a hardware implementation. The instruction ORs the contents of location 500 with the accumulator, the results remain in the accumulator.

16-9. The 16-bit computer discussed in section 16-8 must be able to execute the following instructions:

Add
Subtract
And
Or
Increment
Shift (1 Bit)

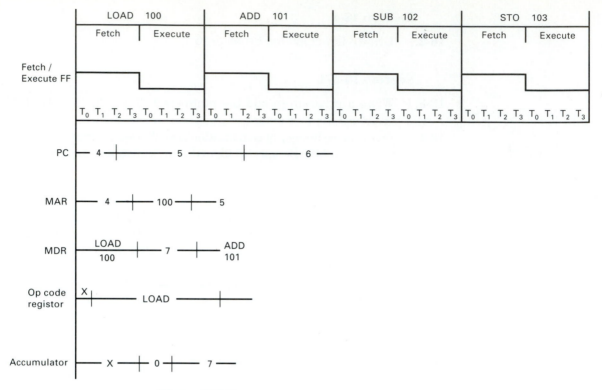

Figure P16-7

Draw the memory data outputs, the ALU, the accumulator, and the Op code decoder. Show the connections between them to execute these instructions.

16-10. The ALU consists of four **74x181**s. What is a CARRY-IN and CARRY-OUT of each **74181** if S = 15, CARRY-IN to the first IC is 1, M = 0, and the sixteen A inputs are $(0000)_{16}$. Repeat if the A inputs are $(0B00)_{16}$.

16-11. If the computer of Fig. 16-16 is executing the instruction 8 JMP 500, sketch the contents of the PC and the MAR at each cycle in the FETCH and EXECUTE parts of the instruction.

16-12. Add a JNZ (Jump on Not Zero accumulator) to the computer of Fig. 16-16.

16-13. Reduce the size of the computer of Fig. 16-16 by replacing the **74x174**s by **74x374**s and the **2102**s by **2114**s where they will save gates.

After attempting to solve the problems, try to answer the self-evaluation questions in section 16-2. If any of them still seem unclear, review the appropriate sections of the chapter to find the answers.

Input/Output
and Computer Interfacing

17-1 Instructional Objectives

This chapter considers how computers and microprocessors (μPs) perform input/output (I/O) functions to communicate with peripheral devices such as printers and MODEMs. Port I/O, memory-mapped I/O, parallel and serial interfacing are discussed. Finally, the use of these devices in personal computers (PCs) is considered.

After reading the chapter, the student should be able to

1. Interface a μP to a printer.
2. Control the direction of data flow in an interface IC.
3. Design decoders to select a particular interface IC.
4. Set up the proper modes in an **8255** Programmable Peripheral Interface and use it for input or output.
5. Use handshaking to coordinate data transfers on input and output.
6. Use a UART to convert serial data to parallel data.
7. Convert digital levels to RS-232C levels.
8. Connect to a MODEM and design a system for transmitting data over telephone lines using UARTs and MODEMs.

17-2 Self-Evaluation Questions

Watch for the answers to the following questions as you read the chapter. They should help you understand the material presented.

1. What is the function of a data direction register? What is the difference between a data direction register and a data direction bit?
2. What is a time-multiplexed bus? What advantage do we gain from the more complex circuitry required?
3. What is the difference between port I/O and memory-mapped I/O? Why is an $M/\overline{IO}$ signal needed in an Intel system?
4. What are the advantages of handshaking? Where should it be used?
5. What is the difference between serial and parallel transmission? Which is faster? When should each be used?
6. What are the functions of the START and STOP bits? What logic levels are they?
7. What is the difference between synchronous and asynchronous transmission? Which is faster?
8. What are RS-232C levels? Why are they used? What else does RS-232C specify?
9. What are the essential peripherals that must be connected to a PC?

17-3 An Overview of Peripheral Devices and Interfacing

A computer or microprocessor (μP) must be able to communicate with the external world; it must be able to send and receive information. A personal computer (PC) is an example of a μP system. It consists of the basic computer, a μP in this case, as described in Chapter 16, and external peripheral devices that make it useful. Some of the peripheral devices that are necessary to a PC's operation are the keyboard, the printer, the video monitor, and the disk. Other peripheral devices that are used on some computers include modems, analog-to-digital converters, optical scanners, and a mouse.

17-4 Interface ICs

μPs do not communicate directly with any of the peripherals mentioned in the previous paragraph. Instead they use special purpose *interface ICs,* such as the **8255** Programmable Peripheral Interface, which is discussed in section 17-7. Figure 17-1 shows the general diagram of such an IC. These ICs act as *intermediaries between the μP and the peripheral;* they communicate with both the peripheral and the μP. The interface IC is at the center of the figure. The bus between the μP and the interface IC is on the left side. The communications between the peripheral and the interface IC is shown on the right side.

17-4-1 The μP Bus

The interface IC is a slave of the μP, because it has no computational power (no brains) and cannot issue commands on its own. It accepts commands from the

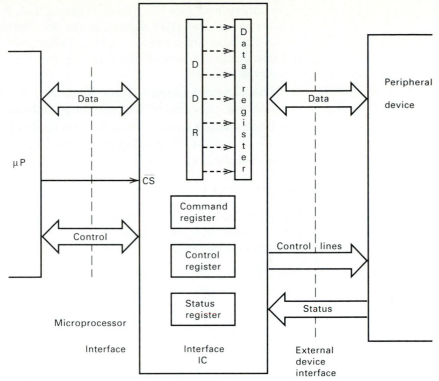

Figure 17-1 A basic Interface IC.

μP and transmits them to the peripheral. It is also a conduit for data between the μP and the peripheral.

The bus between the μP and the interface IC is shown on the left side of Fig. 17-1. It consists of three parts:

- **The Data Bus**—This is the 8- or 16-bit bidirectional data bus connected directly to the μP. It transfers data between the μP and the interface IC.
- **Chip Select (CS̄)**—Unless C̄S̄ is LOW, the IC is not selected. When deselected, it will not respond to any commands from the μP or transfer any data.
- **Control**—These are the signals that control the operation of the μP and the interface IC. They include the READ and WRITE commands and perhaps the M/ĪŌ signal (for Intel μPs). Interrupts from the peripheral to the μP, which are not discussed in this book, also use these lines.

17-4-2 The Peripheral Bus

The bus between the interface IC and the peripheral is shown on the right side of Fig. 17-1. It consists of three parts:

- **The Data Bus**—This bus carries data, usually byte-sized, between the interface IC and the peripheral. Although it is shown as a bidirectional data bus, the direction of data flow (IN if coming from the peripheral, OUT if going to the peripheral) is usually declared at the start of operation and not changed. Thus, data generally flows only one way on the data bus, but the user can decide which way it will flow.

- **The Control Lines**—These lines transmit commands to the peripheral; they control the way the peripheral operates.
- **The Status Lines**—These lines transmit status information from the peripheral. *Status information tells the μP what is happening within the peripheral.*

EXAMPLE 17-1

The temperature of the liquid in a vat is to be kept between 70°C and 80°C. There is an oven to heat the liquid, and it will cool if the oven is off. Conceptually design the circuit.

SOLUTION   The peripheral device, which in this case consists of the oven, must have one control line and two status lines. The control line simply turns the oven on. One status line indicates the temperature is below 70°C and the other status line indicates whether the temperature is above 80°C.

  The computer or μP need only monitor the status lines and issue a command to turn the oven on if the temperature is less than 70°C and turn the oven off if the temperature is above 80°C. This command will be sent to the interface IC, and then to the oven via the control line from the interface IC.

17-4-3   A Printer Interface

The so-called Centronix interface is the de facto standard interface between a μP or PC and a printer. A simplified version of this interface and its timing are shown in Fig. 17-2. The simplified interface consists of the data bus (8 lines), one control line (DATA-STROBE), and three status lines, BUSY, ACKNOWLEDGE, and FAULT. A HIGH on BUSY indicates the printer is busy and the μP must wait before sending a character. ACKNOWLEDGE pulses LOW at the end of BUSY to indicate the printer is ready to receive another character. A HIGH on FAULT indicates some fault condition, and the printer may not receive characters to be printed until that condition is corrected.

  To send a character to the printer, the μP must first check FAULT and BUSY. If they are both CLEAR, it can send the character to the interface IC, which puts the character on the data lines and causes the DATA-STROBE control line to go low. The printer then raises BUSY and prints the character. When it is finished, it clears BUSY and is ready for another character.

EXAMPLE 17-2

The FAULT line on the printer must go HIGH if any one of the following conditions are met:

(a)   There is no paper in the printer.
(b)   The printer is off-line.
(c)   There is no power to the printer.

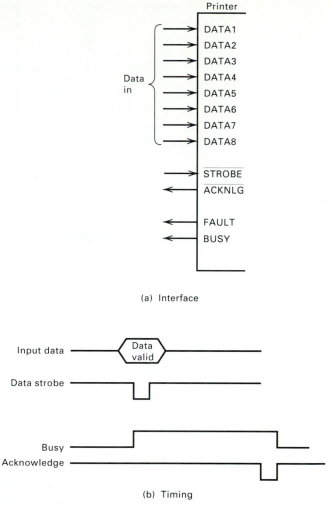

(a) Interface

(b) Timing

Figure 17-2 The Centronix Interface. (John Uffenbeck, *THE 8086/8088 FAMILY: Design, Programming, and Interfacing,* © 1987. Reprinted by permission of REGENTS/PRENTICE HALL, Englewood Cliffs, New Jersey.)

Design the circuit.

SOLUTION   The solution is shown in Fig. 17-3. If the printer is either off-line or there is no paper, the FAULT line will go HIGH. If there is no power to the printer, the lower totem-pole output transistor on the OR gate will be open-circuited and the pull-up resistor will cause the FAULT line to go HIGH.

17-4-4   The Interface IC Registers

Figure 17-1 shows that there are five registers in a typical interface IC. They are:

• **The Command Register**—This register is loaded by the μP and determines how the interface IC functions.

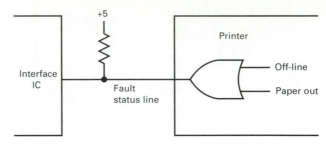

Figure 17-3   Solution to Example 17-2.

- **The Control Register**—This register retains the commands the μP issues to control the peripheral. The commands sent out on the control lines to the peripheral are taken from this register.
- **The Status Register**—This register holds the status information sent back by the peripheral. The μP can examine this register to find the status and then determine how it should respond.
- **The Data Register**—This register temporarily holds the data going to or coming from the peripheral.
- **The Data Direction Register (DDR)**—The DDR determines the *direction* of the data flow between the data register and the peripheral. It is written by the μP. In most Motorola products, each line on the data register can be either input or output and each bit in the DDR controls the direction of the corresponding bit in the data register. For Motorola, a 1 in the DDR means that the data flow in the corresponding bit is OUT, and a 0 means it is IN.

EXAMPLE 17-3

The μP writes the hexadecimal number FO into a DDR. What is the direction of the data for each bit of the data register?

SOLUTION   Bits 0, 1, 2, and 3 have 0s in the DDR. They are set for *input*. Therefore, lines 0–3 of the data register will only receive data from the peripheral. Lines 4, 5, 6, and 7 are set for *output*. They can only transmit data to the peripheral, because the DDR bits for these lines are 1s.

Many interface ICs are designed so that the entire data register can either send or receive data, as opposed to the bit-by-bit control discussed in Example 17-3. In this case, only a single DDR bit is needed to control the direction of the data register.

17-4-5   Serial and Parallel Data Transmission

Parallel data transmission means the transmission of several data bits over a bus simultaneously. It is rapid, but can only be used for data transmission over short distances. Serial data transmission transmits data over one line. It is slower than parallel data transmission, but it is used when data must be transmitted over long distances.

Interface ICs always communicate with the μP over its data bus, which is parallel transmission. Interface ICs can be classified as serial or parallel by considering their outputs to the peripherals. On a PC, connections to the printer and the disk controller are parallel, whereas connections to the keyboard or a modem (see section 17-10) are serial. On the IBM-PC, the keyboard is connected to the computer by a 5-wire cable, but only one wire transmits data. This is an example of serial communications.

17-5 Port Addressing and Intel μPs

Interface ICs are selected by one of two methods: *port I/O,* which is used by Intel, or *memory-mapped I/O,* which is used by Motorola. Both of these concepts are explained in the following paragraphs. Before discussing them, however, bus timing for μPs should be considered.

17-5-1 Memory Cycles and T-States

As stated in section 16-7-1, the basic timing of all computers is determined by a fixed-frequency, crystal-controlled clock. Indeed, this has become part of the specifications for a personal computer (PC). For example, a 25 MHz **386 PC** means the PC uses an Intel **80386** μP that is driven by a 25 MHz crystal. Higher frequency basic clocks allow the μP to execute its instructions faster, which increases its computer power.

Intel uses the concepts of *memory cycles* and *T-states*. A T-state is simply one complete cycle of the clock. Thus, a T-state for a 25 MHz clock would take 40 ns; it would be HIGH for 20 ns and LOW for 20 ns. A memory cycle is the time it takes to access memory and read or write it. *A memory cycle requires several T-states.*

17-5-2 Address/Data Buses for Intel μPs

Figure 17-4 is a simplified timing diagram for the Intel **8086** 16-bit μP. The **8085** and **8088** μP operate similarly. All of these μPs *time-multiplex* their address and data buses to minimize the number of lines required. This means that both data and addresses occupy the same lines, but at different times. When the ALE (Address Latch Enable) signal is HIGH, addresses are on the lines and must be latched into a set of FFs to be retained throughout the cycle. During the later T-states of each memory cycle, these addresses are replaced with the data being transferred between the μP and memory.

The address bus differences between the **8085**, **8088**, and **8086** are as follows:

- **8085**—The **8085** is an 8-bit μP with an 8-bit data bus and a 16-bit address bus. The 8 data bits occupy the lower 8 lines of the bus, which are labeled AD0–AD7 (AD stands for either address or data).
- **8088**—The **8088** is a 16-bit μP with an 8-bit data bus and a 20-bit address bus. Addresses AD0–AD7 form the time-multiplexed address/data bus and must be latched by ALE. Addresses A8–A15 do not have to be multiplexed, but often are. Addresses A16–A19

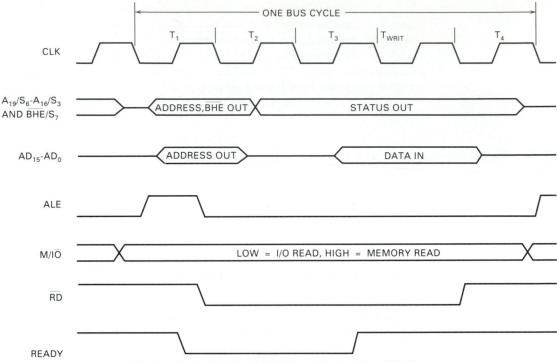

Figure 17-4 The Intel 8086 bus timing (simplified).

must also be multiplexed because these addresses are replaced by *status information*[1] during the later T-states of a memory cycle.

- **8086**—The **8086** is a 16-bit μP with a 16-bit data bus and a 20-bit address bus. AD0–AD15 form the multiplexed address/data bus. All lines must be multiplexed.

The IBM-PC uses an **8088** and part of its schematic is shown in Fig. 17-5. The addresses are latched by ALE. They use the three **74LS373** 8-bit latches in U10, U9, and U7. The **74LS245** is an 8-bit transceiver for the data lines connected to U8. On the **8088** side of the **74LS245**, the data lines are the same lines as the address lines.

17-5-3   The Intel Memory Cycle

Referring back to Fig. 17-4, we can follow the memory cycle for the Intel **8086**. An **8086** memory cycle takes at least four T-states (the figure shows five). During T_1, ALE is HIGH and the addresses must be latched. The ADDR/DATA and ADDR/STATUS buses contain addresses at this time. During T_2 the $\overline{RD}$ line goes low, indicating a READ cycle. The ADDR/DATA bus is reserved for data at this time, and the data from the memory must be valid during T_4. The data is clocked in on the rising edge of $\overline{RD}$. Figure 17-4 also shows a READY line and a *wait state* (T_{WAIT}). If the computer is too fast for the memory, meaning the computer needs the memory data before the memory access time has expired, the

---

[1]Status information indicates what type of memory cycle is being executed and gives other information. The user should refer to books on the **8086** or **8088** for further information.

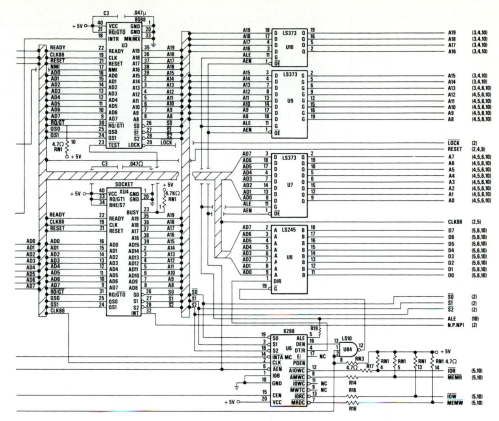

Figure 17-5   Control of the address and data buses on the IBM-PC.
(Courtesy of International Business Machines Corporation.)

computer may be forced to wait. This can be done by lowering the READY line. If the READY line is LOW at the start of T_3, the μP will insert one or more wait states between T_3 and T_4. Each wait state forces the μP to be idle for one T-state. Operation resumes when READY goes HIGH.

EXAMPLE 17-4

An **8086** is operating with a 25 MHz crystal. What is the minimum access time the memory must have if it must operate with no wait states?

SOLUTION   At 25 MHz, each T-state takes 40 ns. There are two T-states or 80 ns between the end of T_1, when the addresses are latched, and the start of T_4, when the data must be available. Thus, no wait states should be required if the memory's access time is less than 80 ns.

EXAMPLE 17-5

Assume the memory for Example 17-4 has a 100 ns access time. Design a circuit to provide the necessary wait states.

SOLUTION   One wait state will extend the time between the end of T_1 and the start of T_4 to 120 ns, so a single wait state in every memory cycle should be sufficient. The circuit of Fig. 17-6 solves the problem. ALE causes both the TOGGLE and READY FFs to clear, placing a LOW on the READY line. The rising edge of the CLK signal during T_2 sets the TOGGLE FF, but does not affect the READY FF. The next rising edge of CLK, during T_3, clears the TO-GGLE FF, which produces a rising edge on its $\overline{Q}$ output. This rising edge sets READY. Thus, READY is LOW at the start of T_3 so a wait state will be inserted, but READY will also go HIGH and no additional wait states will be inserted.

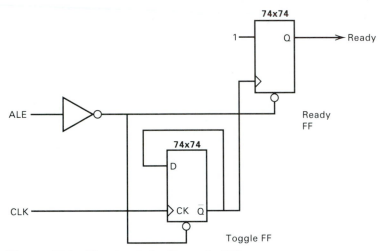

Figure 17-6   The circuit for Example 17-5.

Most **8086** or **8088** systems, including the IBM-PC, use the Intel **8284** IC to control timing. It provides the wait states as required.

17-5-4   Port Addressing

Intel uses the concept of port addressing for its I/O. Each interface IC is assigned a *port number,* which distinguishes it from other interface ICs. On the IBM-PC, for example, ports 0378–037F are reserved for the printer(s), and ports 0320–032F are reserved for the fixed disk.

Intel μPs have a set of ports and a set of memory locations. The concept is shown in Fig. 17-7 for the **8086** and **8088**, which can address 1 Mbyte of memory or 64K ports. These μPs have a 20-bit address bus, but the 16 LSBs of this address bus are also used for the port addresses. This means the memory and interface ICs must be able to distinguish between a memory address and a port address. On the simpler systems this is done by using the M/$\overline{\text{IO}}$ signal, which is a control line on the **8086** and **8088**. If M/$\overline{\text{IO}}$ is LOW, port addresses are on the line; the selected interface IC must respond and may put data on the data bus, but the memory must ignore the cycle. Conversely, when M/$\overline{\text{IO}}$ is HIGH, only memory addresses are selected and all interface ICs must be deselected and not respond. Thus, M/$\overline{\text{IO}}$ must be an input to the decoders that selects a particular port IC or a particular memory IC.

A single interface IC usually contains several ports. These ports must have *contiguous addresses*. These interface ICs have address inputs that are connected to the LSBs of the address bus and determine which port is selected.

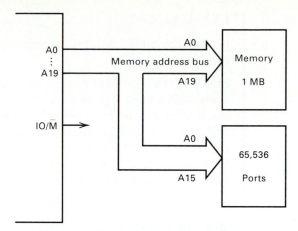

Figure 17-7   Ports and Memory on an **8086** or **8088**.

The Intel **8255** parallel interface IC, discussed in section 17-7, has four ports. The one port of the four on the **8255** that is to be addressed is selected by bringing the two LSBs of the address bus into the **8255**.

EXAMPLE 17-6

An **8255** has four ports that occupy addresses 8000–8003. Design a decoder to select the **8255** and show what must be on the address bus to select port 2 on this interface IC.

SOLUTION   The solution is shown in Fig. 17-8. The decoder is shown as a 15-input NAND gate connected to the $\overline{\text{CHIP SELECT}}$ ($\overline{\text{CS}}$) input of the **8255**. The **8255** will only be selected if $\text{M}/\overline{\text{IO}}$ is LOW, indicating a port address for an interface IC, A15 is HIGH, and A14 . . . A2 are low. Thus $\overline{\text{CS}}$ will go LOW for any address between 8000 and 8003. To select port 2, A1 must be 1 and A0 must be 0.

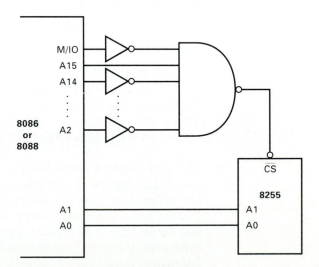

Figure 17-8   Solution for Example 17-6.

EXAMPLE 17-7

Repeat Example 17-6, but assume there are no other ports connected between 8000–8FFF.

SOLUTION   In this example the decoder is much simpler, because any address with a most significant nibble of 8 can be used to select the **8255**. Thus, the NAND gate of Fig. 17-8 only needs five inputs. A15 must be 1, and M/$\overline{\text{IO}}$ and A14, A13, and A12 must be 0 to select the **8255**. This decoding, however, causes *redundant addressing;* any address in the range will cause the **8255** to respond. Port 0, for example, will respond to any port address that starts with 8 and has two LSBs of 0, such as 8000, 82CC, or 8334.

17-5-5   IN and OUT Instructions

The M/$\overline{\text{IO}}$ line only goes LOW during two **8086** or **8088** instructions, IN and OUT. An IN instruction takes two bytes; it is coded as the Op code for IN and the port address. It reads data (one or two bytes) from the interface IC into the accumulator. IN instructions are also referred to as $\overline{\text{IOR}}$ (I/O Read) instructions.

An OUT instruction is coded similarly; it has the Op code for OUT followed by the port address. OUT instructions are also referred to as $\overline{\text{IOW}}$ (I/O Write) instructions, because they write data from the accumulator into the selected port.

17-5-6   IN and OUT Timing

The Intel **8085** is a simpler 8-bit μP and the timing is easier to understand. The IN and OUT instructions are similar to the 16-bit μPs, with the following differences:

1. The **8085** can only address 256 ports. This means, however, that the port address can be limited to one byte. The **8086/8088** port addresses can be either 1 or 2 bytes long.
2. The M/$\overline{\text{IO}}$ line is inverted; In the **8085** this line must be HIGH for an I/O cycle and is called IO/$\overline{\text{M}}$.
3. The **8085** executes many memory cycles using only three T-states instead of 4.

The timing for an OUT instruction for the **8085** is shown in Fig. 17-9. IN and OUT instructions take three Machine Cycles consisting of 10 clock cycles or T-states. Figure 17-9 shows the relationship of the clock, ALE $\overline{\text{RD}}$, $\overline{\text{WR}}$, A8, and the lower 8 bits of the multiplexed address-data bus for an OUT 15 instruction. The instruction proceeds as follows:

1. The first Machine Cycle is an Op code fetch. This requires four clock cycles. During the first clock cycle, the instruction address (N) is on the address-data (AD) bus and ALE goes HIGH to latch the address. During the second and third cycles, $\overline{\text{RD}}$ goes LOW while the Op code (D3 for an OUT instruction) is placed on the AD bus. The fourth clock cycle gives the **8085** time to decode the instruction and prepare for the following cycles.

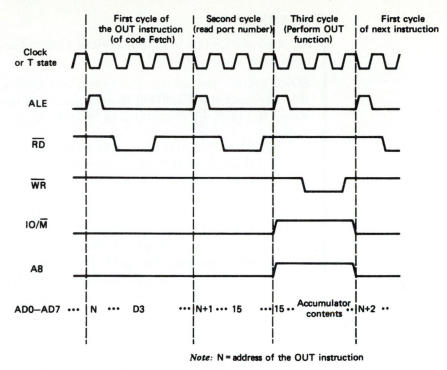

Figure 17-9   Timing for an OUT instruction.

2. The second cycle of the OUT instruction is a memory read of the second byte. The memory address is (N + 1), the location of the second byte, and the data read from memory is the port address (15 in this example).

3. The data transfer takes place during the third Machine Cycle. During this cycle IO/$\overline{\text{M}}$ is HIGH. The port address is placed on the address-data bus, but is quickly replaced by the accumulator data. At this time the $\overline{\text{WR}}$ line goes LOW. Address bits 8–15 will also contain the port address (15), but they will remain there throughout the entire cycle as the behavior of A8 shows.

The timing for an IN instruction is identical except that an additional pulse occurs on the $\overline{\text{RD}}$ line instead of on the $\overline{\text{WR}}$ line. All timing can be checked on an oscilloscope by using the simple program:

    2010   OUT   15   (or IN)
    2012   JMP   2010

17-5-7   Execution of the OUT Instruction

The OUT instruction (I-O WRITE) can be executed by tying the AD bus to the inputs of the various peripherals and then strobing the data in with a pulse that gates IO/$\overline{\text{M}}$ and the $\overline{\text{WR}}$ line.

EXAMPLE 17-8

An **8085** system has 16 output devices. Design the circuitry so the **8085** can send data to any of these devices.

SOLUTION The output devices can be set up so they use ports 0–15. In this way, only the 4 LSBs of the address lines need be used. The circuit is shown in Fig. 17-10. It operates as follows:

1. The AD0–AD7 lines go to all devices. These lines are shown going through a **74LS241**, 8-input, noninverting, buffer/line driver. If the data input to all devices is CMOS, the **74LS241** might be omitted, but it has two advantages: it alleviates *fanout or loading problems,* and it *isolates* the μP from the external devices. Isolation is important. Otherwise, a miswire or other problem on an external device could disable the μP. The buffer is enabled by IO/M̄ so that it transmits data only during OUT or IN instructions.

2. The 4-bit port address is sent to a **74154** decoder (see section 11-4-2). Addresses A8 through A11 are used (instead of AD0–AD3) because the higher address lines are not multiplexed and this eliminates the need for an address latch.

3. The decoder is enabled by the inverse of IO/M̄ and by W̄R̄. Thus, it will only function during the write portion of an OUT instruction. When enabled, the decoder provides a LOW input to the DEVICE-SELECTED line of the addressed peripheral. The peripheral must use this signal to strobe in the data. Note that the AD lines contain data at this time, not addresses, because DEVICE-SELECTED is synchronized with the W̄R̄ signal.

4. More than 16 (up to 256) peripherals may be accommodated by enlarging the decoder and adding more drive capability if necessary.

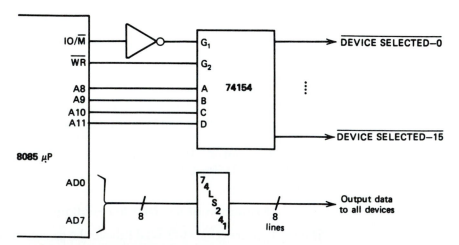

Figure 17-10 Sending data from an **8085** to 16 peripherals.

17-5-8 Execution of the IN Instruction

The IN instruction (I-O READ) brings data in from the peripherals and enters them into the accumulator. The IN instruction can be executed by a circuit similar to Fig. 17-10. The major differences are:

1. The $\overline{RD}$ line would enable the decoder instead of $\overline{WR}$.
2. The peripherals must all put their data on the AD lines of the μP at the proper time. If these data are present at the wrong time (when the μP is executing other instructions, for example), they will cause a conflict and cripple the μP. Therefore, all peripherals that are inputs must be tied to the address-data bus via 3-state gates that are only enabled during the third Machine Cycle of an IN instruction. Fortunately, the DEVICE-SELECTED outputs from the decoder only go LOW at this time, so DEVICE-SELECTED provides an ideal signal for enabling the 3-state drivers going from the peripherals to the μP.

17-5-9 IN and OUT Instructions on the IBM-PC

The IBM-PC uses the **8088** in its so-called *maximum mode*. This mode does not use $M/\overline{IO}$ directly, but requires an Intel **8288** IC to set the I/O. It is shown in the bottom right part of Fig. 17-5. The four outputs are $\overline{IOR}$, $\overline{MEMR}$, $\overline{IOW}$, and $\overline{MEMW}$. One of them must be low for each type of cycle.

17-6   Motorola and Memory-Mapped I/O

Unlike Intel, Motorola μPs use *memory-mapped* I/O. Each interface IC is assigned a set of locations in the address space, and the registers for that interface IC are accessed by reading or writing to the register's address. Thus, if there are four registers on an interface IC, and they are assigned to locations 8008–800B, there can be no memory at those addresses; they are reserved strictly for the interface IC. The concept is shown in Fig. 17-11, which shows two interface ICs, each with its reserved area in the memory map.

17-6-1   Selecting an Interface IC

An interface IC is selected or accessed when its $\overline{CS}$ (Chip Select) line goes low. Therefore, there must be a decoder on the address lines to determine when the address is for the interface IC. Many Motorola interface ICs have several CS lines, some responding to positive levels and some responding to negative levels. In these cases, all CS lines must be at their proper level for the IC to respond. The advantage of having several CS lines is that it makes external decoding easier.[2]

---

[2]For a more thorough discussion of this topic, see *Greenfield and Wray, Using Microprocessors and Microcomputers: The Motorola Family,* latest edition, published by Prentice Hall, Englewood Cliffs, N.J.

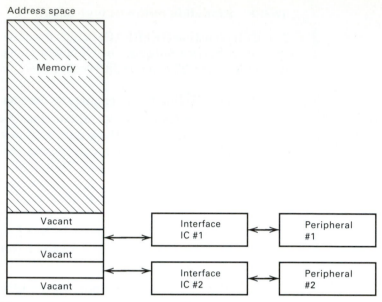

Figure 17-11 Memory mapped I/O.

─ **EXAMPLE 17-9**

An interface IC has four registers that occupy locations 8008–800B in a 16-bit address space. Assume the IC has only one $\overline{CS}$ line. Conceptually describe the decoder needed to select this IC.

SOLUTION Because the interface IC has four registers, addresses A0 and A1 must be used to specify the register to be accessed. The other addresses must be decoded to select the IC. For the given addresses, A15 and A3 must both be 1 and addresses A14–A4 and A2 must be 0. Note that there is no M/$\overline{IO}$ signal in this system.

17-6-2   Advantages and Disadvantages of Memory/Mapped I/O

The obvious disadvantage of memory-mapped I/O is that it takes pieces out of the address space, and these pieces cannot also contain memory. The advantage is that the interface registers can respond to all the normal memory commands. A LOAD will read from a register into the accumulator, a STORE will write from the accumulator to a register, but the registers can also respond to other commands that can access memory, such as INCREMENT, DECREMENT, and SHIFT.

17-7   The Intel 8255

The Intel **8255** is called a Programmable Peripheral Interface. It is one of the most commonly used parallel interface ICs. Its diagram, Fig. 17-12, shows that it contains 24 lines to connect to external devices and an interface to a μP. It also contains four ports as explained in section 17-5-4. The **8255** is used in the

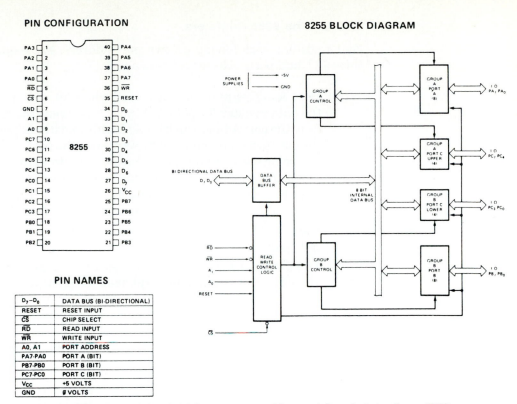

PIN NAMES

D_7-D_0	DATA BUS (BI-DIRECTIONAL)
RESET	RESET INPUT
$\overline{CS}$	CHIP SELECT
$\overline{RD}$	READ INPUT
$\overline{WR}$	WRITE INPUT
A0, A1	PORT ADDRESS
PA7-PA0	PORT A (BIT)
PB7-PB0	PORT B (BIT)
PC7-PC0	PORT C (BIT)
V_{CC}	+5 VOLTS
GND	Ø VOLTS

Figure 17-12 8255A programmable peripheral interface (PPI). (Courtesy of Intel Corporation.)

IBM-PC to read in the status of the switches and for other functions. Its ports occupy locations 60–63.

17-7-1 The External Interface

The interface to external devices, shown on the right side of Fig. 17-12, consists of three 8-bit ports. Port A and Port B can both be either input or output. Port C can be subdivided into two 4-bit ports: a lower port containing PC0–PC3 and an upper port containing PC4–PC7. All connections to external devices are made through connections to these ports.

17-7-2 The μP Interface

The connections between the **8255** and the μP are shown on the right side of Fig. 17-12. They are:

- The 8-bit bidirectional data bus.
- **$\overline{RD}$ and $\overline{WR}$**—Signals from the μP that indicate whether the **8255** is being read or written.
- **A0 and A1**—The two LSBs of the address bus. They select one of the four registers in the **8255**.
- **$\overline{CS}$**—Chip Select. This pin determines when the port address for the **8255** is present and the **8255** should be accessed (see section 17-5-4).

17-7-3 The 8255 Registers

Four registers, each having its own port address, are required to control the three I/O data ports. These registers are:

- **Port A**—This register is accessed when A0 = 0 and A1 = 0. It is the data register for port A. If port A is input, the data coming in from the port A lines can be read in this register. If port A is output, the data to be sent out must be written to the port A register. If port A is being used for input, it makes no sense to write to it, and if port A is being used for output, it makes no sense to read it.
- **Port B**—This register is accessed when A0 = 1 and A1 = 0. It controls port B, but functions in the same way as the port A register.
- **Port C**—This register is accessed when A0 = 0 and A1 = 1. It controls port C.
- **The Control Register**—This register is accessed when A0 = 1 and A1 = 1. Its function is explained in the next paragraph.

EXAMPLE 17-10

The **8255** on the IBM-PC uses ports 60–63. Which port address accesses which register?

SOLUTION   Port A is accessed when A0 and A1 are both 0. This is only true for address 60. Similarly, port B is accessed by port address 61, port C by address 62, and the control register by address 63.

EXAMPLE 17-11

What instruction must be given to the **8255** to

  (a)   Write data into port A?
  (b)   Read data from port B?

SOLUTION

  (a)   The data to be written must be in the accumulator. Then the instruction OUT 60 will write it into port A and send it out on the port A lines.
  (b)   The instruction IN 61 will bring data from the port B lines into the accumulator.

17-7-4 The Control Register for the 8255

The control register for the **8255** is shown in Fig. 17-13. The **8255** operates in one of three modes, 0, 1, or 2. Mode 0 is the basic I/O mode, where all ports are either input or output. Mode 1 is the *handshaking* mode, and some of the port C lines are reserved for control functions. Mode 2 is the bidirectional data mode; it is rarely used and will not be discussed further.

Commands are given to the **8255** by writing to the control register. It cannot be read. If bit 7 of the command is a 1, as shown in Fig. 17-13, the command is

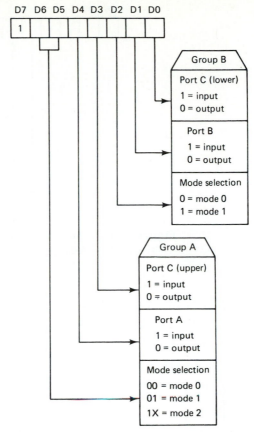

Figure 17-13 The Control Register for the **8255**. (Courtesy of Intel Corporation.)

mode set. The rest of the bits in the command set the mode of the **8255**. They function as listed below:

- **Bits 5 and 6**—These determine the mode of port A. Because the **8255** can operate in one of three possible modes, two bits are necessary to determine the mode.
- **Bit 4**—This is the direction bit for port A. It specifies whether the data at port A flows out to the external peripheral (Bit 4 = 0) or comes in from the external device (Bit 4 = 1).
- **Bits 3 and 0**—These are the direction bits for port C, which is subdivided into an upper half (bits C7–C4) and a lower half (C3–C0). Bit 3 controls the direction of data flow for the upper half of the port and bit 0 controls the lower half.
- **Bit 2**—This is the mode bit for port B, which can only operate in modes 0 or 1.
- **Bit 1**—This is the direction bit for port B.

If bit 7 of the control register is a 0, the command is a *bit set command*. It sets or clears a bit in the port C register, as shown in Fig. 17-14. This feature can be used in mode 1 operation.

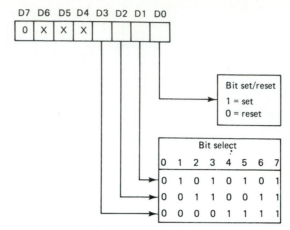

D7 D6 D5 D4 D3 D2 D1 D0

| 0 | X | X | X | | | | |

Bit set/reset
1 = set
0 = reset

Bit select

0	1	2	3	4	5	6	7
0	1	0	1	0	1	0	1
0	0	1	1	0	0	1	1
0	0	0	0	1	1	1	1

Figure 17-14 The configuration of a bit set command in the 8255. (Courtesy of Intel Corporation.)

EXAMPLE 17-12

What are two ways to set bit 5 of port C? Assume the **8255** is assigned to ports 6C–6F.

SOLUTION One way would be to write a 1 into bit 5 of port C. For example, we could load the accumulator with 20 and then issue an OUT 6E command. This will affect all the bits in port C.

Another way is to load the accumulator with 0D and issue an OUT 6F command. This clears bit 7 of the control register to identify the command as a bit set/reset command. Bits 3–1 are 101, which identify bit 5, and bit 0 is a 1, which indicates that the bit must be set.

17-7-5 Mode 0 Operation

Mode 0 operation is simple I/O. All lines on all ports are either input or output, as determined by the word written to the command register. This is the most commonly used mode and is the mode used by the IBM-PC.

EXAMPLE 17-13

The **8255** in the IBM-PC uses ports 60–63 and is set up so that ports A and C are used for input and port B is used for output.

(a) What control word must be sent to the **8255** to set the ports for this operation?

(b) What instructions must be issued to do this?

SOLUTION

(a) The command word is determined as follows:

- **Bit 7**—The command must be a mode set word, not a bit set word. Bit 7 must be 1.

- **Bits 5 and 6**—These bits set the mode of port A to 0. They must both be 0.
- **Bits 4 and 3**—These two bits set the direction for port A and the upper half of port C. Because the direction is IN, both bits must be 1.
- **Bit 2**—This bit sets the mode for port B. It is also operating in mode 0 so the bit must be 0.
- **Bit 1**—This is the direction bit for port B. Because port B is to be output, the bit must be 0.
- **Bit 0**—This is the direction bit for the lower half of port C. Port C is input so this bit must be 1.

In accordance with the above, a command word of 99 must be sent to the control register in the **8255**.

(b) Because the control register must be at port 63 (address bits A1 and A0 must both be 1 to access the control register), the instructions LOAD 99–OUT 63 will send the proper word to the proper port.

In the BIOS (Basic I/O System) code for the IBM-PC, which is in ROM, the following code is found at location FE0BB:

$$B0 \ 99 \ E6 \ 63$$

The first byte, B0, is the Op code for a LOAD (or MOVE) instruction, and the next byte, 99, is the data to be moved into the accumulator. The next byte, E6, is the Op code for the OUT instruction, and the last byte, 63, is the port address.

17-7-6   Mode 1 Operation—Input

Mode 1 operation is often called the *handshaking mode*. Either port A or port B (or both) can operate in this mode. If they do, certain bits of port C are reserved for controlling the flow of information between the **8255** and the peripheral device, and may not be used for data transmission. In this paragraph we will consider *mode 1 input on port A* only. Mode 1 input on port B is similar, but uses different pins on port C.[3]

The configuration for port A in input handshaking is shown in Fig. 17-15. It uses three lines of Port C.

- **STBA (STROBE A)**—This signal is sent from the peripheral device to the **8255**. It indicates that the peripheral device has a valid byte for the **8255** and has placed it on the port A lines.
- **IBFA (Input Buffer Full A)**—When IBF is HIGH, it indicates that the **8255** has received data, but has not processed it. The peripheral should not send another byte of data while IBF is HIGH.
- **INTRA (Interrupt)**—This signal allows port C to interrupt the μP when data has been received, so it can read in the byte and allow

[3]A discussion of mode 1 operation for port B is given in the books by Gaonkar and Uffenbeck (see References).

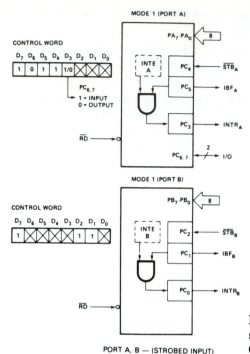

MODE 1 (PORT A)

CONTROL WORD

D_7 D_6 D_5 D_4 D_3 D_2 D_1 D_0

| 1 | 0 | 1 | 1 | 1/0 |✕✕✕|

$PC_{6,7}$
1 = INPUT
0 = OUTPUT

MODE 1 (PORT B)

CONTROL WORD

D_7 D_6 D_5 D_4 D_3 D_2 D_1 D_0

| 1 |✕✕✕✕✕| 1 | 1 |✕|

PORT A, B — (STROBED INPUT)

Figure 17-15 Port A set up for input handshaking on the **8255**. (Courtesy of Intel Corporation.)

the next byte to be sent. A thorough discussion of interrupts is beyond the scope of this book.

Figure 17-15 shows that these pins are on PC4, PC5, and PC3 respectively. The pins for mode 1 input to the B port, $\overline{STBB}$, IBFB, and INTRB, are on pins PC2, PC1, and PC0, respectively.

EXAMPLE 17-14

Explain each bit in the control word shown in Fig. 17-15.

SOLUTION   Bit 7 is 1 because a mode set word is required. Bits 6 and 5 are 0 and 1 to set the A side of the **8255** for mode 1 operation. Bit 4 is 1 to set port A for input. In this mode bits 6 and 7 of port C are not used in the handshaking and can be used for any other I/O. Bit 3 is a direction bit for both of these bits (PC6, PC7). Bits 2 and 1 pertain to port B and are not relevant here. Bit 0 is not used.

The timing for mode 1 operation is shown in Fig. 17-16. It starts with IBF low, which allows the peripheral device to send information. When the peripheral has valid data, it places it on the input data lines and sends a $\overline{STB}$. The leading edge of $\overline{STB}$ causes IBF to go HIGH. The peripheral can use this signal as an indication that it cannot send any more data.

In normal operation, the μP must periodically read IBF to determine if it has gone HIGH, indicating that data is available on port A. This is called *polling*. The μP does this by periodically reading port C (using an IN port C instruction), and examining bit 5. When it finds data is available, it reads the port. This is shown by the low pulse on the $\overline{RD}$ line in Fig. 17-16. This pulse

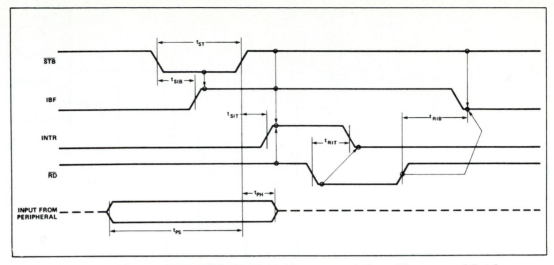

Figure 17-16 8255 mode 1 input port timing. (Courtesy of Intel Corporation.)

shown in Fig. 17-16 *only occurs on an IN instruction to port A.* This same IN instruction will read the data in port A's latches into the μP's accumulator, where the μP can access it. The trailing edge of $\overline{RD}$ causes IBF to go LOW, which allows the peripheral to send the next byte of data along with the next strobe.

The μP system can also be configured to use interrupts when a byte is received instead of polling, but this is beyond the scope of this book.

EXAMPLE 17-15

How can an **8255** be monitored in the laboratory to show its operation in mode 1?

SOLUTION One method is shown in Fig. 17-17. It assumes that the **8255** is the only interface IC to be accessed while the μP program that controls the operation is running. This allows us to select the **8255** by directly connecting its $\overline{CS}$ lines to M/$\overline{IO}$ for an **8086/88** or to complement IO/$\overline{M}$ for an **8085**. The **8255** will be selected only during IN and OUT instructions.

Figure 17-17 shows that a periodic strobe is provided by the pulse generator and the $\overline{CS}$, $\overline{STB}$, and IBF lines are connected to an oscilloscope to show the relationship between them. The timing should be similar to Fig. 17-16.

The control program in the μP should do the following:

1. Write the proper control word to the **8255**.
2. Constantly monitor IBF by issuing IN port C instructions. This will produce a steady stream of pulses on $\overline{CS}$.
3. When IBF goes HIGH, in response to a $\overline{STB}$, go into a time delay loop. At this time the stream of pulses on $\overline{CS}$ will stop.
4. When the time delay expires, read port A. This should reset IBF.
5. Return to step 2, causing the process to repeat. In this way the waveforms can be observed on an oscilloscope.

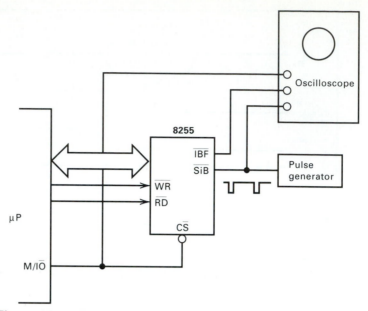

Figure 17-17 Demonstrating handshaking in the laboratory.

Of course, this procedure requires a program to be written into the μP. Unfortunately, space in this book precludes a discussion of μP programming.

17-7-7 Mode 1 Operation—Output

Handshaking on output can occur when the μP is sending data to a peripheral IC via the **8255** interface IC. Mode 1 operation for port A is shown in Fig. 17-18 and its timing is shown in Fig. 17-19. Again three lines on port C are used to control the data transmission. They are:

- **OBFA (PC7)**—A write to port A in mode 1 causes $\overline{OBF}$ to go LOW. This indicates to the peripheral that data is available.
- **ACKA (PC6)**—A LOW pulse on the $\overline{ACK}$ line, which is controlled by the peripheral, indicates that it has accepted the data, and causes $\overline{OBF}$ to go HIGH. It will remain HIGH until the μP writes the next byte to the port.
- **INTRA (PC3)**—If interrupts are used, the trailing edge of $\overline{ACK}$ will cause an interrupt. This interrupt will then cause the μP to write the next data byte to the output port.

The timing chart of Fig. 17-19 clearly shows the sequence of these events.

EXAMPLE 17-16

A printer using a Centronix interface (see Fig. 17-2) is to be connected to a μP via port A of an **8255** operating in mode 1. Show the hardware connections and describe the program.

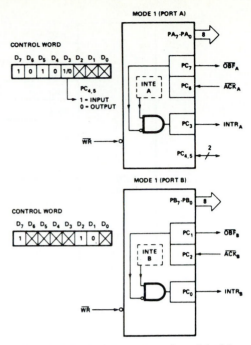

Figure 17-18 The **8255** set up for output handshaking. (Courtesy of Intel Corporation.)

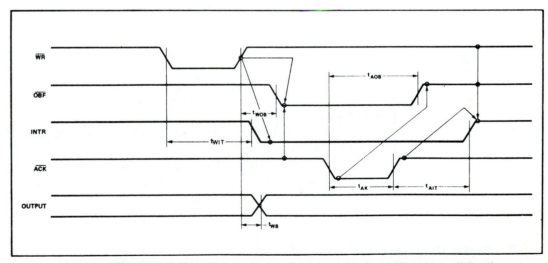

Figure 17-19 8255 mode 1 output port timing. (Courtesy of Intel Corporation.)

SOLUTION The connections are shown in Fig. 17-20. The data flows from port A to the printer. $\overline{\text{OBF}}$ can be connected to the $\overline{\text{STROBE}}$ input to the printer; when it goes LOW data is available. The $\overline{\text{ACKNLG}}$ line on the printer, which goes LOW at the end of BUSY, can be connected to the $\overline{\text{ACKA}}$ line on the **8255**.

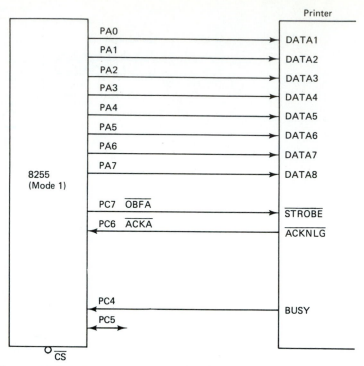

Figure 17-20 Connecting a printer to an **8255**. (John Uffenbeck, *THE 8086/8088 FAMILY: Design, Programming, and Interfacing*, © 1987. Reprinted by permission of REGENTS/PRENTICE HALL, Englewood Cliffs, New Jersey.)

The control program in the μP must do the following:

(a) Issue an OUT instruction to the control port to set up the proper control word.

(b) Issue an OUT command to port A with the first data byte. This causes $\overline{OBF}$ to go LOW.

(c) Monitor $\overline{OBF}$ by looping and reading port C with IN commands. When PC7 goes HIGH, in response to the LOW pulse on $\overline{ACK}$, the program can send the next byte to the printer.

17-8 Serial Communications

Parallel Communications, which involves transferring several bits of data over several lines (typically 8 bits as in an **8255** port), is the fastest way to communicate with a μP. When data is to be transmitted over *long distances,* however, it is neither cost effective nor technically feasible to use several lines. In this case the parallel data must be converted to serial data and sent out one bit at a time on a single line. This is called *serial data transmission.*

A typical communications system is shown in Fig. 17-21, where the μP or computer on the left side must communicate with a *terminal* on the right side of

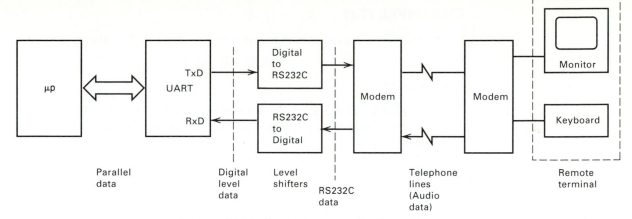

Figure 17-21 A serial communication path.

the figure. The terminal consists of a *video monitor* for displaying computer information to the user, and a keyboard that allows the user to enter commands and data. Many offices have terminals on each employee's desk, and many colleges use terminals extensively. At the Rochester Institute of Technology, for example, there were terminals on each professor's desk, and user centers with many terminals for the students. These were connected to a bank of VAX computers, made by Digital Equipment, Inc., that performed the major computing for the entire campus.

In an office or campus, cables are often strung throughout the facility to connect the computer to the various terminals. These cables contain a single line for transmission, a single line for reception, and perhaps their associated ground wires, but the data transmission is serial.

In many cases, however, people want to work at home or at a remote location. Then the terminal must be connected to the computer via the existing telephone network. MODEMs (see section 17-10) are required as shown in Fig. 17-21. Computers can also be connected to other computers or bulletin board systems using the phone lines. Dialing up over telephone lines also uses serial transmission.

17-8-1 Data Characters

Data characters that are sent from a μP to a terminal, or received by a μP from a keyboard, are always retained in memory in the ASCII (American Standard for Communications Information Interchange) code. These characters are mainly *alphanumeric;* they consist of the standard alphabetics, numbers, and punctuation marks.

The ASCII code is given in Appendix B. It is a 7-bit code. The most significant hex digit is given along the top or heading of each column of the table. The least significant digit is given along the side or row. The character referred to is at the intersection of the row and column. The characters in the first two columns are for communications control and are used in special circumstances. The alphanumeric characters occupy the remaining six columns of the table.

> **EXAMPLE 17-17**
>
> The following urgent message has been received in ASCII:
>
> 48 45 4C 50 21
>
> What is it?
>
> **SOLUTION**   The first character can be found in the ASCII table at the intersection of column 4 and the row for 8. It is the character H. Continuing with the rest of the characters, the message reads:
>
> H E L P !

The 7-bit ASCII characters are always allocated one byte in a computer's memory. The eighth bit is either not used or used for parity. The ASCII equivalent of the characters appearing on a computer screen can often be found in an area of memory called the *screen image*. The screen image for the IBM-PC can be found starting at memory location B0000.

17-8-2 Serial Data Transmission

Serial data transmission is shown in the upper lines of Fig. 17-21. It can be visualized as the μP or computer sending a character which is to be displayed on the screen of the monitor. It involves several devices that will be explained in more detail in later paragraphs.

Referring to Fig. 17-21, serial data transmission requires the following steps:

a. The computer or μP sends an ASCII character in *parallel* to the UART (see section 17-9).

b. The UART converts the character to *serial* and sends it out on its TxD line in asynchronous format (see section 17-9-4).

c. Level shifters change the voltage levels to RS-232C levels (see section 17-10-5). These levels are compatible with MODEMs and terminals.

d. The MODEM (see section 17-10) on the left converts the digital inputs to audio tones and sends them out on the telephone.

e. The MODEM on the right reconverts the audio tones it receives on the telephone lines to digital and sends the bits to the monitor.

17-8-3 Serial Data Reception

Data sent by the keyboard must be received at the μP. This transmission path is shown by the lower line in Fig. 7-21. A character entered at the keyboard is sent to the MODEM, where it is transmitted over the telephone lines. The MODEM on the left receives audio telephone signals, which it converts to digital at RS-232C levels. The level shifters convert the RS-232C to standard logic levels (0 V and +5 V) and send them into the UART. The UART converts the character to parallel and sends it to the μP.

17-8-4 Asynchronous Transmission

Low-speed MODEMs generally utilize *asynchronous* (unclocked) *data transmission*. The word asynchronous means without a clock, or unclocked. The most common code for asynchronous data transmission is the 10-bit START-STOP code that is described here. Some systems and devices use other codes that are variants of the 10-bit START-STOP code.

The basic pattern for asynchronous data transmission is shown in Fig. 17-22. Transmission engineers often use the term *mark* for the logic 1 state and *space* for the 0 state. When the line is quiescent (transmitting no data), it is constantly in the mark or 1 state. The start of a character is signaled by the START bit, which drives the line to the 0, or space state, for one bit time. A *bit time* is the time required to serially transmit a single bit of data. It is determined by the speed of the terminal or MODEM receiving the data.

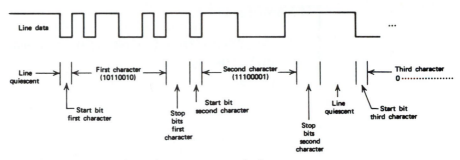

Figure 17-22 Asynchronous transmission.

EXAMPLE 17-18

A terminal is set to run at 2400 Bits Per Second (BPS).[4] What is its bit time?

SOLUTION The bit time is simply the time of a single bit. Here it is $\frac{1}{2400}$ BPS = 416.67 μs.

The 8 bits immediately following the START bit are the *data bits* of the character. The ASCII code is used. The 8 data bits typically consist of 7 bits that determine the ASCII code for the character, plus a parity bit to insure the integrity of the data. The LSB of the character is transmitted first, right after the START bit. The character shown in Fig. 17-22 is 4D or the code for the letter M.

After the last data bit, the transmission line must go HIGH for at least one bit time. This is called the STOP bits. If no further data are to be transmitted, the line simply stays HIGH (marking) until the next START bit occurs.

This data pattern requires 10 bits:

1. One START bit (always a space or 0).
2. Eight data bits.

---

[4]The term BAUD is often used synonymously with the term bits per second. There is a technical difference between the terms, but it is unimportant here.

3. One STOP bit (always a mark or 1). Some older systems use two STOP bits. Two STOP bits were actually used in Fig. 17-22.

EXAMPLE 17-19

How many characters per second can a 300-BPS line transmit if it is running at its maximum rate?

SOLUTION   Because each character requires 10 bits, the maximum data transmission rate is

$$\frac{300 \text{ bits/second}}{10 \text{ bits/character}} = 30 \text{ characters/second}$$

EXAMPLE 17-20

    (a)   What is the bit pattern of the character shown in Fig. 17-23?
    (b)   Is the parity odd or even?
    (c)   What ASCII character is it?

SOLUTION

    (a)   After the initial quiescent period, the waveform bits are 00001001011. The first 0 is the START bit and the next 8 are the character. Therefore, the character is 00010010. The two 1s following the last datum bit are the STOP bit and an idle bit.

    (b)   Since the characters contain an even number of 1s, this is an *even* parity character.

    (c)   The character bits are 00010010. Remembering that the LSB is transmitted first, this is a $(48)_{16}$. Appendix B shows that 48 is the ASCII code for the letter "H."

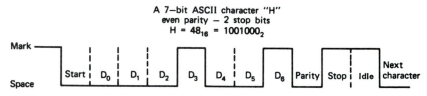

Figure 17-23   Character pattern for Example 17-20.

17-9   UARTS

The term UART stands for *Universal Asynchronous Receiver Transmitter*. Its purpose is to transform parallel data into asynchronous serial data at a speed compatible with the receiving device. It performs all of the following functions:

1. Convert parallel data from the μP to serial data.
2. Produce the proper pulse width (416 μs per bit for a 2400 bps terminal).

3. Add START and STOP bits.

4. Leave the output in the mark state when no data are being transmitted.

UARTs are also used to receive serial data. They strip off the START and STOP bits, check for errors, and send the data to the μP in parallel form.

17-9-1 The AY-5-1013

The **AY-5-1013**[5] is one of the most popular and commonly used UARTS. Its pinout and block diagrams are shown in Fig. 17-24. For transmitting, the UART accepts 8 data bits on its DB lines and sends out the asynchronous character on its SO (Serial Output) pin. It receives data on its SI (Serial Input) pin and presents the data to the computer on its eight RD lines. All signal levels are TTL.

A list of the pins and the function of each pin is given in Fig. 17-25. The pins can roughly be divided into three groups: those involved with data transmission, those involved with data reception, and those involved with the *control* of the UART.

The basic timing for the UART is generated by external clocks applied to pin 17, a control pin, on the receive side and pin 40 on the transmit side. The frequency of these clocks must be 16 times the bit rate of the external device.

EXAMPLE 17-21

A UART must interface with a printer that has a serial interface that accepts data at 300 bps. What frequency clock must be applied to pins 17 and 40?

SOLUTION Since both the transmit and receive sides of the TTY use the same frequency (300 BPS), the same clock can drive pins 17 and 40. Its frequency must be 16×300 BPS, or **4800** Hz.

17-9-2 The Control Pins on the AY-5-1013

The characteristics of the asynchronous data for both transmission and reception are determined by the levels on control pins 35–39, and are explained in Fig. 17-25. The characteristics include the number of bits in each character (pins 37–38), the number of STOP bits (pin 36), and parity (pins 35 and 39). The levels on these pins are set whenever CONTROL STROBE (CS–pin 34) is HIGH. For laboratory experiments CS is usually left HIGH and the levels are hard-wired to the respective pins. They can also be entered from a computer by pulsing CS HIGH.

EXAMPLE 17-22

A standard teletype (TTY) runs at 110 BPS and uses 7 bits plus even parity and two STOP bits. How would pins 34–40 be connected? Assume hard-wire connections.

---

[5]Many engineers prefer the **AY5-1015**. This UART is slower and more expensive than the **AY5-1013**, but only requires a single +5 V power supply. Otherwise the two UARTs operate identically.

Pins

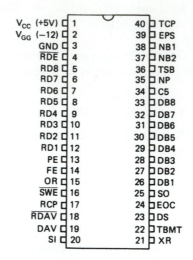

PINOUT FOR AY5-1013 UART

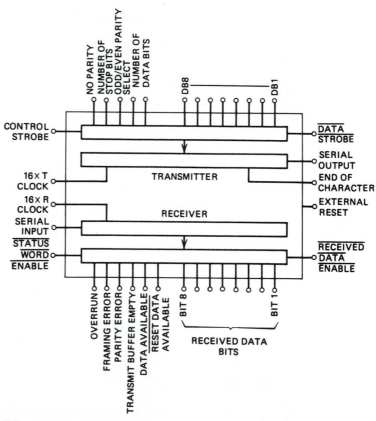

Figure 17-24 Pinout and block diagram of the **AY-5-1013**. (Courtesy of General Instruments Microelectronics.)

Pin No.	Name (Symbol)	Function
1	V_{CC} Power Supply (V_{CC})	-5V Supply
2	V_{GG} Power Supply (V_{GG})	-12V Supply (Not connected for AY-3-1014A 1015)
3	Ground (V_{GND})	Ground
4	Received Data Enable (RDE)	A logic "0" on the receiver enable line places the received data onto the output lines
5-12	Received Data Bits (RD8-RD1)	These are the 8 data output lines. Received characters are right justified, the LSB always appears on RD1. These lines have tri-state outputs, i.e., they have the normal TTL ouput characteristics when RDE is "0" and a high impedance state when RDE is "1". Thus, the data output lines can be bus structure oriented
13	Parity Error (PE)	This line goes to a logic "1" if the received character parity does not agree with the selected parity. Tri-state.
14	Framing Error (FE)	This line goes to a logic "1" if the received character has no valid stop bit. Tri-state.
15	Over-Run (OR)	This line goes to a logic "1" if the previously received character is not read (DAV line not reset) before the present character is transferred to the receiver holding register. Tri-state.
16	Status Word Enable (SWE)	A logic "0" on this line places the status word bits (PE, FE, OR, DAV, TBMT) onto the output lines. Tri-state.
17	Receiver Clock (RCP)	This line will contain a clock whose frequency is 16 times (16X) the desired receiver baud.
18	Reset Data Available (RDAV)	A logic "0" will reset the DAV line. The DAV F/F is only thing that is reset. Must be tied to logic "1" when not in use on the AY-3-1014A.
19	Data Available (DAV)	This line goes to a logic "1" when an entire character has been received and transferred to the receiver holding register. Tri-state. Fig. 12, 34.
20	Serial Input (SI)	This line accepts the serial bit input stream. A Marking (logic "1") to spacing (logic "0") transition is required for initiation of data reception. Fig. 11, 12, 33, 34.
21	External Reset (XR)	Resets all registers except the control bits register (the received data register is not reset in the AY-5-1013A and AY-6-1013). Sets SO, EOC, and TBMT to a logic "1". Resets DAV, and error flags to "0". Clears input data buffer. Must be tied to logic "0" when not in use.
22	Transmitter Buffer Empty (TBMT)	The transmitter buffer empty flag goes to a logic "1" when the data bits holding register may be loaded with another character. Tri-state. See Fig. 18, 20, 40, 42.
23	Data Strobe (DS)	A strobe on this line will enter the data bits into the data bits holding register. Initial data transmission is initiated by the rising edge of DS. Data must be stable during entire strobe.
24	End of Character (EOC)	This line goes to a logic "1" each time a full character is transmitted. It remains at this level until the start of transmission of the next character. See Fig. 17, 19, 39, 41.
25	Serial Output (SO)	This line will serially, by bit, provide the entire transmitted character. It will remain at a logic "1" when no data is being transmitted. See Fig. 16.
26-33	Data Bit Inputs (DB1-DB8)	There are up to 8 data bit input lines available.
34	Control Strobe (CS)	A logic "1" on this lead will enter the control bits (EPS, NB1, NB2, TSB, NP) into the control bits holding register. This line can be strobed or hard wired to a logic "1" level.
35	No Parity (NP)	A logic "1" on this lead will eliminate the parity bit from the transmitted and received character (no PE indication). The stop bit(s) will immediately follow the last data bit. If not used, this lead must be tied to a logic "0".
36	Number of Stop Bits (TSB)	This lead will select the number of stop bits, 1 or 2, to be appended immediately after the parity bit. A logic "0" will insert 1 stop bit and a logic "1" will insert 2 stop bits. For the AY-3-1014A/1015, the combined selection of 2 stop bits and 5 bits/character will produce 1½ stop bits.
37-38	Number of Bits/Character (NB2, NB1)	These two leads will be internally decoded to select either 5, 6, 7 or 8 data bits/character. NB2 NB1 Bits/Character 0 0 5 0 1 6 1 0 7 1 1 8
39	Odd/Even Parity Select (EPS)	The logic level on this pin selects the type of parity which will be appended immediately after the data bits. It also determines the parity that will be checked by the receiver. A logic "0" will insert odd parity and a logic "1" will insert even parity.
40	Transmitter Clock (TCP)	This line will contain a clock whose frequency is 16 times (16X) the desired transmitter baud.

Figure 17-25 Pin functions of the **AY-5-1013**. (Courtesy of General Instruments Microelectronics.)

SOLUTION For hard-wire connections the CS input must be tied to 1 to allow the levels on pins 35–39 to enter the UART. It would be wired as follows:

Pin 35 tied to 0. This indicates parity is present. This is really for the receive side. The TTY does not check parity in the data sent to it, but the UART can now check parity on the data it receives from the TTY.

Pin 39 tied to 1. This selects even parity when parity is used.

Pin 36 tied to 1. This inserts two STOP bits in each character.

Pins 37 and 38. They must be set up for a character comprised of 7 data bits and one parity bit. Figure 17-25 indicates that pin 37 should be connected to a 1 and pin 38 to a 0 for a 7 *data bit* character.

Pin 40. The transmit clock should be connected to a 1760-Hz clock for 110 BPS TTY.

17-9-3   The Receive Side of the AY-5-1013

On the receive side of the UART, the data from the keyboard or a MODEM enters on the serial input (SI) pins. It is sent to the computer using the 8 data outputs (RD8–RD1) pins. The other pins on the receive side include three error condition detectors[6] (parity error, framing error, and over-run), a DATA AVAILABLE signal that goes HIGH when the UART has received a character and it is ready for the computer, and a RESET DATA AVAILABLE line that will reset the DATA AVAILABLE (DAV) line when it goes LOW. There are also two 3-state enable lines. A LOW on the $\overline{RDE}$ line puts the received data on the eight RD lines while the $\overline{SWE}$ puts the error indicators and DAV on the lines.

EXAMPLE 17-23

A computer must read both data and status from a UART over the same 8-bit data bus. How can this be done?

SOLUTION   The error outputs and DAV can each be connected to one of the RECEIVED DATA BIT lines. The computer can request data by sending out a DATA REQUEST signal that causes $\overline{RDE}$ to go LOW, enabling the DATA outputs to be placed on the lines. If the computer sends out a STATUS REQUEST, $\overline{SWE}$ goes LOW and only the error status and DAV are placed on the output lines. Note how the use of 3-state outputs and separate enables allows data and status to be multiplexed onto the same lines.

When a character is being received by the UART, its bits are entered into a Shift Register at the clock rate. When it is fully received, and checked for parity and proper STOP bits (no FRAMING ERROR), it is transferred from the shift register to a holding register. The UART then raises DAV to inform the computer that a character is ready. The computer should then read the character and pulse $\overline{RDAV}$ LOW to acknowledge receipt of the data.

The character can remain in the holding register while the next character is being shifted into the shift register. This is known as *double-buffering* because one character is being held in the holding register, or buffer, while a second character is being shifted into the shift register (which is considered as the second buffer). If the second character gets completely in before receipt of $\overline{RDAV}$, the UART has more data than it can store because the holding register still contains the first character. In this case the UART raises the OVERRUN error to indicate that the computer did not accept the data fast enough, and the UART lost data that it could not store.

[6]These errors are explained in Fig. 17-25.

EXAMPLE 17-24

A computer is connected to a 1200 bps modem. Assume each character is transmitted with only one STOP bit. In worst case, how much time does the computer have to respond to DAV?

SOLUTION   The computer must read the data before the next character is completely entered. Because each bit requires 833 μs and there are 10 bits per character (START, STOP, and 8 data bits), the computer has 8.33 ms to respond to DAV before an OVERRUN error occurs.

17-9-4   The Transmit Side of the AY-5-1013

On the transmit side, the **AY-5-1013** has a transmit buffer register and a transmit shift register. These registers are shown as the upper two registers in Fig. 17-24. The computer sends a character to the transmit buffer register. If the transmit shift register is empty, it will transfer the character to the transmit shift register, where it will be shifted out at the clock rate determined by the clock on pin 40. It will appear at the SO (Serial Output—pin 25) line.

   When a computer is transmitting, it must monitor either Transmitter Buffer Empty (TMBT, pin 22), or End of Character (EOC, pin 23). EOC is HIGH whenever no character is being transmitted. TMBT indicates the transmit holding register is empty and a character may be sent to the UART. Computers usually monitor TMBT, which is also enabled by the $\overline{\text{SWE}}$ strobe. If TMBT is HIGH, the computer sends a character to the UART and sends out DATA STROBE to inform the UART that a character is available on its transmit data input lines and should be strobed into its Transmit Holding Register. The UART then serializes the character and sends it out via the SO line (Pin 25).

17-9-5   Connecting a Computer to a UART

A computer can be directly connected to a UART, but this may require some sophisticated decoding and other circuitry. It may also be connected via a parallel interface IC, such as an **8255**.

EXAMPLE 17-25

Show how to connect the receive side of an **AY-5-1013** to an **8255**.

SOLUTION   The connections are shown in Fig. 17-26, where port A of the **8255** has been set for input and the upper half of port C has been set for output. The **8255** should put a LOW on $\overline{\text{SWE}}$ (PC7) and read the status in. DAV will be on PA0. When it is a 1, the error inputs on PA1, PA2, and PA3 should be checked by the μP, which will take appropriate action in the event of an error.

   If there are no errors, $\overline{\text{SWE}}$ should be made HIGH and $\overline{\text{RDE}}$ should be made LOW. Then the character can be read into the **8255** via port A. Then PA5 should be pulsed LOW to provide an $\overline{\text{RDE}}$ signal to reset the data available signal.

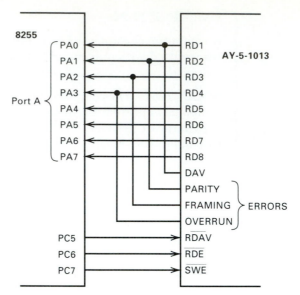

Figure 17-26 Solution to Example 17-25.

17-9-6   Intel and Motorola UARTs

Intel manufactures an IC called the **8251** Programmable Communications Interface, and Motorola manufactures the **6850** Asynchronous Communications Interface Adapter (ACIA). Both of these ICs are essentially UARTs, but they are designed so they can be easily placed on the μP's bus. The **8251**, for example, can have its own port addresses and will not need an **8255** between it and the μP. Both of these ICs are more complex than the **AY-5-1013**, but more information is available on them from Intel and Motorola.

17-10   MODEMs and RS-232C

As shown in Fig. 17-21, digital data is often transmitted via the already existing telephone network. This requires that the *digital data be transformed into an audio tone that the telephone lines can transmit, and be demodulated back to the original bit stream at the receiving end*. This is accomplished by a MODEM. The term MODEM is an acronym for MODULATOR/DEMODULATOR.

A modem can range from a single PC card to a highly sophisticated prepackaged device. The function of a modem is to transform a digital bit stream into audio signals compatible with transmission lines on the transmitting end, and reconstruct the original digital bit stream at the receiving end. A typical modem transmission system is shown in Fig. 17-27.

17-10-1   Modes of Data Transmission

Figure 17-27 shows the most common mode of data transmission, *full duplex*. In full duplex operation, there is a signal line in each direction so that *simultaneous data transmission* can take place. Other modes of operation are *half duplex,* where one side sends and one side receives at any one time (the sides take turns

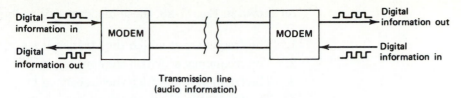

Figure 17-27 A full-duplex data transmission system.

sending and receiving), and *simplex,* where one side is always sending and the other side is always receiving. Simplex is one-way data transmission.

17-10-2   Speed of Modems

Modems are generally classified by the number of *bits per second* (BPS) they can transmit. High-speed modems transmit between 2400 and 9600 BPS. Low speed modems generally handle data rates between 300 and 2400 BPS. Most modems currently available in computer stores run from 2400 BPS to 9600 BPS.

As the speed of a modem increases, its error rate also increases. This adversely affects the actual data rate because errors can cause the retransmission of large quantities of data.

17-10-3   Low Speed Modems

Low-speed modems with speeds from 300 BPS to 2400 BPS are attractive for many applications because they can be used on directly dialed telephone lines. The 103A modem, manufactured by the Bell System, is a popular low-speed modem. These modems often are equipped with cradles for a telephone, so the user can dial the proper phone number, place the handset in the receiver, and start transmitting digital data. A *voice coupler,* often used with teletypes, is a form of low-speed modem.

The operation of a telephone-transmission system is shown in Fig. 17-28. The method of modulation is known as *frequency shift keying* (FSK), where the 0s and the 1s are sent out at different frequencies. All frequencies must be below the 3000 cycle limit on the bandwidth of a telephone line.

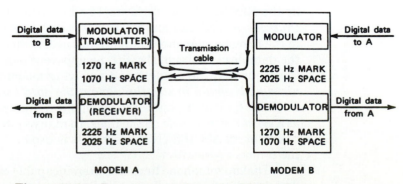

Figure 17-28 Data transmission using FSK modems.

As shown in Fig. 17-28, these modems utilize four different frequencies:

1. The frequency at which they transmit a 1.
2. The frequency at which they transmit a 0.
3. The frequency at which they receive a 1.
4. The frequency at which they receive a 0.

Because the modems operate in pairs, modem A must transmit at the frequency modem B receives. The receiver for modem A and the transmitter for modem B work on the alternate pair of frequencies. The standard frequencies are shown in Fig. 17-28. Under this scheme, data can be transmitted over the telephones in both directions (full duplex) without interference, because the data from A to B use different frequencies than the data from B to A.

EXAMPLE 17-26

What occurs when a 1 is being transmitted from modem A?

SOLUTION   The logic 1 is applied to the digital input of modem A, which modulates it and sends out a 1270-Hz signal on the telephone line. The receiver of modem B detects the 1270-Hz note and demodulates it, producing a logic 1 on its digital output.

17-10-4   High-Speed Modems

Modems that operate at speeds between 2400 BPS to 9600 BPS are considered to be high-speed modems. These modems must compress their high speeds onto a telephone transmission line that has a bandwidth of about 3000 Hz. They use *phase-shift modulation and dibits* instead of FSK. The theory is covered in more advanced texts (see references).

High-speed modems provide a TRANSMIT CLOCK and a RECEIVE CLOCK. These clocks are used to synchronize the digital data. *The transition from 0 to 1 is the center of the datum bit.* When transmitting a stream of bits, data should change when the TRANSMIT CLOCK makes a 1 to 0 transition. On the receive side, data should be sampled when the RECEIVE CLOCK makes a 0 to 1 transition because this is the center of the datum bit.

Data transmission on high-speed modems is synchronized and controlled by the RECEIVE and TRANSMIT CLOCKS. This is called *synchronous data transmission*. START and STOP bits (which transmit no data and therefore add to the transmission overhead) are not used. Once data transmission is started, the data flow continuously with no apparent demarcation between characters. Consequently, the sending and receiving modems must be carefully *synchronized and remain in step throughout the data transmission*. When no data are available for transmission, the transmitter must send out a special byte, known as a *synchronizing byte* on the lines. In many systems, the synchronizing byte is hex 16 or an ASCII SYN character. It is used to synchronize the receiver with the remote transmitter.

Dial up telephone lines are often incapable of handling high-speed modem data transfers. Generally special lines are leased from the telephone company and these lines must be *conditioned,* which means that special circuitry is added to monitor and improve the characteristics of the leased lines, so that they can operate at the required speeds. Privately leased lines are an additional expense that must be recovered by the increased rate of data transmission.

17-10-5 RS-232C—Level Shifting[7]

Data at digital logic levels (0 V and +5 V) cannot be transmitted over long distances. The data must be transformed to *RS-232C* levels for transmission. RS-232C is a widely used engineering specification that specifies both voltage levels and pin numbers for digital signals that are being sent to peripheral devices such as MODEMs and terminals.

The RS-232C voltage levels are:

- **Logic 0**—Any voltage between +3 V and +25 V.
- **Logic 1**—Any voltage between −3 V and −25 V.

Typical RS-232C levels are +12 (logic 0) and −12 (logic 1). Note that RS-232C inverts because logic 0 is a positive voltage. Also voltages between +3 V and −3 V should not occur. This separation between a logic 0 and a logic 1 is called the *noise margin.* It is greater for RS-232C than for standard voltage levels. Figure 17-29 shows the letter E in asynchronous, odd parity format is both its digital and RS-232C levels.

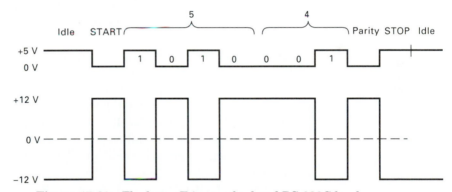

Figure 17-29 The letter E in standard and RS-232C levels.

Most modems and terminals will only accept RS-232C levels. Standard digital levels (0 V and +5 V) can be converted to and from RS-232C levels using a *level converter.* The **MC145407**, is a new level converter manufactured by Motorola, Inc. for this purpose.

Figure 17-30 is a page from the Motorola Data Book describing the **MC145407**. The pin assignment shows that the IC contains three drivers that convert TTL levels to RS-232C, and three receivers that receive RS-232C levels from a modem or other device and convert them to TTL levels. The **MC145407** can be driven by a standard +5 V supply. It contains a *charge pump* that converts the +5 V power input to ±10 V, which can be used to provide RS-232C outputs.

[7]Several books have been written on RS-232. Joe Campbell's book (see References) is one of them.

The receiver diagram in Fig. 17-30 shows that voltage or the received data (Rx) is confined by the action of the diodes and then becomes the output voltage D0, which is TTL. The driver diagram shows that the input TTL voltage goes through a level shift, and becomes the transmitted data (TxD) that is sent out at RS-232C levels.

Motorola also manufactures a highly popular RS-232C to TTL converter, the **1489**, and a TTL to RS-232C converter, the **1488**. These ICs are both quads and have been available for at least 15 years. The **1488** requires external ± 12 V power supplies to develop the necessary RS-232C levels.

17-10-6 Connecting to a MODEM

Almost all terminals and MODEMs use a type DB-25, 25 pin connector as shown in Fig. 17-31. RS-232C specifies both the levels and the functions of the pins to be connected to the MODEM.

The important signals between a MODEM and a digital device, a μP or interface IC, are shown in Fig. 17-32. They are:

TRANSMITTED DATA (Pin 2)—This is data flowing from the μP or UART to the MODEM. It is the data to be transmitted.

RECEIVED DATA (Pin 3)—Data from the MODEM to the digital device. This is the data received from the remote MODEM.

REQUEST TO SEND (Pin 4)—A line from the digital device to the MODEM. This signal should be a 0 whenever data is to be transmitted. It is important in half duplex transmission where it is a 1 when data are being received. In full duplex operation, it generally remains in the 0 state.

CLEAR TO SEND (Pin 5)—A line from the MODEM to the digital device. This signal is a response to REQUEST TO SEND and indicates that the MODEM can accept data for transmission. In full duplex operation, it is normally always active and presents a 0 level to the digital device.

DATA SET READY (Pin 6)—A line from the MODEM to the μP or UART. A 1 on this line indicates that the data set is not ready, usually because of an abnormal condition. This line must be in the 0 state for transmission or reception.

TRANSMIT CLOCK (Pin 15)—A clock from the MODEM to the μP or UART. This clock is provided to synchronize the data to be transmitted.

RECEIVE CLOCK (Pin 17)—A clock from the MODEM to the μP or UART. This clock is provided to synchronize the data being received.

All of these signals are digital and use RS-232C levels. The transmit and receive clocks are not used for asynchronous data. UARTs such as the Intel **8251** or Motorola **6850** (see section 17-9-6) have pins on them to accommodate signals such as RTS, CTS, and so on.

Additional signals such as ring indicators are available for use in special situations. A complete listing and specifications of all the available signals are given in specification RS-232C. Users can provide their own clock to control the bit rate of the modem as an option. Most users, however, prefer to synchronize their data with the clocks provided by the modem.

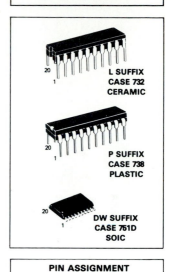

MC145407

Advance Information

5-Volt-Only Driver/Receiver
EIA-232-D and CCITT V.28

The MC145407 is a silicon-gate CMOS IC that combines three drivers and three receivers to fulfill the electrical specifications of EIA-232-D and CCITT V.28 while operating from a single +5 volt power supply. A voltage doubler and inverter convert the +5 volts to ±10 volts. This is accomplished through an on-board 20 kHz oscillator and four inexpensive external electrolytic capacitors. The three drivers and three receivers of the MC145407 are virtually identical to those of the MC145406. Therefore, for applications requiring more than three drivers and/or three receivers, an MC145406 can be powered from an MC145407, since the MC145407 charge pumps have been designed to guarantee ±5 volts at the output of up to six drivers. Thus the MC145407 provides a high-performance, low-power, stand-alone solution or, with the MC145406, a +5 volt-only, high-performance two-chip solution.

Drivers
- ±7.5 Volt Output Swing
- 300 Ohms Power-Off Impedance
- Output Current Limiting
- TTL and CMOS Compatible Inputs
- Slew Rate Range Limited from 4 V/μs to 30 V/μs

Receivers
- ±25 Volt Input Range
- 3 to 7 Kilohms Input Impedance
- 0.8 Volt Hysteresis for Enhanced Noise Immunity

Charge Pumps
- +5 Volts to ±10 Volt Dual Charge Pump Architecture
- Supply Outputs Capable of Driving Three On-Chip Drivers and Three Drivers on the MC145406 Simultaneously
- Requires Four Inexpensive Electrolytic Capacitors
- On-Chip 20 kHz Oscillator

L SUFFIX
CASE 732
CERAMIC

P SUFFIX
CASE 738
PLASTIC

DW SUFFIX
CASE 751D
SOIC

PIN ASSIGNMENT

C2+	1	20	C1+
GND	2	19	V_CC
C2−	3	18	C1−
V_SS	4	17	V_DD
Rx1	5	16	DO1
Tx1	6	15	DI1
Rx2	7	14	DO2
Tx2	8	13	DI2
Rx3	9	12	DO3
Tx3	10	11	DI3

D = DRIVER
R = RECEIVER

FUNCTION DIAGRAM

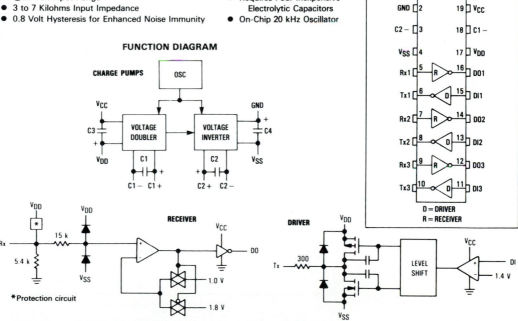

Figure 17-30 The **MC145407** level shifter. (Courtesy of Motorola, Inc.)

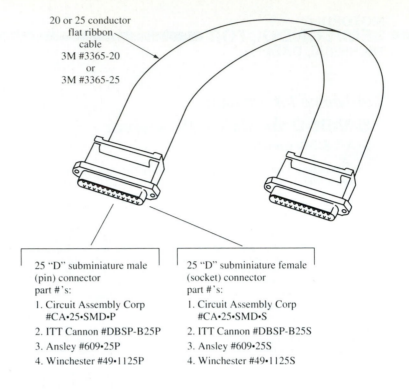

20 or 25 conductor
flat ribbon
cable
3M #3365-20
or
3M #3365-25

25 "D" subminiature male
(pin) connector
part #'s:

1. Circuit Assembly Corp
   #CA•25•SMD•P
2. ITT Cannon #DBSP-B25P
3. Ansley #609•25P
4. Winchester #49•1125P

25 "D" subminiature female
(socket) connector
part #'s:

1. Circuit Assembly Corp
   #CA•25•SMD•S
2. ITT Cannon #DBSP-B25S
3. Ansley #609•25S
4. Winchester #49•1125S

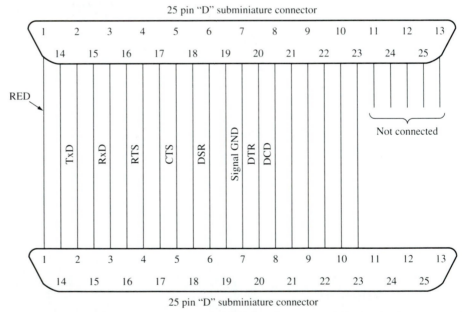

Figure 17-31 A communications cable. (Courtesy of Motorola, Inc.)

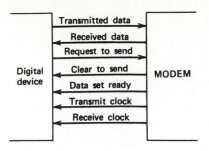

Figure 17-32 Interconnections between a MODEM and a digital device.

17-10-7 Data Transmission—A Summary

To transmit data using a modem, the computer should

1. Check the control signals from the modem to be sure the modem is up and ready to transmit data.
2. Send the data to a UART or Communications controller.
3. The data must now be translated to RS-232C levels using a **1488** or **MC145407**.
4. The output of the **1488** must be placed on a DB-25 cable and sent to the modem.
5. The output of the modem is typically a telephone jack that connects to the phone system.

Data Reception is basically the same process in reverse.

17-11 An Overview of PCs

A personal computer is basically a large, sophisticated digital circuit. At this time, the student should be able to start to understand how PCs work.

Figure 17-33 is the block diagram of basic PC. We have already covered most of the components in the previous sections of this book. The components are:

1. **The CPU and Support Circuitry**—CPU stands for Central Processing Unit. It is an elegant term that really means the μP in most cases.
2. **The Memory Subsystem**—This includes the RAM, ROM, and boot ROM. The boot ROM is accessed when power is first turned on. It checks out the RAM memory, performs other diagnostics, and allows the PC to start properly. In the IBM-PC the boot ROM contains the BIOS (Basic Input/Output System), a program which is accessed when power is applied.
3. **The Keyboard System**—This is the keyboard connected to the PC. It also includes the keyboard interface, which is part of the motherboard.
4. **The Disk Interface and Disks**—This includes both hard drives and floppy drives and their associated disk controllers.
5. **The Video Circuitry and Monitor**—This includes the circuitry needed to drive the monitor display. It usually includes memory to hold the screen image (see section 17-8-1).

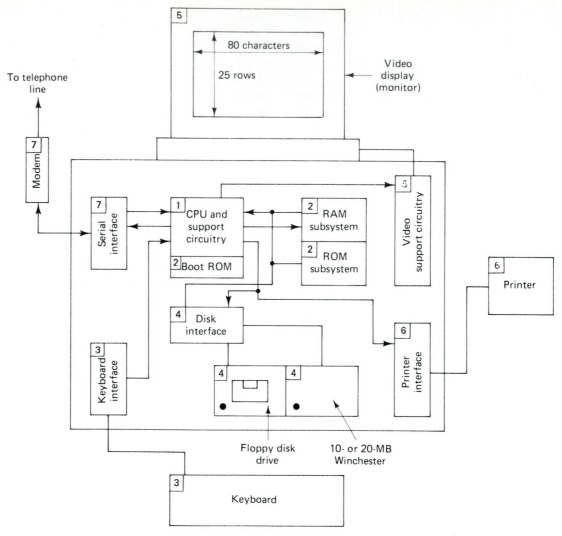

Figure 17-33 Block diagram of microcomputer. (Bryon Putman, *DIGITAL AND MICROPROCESSOR ELECTRONICS: Theory, Applications and Troubleshooting,* © 1986. Reprinted by permission of REGENTS/PRENTICE HALL, Englewood Cliffs, New Jersey.)

6. **The Printer Interface and Printer**—This is usually a parallel printer interface, similar to the interface discussed in section 17-7-7.

7. **The Serial Interface and MODEM**—This has been discussed in the previous paragraph.

The numbers in the boxes correlate with the numbers in the list.

17-11-1   The IBM PC

Figure 17-34 is a diagram of the system board of the IBM-PC. The **8284** at the upper left corner is the clock generator that generates the basic timing of the PC. It feeds the **8088** μP. There is also a socket provided for a coprocessor, if

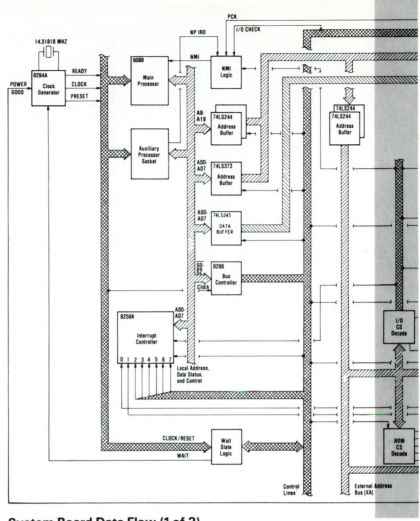

System Board Data Flow (1 of 2)

Figure 17-34 A block diagram of the IBM-PC. (Courtesy of International Business Machines Corporation.)

needed. The **8259** controls interrupts to the **8088**. The rest of the left side consists of the address and data buffers that allow the μP to access memory.

In the center right there is the **8237**, which is a Direct Memory Access (DMA) controller. Below it is the **8255**, used to monitor the switches and keyboard and control the speaker. Below the **8255** is the **8253** timer IC, which develops timing pulses needed by the PC. More information on the DMA controller, the interrupt controller, and the timer can be obtained from Intel, Inc.

The ROM memories are at the bottom right and the RAM memories are in the upper right. At the extreme right are the *slots* that the user needs for peripheral IC cards. These slots are used for the disk controller, the video controller, the printer, and serial output to a communications channel.

Although some of these devices, like the interrupt controller, the DMA controller, and the Timer Interface ICs, are beyond the scope of this book, the student should be able to understand most of what is happening in the IBM-PC.

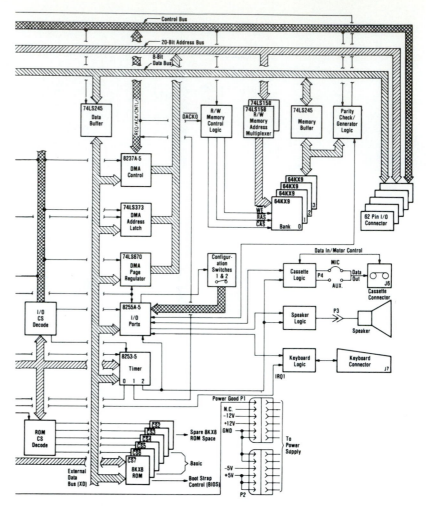

System Board Data Flow (2 of 2)

SUMMARY

Microprocessors and PCs communicate with their peripheral devices using interface ICs. One side of an interface IC communicates directly with the μP's bus, and the other side is connected to the peripheral.

This chapter started with an explanation of the basic functions of an interface IC, and described methods for addressing them. Port I/O and memory-mapped I/O were explained. Then the **8255**, a commonly used interface IC for parallel communications, was discussed.

The second part of the chapter was concerned with serial data transmission. UARTs and Communications Interface Adapters, ICs that change parallel data to serial data, were discussed. The **AY-5-1013** was discussed because this is one of the simplest and most commonly used UARTs. Modems that translate digital data to audio tones so they can be placed on telephone lines were explained. The RS-232C specifications for both digital levels and MODEM control were also discussed.

The chapter concluded by presenting an overview of PC systems in general and the IBM-PC in particular.

GLOSSARY

Alphanumeric: A character that consists of an alphabetic, a number, or a punctuation mark.

ASCII (American Standard for Communication Information Interchange): The standard code for alphanumeric characters.

Charge Pump: A circuit within an IC that increases the supply voltage. It supplies voltages which are not limited to the 0 V to +5 V range.

Communications Interface: An Interface IC designed to work with a μP and control a serial communications channel. It is similar to a UART.

Handshaking: A system of coordinating I/O between the transmitting and receiving devices. The receiving device must first signal that it can accept data, and the transmitting device must then strobe to indicate that valid data is on the lines.

Mark: A communications term for a logic 1.

MODEM: A modulator/demodulator used to convert digital signals to audio tones (modulation) and to convert audio tones to digital (demodulation).

Monitor: A name for the CRT screen or display part of a terminal or computer.

Polling: Periodically examining a control line to determine if data is available.

Screen Image: An area of memory that holds the ASCII characters that are being displayed on a monitor.

Space: A communications term for a logic 0.

START bit: A logic 0 that indicates the start of an asynchronous character.

STOP bit: A logic 1 for one bit time. A STOP bit must appear at the end of each asynchronous character.

Terminal: A combination of a CRT screen and a keyboard.

UART: Universal Asynchronous Receiver-Transmitter. An IC that transforms parallel data to serial in the asynchronous format and vice versa.

PROBLEMS

Section 17-4-2.

17-1. Design a simple IC circuit to control the oven of Example 17-1.

Section 17-5-3.

17-2. An **8086/8088** uses a 33 MHz clock and has a memory with 150 ns access time.
    a. How many wait states are required?
    b. Design a circuit to provide these wait states.

17-3. Using a **16R4** write the PALASM equations for the circuit of Fig. 17-6.

17-4. Design a decoder circuit for an **8255** if it is to respond to addresses 9238-923B.

17-5. Repeat problem 17-4 if there are no other ICs assigned to addresses between 9200 and 92FF.

17-6. In Example 17-7, what port is selected by address 8FFD?

17-7. Repeat Example 17-6 using **74x138** decoders.

Section 17-6-1.

17-8. Design the decoder for Example 17-8.

Section 17-7-4.

17-9. An **8255** is in ports 8008–800B. Issue commands to clear bit 3 of port C and set bit 6 without disturbing the other bits of the port.

Section 17-7-5.

17-10. What command word must be sent to the control register of an **8255** to set it up as
   a. All 24 lines output
   b. All 24 lines input
   c. Port A and the upper half of port C input. Port B and the lower half of port C output.

17-11. Figure P17-11 shows an **8255** interfaced to a small keyboard. Assume only one switch at a time is depressed. The depressed key can be detected by driving the appropriate row line LOW and examining the appropriate column line.

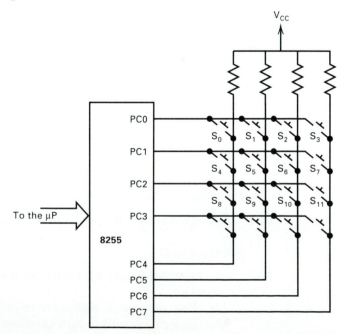

Figure P17-11

a. The **8255** is to be selected only when A15 = A14 = 0 and A13 = 1. Design the address decoder.

b. Write the mode set word to properly initialize the **8255**.

c. Describe the program that must be written to detect when a switch is depressed.

Section 17-7-6.

17-12. Write the control word and list the function of the port C pins if handshaking on input is to use port B.

17-13. In Example 17-15 there are IN instructions to read both ports A and C. What additional line must be connected to the oscilloscope to allow the user to distinguish an IN port A from an IN port C?

Section 17-7-7.

17-14. Explain each bit of the control word shown in Fig. 17-18.

17-15. Explain how to set up an **8255** in the laboratory to monitor the activities on the lines using an oscilloscope.

Section 17-8-1.

17-16. The following word is in ASCII. What is it?
53 74 6F 70.

17-17. Write the message "Hello World" in ASCII. Remember to include the space.

Section 17-8-4.

17-18. A terminal receives the bit stream shown in Fig. P17-18.
a. What characters are they?
b. Is the parity odd or even?
c. How does the terminal respond to them?

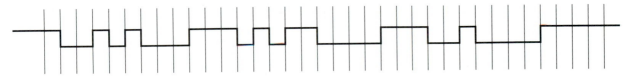

Figure P17-18

Section 17-9-2.

17-19. A terminal is connected to an **AY-5-1013** and operates at 300 BPS.
a. What frequency must be supplied to the UARTs clock inputs?
b. How long does the computer have to respond to DAV?
c. If the character is 8 bits long, must have odd parity appended, and uses one STOP bit, how must pins 34–40 of the UART be wired?

Section 17-9-3.

17-20. A UART or Communications controller is receiving characters at 600 BPS. How long does the μP have to respond to the character before an overrun error occurs? Assume each character has two STOP bits.

Section 17-9-5.

17-21. Show how to connect the transmit side of an **AY-5-1013** to an **8255** and describe the operation.

17-22. A UART is to be set up so that:
OUT 0 provides data and a data strobe.
IN 0 reads the status word with
   Bit 7   DAV
   Bit 6   TBMT
   Bit 5   Overrun
   Bit 4   Parity Error
   Bit 3   Framing Error
IN 1 reads data.

Show the hardware connections between an **8255** and the **AY-5-1013**.

Section 17-10-3.

17-23. Draw a MODEM transmission system using FSK and frequencies of 1000, 1100, 1200, and 1300 Hz.

Section 17-10-6.

17-24. A Communications Interface IC uses all of the signals shown in Fig. 17-32. Show how it must be connected to a MODEM using **MC145407**s.

Section 17-10-7.

17-25. Describe the steps for data being received by a computer. It should be similar to the description given in section 17-10-7.

After attempting the problems, reread the questions in section 17-2 to be sure you understand the chapter thoroughly.

Analog-to-Digital Conversion

18-1 Instructional Objectives

This chapter describes methods of converting analog signals, such as the voltage across a thermistor or the voltage generated by a transducer, into digital signals that may be brought into a computer or microprocessor (μP). It also describes methods of converting digital inputs to analog voltages so a μP can control external systems.

After reading the chapter, the student should be able to:

1. Use the voltage output of a transducer to determine the physical quantity it represents.
2. Build a circuit, using weighted resistors, to convert digital inputs to analog voltages.
3. Build a simple A/D converter.
4. Interface an A/D or D/A converter to a μP.
5. Use commercially available A/D and D/A converters.
6. Test A/D and D/A converters in the laboratory.
7. Explain how a digital voltmeter works.

18-2 Self-Evaluation Questions

Watch for the answers to the following questions as you read the chapter. They should help you to understand the material presented.

1. What is a transducer? Why is it used?
2. What are weighted resistor networks? Give two examples. What is their function?
3. Why is there no need for an SOC command to a D/A converter?
4. What is the difference between an A/D converter using successive approximations, a converter using an SAR, and a flash converter?
5. How are SOC and EOC generated for an **ADC0804**? How can continuous conversions be performed?
6. What is the function of a scalar in a digital voltmeter?

18-3   Communication Between the Analog and Digital Worlds

Microprocessors (μPs) or other computers are often used to monitor and control processes rather than to solve problems. They are being used in the Electronic Control Modules in automobiles, for example, where they monitor the engine vacuum and temperatures, RPM, and so on, and control the action of the car by delivering the optimum mixture of air and gas to the carburetor.

18-3-1   Transducers

Typically, the variables that must be measured are *analog and not electrical*. These physical quantities, such as temperature or pressure, must first be converted to an analog voltage by a *transducer*, which is an electronic circuit that converts physical quantities into voltages. Perhaps the simplest transducer is a *thermistor*, a resistor whose resistance varies with temperature. This variation of resistance can be used to generate a voltage that depends on the temperature of the air surrounding the resistor, its *ambient* temperature. Figure 18-1 shows how the resistance varies with temperature for a typical thermistor.

Perhaps the simplest temperature transducer is shown in Fig. 18-2, where R_1 and V_{in} are fixed. As the temperature at the thermistor varies, its resistance changes, which causes changes in V_{out}. These changes are not *linear*, or directly proportional to the thermistor's resistance, both because the thermistor's resistance is not directly proportional to its ambient temperature, and because the output voltage of the circuit itself is not directly proportional to R_2. However, V_{out} will eventually become the input to a computer that can calculate the temperature corresponding to each value of V_{out}.

EXAMPLE 18-1

The thermistor of Fig. 18-1 is used in the circuit of Fig. 18-2 where $V_{in} = 10$ V and $R_1 = 100$ ohms. V_{out} is found to be 6 V. What is the temperature of the thermistor?

SOLUTION   From circuit theory, the resistance of the thermistor can be calculated to be 150 ohms. This is slightly in excess of 100 or 10^2. From the curve of Fig. 18-1, the temperature corresponding to this resistance is about 100°C.

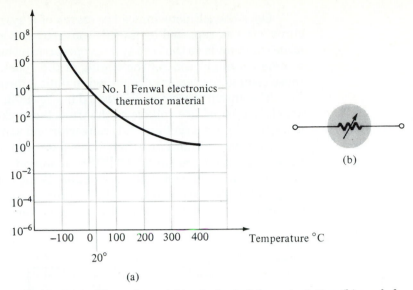

(b)

(a)

Figure 18-1 Thermistor: (a) typical set of characteristics; (b) symbol. (Robert Boylestad and Louis Nashelsky, *ELECTRONIC DEVICES AND CIRCUIT THEORY, 3e,* © 1982. Reprinted by permission of REGENTS/PRENTICE HALL, Englewood Cliffs, New Jersey.)

Other transducers are flow meters, tachometers, and pressure sensors. They measure the rate of flow of a liquid, the speed of a rotating shaft, and the pressure at various points in a physical system, respectively.

18-3-2 Computer-Controlled Systems

Transducer outputs are voltages, but they are analog. They must be converted to digital data before they can be used by a μP or digital controller. This requires an *analog-to-digital* (A/D) converter. We have already seen one A/D converter in the code wheel (section 14-7-2) that converted a shaft position to digital information.

The digital outputs of a μP must often be converted to analog quantities to control a process. Voltages and currents are typical analog quantities. It requires a *digital-to-analog* (D/A) converter to convert the μP's digital outputs to the proper voltages and currents.

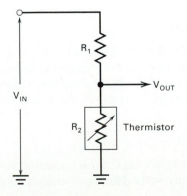

Figure 18-2 A simple temperature transducer.

Once the μP determines the *status* of a system, (its temperatures, pressures, etc.) as reported by its A/D converters, it can operate on this data and issue commands so that the system operates optimally. In an automobile these commands might control the timing of the spark or the flow of gasoline. But these commands can only be issued in digital form.

The entire process is illustrated in Fig. 18-3, where a physical input variable is converted to a voltage by the transducer, then converted from an analog voltage to digital data by an A/D converter and sent to the computer for μP. The μP processes this data and then sends out commands to control the system. If these commands are simply ON/OFF commands, such as opening or closing a switch, they can be sent out via an **8255**. Some commands, such as regulating the gasoline flow in an automobile, require an analog voltage to control the flow by partially opening a valve. Here a D/A converter would be required.

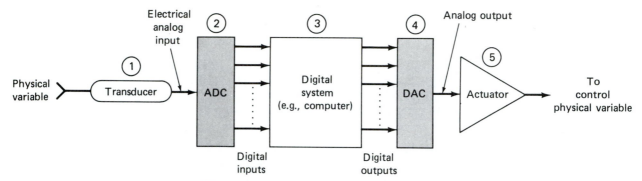

Figure 18-3 Analog-to-digital converter (ADC) and digital-to-analog converter (DAC) are used to interface a computer to the analog world so that the computer can monitor and control a physical variable. (Ronald J. Tocci, *DIGITAL SYSTEMS: Principles and Applications, 5e,* © 1991. Reprinted by permission of REGENTS/PRENTICE HALL, Englewood Cliffs, New Jersey.)

18-4 Digital-to-Analog (D/A) Converters

Because A/D converters often depend on D/A converters, they are considered first. A D/A converter accepts a digital input and produces an analog output equal, as nearly as possible, to the value of the digital input. The analog output of a D/A converter is shown in Fig. 18-4.

18-4-1 Resolution

One of the most important characteristics of D/A converters is its *resolution*. Resolution can be defined as *the change in the output analog voltage corresponding to each increment of the digital number* coming in. In Fig. 18-4, each digital increment results in an increase in the analog output voltage.

EXAMPLE 18-2

A D/A converter has a maximum output of 10 V and accepts 6 binary bits as inputs. How many volts does each analog step represent?

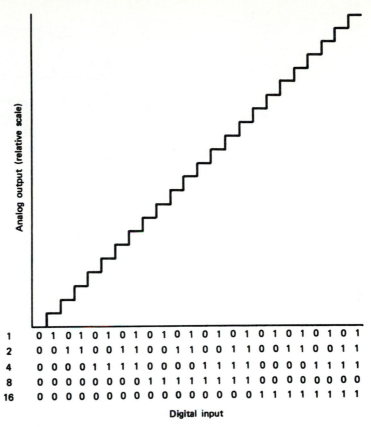

1	0 1 0 1 0 1 0 1 0 1 0 1 0 1 0 1 0 1 0 1 0 1 0 1
2	0 0 1 1 0 0 1 1 0 0 1 1 0 0 1 1 0 0 1 1 0 0 1 1
4	0 0 0 0 1 1 1 1 0 0 0 0 1 1 1 1 0 0 0 0 1 1 1 1
8	0 0 0 0 0 0 0 0 1 1 1 1 1 1 1 1 0 0 0 0 0 0 0 0
16	0 0 0 0 0 0 0 0 0 0 0 0 0 0 0 0 1 1 1 1 1 1 1 1

Digital input

Figure 18-4 Analog output versus digital input in a D/A converter. (From George K. Kostopoulos. *Digital Engineering.* Copyright 1975, John Wiley & Sons, Inc. Reprinted by permission of John Wiley & Sons, Inc.)

SOLUTION With a 6-bit input, the 10 V range can be divided into 64 (2^6) levels. Therefore, each analog step represents 10/64 = 0.15625 V. This is the *resolution* of the output. Greater resolution can be obtained by using more binary bits. If a staircase output, like that of Fig. 18-4, were plotted for this circuit, each step would be 0.15625 V.

18-4-2 Weighted Resistor Networks

A D/A converter operates basically as an adder. *It converts a digital byte to its equivalent analog signal by adding all the 1s in the digital signal, properly weighted, together.* Weighting means giving each bit its proper weight or value. For binary numbers, the MSB has the highest weight, the next bit has half this weight, and so on. When a digital input is applied to a D/A converter, all inputs that correspond to 1s in the digital number receive the same voltage input, typically +5 V, and all inputs that correspond to 0s in the digital number receive an input of 0 V.

D/A converters are usually built using *weighted resistor networks* feeding an operational amplifier (op-amp). The simplest D/A converter is shown in Fig. 18-5. This is a *summing network* where *each resistor is twice as large as the pre-*

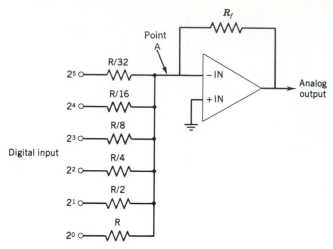

Figure 18-5 A weighted resistor D/A converter. (From George K. Kostopoulos. *Digital Engineering.* Copyright 1975, John Wiley & Sons, Inc. Reprinted by permission of John Wiley & Sons, Inc.)

ceding resistor. Consequently, the current that flows in R1 is twice as large as the current that flows in R2, and so on. Because each binary bit is half the value (or weight) of the preceding bit, the currents are in proportion to the value of the bits. The op-amp sums these currents and develops the correct analog output. It also keeps the voltage at point A at 0 V due to the *virtual ground* that always occurs when an op-amp is used. This prevents the currents in the various resistors from interacting with each other. All of the currents that flow through all of the resistors also flow through R_f, and establish the correct output voltage.

EXAMPLE 18-3

Assume the digital input to the summing network of Fig. 18-5 is 100010, where a logic 1 = + 5 V and 0 = ground. How much current flows into R_f?

SOLUTION With the given digital input, current flows in the 2^5 and 2^1 branches and their sum flows into the analog output.

$$I = \frac{5}{R/32} + \frac{5}{R/2} = \frac{170}{R}$$

The output current is proportional to the binary input number. If R is chosen as 10000 Ω, I = 17 mA. If R_f is selected to be 1000 Ω, the output voltage will simply be this current times R_f or −17 V. Note that the output voltage equals the binary value of the input bits.

18-4-3 Ladder Networks

Another resistor network that is used for D/A conversion is the R-2R ladder network shown in Fig. 18-6. It only uses two values of resistance, R and 2R. The results are the same as the weighted resistor network; the current that flows in node A due to each 1 in the digital input is half of the current due to the preceding bit.

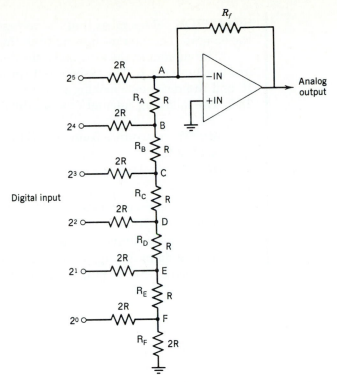

Figure 18-6 The R-2R network D/A converter. (From George K. Kostopoulos. *Digital Engineering*. Copyright 1975, John Wiley & Sons, Inc. Reprinted by permission of John Wiley & Sons, Inc.)

EXAMPLE 18-4

The ladder network of Fig. 18-6 operates on the principle that the resistance looking downward toward ground *from any node* is always 2R Ω. Show that this is true for nodes F, E, and D.

SOLUTION For node F, the impedance looking toward the lower ground is simply the resistor of R_F. To find the resistance looking toward ground from node E, we must assume the voltage at the 2^0 node is 0 V, because of the superposition theorem. Thus, the resistance looking down from point E is the resistance from node F to ground (two resistors, whose value is 2R ohms, in parallel or R Ω) plus the resistance from between node E and node F, which is also R. Thus, the resistance looking downward from node E is also 2R Ω.

The resistance looking downward from node D is R_D plus the resistance from node E to ground. This has already been found to be 2R Ω. It is in parallel with the 2R resistor between node E and the 2^1 terminal. This resistance is R, and when added to R_D, the resistance to ground is again 2R Ω.

EXAMPLE 18-5

In Fig. 18-6 show that the current flow in R_f due to a voltage V at the 2^4 terminal is half that due to the same voltage applied at the 2^5 terminal.

SOLUTION   Remember that the voltage at node A must be 0 V because of the op-amp. The current flowing from the top node is then simply V/2R.

The current in the next node, the 2^4 node, flows through the 2R resistor and reaches node B. The resistance to ground from node B looking upward is R, and the resistance looking downward is 2R. The circuit is shown in Fig. 18-7. The following calculations can then be made:

(a)   The resistance from node B to ground is R Ω in parallel with 2R Ω, or 2R/3 Ω.

(b)   The resistance from the 2^4 terminal to ground is 2R + 2R/3 = 8R/3 Ω.

(c)   The current flowing out of the 2^4 terminal, due to a voltage V, is therefore 3V/8R.

(d)   At node B this current divides. By circuit theory, two thirds of this current flows through resistor R_A and into the op-amp. This calculates to exactly V/4R, or half the current caused by the previous input.

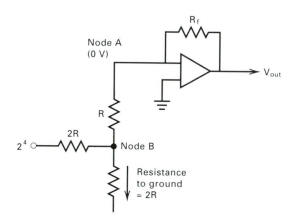

Figure 18-7   Solution to Example 18-5.

18-5   Analog-to-Digital (A/D) Converters

A/D converters are the complement of D/As. They convert an analog quantity (typically a voltage) to digital. There are two popular methods of building an A/D converter. Both utilize comparators. A *comparator* (shown in Fig. 18-8a) has two inputs and one output. The output is essentially digital; it is HIGH if input A is larger than input B, and LOW otherwise. Figure 18-8b shows that if the voltage difference between the inputs is −2 mV or more, the output is +3 V, and if the input voltage difference is +2 mV or more, the output is −0.5 V.[1]

One method of performing A/D conversions is *successive approximations*. To perform the conversion, a D/A converter and a comparator are used. The in-

[1]More modern comparators are superior to the comparator of Fig. 18-8. Texas Instruments, Inc., for example makes the **LM111**, **LM211**, and **LM311**. The specification for these ICs can be found in the TI Linear Circuits Data Book, Volume 1 (see References). These are comparators that switch when in less than 0.1 mV difference between the inputs. The comparator of Fig. 18-8 was used, however, because its explanation is very simple.

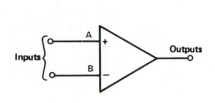

(a) Symbol

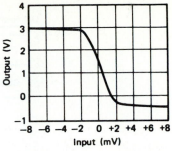

(b) Output voltage vs. input voltage

Figure 18-8 An analog voltage comparator. (From Barna and Porat, *Integrated Circuits in Digital Electronics.* Copyright 1973, John Wiley & Sons, Inc. Reprinted by permission of John Wiley & Sons, Inc.)

puts to the comparator are the unknown voltage, which is to be converted to digital, and the output of the D/A converter. Successively larger digital inputs are applied to the D/A converter until its output most closely approximates the analog voltage. Then the comparator stops the digital inputs from increment- ing. The digital number at this point represents the analog input voltage.

The simplest way to build this converter is to use a counter to increment the digital inputs. The counter stops when the comparator changes states.

EXAMPLE 18-6

(a)   Design an A/D converter. Assume the input voltage and the input of the D/A converter can vary between 0 V and +10 V. Use the compara- tor of Fig. 18-8 and a pushbutton to clear the counter.

(b)   How long does the conversion take if the analog input voltage is 6 V and the clock frequency is 1 MHz?

SOLUTION

(a)   The design is shown in Fig. 18-9. Depressing the pushbutton clears the counter, causing the output of the comparator to be HIGH. When the pushbutton is released, the counter increments. The output of the D/A converter increases until it is greater than the analog input. At this time, the output of the comparator goes LOW and stops the clocks to the counter by cutting off the AND gate. The number locked into the counter is the digital equivalent of the analog voltage.

(b)   The resolution of this counter is 0.15625 V. If the analog input is 6 V, the counter stops when the A/D converter output becomes greater than 6 V, or after 39 counts ($39 \times 0.15625 = 6.09375$). Therefore, it requires 39 μs to make the conversion.

18-5-1   Conversion Speed

Conversion speed is the time a converter requires to produce its result. D/A con- verters are relatively fast, because they are basically adders. The digital input is applied to the resistive ladder network and the sum of the currents is avail-

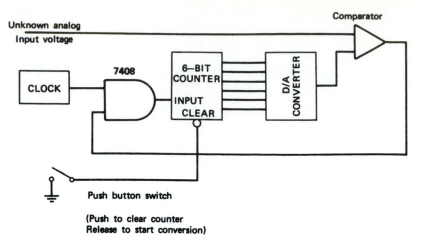

Figure 18-9 Design of an A/D converter.

able almost instantly. Most D/A converters can produce their outputs in less than 1 μs.

A/D converters are generally slower than D/A converters because they must perform a series of approximations before arriving at the digital result. A/D converters depend on two signals:

- Start of Conversion (SOC)—This is a signal from the computer to the A/D converter. It commands the converter to start the conversion process—to start the approximations to the analog input that will result in a valid digital output.
- End of Conversion (EOC)—This is a signal from the A/D converter to the computer. It indicates that the conversion process is complete, and the digital output of the converter is valid.

Thus, conversions are initiated by an SOC pulse from the computer. After the A/D finishes its approximations it provides EOC to tell the computer to read the digital results.

The A/D converter of Fig. 18-9 is simple, but it is relatively slow. If an 8-bit converter is used, for example, it will require 255 clock cycles, in worst case, to increment the counter. Faster A/D converters can be built using a *successive approximation register*.

18-5-2 Successive Approximation Registers

Successive approximation registers (SARs) work by first applying a logic 1 to the MSB of a D/A converter. The output of the D/A is compared to the analog input voltage. If it is higher, the logic 1 is removed, and the binary MSB is 0. If it is less, the input to the MSB remains, and the MSB is a 1. Then the SAR proceeds to the next bit and repeats the procedure. An 8-bit converter would only require 8 cycles of an SAR, so the speed advantage is apparent.

EXAMPLE 18-7

Show how an SAR could provide an A/D converter by using the weighted-resistor D/A converter of Fig. 18-10. Assume, for simplicity, that a logic 1 is -8 V, and that the analog input voltage is $+6$ V.

SOLUTION The SAR would first apply -8 V to point D, leaving inputs A, B, and C at 0 V. Remember that the voltage at the input to the op-amp is always 0 V because of its virtual ground. Then 8 mA would flow through the 1 KΩ resistor and also through R_F. The output voltage, V_{OUT}, would be $+8$ V. The comparator would detect that this output is too high and cause the logic 1 to be removed, returning point D to ground.

Next the SAR should apply -8 V to point C causing 4 mA to flow in the 2 KΩ resistor. V_{OUT} becomes $+4$ V. This is not greater than $+6$ V, so the -8 V input remains on point C.

The SAR now applies -8 V to point B. The 2 mA flowing in the 4 KΩ resistor plus the 4 mA still flowing in the 2 KΩ resistor cause the output to be $+6$ V. This does not exceed the analog input voltage, so it remains.

The SAR now applies -8 V to point A. This adds another 1 mA to the current and makes the output voltage $+7$ V. This is too much, so the logic 1 at point A is removed. This finishes the conversion. The logic levels at points D, C, B, and A are 0110 respectively, or binary 6. The binary output is equal to the analog input voltage.

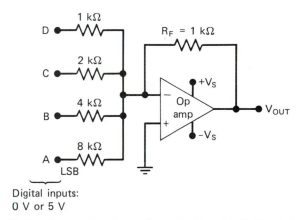

Figure 18-10 Circuit for Example 18-7. (Ronald J. Tocci, *DIGITAL SYSTEMS: Principles and Applications, 5e,* © 1991. Reprinted by permission of REGENTS/PRENTICE HALL, Englewood Cliffs, New Jersey.)

18-5-3 Flash Converters

The flash converter is the fastest A/D converter, but also the most complex. A flash converter for an n-bit output requires $2^n - 1$ comparators, but does the conversion in one cycle. Thus, a converter with an 8-bit output would require 255 comparators.

The circuit of Fig. 18-11 is a flash A/D converter with a 3-bit output. It requires 7 comparators and uses a **74LS148** (see section 11-9-1) to determine its output.

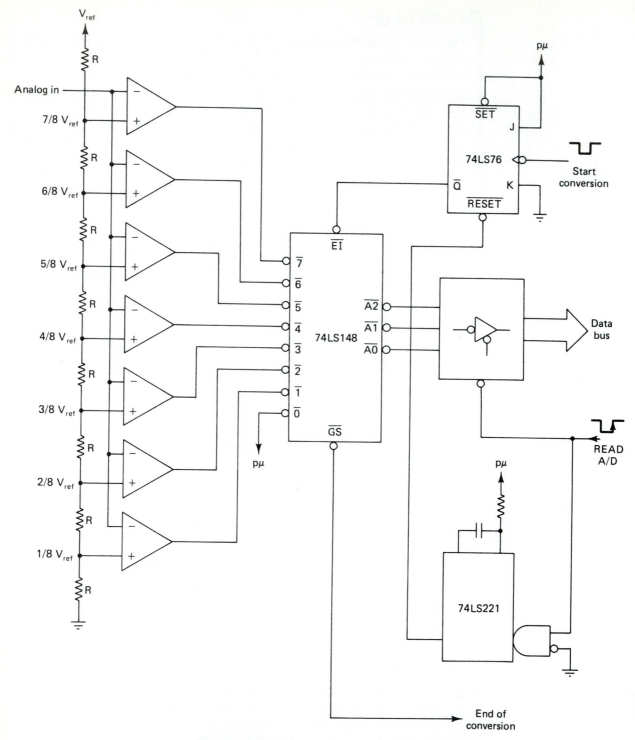

Figure 18-11 Flash-type A/D converter. (Bryon Putman, *DIGITAL AND MICROPROCESSOR ELECTRONICS: Theory, Applications and Troubleshooting*, © 1986. Reprinted by permission of REGENTS/ PRENTICE HALL, Englewood Cliffs, New Jersey.)

EXAMPLE 18-8

In Fig. 18-11, assume V_{ref} is 8 V, and the analog input voltage is 5.5 V. Explain how the converter works.

SOLUTION Because of the resistor-divider network, the input to the + terminal of the top comparator is 7 V, the input to the next comparator is 6 V, and so on. If the analog input voltage, which is applied to the − terminal of all the comparators, is 5.5 V, the lower five comparators will produce LOW outputs and the top two comparators will have HIGH outputs. Thus, input 5 to the **74LS148** will be the most significant LOW input, and the outputs $\overline{A2}$, $\overline{A1}$, and $\overline{A0}$ will be 010, the inverse of 5. This can indicate to the computer that the analog input voltage is 5 V, which is as near as we can come with this resolution.

18-5-4 Ramp Converters

Another method of A/D conversion, sometimes used in digital voltmeters, is to apply the unknown voltage to one input of a comparator and a *ramp voltage*, which increases linearly with time, to the other input. The higher the unknown voltage, the longer the RAMP voltage takes to reach it and trigger the comparator. While the RAMP voltage is rising, a counter is being incremented by a fixed frequency clock. Higher voltages allow more clock pulses before the comparator fires and blocks them. Thus, the final count is proportional to the unknown analog voltage.

18-5-5 A Homemade A/D and D/A Converter

Figure 18-12 shows a "homebrew" circuit that can function as both a D/A and an A/D converter.[2] It contains all of the elements of a converter described in the preceding paragraphs.

To function as a D/A converter, the digital input is placed on the D7-D0 lines and the LOAD input to the **74x193**s (see section 8-10-3) is held LOW. Thus, the digital inputs pass through the **74x193**s and onto the R-2R ladder network. The output of this network appears at point A. It should be buffered. We recommend using an op-amp.

The circuit can also function as a successive approximations A/D converter like Fig. 18-9. It is used as an A/D converter as follows:

1. The switch is thrown to its A/D position. This disables the LOAD to the **74x193**, but enables the 3-state outputs on the **74173**s.
2. The **74x193**s now function as an up-counter driven by IC1, an oscillator. As in the D/A, its outputs feed the R-2R network.
3. When the output at point A exceeds the analog input voltage, the comparator flips. This clocks the **74x193**'s output into the **74173**'s FFs. Thus, the **74173**s retain the digital output that represents the analog input.

[2]This circuit is described by its inventor, Roger W. Mikel, in an article in the February 1981 issue of *BYTE* magazine.

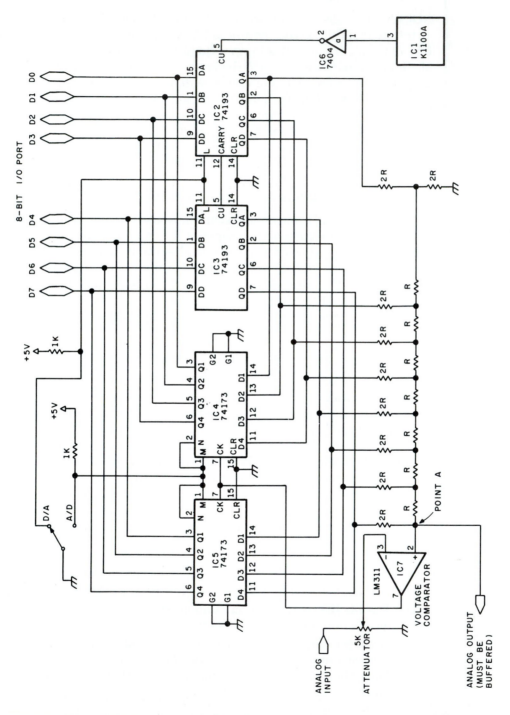

Figure 18-12 A "homebrew" A/D and D/A converter. (Courtesy of Roger W. Mikel.)

EXAMPLE 18-9

Assume a computer generates a low-going SOC pulse to the circuit of Fig. 18-12 and waits for an EOC pulse. Where should the SOC pulse be placed and how would the EOC pulse be generated?

SOLUTION The $\overline{\text{SOC}}$ pulse should be inverted and applied to the CLEAR inputs of the **74x193**s. Thus, the **74x193**s will be cleared and start to count on the trailing edge of $\overline{\text{SOC}}$.

EOC can be generated by sensing when the comparator flips. This is the end of the conversion process.

18-6   Commercial A/D and D/A Converters

Almost all engineers buy A/D and D/A converters rather than building them. In this section we will discuss a popular D/A converter and a popular A/D converter.

18-6-1   Converter Parameters

Before discussing actual converters, their *parameters* must be understood. The important parameters of an A/D converter are:

1. **Resolution**—The change in analog voltage reflected in a 1-bit change in the digital output.
2. **Bit size**—The number of binary bits of output.
3. **Maximum conversion time**—The worst case time of a conversion.
4. **Output codes**—A/D converters are available with either binary or BCD outputs.
5. **Analog input range**—The range of analog input voltages that are converted.
6. **Input impedance**
7. **Power dissipation**
8. **Price**

The parameters that apply to a D/A converter are:

1. **Resolution**
2. **Bit Size**
3. **Error**—The worst case error output as a percent of full scale.
4. **Output mode**—Current or voltage outputs are available in D/A converters.
5. **Settling time**—The time required for the output analog voltage to become firm.
6. **Input coding**—D/A converters may accept binary or BCD inputs.
7. **Analog output range**—The range of the analog output voltage or current.
8. **Power**
9. **Price**

The engineer should consult the manufacturer's catalogs to determine the best D/A or A/D converter for the particular application.

18-6-2 The DAC 0800 D/A Converter

The **DAC 0800** is a readily available D/A converter manufactured by the National Semiconductor Corporation (see References). It is a very simple converter that accepts its digital inputs with no buffering and converts them to the analog output. The digital inputs could come from a set of switches in a laboratory, or from a computer connected to an output port on an **8255**, for example. It is shown in Fig. 18-13 as set up in the laboratory to provide approximately a ± 5 V output.

The **DAC 0800** provides complementary *current* outputs on pins 2 and 4, but the output on pin 2 in Fig. 18-13 was connected to ground, and not used. The power supply voltages were ± 10 V. The eight TTL digital inputs were applied to pins 5 to 12. With pin 4 connected to $+5$ V through a 10-KΩ resistor, the output was found to vary between ± 4.9 V. This limit can be adjusted by changing the voltage on the 10-KΩ resistor.

The positive reference voltage was applied to pin 14, and the negative reference was applied to pin 15 via the 5KΩ resistors, as the manufacturer recommends. Pin 15 was grounded and the potentiometer connected to pin 14 was adjusted so that the mid-scale voltage (with the digital inputs at 10000000 or 01111111) was as close to zero as possible. The analog output at pin 4 then precisely followed the digital input.

In this and in other DACs, the sum of the currents flowing into pins 2 and 4 is constant and equal to the current flowing into pin 14. If the digital input is all 1s, 1 mA flows into pin 2 and no current enters pin 4. In this circuit the output would be $+5$ V. If the digital input were all 0s, the entire current, 1 mA, flows into pin 4 and the V_{OUT} would be -5 V.

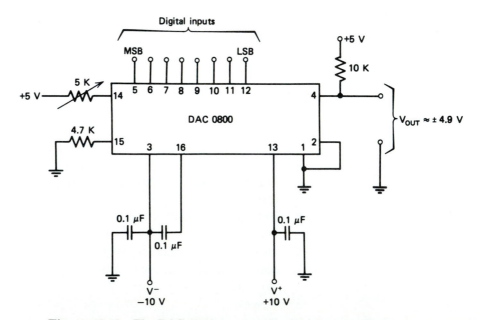

Figure 18-13 The **DAC 0800** as set up in the laboratory.

EXAMPLE 18-10

Assume the **DAC 0800** is adjusted for a ±5-V output. What output voltage would we expect if the digital inputs are 10010110 or (96)16?

SOLUTION   The 10-V analog range is divided into 256 intervals by the 8 digital inputs, so each digital increment corresponds to 10/256 or 0.0390625 V. The digital input corresponds to the decimal number 150 so the voltage output is 150 times 0.0390625V or 5.86 V. This is 5.86 V above the − 5 V that occurs when the digital inputs are all 0, and the output voltage should be 0.86 V.

EXAMPLE 18-11

Modify the circuit of Fig. 18-13 so that the analog output voltage ranges for + 10 V to 0 V.

SOLUTION   There are several ways to adjust the currents and voltages to do this. Perhaps the simplest way is to connect the 10 KΩ resistor to a source of + 10 V instead of + 5 V.

18-6-3   The ADC0804 A/D Converter

The **ADC0804** is an inexpensive, commonly used A/D converter. It is available from National Semiconductor, Texas Instruments, and others. It uses an 8-bit SAR to make its conversions.[3]

Figure 18-14 shows the pins on the '0804 and gives their functions. The left side of Fig. 18-14 is the computer interface. It consists of the 8-bit digital output of the converter and four signals that communicate with a computer or μP and control SOC and EOC. These will be discussed in a later paragraph.

The right side of Fig. 18-14 shows the inputs associated with the analog side of the converter. They are:

- **V_{CC}**—The IC is designed to operate with a power supply of + 5 V.
- **CLK R and CLK IN**—The SAR in the converter must be driven by a clock. These pins are available if the user wishes to connect his own clock. The **ADC0804** has an internal Schmitt Trigger Oscillator (see section 7-6-3) that is activated by connecting a 10 KΩ resistor and 150 pF capacitor to these pins as shown. Most engineers simply use the internal oscillator.
- **$V_{IN}(+)$ and $V_{IN}(−)$**—These are the analog inputs to the converter. Generally $V_{IN}(−)$ is connected to ground, and the voltage to be converted is applied to $V_{IN}(+)$. $V_{IN}(+)$ should not exceed VCC.
- **$V_{REF}/2$**—This pin establishes the *range* of the analog voltage to be converted; see the following paragraph.
- **A GND and D GND**—Separate analog and digital grounds are provided so that any noise on the digital ground caused by rapid switching will not affect the analog ground. Generally these are both connected to separate grounds on the chasis or printed circuit board.

[3]Block diagrams of the **ADC0804** are available with the manufacturer's data sheets.

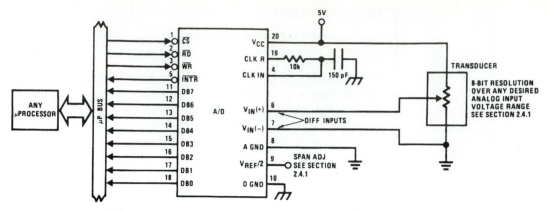

Figure 18-14 The **ADC0804** analog-to-digital converter. (Courtesy of National Semiconductor Corporation.)

The *range* of the analog input signal is determined by the voltages on the V_{CC}, $V_{REF}/2$, and $V_{IN}(-)$ pins. If the $V_{REF}/2$ pin is left open, $V_{IN}(-)$ is always at ground and the range of the analog input voltage is 0 V to +5 V.

If a voltage is applied to $V_{REF}/2$, the voltage will determine the range of the A/D converter. It will be *half the range of the analog voltages to be converted*. In addition, a voltage can be applied to $V_{IN}(-)$. It will function as an *offset*; it will form the lower threshold of the analog range.

National Semiconductor's data sheets give an example where $V_{REF}/2$ is 1.5 V and $V_{IN}(-)$ is 0.5 V. The range of input analog voltages for this converter is twice $V_{REF}/2$, or 3 V, sitting above a 0.5 V offset, so the analog range is from 0.5 V to 3.5 V.

The advantage of setting the range is that the resultant output will have smaller resolution if the input analog voltage is less than 5 V. In the example of the previous paragraph, each digital increment will correspond to 3/256 V instead of 5/256 V if $V_{REF}/2$ were left open. Smaller resolution can result in greater accuracy.

EXAMPLE 18-12

An **ADC0804** is set up so that $V_{REF}/2$ is 2 V and V_{IN} is 0.5 V. What is the digital output if the analog input is 4 V?

SOLUTION The range is from 0.5 V to 4.5 V. The input voltage is 3.5 V above the lower threshold. Because the range is divided into 256 digital increments, the output is 3.5V/4V times 256 or $(224)_{10}$. This number converts to $(E0)_{16}$, so the digital output is 11100000.

18-6-4 The A/D Interface of the ADC0804

The four pins, $\overline{CS}$, $\overline{RD}$, $\overline{WR}$, and $\overline{INTR}$, control the communications between the **ADC0804** and the μP or external world. The function of each of these pins is:

- **Chip Select ($\overline{CS}$)**—This functions like any Chip Select signal. The converter will be dormant unless $\overline{CS}$ is LOW.

- **Read (RD)**—If RD and CS are both LOW, the output of the converter can be read. The internal 3-state drivers on the output latch of the **ADC0804** will be enabled, and it will send its digital data to the μP. A valid READ also resets INTR HIGH.
- **Write (WR)**—The WR pulse with CS LOW acts as SOC. It initiates conversion.
- **Interrupt (INTR)**—INTR goes HIGH on SOC. It goes LOW when conversions are complete. Thus, the trailing edge of INTR can be taken as EOC. The timing for the pulses is shown in Fig. 18-15.

The INTR line can also be inverted and then used to interrupt an **8085** or **80x86** μP. SOC will set INTR HIGH, which is the no interrupt state. After the conversions are complete, INTR will go LOW and this can cause an interrupt. A read of the converter means the μP has read the data. It also sets INTR HIGH so the **ADC0804** will not interrupt the μP until it receives another SOC and finishes the next conversion.

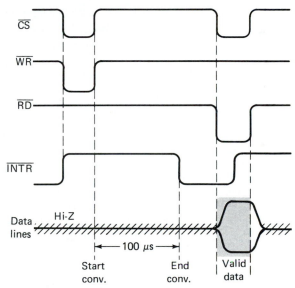

Figure 18-15 The timing for an **ADC0804**. (Ronald J. Tocci, *DIGITAL SYSTEMS: Principles and Applications, 5e,* © 1991. Reprinted by permission of REGENTS/PRENTICE HALL, Englewood Cliffs, New Jersey.)

EXAMPLE 18-13

Explain how to interface an **ADC0804** to a μP using an **8255**.

SOLUTION   One way is to set up the lower half of port C and port A for input and the upper half of port C for output. The CS line on the **8255** could be connected to the CS line on the converter so that the converter is accessed whenever the **8255** is addressed. Then PC7 on the **8255** could be tied to WR on the **ADC0804**, PC6 could be tied to RD, PC0 could be tied to INTR, and the data out of the **ADC0804** could be the input to port A.

Conversions could be initiated by causing the μP to pulse PC7 LOW by an OUT command causing PC7 to go LOW followed immediately by another OUT command causing PC7 to return HIGH. This negative pulse should also cause $\overline{INTR}$ to go HIGH. PC0 could be monitored by a series of IN port C commands to determine when EOC occurs. When PC0 goes LOW, conversions are complete. PC6 should be taken low to provide a read command to the **ADC0804**, the data should be read by an IN port A instruction, and PC6 should then be commanded to return HIGH.

18-6-5 Continuous Conversions

The **ADC0804** can be set up to continuously convert its analog input to digital outputs by making the following connections:

1. Connect $\overline{CS}$ to ground so the converter is always selected.
2. Connect $\overline{RD}$ to ground so the read outputs are always enabled.
3. Connect $\overline{INTR}$ to $\overline{WR}$ so that when $\overline{INTR}$ goes LOW, indicating EOC, it also puts a LOW pulse on $\overline{WR}$, which functions as an SOC, and the conversion process restarts.

Continuous conversions make it much simpler to connect the **ADC0804** to an **8255**. One can simply read the digital output into the **8255**.

18-6-6 Testing D/A and A/D Converters in the Laboratory

One simple way of testing an A/D converter is to apply a known input voltage to V_{IN} and connect the digital outputs to LEDS. As the input voltage varies, the LEDs should indicate a binary number corresponding to the input. The circuit is shown in Fig. 18-16. As the input voltage is varied and read out on the Voltmeter (VM), the LEDs should change accordingly.

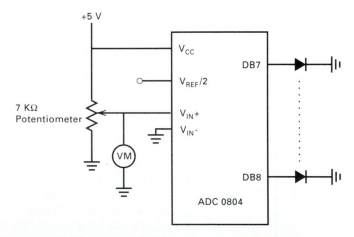

Figure 18-16   Testing an A/D converter in the laboratory.

EXAMPLE 18-14

For the circuit of Fig. 18-16, what LEDs should be lit if the input voltage is 2 V?

SOLUTION Because $V_{REF}/2$ is an open pin, the range of the converter will be 0 V to V_{CC} or 5 V. The decimal number of the output is 2x256/5 or 102.4. This translates into $(66)_{16}$. Therefore, lights DB6, DB5, DB2, and DB1 should be on to display a 66.

D/A converters can be tested by applying a switch bank to the digital inputs and tying a voltmeter to the output. The **DAC0800** of Fig. 18-13 can be tested by connecting the VM to pin 4.

18-7 Applications of Converters

There are many applications of A/D and D/A converters. Two of the most common are presented in this section.

18-7-1 The Digital Voltmeter

A/D converters are used to build digital voltmeters that are found on almost every laboratory bench. A photograph of a digital voltmeter is shown in Fig. 1-3. Figure 18-17 is the block diagram of a typical digital voltmeter (DVM).

In Fig. 18-17, the input analog voltage is first applied to a *sample-and-hold* circuit. This type of circuit prevents the input analog voltage from changing while the A/D is converting it. If the input voltage does not change rapidly, it may not be needed. Inexpensive DVMs omit this circuit.

The analog voltage is next applied to a *scalar*. This circuit assures that the proper analog voltage is applied to the A/D converter. Assume, for example, that the range of the A/D input is 0–10 V. If the analog input is less than 10 V, it can be applied directly to the converter. If the analog input is between 10 V and 100 V, the scalar would insert a 10-to-1 attenuator so the input to the A/D would be within its range. The scalar could determine if the input voltage is greater than 10 V by using a comparator with one input tied to 10 V. The scalar output could also affect the position of the decimal point on the 7-segment displays.

The output of the scalar is then connected to the A/D converter. The A/D can be quite slow in this application, because the output is designed to be read by humans, so ten conversions per second would be more than sufficient.

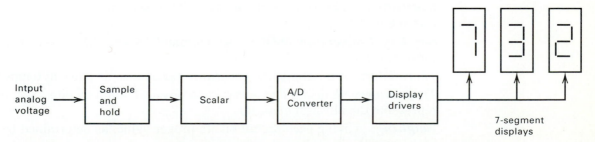

Figure 18-17 A block diagram of a digital voltmeter.

The output of the A/D is connected to a set of display drivers that drive 7-segment displays (see section 11-10). It might be advantageous to select an A/D converter with BCD outputs. Otherwise a binary-to BCD converter must be used.

The output that the human reads is on 7-segment displays. Three displays are shown in Fig. 18-17. The scalar should determine where the decimal point is. If the input voltage is less than 10 V, the output might read 7.32, as shown, but if the input is greater than 10 V, the output must read 73.2 V. Note the shift in the decimal point for these two cases.

18-7-2 Digital Oscilloscopes

Digital oscilloscopes sample the input waveform using an A/D converter and remember it. Thus, the waveform of a single occurrence can be displayed continuously or recorded for posterity. Digital oscilloscopes require very fast A/D converters. They are discussed further in section 19-9.

SUMMARY

This chapter considered the conversion of analog voltages to digital signals (A/D) and digital signals to analog (D/A). It started by introducing transducers, which convert physical quantities such as temperature and pressure into electrical voltages. Next the principles of D/A converters using weighted resistors were explained. Then A/D converters, which often incorporate D/A converters, were considered. Successive approximation, flash, and ramp converters were discussed.

Two examples of commercially available and commonly used converters were then presented. Methods of interfacing them to computers and of testing them in the laboratory were explained. Finally, two examples of common test instruments that use A/D converters were presented.

GLOSSARY

Comparator: An IC that compares two inputs and produces an output at one of two levels, depending on which input is greater.

End of Conversion (EOC):   A signal from an A/D converter indicating it has finished its conversions.

Resolution: The change of analog voltage corresponding to each digital increment.

Start of Conversion (SOC):   A command to an A/D converter to start conversions.

Thermistor: A resistor whose resistance varies with its ambient temperature.

Transducer: A device for converting a physical quantity into a voltage that represents the physical quantity.

Weighing: Giving each digital bit its proper value, as determined by its bit position in the digital word.

PROBLEMS

Section 18-3-1.

18-1. Repeat Example 18-1 if the output voltage is 2V.

18-2. For the circuit of Fig. 18-2, what is the output voltage if the input temperature is 200°C?

Section 18-4-1.

18-3. A D/A converter has 8 input bits. If its output ranges from 0 V to +5 V, how much voltage does each digital step require?

18-4. Repeat Problem 18-3 if the output range is from −7 V to +7 V.

18-5. A D/A converter is required to have an output range of 0 V to +20 V. It must have an input resolution of 0.01 V or better. How many bits are required on the digital input?

Section 18-4-2.

18-6. In the circuit of Fig. 18-10, find the current in each resistor, the current in R_F, and the output voltage if the input is
  (a) 0000
  (b) 0111
  (c) 1000
  (d) 1111
  Assume each digital 1 corresponds to +5 V on its input.

18-7. For the summing network of Fig. 18-5, what is the current output if $R = 10K$ and the binary input is:
  (a) 37
  (b) 55

Section 18-4-3.

18-8. In Fig. 18-6 show that the current in R_A due to a voltage V applied to the 2^3 terminal is V/8R.

18-9. For the ladder network of Fig. 18-6, if the digital input is 0 or +5 V, what value of R is needed to make the output current in mA equal to the binary value of the input? Remember the output impedance is 2R Ω to ground.

18-10. What is the output current for the network of Fig. 18-6 if the resistances are 1000 Ω and the digital input is:
  (a) 20
  (b) 25
  (c) 35
  Assume digital 1s and 0s correspond to +5 V and 0 V, respectively.

Section 18-5.

18-11. What is the output of the comparator of Fig. 18-8 if the a input is greater than the b input by
   (a) 1 V
   (b) 2 mV
   (c) 1 mV
   (d) 0.5 mV

18-12. If the A/D converter of Fig. 18-9 is required to have a worst case settling time of 50 μs, how fast must the clock run?

Section 18-5-2.

18-13. What is the range of the A/D converter of Fig. 18-10? What is its resolution?

18-14. Redraw Fig. 18-10, making it an R-2R network where R = 1K Ω. Find the currents in the network if the inputs are:
   (a) 1100
   (b) 1010
   (c) 1011
   where the input voltage is 5 V.

18-15. Design an A/D converter that first compares the MSB to the input voltage, then the next most significant MSB, and so on. How many steps are involved for a worst case comparison?

Section 18-5-5.

18-16. Interface the converter of Fig. 18-12 to an **8255**.

Section 18-6-2.

18-17. If a **DAC 0800** is set up to provide an analog output of ±10 V and its input is 00110011, what is the analog output voltage?

18-18. Design a **DAC 0800** to provide an output from −5 V to +10 V.

18-19. A **DAC0800** is connected using an 8 KΩ resistor to pin 5 and a 6 KΩ resistor between + 5 V and pin 14 (refer to Fig. 18-13). What is the range of the analog output voltage?

Section 18-6-3.

18-20. If an **ADC 0800** is set up to convert analog voltages between 0 V and +10 V, what is its digital output if the input is 3.5 V?

Section 18-6-6.

18-21. For the circuit of Fig. 18-16, If $V_{REF}/2$ is tied to 2.56 V and V_{IN} is 4V, what lights will be on?

18-22. Sketch a laboratory circuit for testing a D/A converter.

After attempting the problems, reread the questions of section 18-2 to be sure you understand the material presented.

19
Construction and Debugging of IC Circuits

19-1 Instructional Objectives

In this chapter we consider methods of building and debugging IC circuits. Modern equipment for testing and developing IC circuits is also discussed. After reading the chapter, the student should be able to:

1. Write a specification for a circuit.
2. Draw up logicals, wire lists, and module charts.
3. Buzz-out (trace) a circuit.
4. Design and run an acceptance test.
5. Locate faulty ICs and faulty circuit wiring.
6. Start up the timing chain of a circuit.
7. Begin to debug a circuit.
8. Use logic analyzers, digital storage oscilloscopes, and workstations to test and design digital systems.

Watch for the answers to the following questions as you read the chapter. They should help you understand the material presented. When you have finished the chapter, return to this section and be sure you can answer all of the questions.

1. What are the advantages of wire-wrapping construction over printed circuit boards?
2. What are the advantages of printed circuit boards?
3. What criteria are used to determine the best method of building a circuit?
4. What are specifications? Why are they important?
5. Why should wire runs spanning more than one logical be named?
6. What are the typical causes of errors in a combinatorial circuit?
7. Why is it more difficult to find errors in sequential circuits than in combinatorial circuits?
8. Are synchronous or asynchronous circuits generally easier to debug? Explain why.

19-3   Wire Wrapping

The simplest method of constructing a digital circuit so it can be tested is to put it on a superstrip as discussed in Chapter 2. If larger or sturdier circuits are required, however, or if they must be produced in quantity, there are two major methods of building them. The first is to place the ICs in sockets and *wire wrap* the interconnecting wires. The alternate method is to solder the ICs into a *printed circuit board* where the interconnecting lines are already printed in copper on the board. Wire wrapping is more flexible and versatile than printed circuits because it is easier to reroute the wires, when necessary. It is considered first.

For use in wire wrapping, the ICs are plugged into a socket or a panel. Sockets are generally small, most contain only 14 or 16 contacts, and are designed to accommodate a single IC in a dual-in-line (DIP) package. *Panels* are generally larger, contain several rows of contacts, and can accommodate *many* ICs. A wire-wrap panel with several ICs in it is shown in Fig. 19-1.

In both sockets and panels, the contacts that engage the IC pins are connected to *wire-wrap posts*. These posts are usually square, 0.025 in. (0.635 mm) on a side, and long enough to accept three wraps of wire. The wire used is thin, solid wire (No. 30 AWG is the most popular) and is wrapped around the post. The sharp, rectangular edges of the post cut into the wire slightly to make a solid electrical contact. The ideal wrap consists of about six turns of bare wire plus a turn of insulated wire. A wire-wrap socket with three separate wraps of wire on one pin is shown in Fig. 19-2. The wire is generally wrapped by an electrically (or pneumatically) driven tool called a *wire-wrap gun*, shown in Fig. 19-3. A small plastic hand-operated wire-wrap tool, shown in Fig. 19-4, is also available for small jobs. About 1.5 in. (3.8 cm) of insulation is stripped from the solid wire and it is inserted into the smaller hole in the gun or tool. The larger

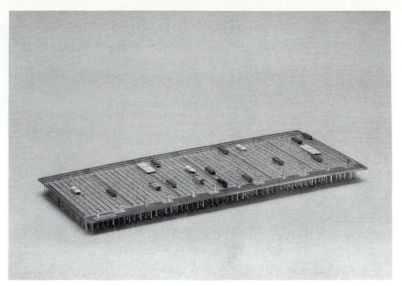

Figure 19-1 An IC panel. (Photo courtesy of Cambridge Thermionic Corp.)

hole in the center of the gun is then placed over the wire-wrap post and the gun spins the wire onto the post to make the connection. The other end of the wire is then brought to the next post in the wire run and the process of stripping and wrapping is repeated to connect the wire.

Figure 19-2 An IC socket with three wraps of wire. (Photo courtesy of Cambridge Thermionic Corp.)

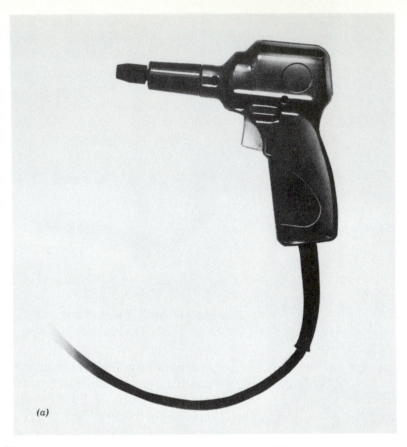

(a)

Figure 19-3 An electric wire-wrap gun. (Photo courtesy of Cambridge Thermionic Corp.)

(b)

Figure 19-4 A hand wire-wrap tool.

19-3-1 Semi-Automatic Wire Wrapping

For commercial production or large quantity runs of electronic equipment, *semi-automatic* wire-wrapping machines are often used. The panel to be wrapped is placed on the machine and the wire-wrap gun automatically positions itself for the first wrap, as shown in Fig. 19-5. The operator then selects the proper wire from a group of prestripped and presized wires and inserts it into the gun. The gun wraps the wire and then moves to the next position to be wrapped.

The motion of the gun is numerically controlled by punched cards or a paper tape prepared from a *wire list* (see section 19-5-5). The device being built must be thoroughly tested and the wire list correct before the punched card deck or paper tape is made, otherwise, identical errors will appear in each piece of the production run.

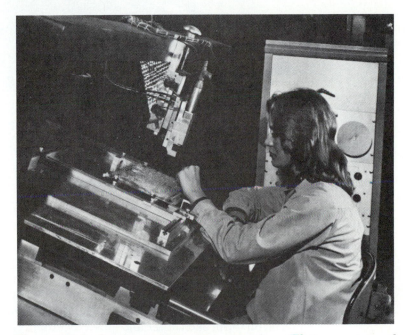

Figure 19-5 Semi-automatic wire wrapping. (Photo courtesy of Cambridge Thermionic Corp.)

19-4 Printed Circuits

The design and construction of a printed circuit (PC) board can start only after a circuit has been thoroughly developed, tested, and debugged. A copper-clad board is used and the circuit is built as follows:

1. The copper is etched off the board by an acid solution, leaving only the pads and copper paths that serve as interconnecting wires.
2. Holes are drilled into the board to accommodate the IC pins and the leads of any discrete components in the circuit.
3. The components are then soldered to the PC board. Usually the entire board is either *wave soldered* or *flow soldered* so that all components are connected simultaneously.

4. If a system is large enough to require several PC boards, fingers (contact strips) are printed on the edge of each board. The boards are inserted into card connectors, which are then wired together to interconnect the boards. Figure 19-6 shows a general purpose printed circuit board with edge contact fingers to plug into a connector.

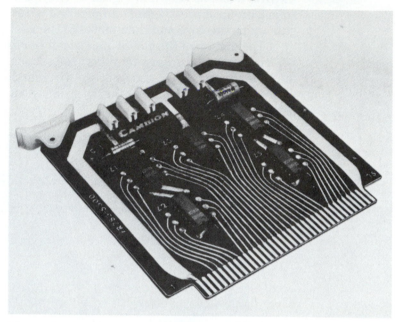

Figure 19-6 A printed circuit board. (Photo courtesy of Cambridge Thermionic Corp.)

19-4-1 Multilayer Boards

In many modern circuits, the wiring is too complex to fit on a single two-sided board. Then several *layers* of a single board must be etched. The layers are placed on top of each other and glued together. Then the holes can be drilled and the components mounted.

19-4-2 Comparison of Printed Circuits and Wire Wrapping

Almost all preproduction testing, or prototyping, is done using wire-wrap connections. The advantages of wire wrapping are:

1. Wire wrapping requires a much shorter lead time. Once the wire list is complete, wire wrapping can start immediately. The production of a PC board, which has been described briefly, is a much more laborious procedure. It requires sophisticated drafting to make a mask, followed by photographing, etching, component insertion, and soldering. Only rarely is a PC board ready in less than two weeks after the circuit has been finalized.

2. It is much easier to correct errors or make changes in a wire-wrap circuit. A faulty wire can be removed simply and quickly using a small unwrap tool. A wire change on a PC board generally requires cutting

the print with a knife or razor blade to break a connection, and soldering a wire bridge between existing printed wires to form new connections.

3. Once the circuit is being tested, it is easier to replace faulty ICs because ICs in a wire-wrap circuit are inserted in sockets, whereas ICs are usually soldered into PC boards.

PC boards are best used when large production quantities are involved. The advantages of PC boards are:

1. **Wiring.** PC boards do not require any wiring, thus eliminating human wiring error in a production run.
2. **Cost.** In large quantity, after the board has been designed, additional boards can be built quickly and economically.
3. **Production speed.** It is much faster to mount components on a PC board and solder them automatically by machine than to build the equivalent wire-wrap circuit.
4. **Repeatability of efforts.** If an error is made in a PC board, its repair is painful, but at least it occurs at the *same* place on every board. When several identical circuits are built using wire wrap, the wiring errors occur in different places for each circuit. This increases the time, cost, and aggravation of debugging wire-wrap circuits.

The foregoing clearly indicates that the wire-wrap method is better for circuits that have not been fully tested and debugged, or for circuits that are to be produced in small quantities. In prototype development, it is almost mandatory. Many military systems use wire-wrap boards because they can be modified more easily than PC boards if the need arises. When a large number of identical circuits are to be produced for commerical use, however, the PC board advantages predominate. Portable radios, television sets, calculators, and other *mass-produced* items use PC boards.

19-5 Construction of Wire-Wrap Circuits

There are five steps in the construction of a wire-wrap circuit described in detail in this section. Each step must be carefully documented with appropriate paperwork. It is almost impossible to debug a prototype circuit *without complete* and *precise documentation*.

19-5-1 Specifications

The *specifications* describe how a circuit is to operate. They are written before work is started and should state exactly how the final circuit will perform. Engineering contracts are generally based on the specifications, which are written by the customer. The vendor, usually an engineering firm, estimates the cost of the circuit from the specifications and bids on the job. It often happens that during the design, building, and testing of a circuit, situations arise that were not anticipated in the original specifications. Often disputes arise between the customer and the vendor as to why the circuit doesn't work and who will bear the cost of repairing or redesigning it. In extreme cases, threats of legal action are heard. It is extremely important, therefore, that *specifications are written very carefully*, and cover as many situations and contingencies as can be envisioned.

19-5-2 Logicals

After the specifications are finished, the engineer designs the circuit by drawing *logic diagrams* (called *logicals*) as we have done in this book. In Chapter 10, Fig. 10-18 is a logical. Generally the engineer roughly sketches the logicals. The drafting department then redraws them professionally. Since a complex circuit may require many pages of logicals, careful labeling is extremely important. To increase the clarity and usefulness of the logicals, the following suggestions are made.

1. Label every FF with a *descriptive name* that relates to *its function*. In Fig. 10-18 the FF was called the DATA AVAILABLE FF, because it controlled the DATA AVAILABLE signal.

2. Each *stage* or *output* of a shift register or counter should also be labeled. If, for example, a **74x164** shift register with outputs A through H is used to control the timing of a circuit, then TIMING SHIFT REGISTER-B is a reasonable way to describe the B output of the **74x164**. Counters should be labeled similarly. If a four-stage counter is built of FFs, WORD COUNTER 3-Q is a way to describe the Q output of the third stage of that counter. The advantage of labeling FFs and registers is that they can be referred to more easily when the documentation (see section 19-5-7) describing the system in written.

3. When a single wire run appears on several pages of logicals, it should leave the right-hand side of the drawing that contains its *source* and enter all drawings that contain *loads* on the left side. At the point it leaves the source drawing, all the logical page numbers where it appears should be in parentheses. When it enters another logical where it is feeding loads only, the source drawing should be referenced in parentheses. It is also useful to name each wire run that appears on several logicals. Sometimes the source location is used for the name of a run. For example, U17-6 means the source is pin 6 of the IC at location U17.

All the above points are illustrated in Fig. 19-7, which shows parts of sheets 4 and 5 of a disk interface. Two signals, PATTERN COUNTER-D and GATED PATTERN COUNTER, leave sheet 4 on the right. The numbers in parentheses indicate that PATTERN COUNTER-D goes to sheets 2, 5, and 7, and GATED PATTERN COUNTER goes to sheets 5 and 6.

The left side of sheet 5 where the signals enter is also shown. Note that on sheet 5, they only reference sheet 4, their source sheet. Note also that these logicals are complete because they contain the pin numbers, module locations, and IC numbers for each gate.

Figure 13-23 is a logical diagram for sheet 6 of the IBM-PC. It conforms to most of the specifications we have listed. Signals coming onto the board are on the left, along with the page numbers they originated on. Signals going off the board are on the right along with the page numbers of the other logicals where they are used.

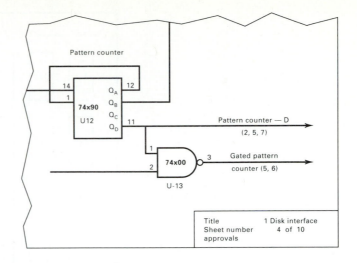

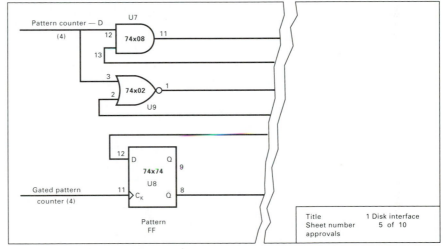

Figure 19-7 Part of a set of Drum Interface logicals, complete with pin numbers and module locations.

19-5-3   Module Placement and Pin Numbering

In the process of drawing the logicals, all gates should be labeled with their type number (i.e., all 2-input NAND gates are labeled **74x00**, all 3- input NAND gates are labeled **74x10**, etc.) Before the logicals are finished, each gate must be assigned a physical location and correct pin numbers.

A *module chart* is used to assign physical locations to the various ICs. All ICs are assigned numbers that start with the letter U, as shown in Fig. 19-7. The module chart should conform as nearly as possible to the panel on which the ICs are to be mounted. For example, if the ICs in a system are to be plugged into 24 socket wire-wrap panels, a module chart, shown in Fig. 19-8, could be used. Each rectangular box represents one IC socket. We recommend subdividing each box into a number of slots. The IC type number is placed in the top slot and an *additional slot is allocated for each gate in the IC*. The drawing or page number of the logical where that gate is used is then placed in the slot.

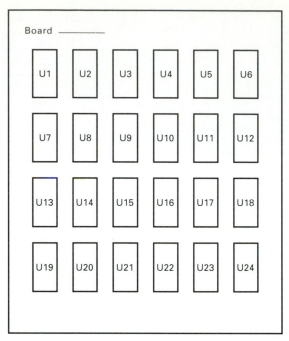

Figure 19-8 A blank module chart for a 24-socket IC panel.

Figure 19-9 is an example of the proper use of a module chart. Assume a 2-input NAND gate appears on page 1 of the logicals. A **74x00** is required, so it is placed in the top slot of an available socket on the module chart (location U1 in Fig. 19-9). The first NAND gate uses pins 1, 2, and 3. These pin numbers are written on the logicals and the number 1 (to indicate sheet 1 of the logicals) is

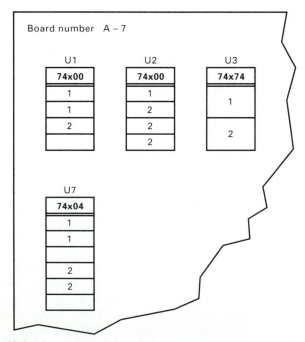

Figure 19-9 A portion of a module chart.

placed in the first of the slots reserved for the 4 gates of the **74x00**. When a second NAND gate is encountered on sheet 1, pins 4, 5, and 6 are assigned and a 1 is placed in the second slot on the module chart. On page 2 a third NAND gate is encountered. A 2 is placed in the third slot of the module chart, and pins 9, 10, and 8 are assigned to the NAND gate. The fourth slot remains vacant, indicating there is a NAND gate available in case one has been overlooked, or a NAND gate is needed to correct an error. If additional gates are needed after all the slots are filled, additional ICs must be used. Figure 19-9 also shows that there are only two slots assigned to the **74x74** in location 3, because a **74x74** contains only two FFs. The **74x04** in location 7, however, has six slots assigned to it, one for each inverter in the IC.

Referring back to Chapter 1, Fig. 1-2 is the module chart for the IBM-PC. The ICs are shown as they appear on the motherboard.

19-5-4 Wire Lists

A wire list is simply a list of the routing or path of each wire in a system. A typical blank wire list form for a small circuit, such as could be built in a college laboratory, is shown in Fig. 19-10. A wire list should contain an entry for each wire in the system. The first column in the list (name) is used only when the wire is designated by a name on the logical. Otherwise it is left blank and the name of the run is taken from the source.

The entry in the *source* column is the module location and pin number of the gate that is the source of the wire run. If a run has multiple sources (3-state or open collector gates are examples), the additional sources could be identified by an asterisk. The entries in the To columns are the various loads or destinations of the wire run. If a run contains more than three destinations, the To columns in the succeeding line are filled, with the Source column left blank. The absence of an entry in the Source column indicates that the source of the run can be found on the line above.

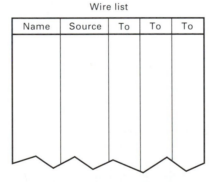

Wire list

Name	Source	To	To	To

Figure 19-10 A blank wire list.

EXAMPLE 19-1

Construct a module chart and wire list for the circuit of Fig. 10-13.

SOLUTION   Looking back at Fig. 10-13, we see that the circuit contains four ICs. They should be assigned U numbers and the module chart of Fig. 19-11 can be drawn. Next pin numbers can be assigned to each IC. We have also labeled the **74x164** shift registers as TC1 and TC2 (Timing Control-1 and

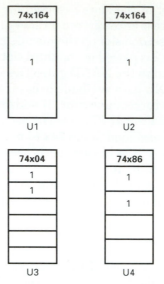

Figure 19-11 A module chart for Example 19-1.

Timing Control-2). The circuit, complete with pin numbers, U numbers, and labels, is shown as Fig. 19-12.

The wire list for the circuit can then be drawn. It is shown in Fig. 19-13. By assigning lables to the shift registers, it is very easy to assign names to some of the wire runs.

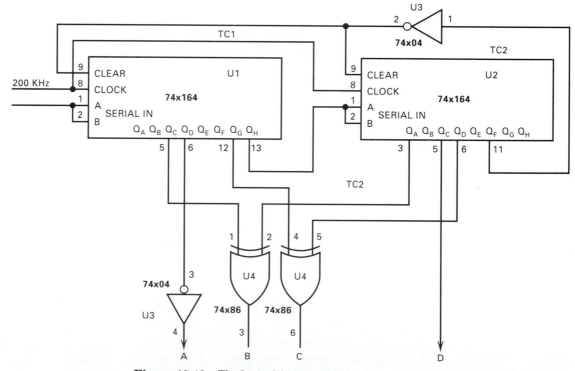

Figure 19-12 The logical for Example 19-1.

Name	Source	To	To	To
200 KHz	200 KHz	U1-8	U2-8	
	VCC	U1-1	U1-2	
	U3-2	U1-9	U2-9	
TC1-C	U1-5	U4-1		
TC1-D	U1-6	U3-3		
TC1-G	U1-12	U4-4		
TC1-H	U1-13	U2-1	U2-2	
TC2-A	U2-3	U4-2		
TC2-D	U2-8	U4-5		
TC2-F	U2-11	U3-1		
TC2-C	U2-5	Output D		
	U3-4	Output A		
	U4-3	Output B		
	U4-6	Output C		

Figure 19-13   The wire list for Example 19-1.

The wire list of the previous paragraph applies to small systems that can be developed in the laboratory. The wire lists for large commercial systems are usually generated by computer. The pins in the circuit are each listed in order and each wire connected to each pin is listed. Table 19-1 shows a sample wire list.

TABLE 19-1

A Sample Commercial Wire List

From	Level	To	Level	Name
U1-1	1	U5-3	1	RUN1
U1-1	2	U6-6	2	RUN1
U1-2	1	U7-5	1	RUN2
.				
.				
.				
U5-3	1	U1-1	1	RUN1
U5-3	2	U8-2	2	RUN1
U5-4	1	U8-5	1	RUN3
.				
.				

The wire list of Table 19-1 starts with the first pin of the first IC (U1-1), lists both wires connected to it, and then progresses to the second pin of the first IC (U1-2) and so on. The name is very important; all wires in the same run must have the same name. In this type of wire list, *each wire is listed twice*. The first wire in the list also appears again in the list when pin U5-3 is encountered. Thus, the list contains *twice as many entries as there are wires in the circuit*, but it makes it easier to find all the wires connected to each pin. In the wire list of Fig. 19-13, if U5-3 was a load, it might be necessary to search the entire list to find the entry for the pin. Here, however, we can go directly to the listing for the pin. Notice also that level 1 wires are always connected to level 1 and level 2 wires are always connected at level 2. This allows the engineer to make changes without tearing out the entire run, if changes become necessary.

19-5-5 Wiring and Buzz-Out

After the logicals, module charts, and wire list are completed, the circuit is ready for wire wrapping. The simplest procedure is to follow the wire list. When wiring from the list, a third wire should *never* be wrapped on a post. If a wire is listed for a post that already has two wires on it, this is surely an indication of an error, probably the same point listed on two different runs, and the wire list should be checked against the logicals before proceeding.

After completing the wiring, the circuit should be "buzzed out" *before* inserting the ICs and applying power. Buzz-out means checking the continuity of each run in the wire list. In industry this check is often made using a buzzer, or a signal injector and tracer, but it can be made with an ohmmeter. As part of the buzz-out, one should always check to be sure *power and ground have not been wired together*. If this error is made, the results, when power is applied, may be spectacular!

19-5-6 Acceptance Test

The *acceptance test* is a series of tests performed on the completed system to determine whether it has met the specifications. Usually if a system passes the acceptance test, it is delivered to the customer. The vendor is then entitled to payment in full, as provided by the contract. Therefore, the customer should document the acceptance test so it is thorough and comprehensive, to be sure that the unit satisfied all requirements. Sometimes partial acceptance tests are made at fixed times after the award of contract so the customer can determine the vendor's progress. These tests, and the penalty for failure, should also be well documented in the specifications (section 19-5-1).

19-5-7 Manuals

For any complex circuit, the vendor must usually write a manual as part of the contract. The manual usually contains a written description of the system, instructions on how to operate it, and a detailed explanation of the systems operation. Most manuals also contain logicals, module charts, wire lists, mechanical drawings, maintenance procedures, and aids for trouble-shooting.

The design engineers are usually responsible for the manuals. Although they may get some help from technical writers, the engineers are the people who really understand how the system operates. Consequently, they must generate at least the first draft of the manual. Engineers are normally used to write specifications, proposals, and instruction sheets, as well as manuals, and *an engineer who writes well is a very valuable employee*.

19-6 Error Detection in Combinatorial Circuits

When a digital circuit or system is being developed and tested problems will arise continuously, and it is extremely important for the engineer to be able to diagnose and fix the problems as they occur. We estimate that a circuit with five ICs has no better than a 50 percent chance of working when power is first applied. As the complexity increases, the probability of the circuit working on the first try decreases rapidly, and it is wildly optimistic to expect a complex circuit to work when first turned on.

The process of locating and correcting the faults in a circuit is colloquially called "*debugging*." The engineers who designed the circuit are often called upon to fix or debug it because they best know how it should work. The principal causes of error are faulty components, faulty wiring, and faulty design. Often all three are present, especially in a new design.

The ability to debug a complex circuit is one of the criteria that separates the competent from the mediocre engineer. Routine debugging is often left to technicians, but when a *new system is being tested the engineer is indispensable because design errors are as prevalent as any other problem*.

To debug successfully, *one must start by knowing exactly what the circuit is supposed to do*. The circuit should then be tested by placing it in the mode where it *makes errors most frequently*, continually if possible. Now, comparing the way the circuit actually works to the way it was designed to work should reveal the source of the problems. In complex circuits there are often many errors of various types working together to befuddle the engineer. *As soon as any part of a circuit is found to be faulty, it should be repaired*. An entire logic system usually will not work in response to a single fix (in fact, it may appear worse than before), but only by killing the bugs one at a time can one hope for eventual success.

19-6-1　Faulty ICs

In combinatorial circuits, errors usually occur for particular values of the input variables. The circuit should be set up using these input variables and then investigated step by step until a gate is found that is not operating as its function table specifies. If a faulty IC is the cause of a problem, its output is generally constant regardless of the inputs. Faulty ICs are usually relatively easy to find, compared to wiring and design errors.

EXAMPLE 19-2

A circuit is constructed to produce the function:

$$f(X,Y,Z) = \Sigma(0,4,5,6)$$

which simplifies, using Karnaugh maps, to:

$$X\bar{Z} + X\bar{Y} + \bar{Y}\bar{Z}$$

The circuit is built as shown in Fig. 19-14 and produces a high output in response to a 1 input ($X = 0$, $Y = 0$, $Z = 1$). Locate the problem.

SOLUTION　These values of *XYZ* are set into the circuit and the logic levels throughout the circuit are checked. The results are also shown in Fig. 19-14. An examination of the circuit reveals that gate 3 is not operating as it should. The 0 input to the top leg of the NAND gate should produce a 1 output.

The problem illustrated in Example 19-2 could be caused by a faulty IC and once the IC is located, it should be replaced. As often as not, however, the problem will persist. These other possible sources of error should then be checked:

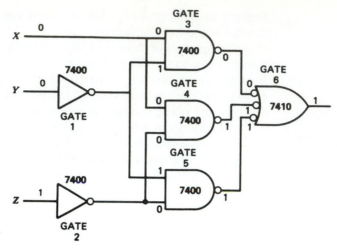

Figure 19-14 Circuit for Example 19-2.

1. The IC is inserted backwards.
2. Power and ground are not wired properly to the IC socket.
3. Wiring is faulty (see section 19-6-2 below).

19-6-2 Faulty Wiring

There are several possible ways in which faulty wiring can cause erroneous operation of a circuit.

1. **Floating inputs.** Floating inputs are gate inputs that are not connected to an output of another gate or a switch. Typically they act like a logic 1, but may appear as a logic 0 or a "maybe" (a voltage between 0.8 V and 2.0 V, which is neither a 1 or a 0, is sometimes called a maybe) on an oscilloscope (CRO). If the top leg of gate 3 is floating, it would cause the error shown. If the CRO indicates a LOW where a floating input is suspected, connect the point to V_{CC} through a 1-kΩ resistor. If the point is actually floating, it will go to V_{CC}, but if the point is genuinely LOW (because of a gate or switch output), it will remain LOW despite the pull-up resistor.

2. **Open wires.** Sometimes wires break or open inside the insulation where a break is not visible. Such breaks act as a floating input. An open wire in the X run could cause the problem of Example 19-2. Open wires can be detected by noting a different voltage level at each end of the same wire.

3. **Parallel sources.** The outputs of two or more TTL ICs should never be tied together, except for open-collector or 3-state ICs. Even when this design rule is followed, a misrouted wire can result in tying two outputs together or tying an output to ground. This can cause an IC to appear faulty and is one of the more difficult errors to detect. A parallel output can cause the line under investigation to be LOW or "maybe" when it should be HIGH. This could be the cause of the error in Fig.

19-14. If the wire between gate 3 and gate 6 were connected to another output, which was LOW, the effect would be as shown. If parallel sources are suspected as the cause of error, and all else fails, unwrap the output. It should go HIGH. Then by reconnecting the wires to the output one at a time and observing which wire causes the output to go LOW, one can begin to trace the faulty wire run and find the other source.

4. **Misrouted wires.** Circuits can be wired too tightly, and this often results in connecting two wire runs together. In a wire wrap board, if an insulated wire is dressed too tightly around a pin, the pin might cut it and connect the pin into the run. Adequate space should be left for wires in such a circuit. Misrouted wires can be checked in the same manner as parallel sources.

The buzz-out procedure (section 19-5-5) checks each point in a run to see if it is connected. The run could be faulty, however, if it is connected to other points in addition to the points on the wire list. One way to check for this is to count the number of *wire terminations*. This should equal $2N - 2$, where N is the number of points on the list. If a run has five points on the wire list (a source and four loads), each pin should be examined and the number of wire wraps counted. The total should be 8. Any other total indicates a faulty wire run.

19-7   Error Detection in Sequential Circuits

Error detection in sequential circuits (circuits containing FFs, one shots, shift registers, etc.) is far more complex than finding errors in combinatorial circuits. Of course, sequential circuits are susceptible to all the problems of combinatorial circuits, but a circuit that fails because of a faulty FF or a missing wire presents a relatively easy debugging problem. The more difficult and more interesting problems occur when the circuit does different things at different times. This *intermittent operation* or *inconsistent behavior* is usually because of FFs being SET or CLEARED at unexpected times, and the problem of determining the cause of these CLEARS or SETS is often quite difficult.

As with combinatorial circuits, the engineer who observes the following principles has the best chance of successfully debugging a sequential circuit.

1. *Thoroughly understand the circuit and know what it should be doing at all times.*
2. *Set the circuit up to operate so that errors occur as frequently as possible.*
3. *Expect multiple errors.*
4. *Compare what the circuit is doing to what it should be doing and correct each specific discrepancy as it is discovered.*

The first principle—thoroughly understanding the circuit—is achieved by studying the specifications and the logicals. The second step must be accomplished by setting the circuit up so that it *does the same thing consistently*, preferably the *wrong* thing, as often as possible. This erroneous behavior gives the engineer an excellent starting point for debugging.

19-7-1   Light and Switch Panels

Light and switch panels, mentioned in section 2-6, help debug complex circuits. These panels, which consist simply of a group of lights and switches, are similar to computer front panels and help the engineer determine how the circuit is functioning. The switches allow him or her to set in various combinations of input variables and the lights monitor the system response. At the start of a project it may seem like extra work to build these panels, but they always pay for themselves by saving debugging time and aggravation.

19-7-2   Start-up Procedures

The operation of a sequential circuit is generally controlled by the basic timing of the circuit, and this should be debugged first. If the circuit is controlled by an oscillator, there is usually little problem in getting the circuit going, but if it is controlled by a one-shot timing chain (see Example 7-10), the one-shots may fail to start. If the timing chain is initiated by a switch, it may be difficult to determine where the problem is. Fortunately, most Cathode Ray Oscilloscopes (CROs) have lights on them that flash whenever the CRO is triggered. By attaching the CRO probe to the output of the first one-shot and causing the output signal to trigger the CRO, we can determine whether the first one-shot is firing, or if there is something wrong with it. If it is firing, the problem is farther down the line. By proceeding through the timing chain, one circuit at a time, the problems can be found and eventually the timing circuit will start.

Once the timing chain is operative, repetitive circuit behavior is best observed on the CRO and the engineer stands a much better chance of debugging a circuit. In Chapter 7, many circuits had both SINGLE/CYCLE and CONTINUOUS modes of operation. It is wise to incorporate the CONTINUOUS mode in circuits that are not designed to work continuously or repetitively. In practice (a computer memory would be an example), the CONTINUOUS mode is used while debugging. It simplifies debugging by causing the circuit to operate repetitively, and is well worth the extra effort required to incorporate it.

19-7-3   Glitch Hunting

Once the basic timing has been debugged and the circuit placed in repetitive operation, most troubleshooting problems are solved fairly easily. Only the hardiest bugs will survive. Unfortunately, these require the shrewdest engineers to exterminate them.

At this stage of debugging, *most problems are caused by FFs being SET or CLEARED at the wrong time because of design errors and/or glitches*. Design errors generally allow pulses to occur at the wrong time. Glitches are often caused by race conditions (see section 6-11) or noise and are shorter pulses that are much more difficult to see on a CRO. Unfortunately, by the time the error is apparent, its causes have generally disappeared. Therefore, if the circuit can be run in a loop and FF inputs observed before the FF erroneously SETS, we may succeed in observing the error and tracing its cause. When building a computer, for example, it is wisest to debug the JUMP instruction first. Then the computer can be set up to execute an instruction or series of instructions, jump back and do it again. This sets up a repetitive loop and the action of the instructions can be observed on a CRO.

Some errors may occur only when the circuit is in a certain mode of operation, or when a certain FF is SET or CLEAR, and these conditions may occur relatively rarely. The frequency of occurrence of these conditions can be increased by the use of temporary jumpers to ground, which can artificially hold FFs CLEARED or SET gate outputs to desired levels. Although the entire circuit will not operate properly with these jumpers in place, they can often help to isolate and detect an error.

Glitches are caused by race conditions, transient circuit outputs (such as the decoder of section 8-5-1), or electrical noise or crosstalk. Careful attention must be paid to the design of the SET, CLEAR, and CLOCK inputs of FFs to prevent spurious SETS. In worst cases, a glitch may occur very infrequently and last for only a few nanoseconds. It is almost impossible to see on a CRO and its cause must be inferred from the action of the circuit. A digital circuit that fails one time in a million is virtually useless! If a computer executes 200,000 instructions in a second and errs once in every million instructions, we can expect it to run for only 5 seconds before chaos sets in! Finding the cause of these invisible and infrequent glitches is perhaps the most difficult and challenging of debugging problems.

When working with a computer, keep in mind that errors can also be caused by the program. Determining whether the source of an error is in hardware or software requires a grasp of programming. This is another problem in advanced debugging that is beyond the scope of this book.

19-7-4   Turn-on Procedures

Every system should incorporate a MASTER CLEAR or POWER ON RESET circuit. It is the function of this pulse to CLEAR or SET all the critical FFs so that a system may start properly. This pulse should also be capable of being generated electronically by a switch or circuitry so that it can be used as a last resort to clear a circuit that "hangs up," or locks into a particular state and will not change. Hang-up conditions are another source of error that must be avoided.

Microprocessors turn on by jumping to a specific memory address when they are powered up and starting to execute the instructions found there. Because no program can be written into RAM before power is applied, this location must be in ROM or nonvolatile memory.

19-7-5   Synchronous and Asynchronous Behavior

In *synchronous circuits*, all events are controlled by a clock and errors or glitches normally occur near the clock edges. In *asynchronous* circuits, events are often sequential; one event starts when the preceding event finishes. Because of the greater variation in time, glitches and races can occur more frequently in asynchronous circuits.

Sometimes even synchronous circuits must communicate with other circuits that send it pulses asynchronously. It is often wise to synchronize these pulses by gating them with the circuit clock, rather than allowing them to enter whenever they are generated. This guards against the possibility of a pulse entering at precisely the wrong time and causing an unforseen race or glitch. The use of synchronizing FFs (see section 6-16) is recommended for troublesome circuits.

19-7-6 Logic Probes

Logic probes are often used in digital testing. A typical logic probe was shown in Fig. 2-15 and is powered from V_{CC} of the system under test. It contains a metal tip that is touched to the point under test. There are usually two indicator lights that indicate a 1 and a 0. If neither indicator glows, the point under test is floating (unconnected). The ability to recognize a floating input is an advantage of logic probes.

Most logic probes also have the ability to detect single pulses. A short single pulse causes a light on the probe to glow for perhaps 200 ms, so that its occurrence is clearly visible. Some probes contain a light that stays on after being triggered by a pulse until a switch is depressed. Logic probes can detect the occurrence of short pulses where they should not normally appear.

The circuit for a logic probe was shown in Fig. 7-6. A 0 on the input lights the top Light-Emitting Diode (LED) and a 1 lights the lower LED. If a short pulse or glitch appears on the input, it triggers the **74x121** one shot, which lights the pulse LED for 100 ms, so the user can see it and is aware of its existence.

19-8   Logic Analyzers

As time progresses, electronic systems become larger and more complex and the problems of debugging them increase. Two devices that help analyze and troubleshoot these systems are the logic analyzer and the digital storage oscilloscope (DSO). Both use the digital technology that has been explained in this book.

The analog oscilloscope, found on the workbench of almost every laboratory, is an indispensible tool for examining electronic circuits. The oscilloscope is the best tool for examining waveforms, and determining voltage levels and rise times. It has two limitations, however:

1. The waveform shown on an oscilloscope must be *repetitive*.
2. Oscilloscopes have a limited number of channels. Four is about the largest number of channels on a commonly available oscilloscope.

Although not as accurate as the analog oscilloscope, the DSO overcomes the first limitation and the logic analyzer overcomes the second.

To debug a modern digital system, it is often necessary to monitor or observe *many constantly changing lines*. If an 8-bit counter (see section 8-10), for example, is not operating correctly, it may be necessary to monitor all 8 outputs and several other points, such as the clock driving the counter, to analyze the problem. A microprocessor (μP) is a second example. Even a small, 8-bit μP, such as an Intel **8085** or a Motorola **6800**, has 8 data lines, 16 address lines, and several control lines, all of which change on every μP clock cycle. To diagnose and fix a fault in these systems, a *logic analyzer* should be used because it can monitor many points in a digital system.

The display in both the analog oscilloscope and the logic analyzer are initiated by a *trigger*. But the oscilloscope can only display events that occur *after the trigger*, whereas the logic analyzer can record events that occurred both *before and after*. This ability is a powerful advantage in debugging a digital circuit.

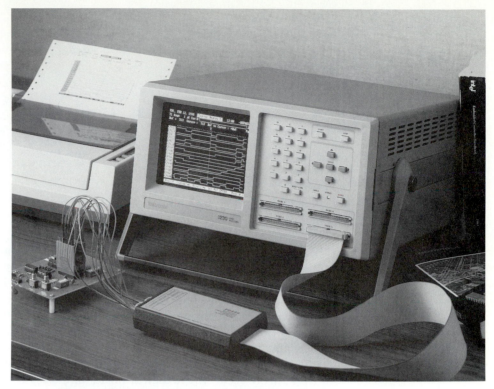

Figure 19-15 The Tektronix **1230** logic analyzer.
(Courtesy of © Tektronix, Inc.)

Figure 19-15 shows a modern logic analyzer, the Tektronix **1230**, which is a relatively inexpensive analyzer and is used in many college laboratories. It comes in a *console* that consists of three parts:

- **The keyboard**—This is used to enter commands and set up the parameters, such as the pattern recognizer or memory timing, that the logic analyzer will use.
- **The CRT display**—The CRT displays the command menu for the operator. It also displays the output data.
- **Pod Slots**—These are the connections for the pods that carry data from the μP or digital circuit into the logic analyzer.

The logic analyzer will be discussed in more detail in the following paragraphs.

19-8-1 Data Sampling

A logic analyzer obtains its information by taking samples of the data throughout the system at *discrete time intervals*. These are called *sample times* or *sample clocks*. The voltage is read at each input on each sample clock. The voltage at each sample, however, is only recorded as a logic 0 or logic 1, depending on whether it is above or below a threshold voltage. In many logic analyzers, the threshold is set to 1.4 V, so the logic 0s and 1s correspond to standard logic levels.

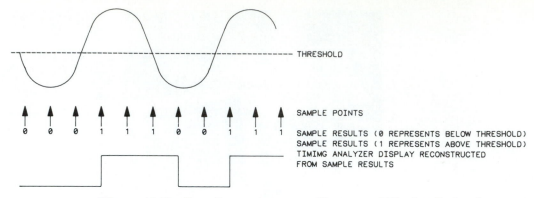

THRESHOLD

SAMPLE POINTS

0 0 0 1 1 1 0 0 1 1 1

SAMPLE RESULTS (0 REPRESENTS BELOW THRESHOLD)
SAMPLE RESULTS (1 REPRESENTS ABOVE THRESHOLD)
TIMIMG ANALYZER DISPLAY RECONSTRUCTED
FROM SAMPLE RESULTS

Figure 19-16 Sampling a sine wave. (Courtesy of Hewlett-Packard Inc.)

The concept is shown in Fig. 19-16, where a channel on a logic analyzer is sampling a sine wave. Observe that the sample points occur at discrete time intervals. Because the logic analyzer can only resolve its inputs to 0 or 1, the output is the square wave shown on the lower trace.

We must concede that the logic analyzer trace of Fig. 19-16 is only a crude approximation to the input sine wave, and that an oscilloscope would show it far more accurately. But a *logic analyzer is not meant to analyze sine wave*. They may be the most commonly used waveforms in transistors and audio circuits, but they rarely occur in the digital world, and a logic analyzer can do what an oscilloscope cannot. It can monitor the digital outputs at many points simultaneously.

Another problem with data sampling is that glitches that occur between data samples could be missed. Glitches were discussed in section 6-11-1 and shown in Fig. 6-20. The Hewlett-Packard Corporation defines a glitch as two transistions of a waveform occuring between sample pulses, as shown in Fig. 19-17,and they can be missed by the analyzer. Fortunately, most logic analyzers have a *glitch-latch* circuit on them that will detect glitches when they occur.

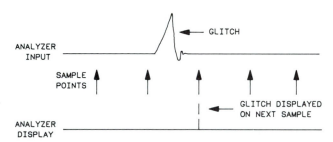

ANALYZER INPUT

GLITCH

SAMPLE POINTS

GLITCH DISPLAYED ON NEXT SAMPLE

ANALYZER DISPLAY

Figure 19-17 A glitch applied to a logic analyzer. (Courtesy of Hewlett-Packard Inc.)

19-8-2   Input Pods

Data is brought into the logic analyzer by an *input pod*. A simple 10-input pod is shown in Fig. 19-18. It consists of 10 *grabbers*, which are probes with clips on the end of them to grab onto various points in a circuit. They are connected to the

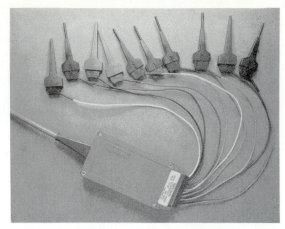

Figure 19-18   A logic analyzer pod. (Photo courtesy of Ed Pickett.)

logic analyzer via the cable. Actually, the pod of Fig. 19-18 consists of only 8 data inputs, a ground input, and an input used for other functions, such as the *system clock*. It is considered to be an 8-channel pod, where each channel is a data point. A modern logic analyzer, such as the Tektronix **1230**, uses 16-input pods and can accommodate up to four of them. Figure 19-15 shows a pod connected between the circuit under test and the **1230**.

EXAMPLE 19-3

A 32-channel logic analyzer is to monitor the buses on an **8085** microprocessor. Where would the inputs be connected?

SOLUTION   Sixteen of the channels should monitor the address bus, and eight channels should monitor the data bus. The remaining eight channels could monitor other points. Typically, $\overline{RD}$, $\overline{WR}$, and IO/$\overline{M}$ would also be monitored. The other five channels can monitor other signals on the **8085** control bus, such as the clock and READY, or can monitor other points in the system.

19-8-3   The Memory in a Logic Analyzer

The memory is the core of any logic analyzer. The signals to be monitored are connected to the pods and then become the data inputs to the memory. This data is written into the memory at the sample clock rate as shown in Fig. 19-19. In the figure the sample clock is assumed to be a square wave and a WRITE pulse is generated on every negative transistion. The WRITE pulses must be as long as the cycle time of the memory.

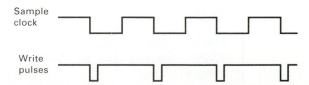

Figure 19-19   Write pulses to a logic analyzer's memory.

EXAMPLE 19-4

The memory in a logic analyzer has a 150 ns cycle time and needs 50 ns to recover. What is the maximum sample rate for this analyzer?

SOLUTION   This memory requires 200 ns for each write. This is its maximum sample rate. For many applications, this is too slow. Most logic analyzers have faster memories that allow them to handle higher clock rates.

Logic analyzers offer the user a choice of many clock rates. The **1230**, for example, has clock rates from 10 ns to 40 ms. The sampling rate can be selected from the menu on the console. The slower clock rates allow for a larger time range, but have less resolution. Glitches are also more probable at slower clock rates.

In addition to selecting specific sampling rates, such as 1 µs, the user can select EXTERNAL. This allows the clock from the system under test to determine the sample rate. This is very important because many systems, including µPs, are *synchronous* and only change their states on a clock edge. If an **8085**, for example, uses a 300 ns clock for its T-states (see section 17-5-1), the clock ouput of the **8085** can be connected to the EXTERNAL input of the logic analyzer. This is called *synchronous sampling* because the analyzer is *synchronized to the clock of the system under test*. Usually a special pod input connects the system clock to the logic analyzer.

19-8-4   Data Acquisition

A logic analyzer operates in one of two modes, data acquisition and data display. During data acquisition, the memory is constantly being written. The Tektronix **1230** has a 2K word memory. If the sample rate is 1 µs, for example, the memory will be filled in 2.048 ms. Acquisition *does not stop* when the memory is filled, however. Instead it continues, with *the new data overwriting the old data*. Thus, at any time during acquisition, the memory will contain the most recent 2K data samples from the external system.

19-8-5   The Trigger

The trigger is an *event that causes the logic analyzer to stop writing data into memory* and leave the acquisition mode. After the trigger, the most recently acquired data remains in memory where it can be examined.

The trigger can be an edge on a line in the external system, but it is usually a *pattern match*.[1] Before entering the acquisition mode, a pattern of 1s, 0s, and Xs is entered on the console. The Xs are don't cares. When the data on the input lines has 1s that match all the 1s in the pattern, and 0s that match all the 0s in the pattern, the trigger occurs.

EXAMPLE 19-5

The logic analyzer is connected to an **8085** µP as in Example 19-4. It should trigger whenever the µP reaches address 200C. How can we make this happen?

---

[1]A pattern match is sometimes called a word recognizer.

SOLUTION With 32 channels available, there would be a 32-bit pattern that could be set into the logic analyzer. The bits corresponding to the 16 address inputs could be set to 200C, and the other bits could be set as Xs (don't cares). The analyzer would trigger whenever it found 200C on the address bus.

When the trigger occurs, the logic analyzer does not stop acquiring data immediately. In most cases it continues to acquire data until it fills half the memory. Thus, a logic analyzer with a 2K word memory will acquire 1024 words after the trigger and then stop. Consequently, the memory will hold 1024 data points that occurred after the trigger and 1024 data points that occurred before the trigger. This helps the user debug because he can examine the system both before and after the trigger.

19-8-6   The Display Mode

After the trigger, the logic analyzer goes from the acquisition mode to the display mode. The information in the memory is no longer changing and can be displayed on the console.

There are two ways to display the information in the memory: a *timing diagram* and a *state diagram*. A typical timing diagram, taken from the Tektronix **1230** manual, is shown in Fig. 19-20. It looks like the output of a multi-trace oscilloscope, except that none of the waveforms have any slope; they are all square. This is because each waveform is sampled at one time slot and converted to a 1 or 0. It retains that value until the next sample. Waveform changes between samples cannot be shown on a logic analyzer.

A simplified, 4-channel timing diagram is shown in Fig. 19-21. While in display mode, the memory data and the trigger are fixed. The *cursor*, however, is a marker that can be moved throughout the memory and allows the user to examine any part of the memory of interest. Figure 19-21 shows three points of particular interest:

1. The cursor position relative to the trigger position. In Fig. 19-21 the cursor is a TRIGGER + 5, or 5 positions after the cursor.
2. The data at the trigger. This should be the same data as entered into the trigger pattern. Here it is shown as 0101.
3. The data at the cursor position. Here it is 1100.

EXAMPLE 19-6

In Fig. 19-21, what would the cursor data be if the cursor position were TRIGGER − 3?

SOLUTION The cursor would be three positions or locations before the trigger. The data in Fig. 19-21 at this position is 1110.

Referring back to Fig. 19-20, the timing diagram for the **1230**, it shows first that the clock rate is 40 ns. The *reference*, the point where the trigger occurred,

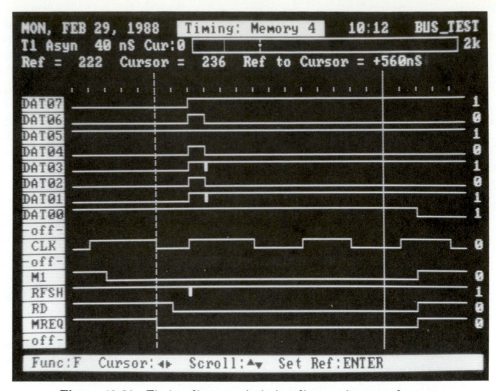

Figure 19-20 Timing diagram. A timing diagram is a pseudo-wave-form display. You can use timing diagrams to look at timing relationships between signals. In these diagrams, the horizontal axis shows samples and the vertical axis shows data channels. (Courtesy of © Tektronix, Inc.)

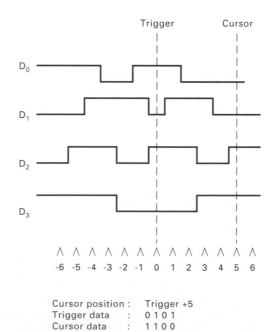

Figure 19-21 A simplified logic analyzer timing diagram.

is at memory location 222, and the cursor is at 236. The cursor is at REF + 14, and shows what occurred 560 ns after the trigger. The 1s and 0s at the right side of the timing diagram show the data at the cursor position. Note that when the data changes at the cursor position, it is the data after the cursor that is reported.

A state table display is shown in Fig. 19-22. It lists the memory data, usually in hexadecimal form, rather than displaying it as a timing chart. Many engineers prefer a state table display, especially when debugging a μP.

EXAMPLE 19-7

Due to a fault, a μP is jumping to a location around 2100. How can a logic analyzer help solve this problem?

SOLUTION Even though the erroneous address is not fully known, the logic analyzer can help. The inputs can be put on the μP's address bus. The trigger pattern can be set to $(21XX)16$, where XX means that eight least significant switches are set to don't cares, and sampling started using the μP's clock. When the μP reaches the first address starting with 21, wherever it is, the logic analyzer will trigger and display not only the first address, but the addresses that preceded the jump to the 2100 area. This should allow the user to examine these locations to discover what caused the erroneous jump.

| THU, FEB 25, 1988 | | | State: Memory 4 | | 10:24 | MICROCHK |

Loc	DAT	ADR	STB	CTL	I/O	COM
	hex	hex	bin	bin	hex	hex
1983	C2	24CA	00000	000	FC6C	FCE3
1984	C7	24CB	01101	111	FC6C	FCE3
1985	20	24CC	01101	111	FE48	FEEE
1986	1B	24C7	01001	111	FE48	FEEE
1987	7A	24C8	01001	111	FE48	FEEE
1988	B3	24C9	01001	111	FE48	FEEE
1989	C2	24CA	01001	111	FFC0	FF95
1990	C7	24CB	01101	111	FFC0	FF95
1991	20	24CC	01101	111	FFC0	FF95
1992	1B	24C7	01001	111	FFC0	FF95
1993	7A	24C8	01001	111	7FC0	7F95
1994	B3	24C9	01001	111	7FC0	7F95
1995	C2	24CA	01001	111	7FC0	7F95
1996	C7	24CB	01101	111	3FC0	3FD5
1997	20	24CC	01101	111	3FC0	3FD5
1998	1B	24C7	01001	111	3FC0	3FD5
1999	7A	24C8	01001	111	3FC0	3FD5
2000	B3	24C9	01001	111	1FC0	1FDE
2001	C2	24CA	01001	111	1FC0	1FDE
2002	C7	24CB	01101	111	1FC0	1FDE

| Func:F | Scroll:▲▼ | Cursor:◄► | Jump:ENTER | Radix:E |

Figure 19-22 A logic analyzer state table. (Courtesy of © Tektronix, Inc.)

EXAMPLE 19-8

Using the circuits developed in this book, conceptually design a logic analyzer.

SOLUTION A high-frequency oscillator, probably crystal controlled for best accuracy, would generate the sample clocks (see section 7-6-5). Various lower output frequencies could be obtained by a divider circuit (see section 8-10) and an EXTERNAL switch should be provided so the clock can be taken from the system under test. The negative edge of the clock could fire a one-shot (see section 7-3) and generate the WRITE pulses for the memory.

The input data to the memory (DRAM or SRAM; see Chapter 13) must come from the connections to the circuit under test, via pods if they are available. The memory addresses are incremented by a counter (see Chapter 8) that simply rolls over. If the logic analyzer has a 2K word memory, an 11-bit counter would be used.

Another register must be available to hold the trigger pattern, and there must be a *comparator* (see section 14-3-3) to compare the input data and the pattern. When they match, a second counter, typically 1024 words, would start. It would also be incremented by the sample clock. When this counter rolls over it would set a flip-flop and stop the counter. The data is now in memory, with 1024 words before the trigger and 1024 words after the trigger. This data can then be displayed. Probably a μP would be needed to coordinate and present the display.

19-8-7 Other Logic Analyzer Features

There are many other features available in logic analyzers. Space limitations preclude a full explanation of these features, but we can list them briefly:

- **Personality modules**—This is a module that is connected between the inputs to a logic analyzer and the buses of a μP. There is a different personality module for each popular μP. These modules tailor a logic analyzer to a specific μP and display the data as that particular μP's instructions. This is called a *disassembler*. Figure 19-23 shows a personality module connected to an analyzer. Figure 19-24 shows the display of such an output. It shows a code segment on the Motorola **68000** μP. In the figure the **Loc** column is the address in the analyzer's memory. The **Addr** column is the 24-bit memory address of the **68000** instruction, expressed as six hex digits. The **Data** column is the 16-bit data found at that location, expressed as four hex digits. In the last columns the disassembled code is displayed.

- **Probe qualification**—Most logic analyzers contain an additional input called a *probe qualifier*, which is also brought in on a pod. This is not a data input (it cannot be displayed), but is used to affect triggering. When used, the pattern recognizer will not trigger unless all data and the probe qualifier match the corresponding switches.

- **Clock qualification**—Some logic analyzers contain a *clock qualifier*. The logic analyzer will not accept a sample clock unless the

Figure 19-23 A personality module. (Courtesy of © Tektronix, Inc.)

```
TUE, FEB 16, 1988   Disasm Memory 1    10:31  -DEFAULT
 Loc   Addr   Data   68000 Disassembly    Md Operation

 1071  0F1FC2  1028  ?MOVE.B 0010(A0),D0    S
 1072  0F1FB0  1210   MOVE.B (A0),D1        S   FFFF03=00C5
 1073  0F1FB2  0201   ANDI.B #10,D1         S
 1076  0F1FB6  6600   BNE     0F1E88        S
-1078--0F1FBA-1210----MOVE.B (A0),D1--------S---FFFF03=00C5-
 1079  0F1FBC  0201   ANDI.B #02,D1         S
 1082  0F1FC0  67EE   BEQ     0F1FB0        S
 1083  0F1FC2  1028  ?MOVE.B 0010(A0),D0    S
 1084  0F1FB0  1210   MOVE.B (A0),D1        S   FFFF03=00C5
 1085  0F1FB2  0201   ANDI.B #10,D1         S
 1088  0F1FB6  6600   BNE     0F1E88        S
 1090  0F1FBA  1210   MOVE.B (A0),D1        S   FFFF03=00C5
 1091  0F1FBC  0201   ANDI.B #02,D1         S
 1094  0F1FC0  67EE   BEQ     0F1FB0        S
 1095  0F1FC2  1028  ?MOVE.B 0010(A0),D0    S
 1096  0F1FB0  1210   MOVE.B (A0),D1        S   FFFF03=00C5
 1097  0F1FB2  0201   ANDI.B #10,D1         S
 1100  0F1FB6  6600   BNE     0F1E88        S
 1102  0F1FBA  1210   MOVE.B (A0),D1        S   FFFF03=00C5
 1103  0F1FBC  0201   ANDI.B #02,D1         S

 Func: F     Scroll: ▼▲    Cursor: ◀▶    Jump: ENTER
```

Figure 19-24 Disassembly of a μP program segment by a logic analyzer. (Courtesy of © Tektronix, Inc.)

clock qualifier is also correct. This is valuable when a μP goes into a loop as it would when awaiting input from a keyboard. The repetitive instructions as the μP loops are of no interest, but would soon fill the logic analyzer's memory. If a clock qualifier is properly set up, the repetitive instructions can be ignored and only the μP's response to the keystrokes reported.

- **Reference memory**—This is an auxiliary memory. The data in the main memory can be transferred into the auxiliary memory and then compared with the data entered into another section of the main memory, or data entered into the main memory at a later time. This is valuable for tracking down intermittents and for comparing the operation of the system at two different times.
- **Glitch latch**—This is an additional circuit used to capture and display any glitches that may not be detected by an ordinary analyzer if the glitch occurs between the sample clocks.

19-8-8 The Tektronix 1230 Logic Analyzer

The **1230** is an inexpensive logic analyzer that is often used in colleges. The user starts by connecting the pods to the system under test (SUT) and setting the menu on the console. The menu will specify at least the following:

- **Synchronous or Asynchronous Clocking**—Asynchronous means the clock is derived from a fixed clock rate. Synchronous means the clock is obtained directly from the μP or SUT.
- **The Clock Rate**—If asynchronous timing is selected.
- **The Display Type**—Timing diagrams or state tables.
- **The Trigger Pattern**—The **1230** gives the user a choice of both simple and complex patterns. A simple pattern was shown in section 19-8-4. An example of a complex pattern might "Trigger on the third occurrence of pattern A followed by the fifth occurrence of pattern B."
- **The Trigger Position**—The position of the trigger need not always be halfway through the memory. If data after the trigger is most important, the analyzer can be set up so that it only remembers a few locations before and many locations after. Conversely, it can be set up to remember and display most of the data before the trigger if that is more important.

Like most modern logic analyzers, the **1230** has many capabilities and is therefore complex. Tektronix provides a 200-page Operator's Manual for its users. We regret that we could only touch on the highlights in this section, but a thorough description would require a small book of its own.

19-9 Digital Storage Oscilloscopes

The Digital Storage Oscilloscope (DSO) is a cross between the analog oscilloscope and a logic analyzer. It acts like an oscilloscope because it accurately displays waveforms on a small number of channels, but it also acts like a logic analyzer because it time-samples the data. The DSO is best used to record and display waveforms that occur only once (are non-repetitive) or have a very low repetition rate.

19-9-1   The Analog Oscilloscope

Before considering the DSO, we will explore the basic operation of an analog oscilloscope, as shown in Fig. 19-25. The signal to be examined enters through an *attenuator*, which adjusts its amplitude by the Volts/Division knob on the front panel, and selects AC or DC coupling. In AC coupling, the input signal is put through a capacitor which strips off its DC component. This allows the engineer to examine a small AC signal riding on a large DC level. With AC coupling, however, the user cannot determine level of the voltage with respect to ground. Many engineers, including the author, use DC coupling wherever possible to eliminate this problem.

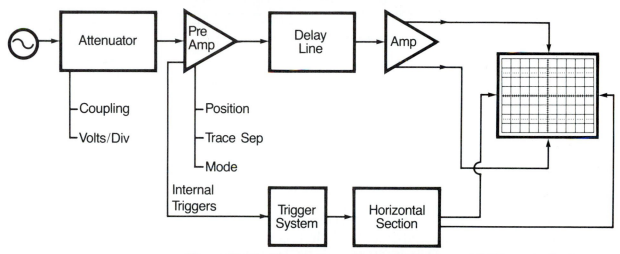

Figure 19-25   The analog oscilloscope. (Courtesy of © Tektronix, Inc.)

The oscilloscope operates by sweeping a beam across the screen and modulating it (moving it up and down) as the amplitude of the input signal varies. The lower path of Fig. 19-25 shows the horizontal section. The centimeters/sec knob (not shown) determines the speed of the beam as it traverses the screen. The trigger system determines when the beam starts. On most oscilloscopes there are two major selections: AUTOMATIC and NORMAL. Automatic means the beam sweep starts at fixed time intervals. NORMAL means the beam starts when the input voltage reaches a particular level. This level is usually set by manipulating the TRIGGER LEVEL knob.

The upper path of Fig. 19-25 is the signal path. The input signal is first delayed for a few nanoseconds to compensate for delays in getting the sweep started. When it is synchronized with the beam, it deflects the beam vertically and produces a representation of the input waveform on the screen.

19-2-2   The Digital Oscilloscope

Figure 19-26 shows the block diagram of a DSO. Like the analog oscilloscope, the signal to be observed enters through a vertical preamp. In the DSO, how-

Analog Oscilloscope

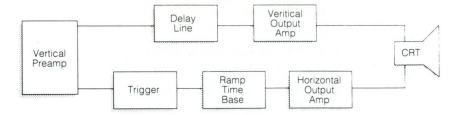

Digitizing Oscilloscope

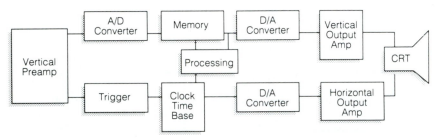

Figure 19-26 A block diagram of a digital oscilloscope. (Courtesy of © Tektronix, Inc.)

ever, the signal goes into an analog-to-digital converter (A/D, see section 18-5), and the output of the converter is written to a memory. Like the logic analyzer, the signal is sampled at discrete time intervals. But the A/D converter preserves its magnitude instead of merely differentiating between a 1 and a 0, as a logic analyzer does. The signal is displayed by reading it out of memory into a digital-to-analog (D/A) converter. The output of the D/A converter determines the vertical deflection of the beam and produces a replica of the input waveform.

The advantage of the DSO is that the waveform is now in memory; it can be displayed as long as needed. A single waveform can be captured by triggering on it, but once it is in the DSO's memory it can be displayed indefinitely. Events before the trigger, or in so-called *negative time*, can also be displayed as in a logic analyzer.

A good DSO must have a very high sampling rate to accurately represent the input waveform. All DSOs specify their maximum sampling rate; this is one of the most important criteria on which DSOs are judged. Many modern DSOs use *charge-coupled devices* (CCDs) on their input. They are like shift registers; they contain many cells, and each cell contains the analog input voltage. But CCDs can sample at a very high rate. The procedure for capturing a single, high-speed event is to shift it into a CCD and a very high sample rate. The CCDs are then shifted out into the A/D converter at a lower rate and brought to memory. In this manner, sample rates that are too high for the A/D converter are slowed down so that the converter can handle them.

19-9-3 The Hewlett-Packard (HP) 54600A Oscilloscope

The **HP54600A** is a two-channel DSO shown in Fig. 19-27. It is a relatively inexpensive DSO that can be used for general purpose laboratory work. It looks and

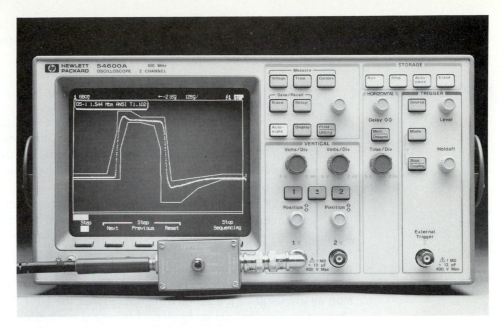

Figure 19-27 The **HP54600** DSA. (Courtesy of Hewlett-Packard Inc.)

feels like an analog oscilloscope, and is often used in the same manner. The buttons in the upper right-hand corner of the panel control digital acquisition and storage.

The major specifications for the **HP54600A** are:

- *Maximum Sampling Rate.* For the **HP54600** this is 20 million samples per second or 20 MSa/s. This means the closest samples can be 50 ns apart, or that the D/A converter within it has a conversion time of less than 50 ns. The more advanced Hewlett-Packard DSOs can sample at a rate of 1 GHz or take one sample per ns. They use flash converters.

- *Vertical Resolution.* This is 255 different levels of amplitude. It shows that there is an 8-bit A/D converter in the **HP54600**. The screen of the DSO is 8 cm. high, so this vertical resolution allows for 32 levels in each centimeter of the screen.

- *Horizontal Resolution.* The **HP54600** can display 500 points on its horizontal axis. Because the screen is 10 cm. wide it is possible to display 50 point per cm. The memory in the **HP54600** can hold 2000 bytes.

The **HP54600**, like many other modern DSOs, do not use D/A converters to display the points. They use a *raster-scan*, similar to the displays on a television set or a computer terminal, to set the position of the dot being displayed. This allows the DSO to perform mathematical operations on the input data and provide some smoothing or averaging, which can minimize the effect of noise on the waveform.

Space in this book allows us to introduce only briefly DSOs. Booklets that explain DSOs in greater detail are available from Hewlett-Packard, Tektronix, and other DSO manufacturers.[2] Engineers should also study the specifications for a DSO or logic analyzer from all the manufacurers before attempting to buy or use this type of test instrument.

19-10   Workstations

A workstation is a computer used to help engineers design products. They are used for design of mechanical devices, where the process is called Computer-Aided Design (CAD), and for the design of electronic circuits. Electronic design is increasingly being created on workstations rather than being sketched out and wired up on a breadboard or superstrip.

Figure 19-28 shows the SPARCstation IPX, a dedicated graphics workstation manufactured by Sun Microsystems, Inc. Sun Microsystems is one of the leading suppliers of workstations; others include Digital Equipment Corporation and Mentor Graphics. Dedicated workstations provide engineers with much capability at a reasonable price.

Software packages exist that will allow personal computers (PCs) to function as workstations. They are available from the OR-CAD corporation of Hillsboro, Oregon; Spectrum Systems, Inc. of Sunnyvale, California and others. They are not as fast or as capable as a dedicated workstation, but are much less expensive.

Figure 19-28 The Sun Microsystems SPARCstation™ IPX™ workstation.

19-10-1   Design Using Workstations

Digital designs can be created directly on the screen of a workstation. For digital design, workstations maintain a library of all the common TTL devices, includ-

[2]A much more thorough discussion of the **HP54600** is given in the February, 1992 issue of the Hewlett-Packard Journal.

ing those discussed in this book. To start a design, the engineer can select a gate he requires from the library, and then place it on the screen by manipulating its position with a mouse and clicking when it is in its proper place. He can then select a second gate or FF, place that on the screen, and specify the wire connections between the gates.

Following this procedure, complicated digital circuits can be constructed on the computer, input points can be specified, and outputs can be labelled. The user can zoom in to get a detailed view of a small part of the circuit or zoom out to see the entire circuit. A complex digital circuit can be created on the screen of the workstation's console.

19-10-2   Testing Using Workstations

Circuits that have been created on workstations can also be tested on workstations. The gate delays of the various components can be selected and input waveforms can be specified. The workstation will then compute a timing chart to show how the outputs react to the specified inputs. This is the modern form of debugging; changes are made electronically rather than changing ICs and re-routing wires. Once the circuit is finalized, a *schematic capture* package can be used to create a printed-circuit board, so that the circuit can be manufactured.

While workstations simplify and speed up testing, they are still only a tool. The engineer must understand the functions and the capabilities of the various ICs before he or she can create a working circuit or system on a workstation. This is what we have endeavored to teach in this book.

SUMMARY

In this chapter, the advantages and disadvantages of wire wrapping and PC board construction were explored. A method of documenting wire-wrap construction was presented. Most companies set their own drafting, wire list, and module chart standards, but usually they are similar to those presented here.

The chapter continued with two sections on debugging. Frankly, debugging is as much art as science and depends on the insight and ingenuity of the engineer. No book can hope to present a specific method for curing the myriad of problems that arise in practice. Here we presented some basic methods used to attack faulty circuits.

Modern, sophisticated test instruments such as logic analyzers and digital oscilloscopes were introduced. They operate using the circuits discussed in this book. Finally, the use of workstations in digital design and testing was explored. These were all described briefly. Their proper use requires a thorough reading of the manual that comes with each of these instruments.

GLOSSARY

Buzz-out: A check of the wiring of a circuit.

Debugging: Correcting the faults in a circuit or system.

Floating input: An input that is not tied to a source.

Hang-up condition: A condition where the state of a system will not change.

Logical: A drawing showing the ICs in a system and their interconnections.

Module chart: A chart showing the locations of the ICs in a system.

Maybe (Colloquial):   A logic level between 0 and 1.

PC board: Printed circuit board.

Prototype: A preproduction model of a system built for testing and debugging.

Wire list: A list of the routing of each wire in a system.

Workstation: A computer used by engineers to create designs.

PROBLEMS

Section 19-5-3.

19-1.   Draw a module chart for the circuit of Fig. 10-18.

Section 19-5-4.

19-2.   Draw a wire list segment for the circuit of Fig. 19-7.

19-3.   Draw a module chart for the circuit of Fig. 10-18. Draw the wire list for the first **74x165** only.

Section 19-8-6.

19-4.   In Fig. P19-4, where is the reference location and the cursor location? What is the data at REF $+$ 35? At REF $-$ 1?

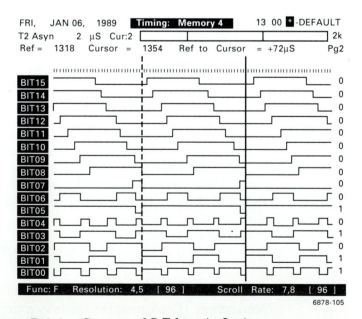

Figure P19-4   (Courtesy of © Tektronix, Inc.)

References

Boolean Algebra

FREDRICK HILL and GERALD PETERSON, *Introduction to Switching Theory,* John Wiley, New York, Latest Ed.

CHARLES H. ROTH, *Fundamentals of Logic Design*, West Publishing Co., St. Paul, MN, Latest Ed.

Digital Electronics

RONALD J. TOCCI, *Digital Systems*, 5th ed., Prentice-Hall, Englewood Cliffs, NJ, 1991.

THOMAS L. FLOYD, *Digital Fundamentals*, 3rd ed., Charles E. Merrill, Columbus OH, 1986.

DOUGLAS V. HALL, *Digital Circuits and Systems*, McGraw-Hill, New York, 1989.

BRYON W. PUTMAN, *Digital and Microprocessor Electronics*, Prentice-Hall, Englewood Cliffs, NJ, 1986.

M. MORRIS MANO, *Computer Engineering*, Prentice-Hall, Englewood Cliffs, NJ, 1988.

WILLIAM KLEITZ, *Digital Electronics*, Prentice-Hall, Englewood Cliffs, NJ, 1990.

RICHARD J. PRESTOPNIK, *Digital Electronics*, Saunders College Publishing, Philadelphia, PA, 1990.

PLDs and Gate Arrays

ROGER C. ALFORD, *Programmable Logic Designer's Guide*, Howard W. Sams and Company, Indianapolis, IN, 1989.

Designing With Programmable Array Logic, Monolithic Memories, 1981. Monolithic Memories has been merged into Advanced Devices, Sunnyvale, CA.

Programmable Array Logic Handbook, Advanced Micro Devices, Sunnyvale, CA, 1984.

The Programmable Gate Array Data Book, XILINX Corp., San Jose, CA, 1989.

PAL Device Data Book, Advanced Micro Devices Corp., Sunnyvale, CA, 1990.

CMOS Gate Arrays Design Manual, 1990, Texas Instruments, Dallas, TX.

Altera Data Book, Altera Corp., San Jose, CA, 1991.

RS-232C

JOE CAMPBELL, *The RS-232 Solution*, SYBEX Computer Books, Berkeley, CA, 1984.

A/D and D/A Converters

National Semiconductor Linear Data Book, Latest Ed.

Datel Databooks (Volumes 1 and 2), Datel Corp., Mansfield, MA.

Microprocessors

Intel 8085.

RAMESH GAONKAR, *Microprocessor Architecture, Programming and Applications*, Charles E. Merrill, Columbus, OH, 1984.

Intel 8086.

JOHN UFFENBECK, *The 8086/8088 Family*, Prentice-Hall, Englewood Cliffs, NJ, 1987.

YU-CHENG LIU and GLENN A. GIBSON, *Microcomputer Systems: The 8086/8088 Family*, Prentice-Hall, Englewood Cliffs, NJ, 1986.

TREIBEL and SINGH have produced several books on 16-bit Intel and Motorola microprocessors, including the **8086** and **80286**. Their books are published by Prentice-Hall. See the PH catalog for the most current information.

Motorola Microprocessors.

W. C. WRAY and J. D. GREENFIELD, *Using Microprocessors and Microcomputers: The Motorola Family*, latest ed. Prentice Hall, Englewood Cliffs NJ.

J. D. GREENFIELD, *The **68HC11** Microcontroller*, Saunders College Publishing, Philadelphia, PA, 1992.

THOMAS L. HARMON has produced several books on the various Motorola 16-bit microprocessors. They are published by Prentice-Hall. See the PH catalog.

IBM. The IBM Corporation has Technical Reference Manuals available for most of their PCs. Information can be obtained from IBM, P.O. Box 1328-C, Boca Raton, FL.

Manufacturer's Data Books

Texas Instruments. Texas Instruments, Inc., Dallas, TX, has published many data books on the electronic devices they manufacture. The data books pertinent to the topics discussed in this text are:

TTL Logic Data Book, 1988

ALS/AS Data Book, 1986

High Speed CMOS Logic, 1989

Advanced CMOS Logic, 1990

Advanced Logic and Bus Interface Logic, 1991

CMOS Gate Arrays, 1990

MOS Memory, 1989

Linear Circuits Data Book (three volumes), 1989

Intel Corporation. The Intel Corporation publishes a series of data books on their products, which are predominantly microprocessors, memories, and peripheral ICs. They can be obtained from INTEL LITERATURE SALES, PO Box 7641, Mt. Prospect, IL, 60056-7641.

Intel Memory Data Book, 1990

Intel Microprocessor Data Book, 1990

Intel Peripherals Data Book, 1990

Motorola. Motorola also publishes a series of data books on their products. They can be obtained from Motorola Literature Distribution, P.O. Box 20912, Phoenix, AZ, 85036.

Motorola Memory Data, 1991

CMOS Logic, 1991

High Speed CMOS Logic Data, 1989

This list of data books is extensive, but not exhaustive. Data books are updated approximately once every three years. For a more complete and up-to-date list of data books, users should contact these companies, or perhaps their local sales offices.

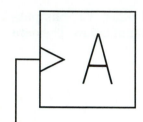

Table of Powers of 2

2^n	n	2^{-n}
1	0	1.0
2	1	0.5
4	2	0.25
8	3	0.125
16	4	0.062 5
32	5	0.031 25
64	6	0.015 625
128	7	0.007 812 5
256	8	0.003 906 25
512	9	0.001 953 125
1 024	10	0.000 976 562 5
2 048	11	0.000 488 281 25
4 096	12	0.000 244 140 625
8 192	13	0.000 122 070 312 5
16 384	14	0.000 061 035 156 25
32 768	15	0.000 030 517 578 125
65 536	16	0.000 015 258 789 062 5
131 072	17	0.000 007 629 394 531 25
262 144	18	0.000 003 814 697 265 625
524 288	19	0.000 001 907 348 632 812 5
1 048 576	20	0.000 000 953 674 316 406 25
2 097 152	21	0.000 000 476 837 158 203 125
4 194 304	22	0.000 000 238 418 579 101 562 5
8 388 608	23	0.000 000 119 209 289 550 781 25
16 777 216	24	0.000 000 059 604 644 775 390 625
33 554 432	25	0.000 000 029 802 322 387 695 312 5
67 108 864	26	0.000 000 014 901 161 193 847 656 25
134 217 728	27	0.000 000 007 450 580 596 923 828 125
268 435 456	28	0.000 000 003 725 290 298 461 914 062 5
536 870 912	29	0.000 000 001 862 645 149 230 957 031 25
1 073 741 824	30	0.000 000 000 931 322 574 615 478 515 625
2 147 483 648	31	0.000 000 000 465 661 287 307 739 257 812 5
4 294 967 296	32	0.000 000 000 232 830 643 653 869 628 906 25
8 589 934 592	33	0.000 000 000 116 415 321 826 934 814 453 125
17 179 869 184	34	0.000 000 000 058 207 660 913 467 407 226 562 5
34 359 738 368	35	0.000 000 000 029 103 830 456 733 703 613 281 25
68 719 476 736	36	0.000 000 000 014 551 915 228 366 851 806 640 625
137 438 953 472	37	0.000 000 000 007 275 957 614 183 425 903 320 312 5
274 877 906 944	38	0.000 000 000 003 637 978 807 091 712 951 660 156 25
549 755 813 888	39	0.000 000 000 001 818 898 403 545 856 475 830 078 125
1 099 511 627 776	40	0.000 000 000 000 909 494 701 772 928 237 915 039 062 5
2 199 023 255 552	41	0.000 000 000 000 454 747 350 886 464 118 957 519 531 25
4 398 046 511 104	42	0.000 000 000 000 227 373 675 443 232 059 478 759 765 625
8 796 093 022 208	43	0.000 000 000 000 113 686 837 721 616 029 739 379 882 812 5
17 592 186 044 416	44	0.000 000 000 000 056 843 418 860 808 014 869 698 941 406 25
35 184 372 038 832	45	0.000 000 000 000 028 421 709 431 404 007 434 844 970 703 125
70 368 744 177 664	46	0.000 000 000 000 014 210 854 715 202 003 717 422 485 351 562 5
140 737 488 355 328	47	0.000 000 000 000 007 105 427 357 601 001 858 711 242 675 781 25
281 474 976 710 656	48	0.000 000 000 000 003 552 713 678 800 500 929 355 621 337 890 625
562 949 953 421 312	49	0.000 000 000 000 001 776 356 839 400 250 464 677 810 668 945 312 5
1 125 899 906 843 624	50	0.000 000 000 000 000 888 178 419 700 125 232 338 905 334 472 656 25
2 251 799 813 685 248	51	0.000 000 000 000 000 444 089 209 850 062 616 169 452 667 236 328 125
4 503 599 627 370 496	52	0.000 000 000 000 000 222 044 804 925 031 308 084 726 333 618 164 062 5
9 007 199 254 740 992	53	0.000 000 000 000 000 111 022 302 462 515 654 042 363 166 809 082 031 25
18 014 398 509 481 984	54	0.000 000 000 000 000 055 511 151 231 257 827 021 181 583 404 541 015 625
36 028 797 018 963 968	55	0.000 000 000 000 000 027 755 575 615 628 913 510 590 791 702 270 507 812 5
72 057 594 037 927 936	56	0.000 000 000 000 000 013 877 787 807 814 456 755 295 395 851 135 253 906 25
144 115 188 075 855 872	57	0.000 000 000 000 000 006 938 893 903 907 228 377 647 697 925 567 626 953 125
288 230 376 151 711 744	58	0.000 000 000 000 000 003 469 446 951 953 614 188 823 848 962 783 813 476 562 5
576 460 752 303 423 488	59	0.000 000 000 000 000 001 734 723 475 976 807 094 411 924 481 391 906 738 281 25
1 152 921 504 606 846 976	60	0.000 000 000 000 000 000 867 361 737 988 403 547 205 962 240 695 953 369 140 625
2 305 843 009 213 693 952	61	0.000 000 000 000 000 000 433 680 868 994 201 773 602 981 120 347 976 684 570 312 5
4 611 686 018 427 387 904	62	0.000 000 000 000 000 000 216 840 434 497 100 886 801 490 560 173 988 342 285 156 25
9 223 372 036 854 775 808	63	0.000 000 000 000 000 000 108 420 217 248 550 443 400 745 280 086 994 171 142 578 125
18 446 744 073 709 551 616	64	0.000 000 000 000 000 000 054 210 108 624 275 221 700 372 640 043 497 085 571 289 062 5
36 893 488 147 419 103 232	65	0.000 000 000 000 000 000 027 105 054 312 137 610 850 186 320 021 748 542 785 644 531 25
73 786 976 294 838 206 464	66	0.000 000 000 000 000 000 013 552 527 156 068 805 425 093 160 010 874 271 392 822 265 625
147 573 952 589 676 412 928	67	0.000 000 000 000 000 000 006 776 263 578 034 402 712 546 580 005 437 135 696 411 132 812 5
295 147 905 179 352 825 856	68	0.000 000 000 000 000 000 003 388 131 789 017 201 356 273 290 002 718 567 848 205 566 406 25
590 295 810 358 705 651 712	69	0.000 000 000 000 000 000 001 694 065 894 508 600 678 136 645 001 359 283 924 102 783 203 125
1 180 591 620 717 411 303 422	70	0.000 000 000 000 000 000 000 847 032 947 254 300 339 068 322 500 679 641 962 051 391 601 562 5
2 361 183 241 434 822 606 848	71	0.000 000 000 000 000 000 000 423 516 473 627 150 169 534 161 250 339 820 981 025 695 800 781 25
4 722 366 482 869 645 213 696	72	0.000 000 000 000 000 000 000 211 758 236 813 575 084 767 080 625 169 910 490 512 847 900 390 625

Table of powers of 2.

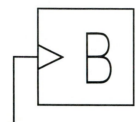

ASCII Conversion Chart

The conversion chart listed below is helpful in converting from a two-digit (two-byte) hexadecimal number to an ASCII character or from an ASCII character to a two-digit hexadecimal number. The example provided below shows the method of using this conversion chart.

Example

				Bits				
		MSB ←					→ *LSB*	
ASCII	*Hex #*	*6*	*5*	*4*	*3*	*2*	*1*	*0*
T	54	1	0	1	0	1	0	0
?	3F	0	1	1	1	1	1	1
+	2B	0	1	0	1	0	1	1

Bits 0 to 3 Second Hex Digit (LSB)	Bits 4 to 6 First Hex Digit (MSB)							
	0	1	2	3	4	5	6	7
0	NUL	DLE	SP	0	@	P		p
1	SOH	DC1	!	1	A	Q	a	q
2	STX	DC2	''	2	B	R	b	r
3	ETX	DC3	#	3	C	S	c	s
4	EOT	DC4	$	4	D	T	d	t
5	ENQ	NAK	%	5	E	U	e	u
6	ACK	SYN	&	6	F	V	f	v
7	BEL	ETB	'	7	G	W	g	w
8	BS	CAN	(	8	H	X	h	x
9	HT	EM	)	9	I	Y	i	y
A	LF	SUB	*	:	J	Z	j	z
B	VT	ESC	+	;	K	[	k	{
C	FF	FS	'	<	L	/	l	/
D	CR	GS	−	=	M	]	m	}
E	SO	RS	.	>	N	∧	n	≈
F	SI	US	/	?	O	—	o	DEL

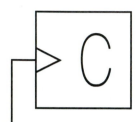

Reduction of Eq. 14-2 to EXCLUSIVE ORs

Reduction of Eq. 14-2 to EXCLUSIVE ORs

$$f(A,B,C,D,) = \bar{A}\bar{B}\bar{C}D + \bar{A}\bar{B}C\bar{D} + \bar{A}B\bar{C}\bar{D} + \bar{A}BCD$$
$$+ A\bar{B}\bar{C}\bar{D} + A\bar{B}CD + AB\bar{C}D + ABC\bar{D}$$

$$= \bar{A}\bar{B}(\bar{C}D + C\bar{D}) + \bar{A}B(\bar{C}\bar{D} + CD)$$
$$+ A\bar{B}(\bar{C}\bar{D} + CD) + (AB)(\bar{C}D + C\bar{D})$$

$$= \bar{A}\bar{B}(C \oplus D) + \bar{A}B(\overline{C \oplus D})$$
$$A\bar{B}(\overline{C \oplus D}) + AB(C \oplus D)$$

$$= (C \oplus D)(AB + \bar{A}\bar{B}) + (\overline{C \oplus D})(\bar{A}B + A\bar{B})$$

$$= (C \oplus D)(\overline{A \oplus B}) + (\overline{C \oplus D})(A \oplus B)$$

$$= A \oplus B \oplus C \oplus D$$

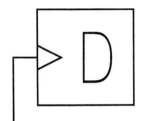

A Proof That XOR Gates Can Be Used As Parity Checkers

To show that if all the outputs of a register are XORed together the resulting output is HIGH on odd parity, we use an induction proof.

Assume the statement is true for an n bit register. We proceed to show that it must be true for an $n + 1$ bit register.

1. If there is an odd number of 1s in the n-bit register, there will be an odd number of 1s in the $n + 1$ bit register only if the $n + 1$ bit is 0. By XORing the n-bit output (HIGH) with the $n + 1$ bit (0) the results will be HIGH.

2. If there are an even number of 1s in the n-bit register, the output of the XOR circuits for the n-bit register is LOW. The $n + 1$ bit register will have an odd number of 1s if the $n + 1$ bit is a 1. By XORing the $n + 1$ bit with the output of the n-bit register, we obtain the correct results.

3. Since the statement is true by inspection for a 2-bit register, it must be true for a register of any size.

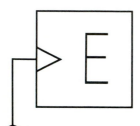

Answers
to Selected Problems

Chapter 1

1-1. 7

1-5. (c) 395
      (d) 27.71875
      (f) 399.453125

1-6. (c) 1011010100011
      (g) 0.110100011

1-7. (c) $A + B = 111101100$
            $A - B = 110100010$

1-8. (b) $A + B = 11011111, A - B = 0100011$

1-9. The numbers 0 through 1023

1-10. (c) 10011010

1-11. Only numbers b, c, and f have 2 LSBs of 0 and are divisible by 4.

1-12. (b)   001010101 $(+85)$
          $+$ 111011011 (-37)
              000110000 $(+48)$

1-13.　(b)　　　 $835 = 01101000011$
　　　　　　　 $214 = 00011010110$
　　　　　 $-214 = 11100101010$

$$
\begin{array}{r}
01101000011 \\
11100101010 \\
\hline
01001101101
\end{array}
\qquad
\begin{array}{r}
835 \\
-214 \\
\hline
621
\end{array}
$$

1-15.　(c)　20FC

1-16.　(c)　0101 1100 1111 0000 0011 0101

1-17.　(c)　6090805

1-18.　(b)　205

1-19.　(c)　19B40

1-20.　(c)　A0A

1-21.　(c)　B00

　　　 (d)　E0CFE

Chapter 2

2-2.

A	B	C	D	G
0	0	0	0	0
0	0	0	1	0
0	0	1	0	0
0	0	1	1	0
0	1	0	0	0
0	1	0	1	0
0	1	1	0	0
0	1	1	1	1
1	0	0	0	0
1	0	0	1	0
1	0	1	0	0
1	0	1	1	1
1	1	0	0	0
1	1	0	1	1
1	1	1	0	1
1	1	1	1	1

2-3.

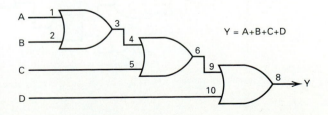

$Y = A+B+C+D$

2-5.   (b)

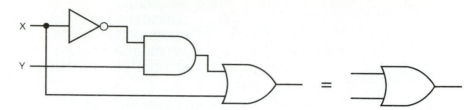

2-7. $P = M + F\bar{A}$ (A policy will be issued to anyone who is married or to a female under 25 years of age.)

2-8.   (c) $X + Z$

       (d)   1

2-10.   (a)   The expression is true.

       (c)   The equation is false for $a = 1, b = 1, c = 0$.

2-12.   (a) $(a + \bar{d} + b\bar{c})(a + d + \bar{b}\bar{c})bc$

2-14.   (b)

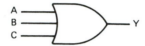

2-15.   (c)

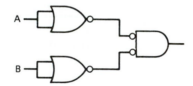

2-16.   (c)

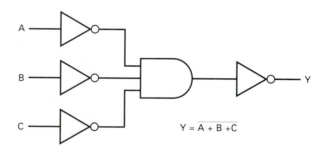

$Y = \overline{A + B + C}$

2-17.   (a)

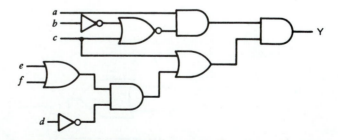

2-17. (d)

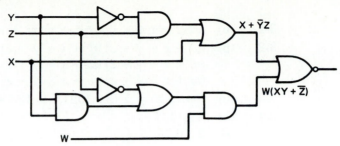

2-17. (f)

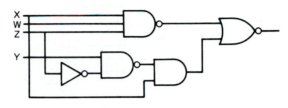

2-18. (d) $\bar{Y} = ABC + A\bar{B} + \bar{A}BC$;

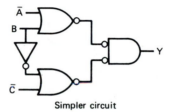

Simpler circuit

2-18. (e) $Y = AB \cdot (\overline{\overline{AB} + \bar{C})\bar{D}} \cdot (\bar{C} + \bar{D})$; simpler circuit

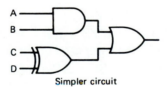

Simpler circuit

2-20. $Y = \bar{C}(AB + CD) = \bar{C}AB$

2-21. (a)

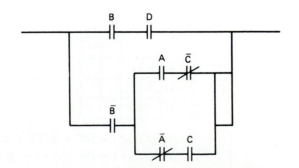

3-1. (b) $w\bar{x}\bar{y}\bar{z} + w\bar{x}\bar{y}z + w\bar{x}y\bar{z} + w\bar{x}yz + wxyz + \bar{w}xyz + \bar{w}\bar{x}\bar{y}z + \bar{w}xy\bar{z}$

3-2. (b) $(\bar{w} + \bar{x} + \bar{y} + z)(\bar{w} + \bar{x} + y + \bar{z})(\bar{w} + \bar{x} + y + z)$
$(w + \bar{x} + \bar{y} + z)(w + \bar{x} + y + z)(w + x + \bar{y} + \bar{z})$
$(w + x + \bar{y} + z)(w + x + y + z)$

3-3. (b) $f(w,x,y,z) = \Sigma\,(8,9,10,11,7,15,1,5) = \pi\,(0,2,3,4,6,12,13,14)$

3-4. (b) 8,9,10,11
(f) 2,3,6,7

3-7. (c)

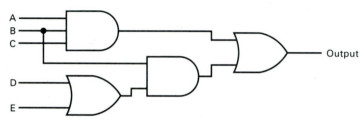

3-8. (a)   (b)

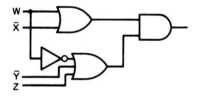   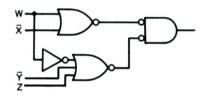

3-10. Subcube
1 $\bar{X}\bar{Y}\bar{Z}$
5 $B\,D$
6 $W + \bar{X} + \bar{Z}$

3-11. $f(W,X,Y,Z) = \bar{W}\bar{Y}\bar{Z} + \bar{X}Y + WYZ + WX$
The subcube composed of the four corner squares is not essential.

3-14. $F(A,B,C,D) = \bar{B}\bar{C} + \bar{A}C + AB$

3-17. (b) $F(W,X,Y,Z) = \bar{X}Z + W\bar{Y}Z + W\bar{X}Y$

3-18. (b) $F(W,X,Y,Z) = (W + \bar{X})(Y + Z)(W + Z)(\bar{X} + \bar{Y})$

3-21. (b)

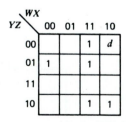

$f(W,X,Y,Z) = W\bar{Z} + WX\bar{Y} + \bar{W}\bar{X}\bar{Y}Z$
$f(W,X,Y,Z) = (W + \bar{X})(\bar{Y} + \bar{Z})(W + Z)(\bar{W} + X + \bar{Z})$

(c)

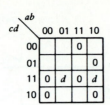

$$f(a,b,c,d) = (\bar{c} + \bar{d})(b + \bar{c})(\bar{a} + \bar{b} + c + d)(\bar{a} + b + \bar{d})$$
$$f(a,b,c,d) = \bar{a}\,\bar{c} + \bar{b}\bar{c}\bar{d} + bc\bar{d} + b\bar{c}d$$

3-22. $f(V,W,X,Y,Z) = \bar{X}\bar{Y} + \bar{V}\bar{X}\bar{Z} + WXY + VWX\bar{Z} + VXY\bar{Z}$

3-23.   Segment $f = A + B\bar{C} + B\bar{D} + \bar{C}\bar{D}$

Chapter 4

4-1.   6

4-3.   5

Chapter 5

5-2.   (a)   Circuit

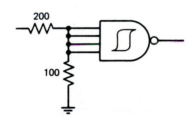

(b)   For minimum pulse width,

$$\sin \theta = \frac{0.9}{1.7} = 0.53$$
$$\theta = 148° - 90° = 58°$$
$$\text{Time} = \frac{58°}{360°} \times 1\ \mu s = 161\ \text{ns}$$

5-4.   (a)   Yes
       (b)   No
       (c)   No
       (d)   Yes

5-8.

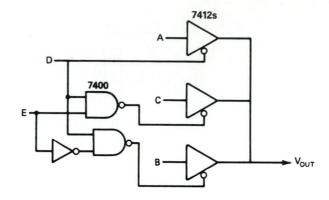

5-12. (a)

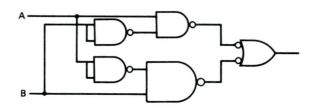

Chapter 6

6-3.   Both outputs would be LOW.

6-5.

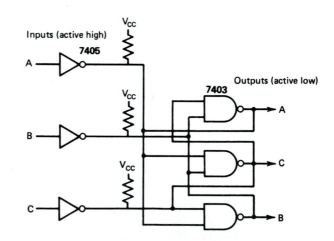

6-6.

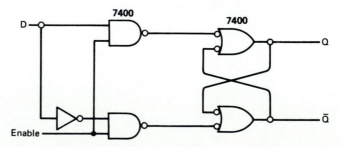

6-9.   When Q_2 SETS, it direct CLEARS Q_1. Q_1 does not toggle on the next pulse because $\bar{Q}_2$ is LOW.

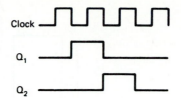

6-13.   (a)   Using **7474**s: Note additional gating to provide the required drive capability. CLEAR inputs also require additional gating.

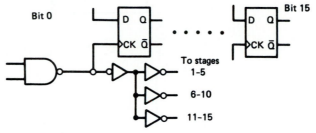

6-13.   (b)   Using **74174**s

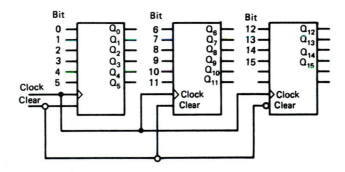

6-14.

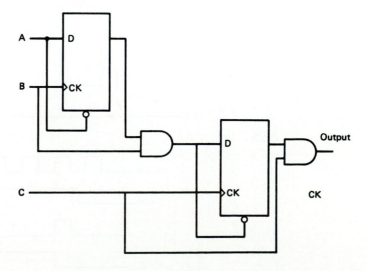

6-17.

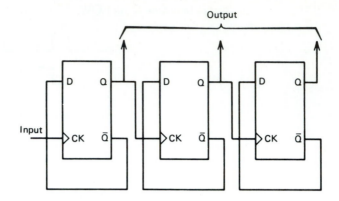

6-18.

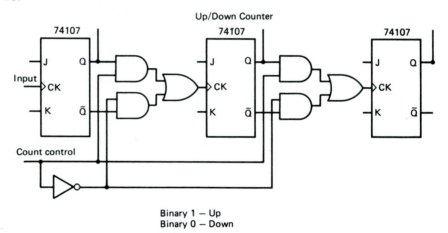

Binary 1 — Up
Binary 0 — Down

6-22.

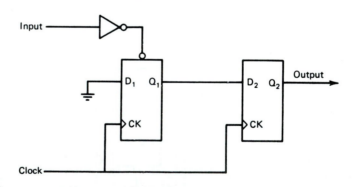

6-25. (a)

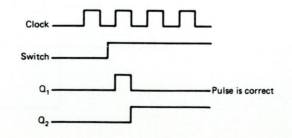

(b)

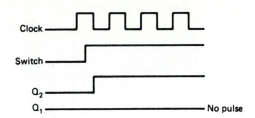

Chapter 7

7-1.   (a) $C = 3.57$ µf. Time between pulses—5 ms.
        (b) $C = 0.175$ µf. Time between pulses—2.57 ms.

7-2.   (a)   3.5 ms
        (b)   17.5 ms
        (c)   The fixed resistor eliminates the possibility of connecting the R_{EXT} input directly to V_{CC}.

7-4b. C for a 10K resistor, $C = 100$ pf

7-5b. C for a 10K resistor, $C = 0.15$ µf

7-6.   (a)   3.5 ms
        (b)   1.65 ms

7-11.   (a)   Each half of the **74LS221** can be set for 10 µs using the 2K internal resistor. $C = 0.00715$ µf
        (b)   Using two halves of a **74LS123** with a 10K external resistor $C = 0.003$ µf
        (c)   For the Schmitt trigger, $R = 330, C = 0.0484$ µf.
        (d)   For the **555**, let $C = 10^{-9}$; then we can choose $R_A = 12$K and $R_3 = 8.2$K.

7-13.   From the chart $f \approx 8$ Hz.

From the formula, $f = \dfrac{1.44}{300 \times 10^3 \times 10^{-6}} = \dfrac{1.44}{0.3} = 4.8$ Hz

Note that the charts give a very approximate reading.

7-15.

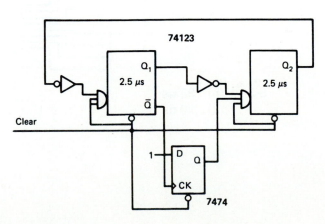

7-16. Circuit shown using all **74123**s. Timing resistors and capacitors not shown for clarity.

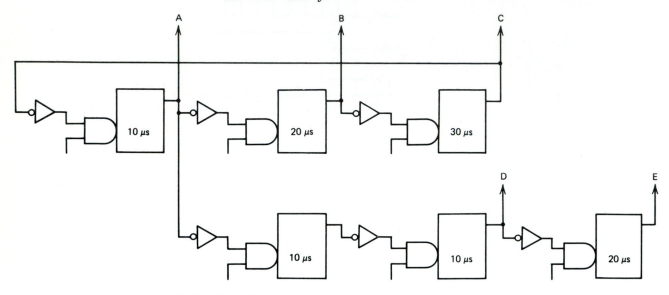

7-18. (a) Yes

(b) Yes

(c) No

Switches generally bounce longer than 500 μs, so bounces could cause additional triggers in parts a and b. Switches never bounce for more than 500 ms, so additional triggers will not be generated in part c. (Note that if a retriggerable one-shot and an undebounced switch are used, the pulse time will be the one-shot time plus the switch bounce time.)

Chapter 8

8-1. Use an ordinary five-stage ripple counter.

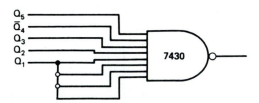

8-3. (b) 95 ns

(d) 90 ns

8-5. (a) Minimum time delay (2 gates max)

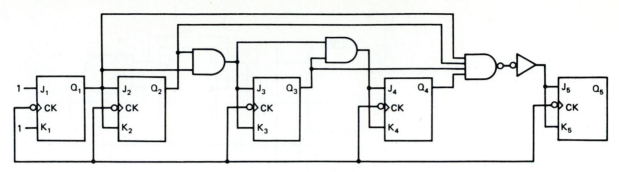

8-7.

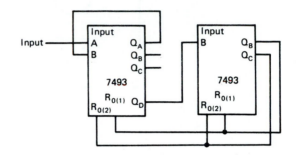

8-11. (a)

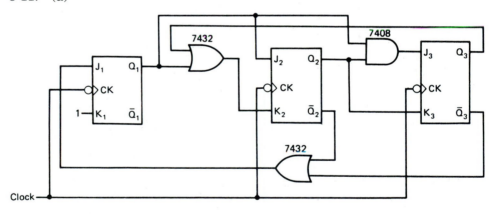

8-11. (b) For **74x74**s:
1. Construct a 3-stage D-type counter.
2. NAND Q_2 and Q_3.
3. AND this output with the signal to every D input.

8-13.

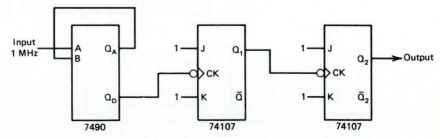

8-19. (a)

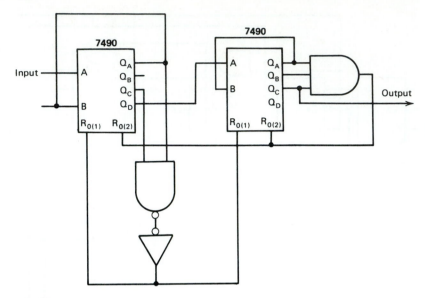

8-20. **7490**s—95
      7493s—149

8-22. (a)

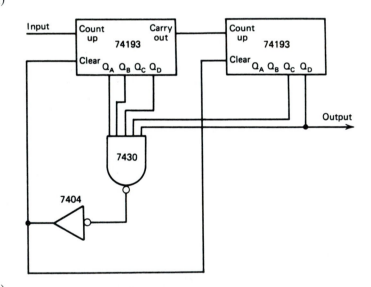

8-22. (b)

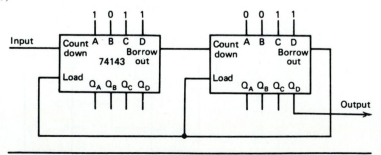

8-25. The counter rolls over on a count of 217.

8-26. (c) For the **7490**s, the output should be HIGH for 830 μs and LOW for 830 μs. For the **74193**s the output should be HIGH for 385 μs and LOW for 1275 μs.

8-28.

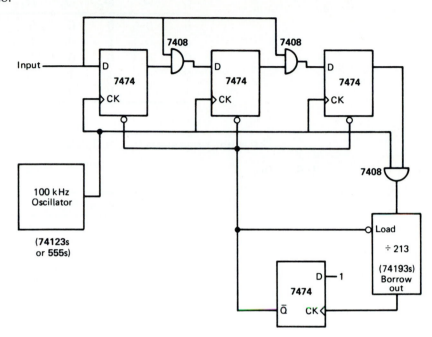

Chapter 9

9-1.

Present state									Next state		
Q_3	Q_2	Q_1	J_3	K_3	J_2	K_2	J_1	K_1	Q_3	Q_2	Q_1
0	0	0	0	0	0	1	1	0	0	0	1
0	0	1	0	0	1	1	1	0	0	1	1
0	1	1	1	1	1	1	1	0	1	0	1
1	0	1	0	0	1	1	1	0	1	1	1
1	1	1	1	1	1	1	1	1	0	0	0

The count progresses 0, 1, 3, 5, 7, 0 . . .

9-4.

Present state			D inputs for next state		
Q_3	Q_2	Q_1	D_3	D_2	D_1
0	0	0	0	0	1
0	0	1	0	1	0
0	1	0	0	1	1
0	1	1	1	1	0
1	1	0	0	0	0

The count progresses 0, 1, 2, 3, 6, 0, . . .

9-6.   (b).

State Tables					
Present state			Next state		
Q_3	Q_2	Q_1	D_3	D_2	D_1
0	0	0	0	0	1
0	0	1	0	1	0
0	1	0	0	1	1
0	1	1	1	0	0
1	0	0	1	0	1
1	0	1	0	0	0
1	1	0	d	d	d
1	1	1	d	d	d

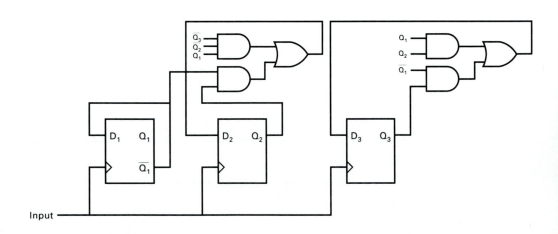

Karnaugh maps

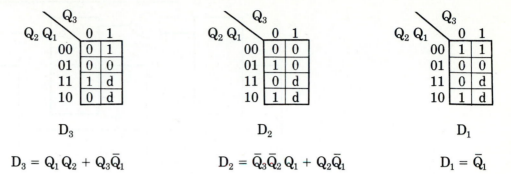

	Q_3 0	1
$Q_2 Q_1$ 00	0	1
01	0	0
11	1	d
10	0	d

D_3

	Q_3 0	1
$Q_2 Q_1$ 00	0	0
01	1	0
11	0	d
10	1	d

D_2

	Q_3 0	1
$Q_2 Q_1$ 00	1	1
01	0	0
11	0	d
10	1	d

D_1

$$D_3 = Q_1 Q_2 + Q_3 \bar{Q}_1 \qquad D_2 = \bar{Q}_3 \bar{Q}_2 Q_1 + Q_2 \bar{Q}_1 \qquad D_1 = \bar{Q}_1$$

9-8.

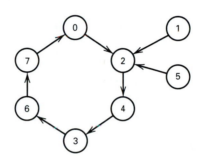

9-9.   (b).

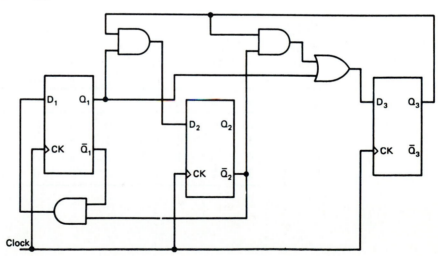

9-10.

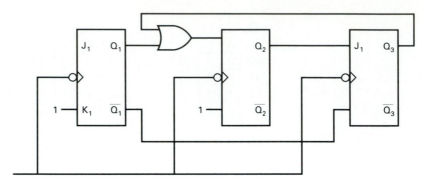

9-13. (a)

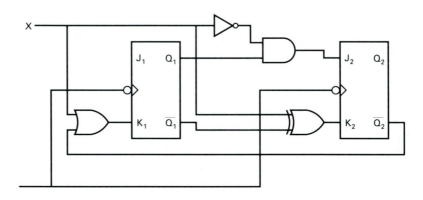

State Tables													
Present State		Next State				$x = 0$				$x = 1$			
		$x = 0$		$x = 1$		J_2	K_2	J_1	K_1	J_2	K_2	J_1	K_1
Q_2	Q_1	Q_2	Q_1	Q_2	Q_1								
0	0	0	0	0	1	0	d	0	d	0	d	1	d
0	1	1	0	0	0	1	d	d	1	0	d	d	1
1	0	0	0	1	1	d	1	0	d	d	0	1	d
1	1	1	1	0	0	d	0	d	0	d	1	d	1

9-13a. (continued)

Karnaugh maps

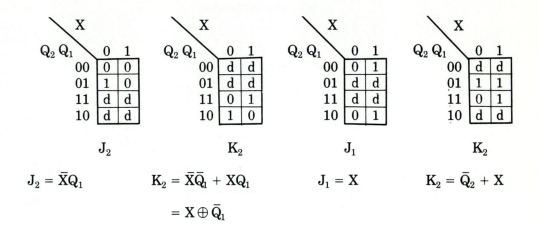

$$J_2 = \bar{X}Q_1 \qquad K_2 = \bar{X}\bar{Q}_1 + XQ_1 \qquad J_1 = X \qquad K_2 = \bar{Q}_2 + X$$

$$= X \oplus \bar{Q}_1$$

9-15c.

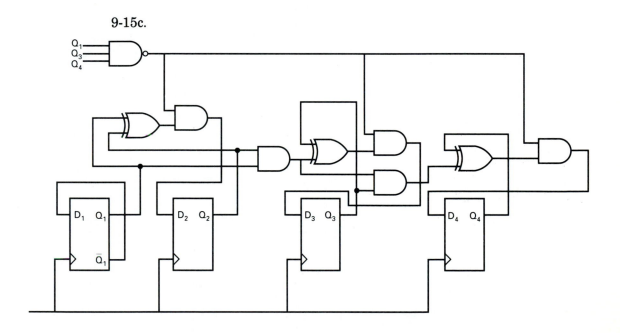

Chapter 10

10-1.　(a)　000100101101
　　　　(b)　011011010000

10-2.

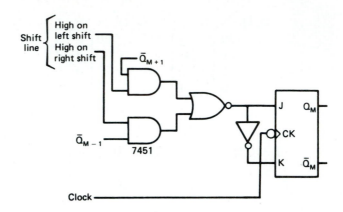

10-6.

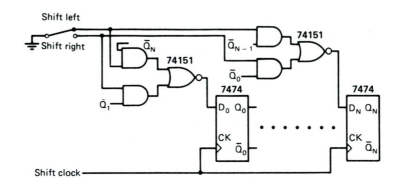

10-8.

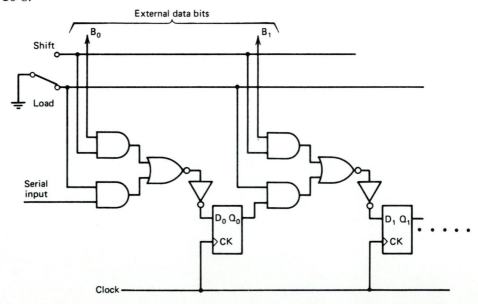

10-12.

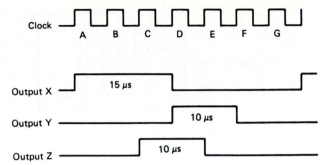

10-13(b).

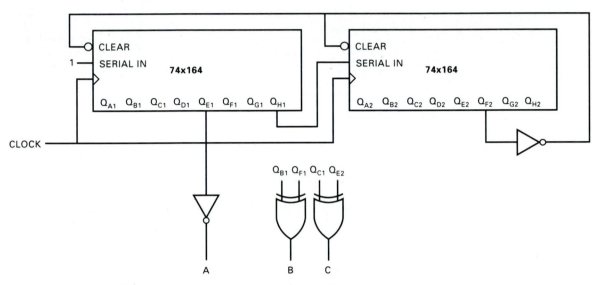

10-15.

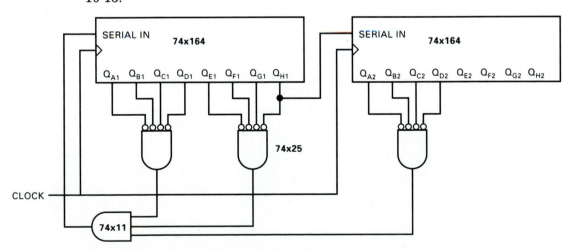

10-17.

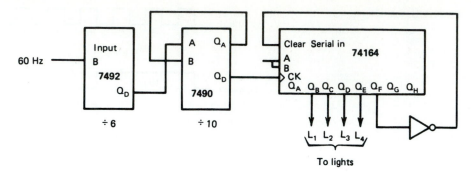

10-20.

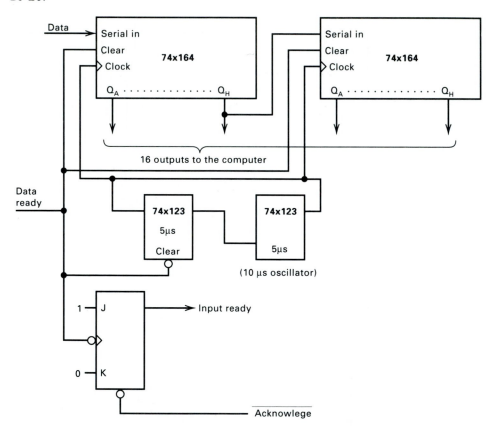

10-22.

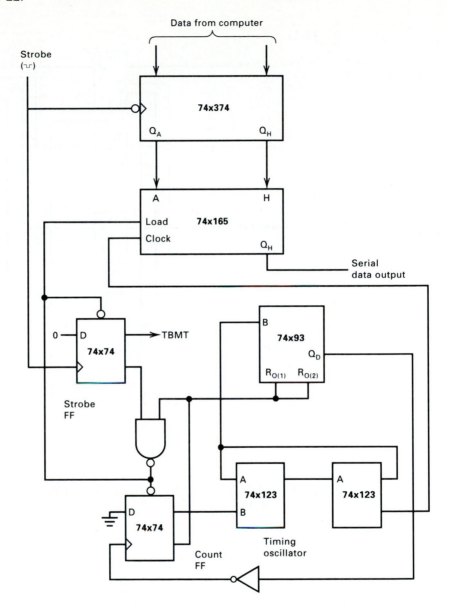

11-1.

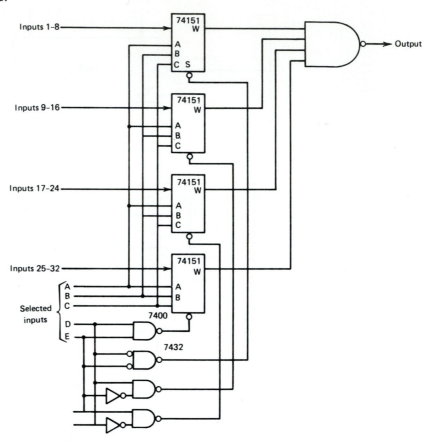

11-2. Delete one IC from the circuit of Problem 11-1 or use a **74150** and a **74151**.

11-6.

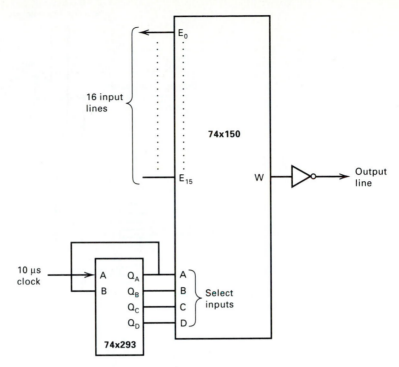

11-9. (a) The decoder requires four **74154**s. Use the select lines as follows:

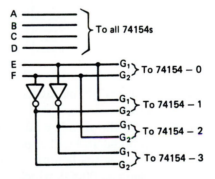

11-10. The circuit requires 11 **7442** 4-line-to-10-line decoders. The select lines of one decoder must be connected to the most significant BCD decade and the select lines of the other decoders must be connected to the least significant BCD decade. Because **7442**s have no strobe, the following circuit should be used to deselect 9 of the multiplexers:

The outputs of the decoder connected to the most significant decade consists of nine 1s and a 0. The 1s cause the C and D inputs of all the corresponding decoders to be HIGH, which causes them to present all HIGH outputs. Only the least significant decoder connected to the LOW output of the most significant decoder receives a genuine BCD input and produces a LOW output.

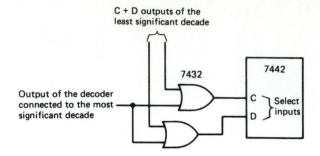

11-11.

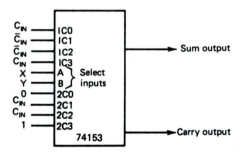

11-13. (a).

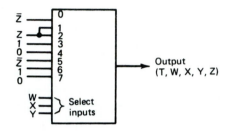

11-15. The key at the intersection of the Y1 and D6 inputs, as shown in Fig. 11-21.

11-20. Note the number 50 will cause the 3 most significant displays to be blanked and will appear as 50, but 100050 will not cause any blanking because the leading 1 is tied to RBI of the MSD. Therefore 100050 appears as 00050, which is distinguishable from 50.

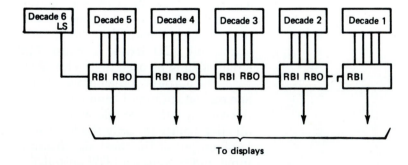

12-1.

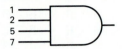

12-3(a).

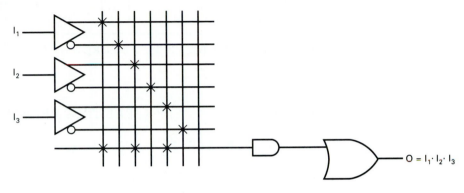

$$O = I_1 \cdot I_2 \cdot I_3$$

12-5.

```
PAL16L8

Skippal PAL
J. D. Greenfield
June 8, 1982
SCS CR SN A7 SP COUT SZ SCC CS GND
E02 NC SA NC NC NC NC FLASH SKIPPOUT VCC
;
; SCS = SKIP ON CARRY SET, CR = CARRY RESET (CARRY FF-0),
; SN = SKIP ON NEG. ACCUM, A7 = ACCUM (ACCUMULATOR) BIT 7,
; SP = SKIP ON POS. ACCUM, COUT = ACCUM CARRY OUT,
; SZ = SKIP ON 0 ACCUM, SCC = SKIP ON CARRY CLEAR
; CS = CARRY SET (CARRY FF-1), SA = SKIP ALWAYS
; FLASH IS AN INTERMEDIATE OUTPUT
; E02 IS A CLOCK
; SKIPPOUT IS THE FINAL OUTPUT. IT SHOULD BE HIGH ONLY
; WHEN FLASH IS LOW AND E02 IS HIGH.
;
/FLASH = /SCS*/CR + /SN*A7 + /SP*/A7 + COUT*/SZ
        + /SCC*/CS + /SA
/SKIPPOUT = FLASH + /E02
```

FUNCTION TABLE

SCS	CR	SN	A7	SP	COUT	SZ	SCC	CS	E02	SA	FLASH	SKIPPOUT
L	L	H	H	H	H	H	H	H	H	H	L	H
L	L	H	H	H	H	H	H	H	L	H	L	L
H	L	L	H	H	L	H	H	H	H	H	L	H
H	H	H	H	H	H	H	H	H	H	L	L	H
H	L	H	L	H	L	H	H	L	H	H	H	L
H	L	H	L	L	H	H	H	H	H	H	L	H

DESCRIPTION PAL IS TO REPLACE SKIPPOUT LOGIC.

12-6.

```
PAL16L8                    PAL TEST PROBLEM
RIT
F1=2,4,5,8,A,B,D     F2=1,5,6,C,D,F
MAY 8, 1986
NC W X Y Z NC NC NC NC GND
NC NC NC NC NC NC NC F1 F2 VCC
;
;
/F1 =/W*/X*/Y*/Z + /W*/X*/Y*Z + /W*/X*Y*Z
     + /W*X*Y + W*/X*/Y*Z + W*X*/Y*/Z + W*X*Y

/F2 =/W*/X*/Y*/Z + /W*/X*Y + /W*X*/Y*/Z + /W*X*Y*Z
     + W*/X + W*X*Y*/Z
```

FUNCTION TABLE

W	X	Y	Z	F1	F2
;					
L	L	L	L	L	L
L	L	L	H	L	H
L	L	H	L	H	L
L	L	H	H	L	L
L	H	L	H	H	H
L	H	H	L	L	L

DESCRIPTION; THIS IS STRICTLY A TEST

12-8.   (a) PAL File

PAL16L8

J. D. GREENFIELD
JUNE 8, 1992
A B C W X Y Z NC NC GND

NC NC NC NC NC NC NC NC M VCC
;
IF (A*/B*C)/M = /W + /X + Y + /Z

FUNCTION TABLE

A	B	C	W	X	Y	Z	M
H	L	H	H	H	L	H	H
H	L	H	H	H	L	L	L
H	L	L	H	H	H	H	Z

DESCRIPTION A PAL USING A
3-STATE GATE

(b) Brief fuse plot.

PAL20 V1.7K – PAL16L8 –J. D. GREENFIELD

```
                      11   1111  1111  2222  2222  2233
          0123  4567  8901  2345  6789  0123  4567  8901
0   -XX-  X---  ----  ----  ----  ----  ----  ----   A*/B*C
1   ----  ----  -X--  ----  ----  ----  ----  ----   /W
2   ----  ----  ----  -X--  ----  ----  ----  ----   /X
3   ----  ----  ----  ----  X---  ----  ----  ----   Y
4   ----  ----  ----  ----  ----  -X--  ----  ----   /Z
```

12-10.

PAL16R4
PROBLEM 12-10
J. D. GREENFIELD
JUNE 8, 1992
CLK A B C D NC NC NC NC GND
GND NC NC NC NC Q2 Q1 NC L1 VCC

; ORIGINAL EQUATIONS
; L1 = (A+B)*C*D
; Q1 :=/A*L1 + B*C
; Q2 := Q1*Q2*/A + /Q1*D

/L1 = /A*/B + /C + /D
/Q1 := A*/B + A*/C + /L1*/B + /L1*/C
/Q2 := /Q1*/D + /Q2*Q1 + /Q2*/D + A*Q1 + A*/D

12-12.

```
/Q1    :=/S1*/S0            +
        /S1*S0*/D1          +
        S1*/S0*/Q1*/Q0      +
        S1*/S0* Q1* Q0      +
        S1* S0*/Q1
```

12-15.

```
PAL16R4
PROBLEM 12-15
J.D. GREENFIELD
JUNE 10, 1992
CLK NC NC NC NC NC NC NC NC GND
GND NC NC Q3 Q2 Q1 Q0 NC NC VCC

/Q0 := Q0 + Q3*Q2
/Q1 := Q0*Q1 + /Q0*/Q1 + Q3*Q2
/Q2 :=/Q2*/Q0 + /Q2*/Q1 + Q2*Q1*Q0 + Q2*Q3
/Q3 :=/Q3*/Q0 + /Q3*/Q1 + /Q3*/Q2 + Q3*Q2
```

FUNCTION TABLE

CLK	Q3	Q2	Q1	Q0
C	L	L	L	H
C	L	L	H	L
C	L	L	H	H
C	L	H	L	L
C	L	H	L	H
C	L	H	H	L
C	L	H	H	H
C	H	L	L	L
C	H	L	L	H
C	H	L	H	L
C	H	L	H	H
C	H	H	L	L
C	L	L	L	L
C	L	L	L	H

12-18.

```
PAL16R4    PROBLEM 12-18
RING COUNTER
J. D. GREENFIELD
JUNE 11, 1992
CLK D   NC NC NC NC NC NC NC GND
GND NC  NC Q3 Q2 Q1 Q0 NC NC VCC

; D IS AN INPUT TO START THE RING COUNTER PROPERLY

/Q0 :=/Q3*/D
/Q1 :=/Q0 + D
/Q2 :=/Q1 + D
/Q3 :=/Q2 + D

FUNCTION TABLE

CLK    D    Q3    Q2    Q1    Q0
------------------------------------
C      H    L     L     L     H
C      L    L     L     H     L
C      L    L     H     L     L
C      L    H     L     L     L
C      L    L     L     L     H
------------------------------------
DESCRIPTION  4-STAGE  RING
COUNTER
```

Chapter 13

13-2. The PC contains 4M bytes or 2^{22} bytes $\times$ 2^3 bits/byte. This equals 2^{25} total bits.

13-3. (a) 12
(b) 12
(c) 13
(d) 1
(e) 8
(f)

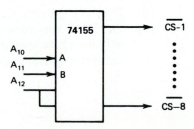

13-6.

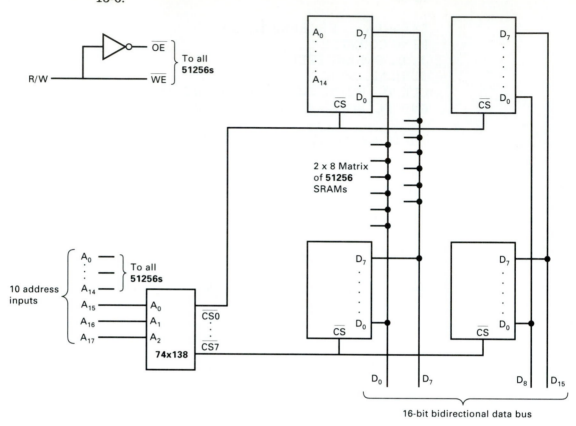

13-9(b). This IC is a <u>1 MB</u> × <u>4 bit</u> DRAM because
   (a) It has $\overline{RAS}$ and $\overline{CAS}$ inputs.
   (b) It has 10 address inputs, for 20 address lines or 1 MB.
   (c) It has 4 data I/O lines.

13-10. (a) 16—They come from the computer system.
   (b) 8—They go to the Dynamic RAM.
   (c) The lower 8 address bits.
   (d) The upper 8 address bits.
   (e) The 8 bits of the Refresh Counter.

13-13. (a) 16
   (b) CSA is selected when $A_{15} = 0$, $A_{14} = 1$, $A_{13} = 0$, and $A_{12} = 1$.
   (c) The **8216**s are transceivers.
   (d) They are synchronizing FFs.
   (e) The **3242** can accommodate a 16K memory IC but is being used only for 4K memory ICs.
   (f) Left for student to draw.

13-14.

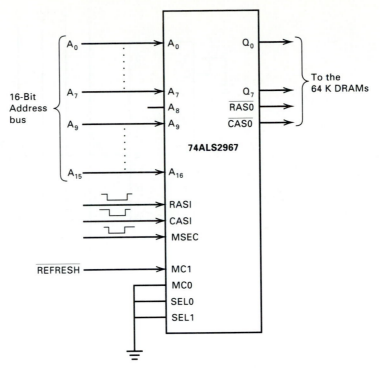

13-17.

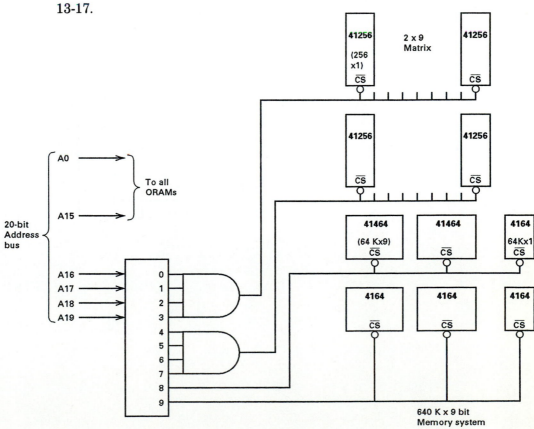

13-19.

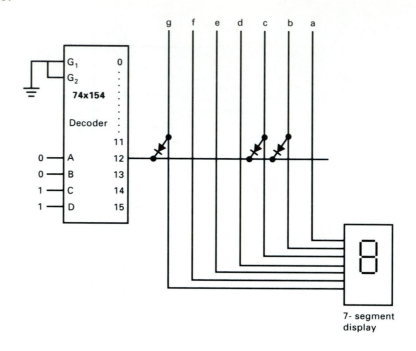

7- segment
display

Chapter 14

14-3.

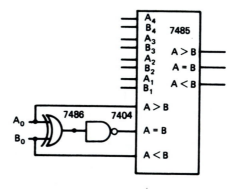

14-5.

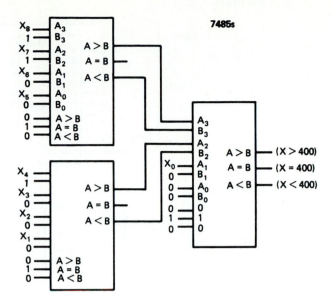

14-8(a).   Even
14-9(a).   1 FF
14-10(a).

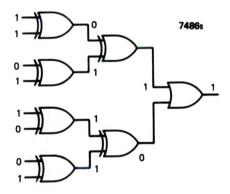

14-12(b).

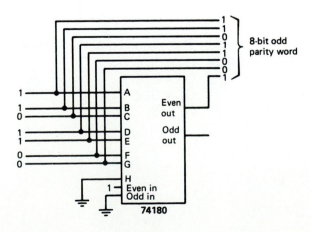

14-14(b).

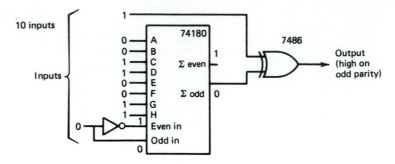

14-17.

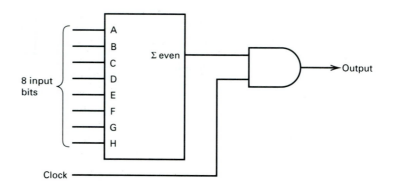

14-18.

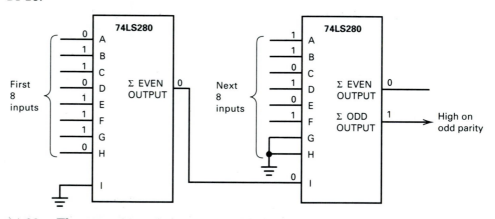

14-22.   The second bit of the third word is wrong.

14-25. $G_0 = B_1 \oplus B_0$
$G_1 = B_2 \oplus B_1$
$G_2 = B_3 \oplus B_2$
$G_3 = B_3$

15-2.

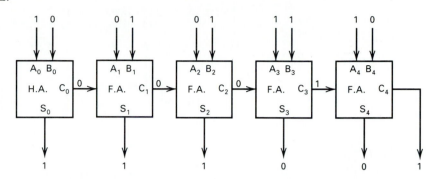

A = 1 1 0 0 1 = 25
B = 0 1 1 1 0 = 14
S = 1 0 0 1 1 1 = 39

15-5(e).

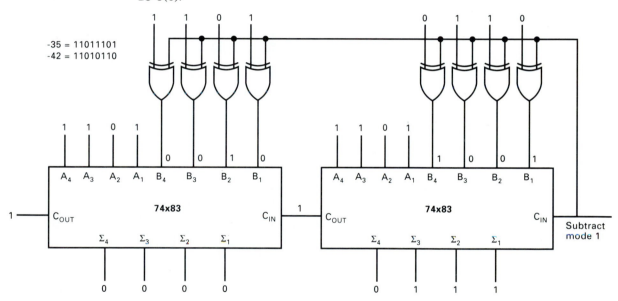

15-6. (a) $(-55) - (-45)$

(b) Gates 3 and 10 are defective.

15-8. Operations a, c, d, and e produce overflow.

15-11. Refer to Fig. 14-10 and start at D_0, where the carry in is 1.

Original numbers	D_3	D_2	D_1	D_0
A	2	4	3	2
B	0	6	7	6
Carry in	0	0	0	1
B	15	9	8	9
Sum of upper **7483** and an input to lower **7483**	1	13	11	12
Carry out	1	0	0	0

B input to lower **7483**	0	10	10	10	
Difference digit	1	7	5	6	Answer

15-14(a).

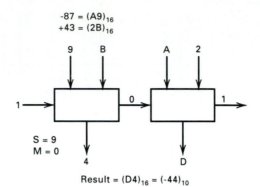

-87 = (A9)$_{16}$
+43 = (2B)$_{16}$

S = 9
M = 0

Result = (D4)$_{16}$ = (-44)$_{10}$

15-14.

(d)

−43 = (D5)$_{16}$
−17 = (EF)$_{16}$

S = 6
M = 0

Result = (E6)$_{16}$ = (−26)$_{10}$

(f)

65 = (41)$_{16}$
(−37) = (DB)$_{16}$

Result = (66)$_{16}$ = (102)$_{10}$

15-16.   Use S = 15 and tie C$_{in}$ of the least significant stage HIGH.

4   15

Result is 0100 1111 = 79

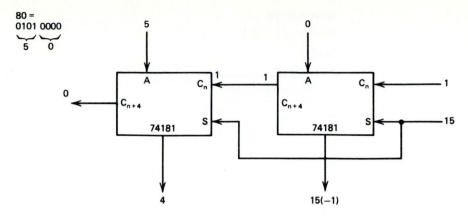

$80 =$
$0101\ 0000$
$\underbrace{}_{5}\ \underbrace{}_{0}$

15-19.

	A_0	B_0	F_0	C_0	A_1	B_1	F_1	C_1
$S = 12$ $M = 0$ $C_{in} = 0$	B	C	7	0	5	3	B	1
$S = 6$ $M = 0$ $C_{in} = 0$	A	D	D	1	6	A	B	1
$S = 9$ $M = 1$ $C_{in} = 1$	A	D	8	X	3	9	5	X
$S = 15$ $M = 0$ $C_{in} = 1$	0	7	F	1	5	7	4	0

15-20.   If A = 0, the carry out of the second stage will be HIGH.

15-22.   Use the multiplexer on the B inputs to the **74LS381** so that when SHIFT is HIGH the A input will also become the B inputs, and S_2, S_1, S_0 become 0, 1, 1 so that the function is A plus B.

15-23.   Multiplier = 45 = 101101

Multiplicand = 37 =   100101

		LSB	
000000			
100101		1	Add
010010	1		Shift
001001	01	0	Shift
101110	01	1	Add
010111	001		Shift
111100	001	1	Add
011110	0001		Shift
001111	0000	0	Shift
110100	00001	1	Add
011010	000001		Shift

(Final answer = 1665)

16-3.

Additional program		
10	LOAD	30
11	ADD	31
12	STORE	30
13	ADD	32
14	STORE	32
15	LOAD	30
16	SUB	33
17	JN	10
18	HALT	

Location 30 contains the number to be added—initially 1
Location 31 contains 5
Location 32 contains the sum—initially 1
Location 33 contains 20,000

16-5.

10	LOAD	100
11	SUBTRACT	22
12	JZ	25
13	LOAD	10
14	ADD	50
15	STORE	10
16	LOAD	51
17	SUB	50
18	STORE	50
19	JP	10
20	HALT	
25	LOAD	52
26	ADD	50
27	STORE	52
28	HALT	

At the start of the program:
    50 contains 1
    51 contains 200
    52 contains 0
It retains the number of 22s in the program.

16-9.

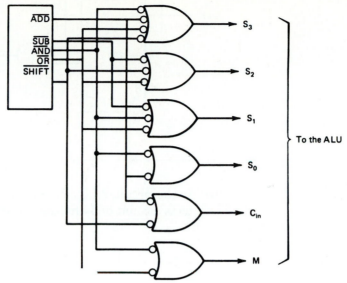

16-11. PC $-8\rightarrow|$ $\leftarrow 9\rightarrow$ $|$ $\leftarrow 500\rightarrow$ $|$ $\leftarrow 501$

MAR $-8\rightarrow|$ $\leftarrow 500\rightarrow$ $|$ $\leftarrow 500\rightarrow$ $|$

$$\underbrace{T_0\,T_1\,T_2\,T_3}_{F}\quad\underbrace{T_0\,T_1\,T_2\,T_3}_{E}\quad\underbrace{T_0\,T_1\,T_2}_{F}$$

16-12. When the JNZ output of the instruction decoder goes LOW, it must set $M = 0, S = 15$, and add the following gate to the JUMP gates.

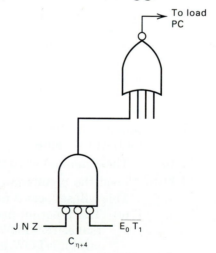

Chapter 17

17-2.

(a) Because each clock takes 30 ns, the circuit must wait for 5 cycles before being ready. Because T2 and T3 must occur before T4, three WAIT states are required.

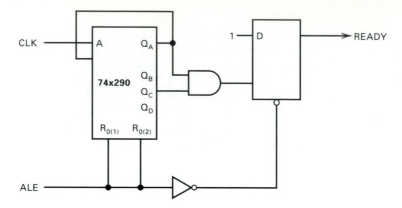

(b) In the circuit, ALE clears the counter and the FF. The **74x290** detects a count of 5 and then resets the FF.

17-4.

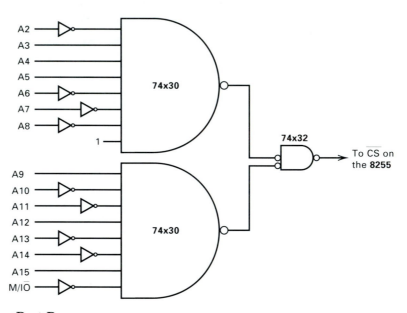

17-6.   Port B.

17-9.   Sending the byte 06 to 800B will clear bit 3. Sending the byte 0D to 800B will set bit 6.

17-10(b).   The command word 9B should set up all lines for input in mode 0.

17-11(b).   Load the accumulator with 81 and issue an OUT 2000 command. This will set port A for output and the lower half of port C for input.

   (c)   The program proceeds as follows: Write 0E into port A. This makes PA0 LOW and PA1, PA2 and PA3 HIGH. Read port C. If any bit is LOW, a switch has been depressed. It will be one of the switches 0–3. Otherwise, write 0D into port A. This will make PA1 LOW. Check port C. Continue as above. The details are left to the student.

17-13.   Address line A1 should be connected to the oscilloscope. If it is LOW during an IN, it is reading port A. If it is HIGH, the μP is reading port C.

17-16. Stop.

17-18. The characters are 0A, 0D, and 42. The parity is odd. The terminal does a Line Feed, a Carriage Return, and types the letter B.

17-19. (a) 4800 Hz.

(b) 33 ms, assuming one STOP bit.

(c) Pin 34 must be HIGH to set the remaining control bits into the UART.

Pin 35 must be LOW if parity is used.

Pin 36 must be LOW for one STOP bit.

Pins 37 and 38 must be HIGH for an 8-bit character.

Pin 39 must be LOW for odd parity.

Pin 40 must have the 4800 Hz clock connected to it.

17-22. The figure shows the connections. To read the status set port A for input and bring PC0 low. Then an IN 0 will read in the status. To send data out, first check the status for TBMT. Then set port A for output, put the data in the accumulator and issue an OUT 0 command. Then pulse PC3 LOW. To read data in, set port B for input. Check the status for DAV. Set PC2 LOW and issue an IN 1 command to read the data. Pulse PC1 to reset DAV.

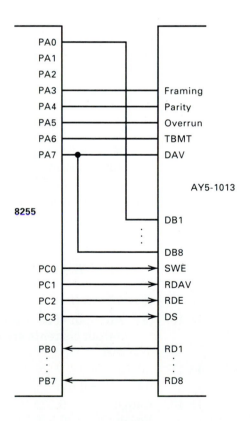

17-23.

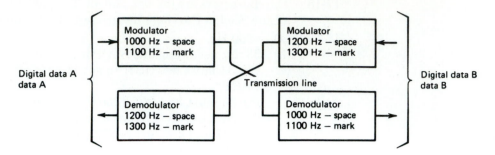

Chapter 18

18-2. If the thermistor temperature is 200°C, the thermistor resistance will be about $10^1 = 10 \, \Omega$.

V_{OUT} will be $\dfrac{10 \, \Omega}{110 \, \Omega} \times 10 \, v = 0.91 \, V.$

18-3. 5 volts ÷ 256 steps = 0.0195 volts/step

18-5. At least 2000 steps are required. Therefore, 11 bits are required on the digital input since $2^{11} = 2048$.

18-6. (a) 0 current, 0 output voltage
(b) R_c = 2.5 mA
R_B = 1.25 mA
R_A = 0.625 mA
Total current = 4.375 mA
V_{OUT} = 4.375 V

18-7. (a) 37 = 100101
Assuming 1 = 5 volts,

$$I = \frac{5}{R} + \frac{5}{R/4} + \frac{5}{R/32} = \frac{185}{R} \, R = 10K, \text{so} \, I = 18.5 \, mA$$

18-9. R = 78.125 Ω

18-10. (a) An input of 20 is 010100. There is $+5V$ on the 2^4 and 2^2 lines. The output currents are 1.25 mA and 0.3125 mA respectively or 1.5625 mA.

18-11. (a), (b) $-0.5 \, V$
(c) $+0.5V$

18-13. Range = $0 - 9.375 \, V$
Resolution = 0.625 V.

18-14(a).

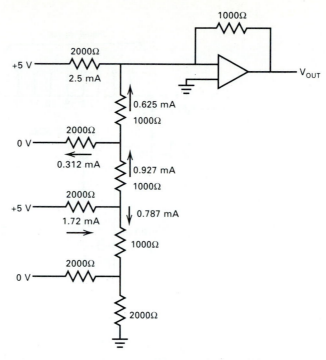

18-15. A single negative pulse is passed down the shift register and sets each FF. If the D/A converter output is too high, the FF just set resets. A worst case comparison requires N steps for an N-bit output.

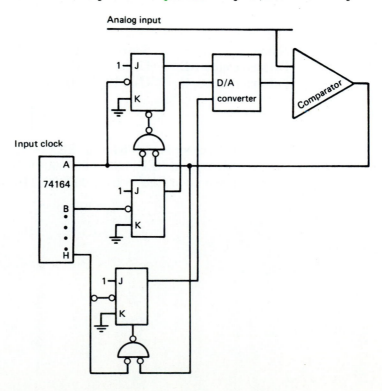

18-17. -5.86 volts

18-18.

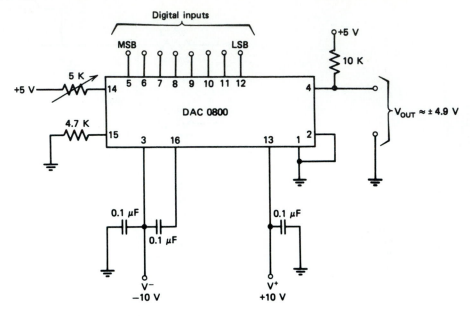

18-20. $(5A)_{16} = (01011010)_2$

18-21. DB7, DB6 and DB3.

Chapter 19

19-2.

Name	Source	To	To	To
Pattern counter-D	U12-11	U13-1		
Pattern counter Q_A	U12-12	U12-1		
	.			
	.			
	.			

19-4. The cursor is at 1354 or at REF $+$ 36.
The data at REF $+35$ is 0000000110011100.
The data at REF -1 is 0000000110011100.

Index

Supplementary Index
of IC Part Numbers

This is an index of the ICs discussed in this book. The numbers are the pages where the ICs are first discussed or are mentioned prominently.